Physics of Matter

The Manchester Physics Series

Properties of Matter	B.H. Flowers and E. Mendoza
Statistical Physics *Second Edition*	F. Mandl
Electromagnetism *Second Edition*	I.S. Grant and W.R. Phillips
Statistics	R.J. Barlow
Solid State Physics *Second Edition*	J.R. Hook and H.E. Hall
Quantum Mechanics	F. Mandl
Computing for Scientists	R.J. Barlow and A.R. Barnett
The Physics of Stars *Second Edition*	A.C. Phillips
Nuclear Physics	J.S. Lilley
Introduction to Quantum Mechanics	A.C. Phillips
Dynamics and Relativity	J.R. Forshaw and A.G. Smith
Vibrations and Waves	G.C. King
Mathematics for Physicists	B.R. Martin and G. Shaw
Particle Physics *Fourth Edition*	B.R. Martin and G. Shaw
Physics of Energy Sources	G.C. King

Physics of Matter

GEORGE C. KING

School of Physics and Astronomy
University of Manchester
United Kingdom

WILEY

This edition first published 2023

Registered Office(s)
John Wiley & Sons, Inc., 111 River Street, Hoboken, NJ 07030, USA
John Wiley & Sons Ltd, The Atrium, Southern Gate, Chichester, West Sussex, PO19 8SQ, UK

For details of our global editorial offices, customer services, and more information about Wiley products visit us at www.wiley.com.

Wiley also publishes its books in a variety of electronic formats and by print-on-demand. Some content that appears in standard print versions of this book may not be available in other formats.

Library of Congress Cataloging-in-Publication Data applied for:

Paper Back ISBN: 9781119468585
ePDF ISBN: 9781119468592
epub ISBN: 9781119468523

Cover Design: Wiley
Cover Image: Courtesy of Ingo Dierking and George C. King

Set in 9.5/12.5pt cmr10 by Straive, Pondicherry, India
Printed and bound by CPI Group (UK) Ltd, Croydon, CR0 4YY

C9781119468585_310723

To my wife Michele

Contents

Editors' preface to the Manchester Physics Series

The Manchester Physics Series is a set of textbooks at first degree level. It grew out of the experience at the University of Manchester, widely shared elsewhere, that many textbooks contain much more material than can be accommodated in a typical undergraduate course; and that this material is only rarely so arranged as to allow the definition of a short self-contained course. The plan for this series was to produce short books so that lecturers would find them attractive for undergraduate courses, and so that students would not be frightened off by their encyclopaedic size or price. To achieve this, we have been very selective in the choice of topics, with the emphasis on the basic physics together with some instructive, stimulating and useful applications.

Although these books were conceived as a series, each of them is self-contained and can be used independently of the others. Several of them are suitable for wider use in other sciences. Each Author's Preface gives details about the level, prerequisites, etc., of that volume.

The Manchester Physics Series has been very successful since its inception over 40 years ago, with total sales of more than a quarter of a million copies. We are extremely grateful to the many students and colleagues, at Manchester and elsewhere, for helpful criticisms and stimulating comments. Our particular thanks go to the authors for all the work they have done, for the many new ideas they have contributed, and for discussing patiently, and often accepting, the suggestions of the editors.

Finally we would like to thank our publishers, John Wiley & Sons, Ltd., for their enthusiastic and continued commitment to the Manchester Physics Series.

J. R. Forshaw
H. F. Gleeson
F. K. Loebinger
June 2023

Author's preface

One of the fundamental discoveries of science is that matter is composed of atoms and molecules. The book describes the nature of matter in terms of these particles and the forces that bind them together. It aims to show how the microscopic properties of the particles can be related to the macroscopic properties of matter. The interatomic forces give rise to the potential energies of the atoms and molecules. These also have thermal energy, and a theme running through the book is the competition between these potential and thermal energies. Whichever dominates determines the state in which the matter exists – the gaseous, liquid, solid, or liquid crystalline state. These states of matter are the subjects of corresponding chapters of the book. The number of molecules involved in any amount of matter is extraordinarily large and, consequently, statistical methods must be used to describe their average behaviour. The book introduces students to these powerful theoretical methods and related concepts such as probability distributions, which are used in many branches of physical science and beyond. The book presents the kinetic theory of gases and its applications. Thus, a unified treatment of transport phenomena of viscosity, heat conduction, and self-diffusion is presented. Classical thermodynamics provides a complementary description of the thermal properties of matter. The book presents the principles of thermodynamics and demonstrates their application. A connection is made between macroscopic quantities such as entropy and the statistical behaviour of the atoms and molecules. Modern experimental techniques provide the ability to 'see' atoms and to determine the structure of highly complex molecules such as biological molecules. Where appropriate, these experimental techniques are described.

The book is based on an introductory 24-lecture course given by the author at the University of Manchester. Chapters on the first and second laws of physics and liquid crystals have been added. The course was attended by first- and second-year undergraduate students taking physics or a joint honours degree course with physics but the book should also be useful to students in chemistry and engineering. Basic knowledge of differentiation and integration is assumed and simple differential equations are used, while undue mathematical complication and detail are avoided.

The organisation of the book is as follows. Chapter 1 deals with the basic properties of atoms, such as their mass and size. The Bohr theory of the atom is described. Although this is a classical model, it can, within certain limits, make useful predictions about atomic energy levels. This chapter also gives an introduction to the quantum mechanical description of atoms. Chapter 2 deals with the forces that bind atoms together and various types of bonding. The general characteristics of atomic forces are discussed and the potential energy between two atoms is described in terms of the Lennard-Jones 6–12 potential. This chapter also gives a first discussion of why matter takes gaseous, liquid, or solid form. Chapter 3 discusses the thermal energy of atoms and molecules including the ideal gas law. It introduces the concept of probability distributions. The law of Boltzmann is described and various examples of its use are discussed including the Maxwell–Boltzmann speed distribution and the isothermal atmosphere. The theorem of the equipartition of energy is introduced and applied to various microscopic and macroscopic systems. This leads to a discussion of the specific heat of gases and the breakdown of the classical theory of specific heats. Chapter 4 introduces and develops the kinetic theory of gases. It shows how this theory, based on molecular collisions and the mean free path of the molecules, describes various transport phenomena such as the diffusion, thermal conduction, and viscosity of gases. This leads to a more general discussion of the random walk problem.

Chapter 5 extends the discussion of ideal gases to real gases. In particular, the van der Waals equation and the virial equation are described, which take account of the finite size of molecules and the interatomic forces of attraction. This leads naturally to a discussion of the phase diagram of a substance. In contrast to the first five chapters, Chapters 6 and 7 take a macroscopic view of matter, i.e. the complementary treatment provided by classical thermodynamics. Chapter 6 deals with the first law of thermodynamics. The relationship between heat and work is discussed, and the concept of internal energy is introduced. Various reversible and irreversible processes are analysed including the expansion and compression of a gas under various conditions. In this chapter, the molar specific heats of an ideal gas are obtained and the concept of enthalpy is introduced. Chapter 7 deals with the second law of thermodynamics. Heat engines and the ideal Carnot engine are discussed along with refrigerators and heat pumps. This naturally leads to a discussion of entropy and various reversible and irreversible processes are described in terms of the change in entropy. The fundamental thermodynamic relationship is presented and phase changes are discussed in terms of the Clausius–Clapeyron equation. The concept and application of Gibbs free energy are also described. Maxwell's relations are obtained and a statistical approach to the second law is presented. In Chapter 8, attention is turned to the solid state of matter. Various types of crystal structures are described and the classification of crystal structure in terms of a crystal lattice, unit cell, and basis is presented. Bragg's law is obtained and various experimental techniques of X-ray crystallography are described along with the complimentary technique of neutron scattering. The macroscopic properties of solids, heat of sublimation, surface energy, and thermal expansion are described in terms of the interatomic forces. Chapter 9 deals with the elastic moduli of solids: Young's modulus, shear modulus, and bulk modulus and also Poisson's ratio. Connections between the three moduli are established, and the relationship between elastic moduli and interatomic forces is obtained. Torsional stress and strain are also described. Chapter 10 describes the thermal and transport properties of solids. In particular, the Einstein model of specific heat of solids is presented. The transport properties of diffusion, thermal, and electrical conductivities are also discussed. Chapter 11 deals with the physical properties of liquids, including the latent heat of evaporation, vapour pressure surface energy, and diffusion. It describes the flow of ideal liquids including the continuity equation and Bernoulli's equation and also the viscous flow of real liquids. Finally, Chapter 12 deals with the liquid crystal phase of matter. The physical features of the liquid crystal phase are described together with various types of liquid crystals. Practical liquid crystal devices usually exploit the birefringence of the liquid crystal material and the polarisation properties of light, and so, these topics are also included.

Worked examples are provided in the text. In addition, each chapter is accompanied by a set of problems that form an important part of the book. These have been designed to deepen the understanding of the reader and develop their skill and self-confidence in the use of physics. Hints and solutions to these problems are given at the end of the book. It is, of course, beneficial for the reader to try to solve the problems before consulting the solutions.

I am particularly indebted to Fred Loebinger who was my editor throughout the writing of the book. He read the manuscript with great care and physical insight and made numerous valuable comments and suggestions. I am grateful to the editors of the Manchester Physics Series for helpful suggestions regarding the content of the book and to Jenny Cossham, Martin Preuss, and Lesley Fenske of Wiley for their valuable assistance. I am grateful to Helen Gleeson for her comments and suggestions regarding the chapter on liquid crystals and to my wife, Michele Siggel-King for her encouragement and patience throughout the writing of the book.

George C. King
June 2023

1

Atoms, the constituents of matter

We now take it for granted that matter is made of atoms. Indeed, using modern imaging techniques, we can even 'see' individual atoms in matter, as we will describe in this chapter. But this fact is of enormous importance. Richard Feynman has said that if just one sentence of scientific knowledge were to survive to be passed to future generations, it would be that *all things are made of atoms – little particles that move around in perpetual motion, attracting each other when they are a little distance apart, but repelling upon being squeezed into one another.* And so, an understanding of matter begins with a study of atoms. In this chapter, we describe the main properties of atoms and their basic electronic structure; their mutual attraction and repulsion will be discussed in Chapter 2. We will see that atoms are extremely small and that any ordinary amount of matter contains an enormous number of atoms. And in this lies the apparently continuous nature of matter on an ordinary scale.

1.1 The mass of an atom

An atom has a well-defined mass and its mass can be determined with high accuracy. In practice, however, it is not the mass of a neutral atom that is measured, but the mass of the *ionized* atom. This is because the motion of a positively charged ion can be readily manipulated by electric or magnetic fields, which is done in a *mass spectrometer*. The atom is ionized by removing one of its electrons and, of course, the mass of the missing electron can be readily taken into account.

One type of mass spectrometer is the *time of flight mass spectrometer*. Its principle of operation is illustrated schematically in Figure 1.1. The atoms (or molecules) in a gas or vapour of the sample are ionized by a short pulse of energetic electrons or photons. The action of the electrons or photons is to knock out one of the atomic electrons, producing a positive ion. The bunch of ions that is produced passes into an *acceleration region*. A potential difference, V_{acc}, is maintained across this region that accelerates the ions, and they leave the acceleration region with kinetic energy given by

$$\frac{1}{2}Mv^2 = V_{\text{acc}}e, \qquad (1.1)$$

Physics of Matter, First Edition. George C. King.

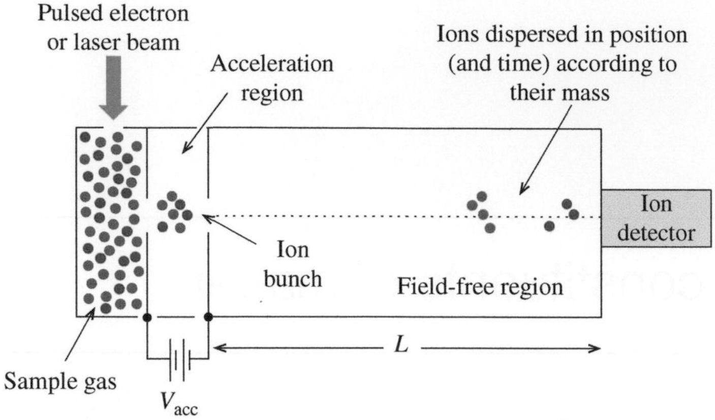

Figure 1.1 Schematic diagram of a time of flight mass spectrometer, illustrating its principle of operation. The atoms (or molecules) in a gas or vapour of the sample are ionized by a short pulse of energetic electrons or photons. The bunch of positive ions that is produced passes into the acceleration region where they are accelerated to a final kinetic energy $V_{acc}e$. The ions then travel through a field-free region, with individual ions travelling at constant velocity. The lighter ions in the bunch travel at a higher velocity than the heavier ones and strike the detector before the heavier ions. Hence, the ions are dispersed in time (and position) according to their mass; the variation in mass has been translated into a variation in arrival time at the detector.

where M and v are the mass and velocity of the ion, respectively, and e is the electronic charge. The ions then travel through a *field-free region* of length L. Because this region is field-free, i.e. there is no potential difference across it, the individual ions travel at constant velocity in this region. The ions then strike an ion detector at the far end of the region. The lighter ions in the bunch travel at a higher velocity and strike the detector before the heavier ions. Hence, the ions are dispersed in time (and position) according to their mass; their variation in mass has been converted into a variation in arrival time at the detector. The time t that the ions take to transverse the field-free region is recorded. Since $v = L/t$, we have

$$t = L\left(\frac{M}{2V_{acc}e}\right)^{1/2}, \tag{1.2}$$

and

$$M = \frac{2V_{acc}e}{L^2}t^2. \tag{1.3}$$

A mass spectrum is obtained by plotting the yield of detected ions versus t^2. Such a mass spectrum is shown schematically in Figure 1.2 for the example of a sample of chlorine gas. This element has two main isotopes ^{35}Cl and ^{37}Cl, where the superscript is the *mass number* of the isotope. The element has *isotopic abundances* in the ratio 3 : 1 and two peaks are observed in the mass spectrum. Indeed, the existence of isotopes was first discovered in a mass spectrometer by J.J. Thompson. If there are different gases in a sample, this will give rise to different peaks in the mass spectrum. In this way, the composition of the sample gas can be determined.

Mass spectrometers are used extensively in research and industry and for a variety of purposes. For example, a time of flight mass spectrometer was one of the instruments on board the *Rosetta* spacecraft,

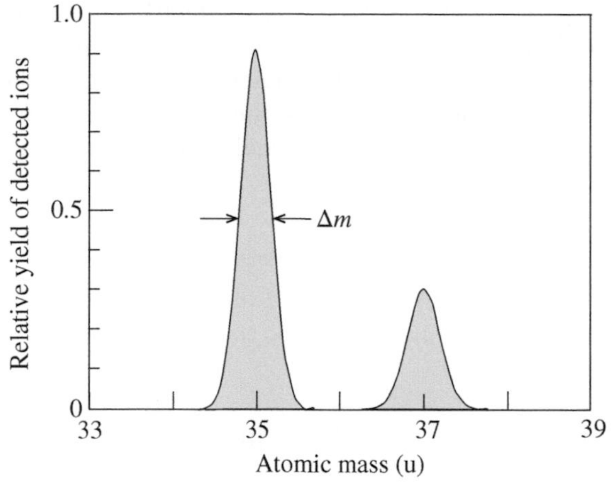

Figure 1.2 A mass spectrum of chlorine gas. Chlorine has two main isotopes ^{35}Cl and ^{37}Cl with isotopic abundances in the ratio 3 : 1 and two peaks are observed in the spectrum. Δm is the width of each peak in the spectrum, measured at half peak height, and is a measure of the ability of the mass spectrometer to resolve individual mass peaks.

which landed on comet Churyumov-Gerasimenko in 2014. The primary role of the spectrometer was to identify species present in the coma of the comet, which is the nebulous envelope of atoms and ions that surrounds the comet. This coma varies with distance of the comet from the sun and also the comets' rotation. A photograph of the time of flight spectrometer is shown in Figure 1.3. The relatively low weight of a time of flight spectrometer is an important advantage for rocket flights. A different type of spectrometer, which allows atomic masses to be determined with very high precision, is the *magnetic mass spectrometer*. In this type, the ions are passed between the poles of a large magnet. The magnet field disperses ions according to their mass; analogous to the way a glass prism disperses light according to wavelength. Using this

Figure 1.3 A photograph of the time of flight mass spectrometer that was aboard the Rosetta spacecraft that landed on comet Churyumov–Gerasimenko. The primary role of the spectrometer was to identify species present in the coma of the comet. The relatively low weight of a time of flight mass spectrometer is an important advantage for space flights. Source: ESA/Rosetta/ROSINA/UBern/BIRA/LATMOS/LMM/IRAP/MPS/SwRI/TUB/UMich/European Space Agency.

magnetic type of spectrometer, atomic masses can be measured with a precision of 1 part in 10^7 or 10^8. It is valuable to have this high precision when, for example, it is required to predict the amount of energy that would be produced in a particular nuclear reaction. If we know the total mass of the interacting nuclei before the reaction and the total mass of the reaction products, the mass difference Δm can be converted into the released energy via Einstein's equation of mass-energy equivalence $E = \Delta mc^2$.

1.1.1 Atomic masses

Precise measurements give the mass of a ^{12}C atom to be 1.992647×10^{-26} kg. A more natural and convenient unit is the *atomic mass unit* (u). This unit is defined so that the mass of a ^{12}C atom is taken to be exactly 12 u. In terms of the kilogram:

$$1\,\mathrm{u} = 1.660539 \times 10^{-27}\,\mathrm{kg}.$$

Other atomic masses are measured on this relative scale. For example, the atomic mass of ^{20}Ne, the most abundant isotope of neon, is 19.992436 u. We can also have the *relative atomic mass* A_r of an atom, which is given by

$$A_r = \frac{\text{Mass of the atom}}{\text{Mass of a }^{12}\text{C atom}} \times 12. \tag{1.4}$$

Clearly, being a ratio, A_r is a number and has no units. So, for example, the relative atomic mass of ^{20}Ne is 19.99, rounded to two significant figures.

Most elements in their natural state consist of several or more isotopes. In that case, we take a weighted mean of the atomic masses of the individual isotopes according to their natural abundance. This is called the *mean atomic mass*. Carbon has two stable isotopes: ^{12}C and ^{13}C with atomic masses of 12 u and 13.003355 u, respectively, and relative natural abundances of 98.89% and 1.11%, respectively. Hence, the mean atomic mass of carbon is

$$\left(\frac{98.89}{100} \times 12\right)\mathrm{u} + \left(\frac{1.11}{100} \times 13.003355\right)\mathrm{u} = 12.011\,\mathrm{u}.$$

Worked example

In a particular time of flight mass spectrometer, the length L of the field-free region is 50 cm, the acceleration voltage V_{acc} is 100 V, and the pulse width Δt of the electron beam that is used to ionize the atoms is 0.1 μs. Would this mass spectrometer be able to resolve the two main isotopes of chlorine, ^{35}Cl and ^{37}Cl?

Solution

The finite width of the electron pulse results in an uncertainty Δt in the arrival time of the ions at the detector, and, in turn, this leads to an uncertainty Δm in the measured mass of the isotopes. From Equation (1.3), we have

$$M = \frac{2V_{acc}e}{L^2}t^2.$$

$$\frac{\Delta M}{\Delta t} \approx \frac{\mathrm{d}M}{\mathrm{d}t} = \frac{2V_{\mathrm{acc}}e}{L^2}2t = \frac{M}{t^2}2t.$$

Hence,

$$\frac{\Delta M}{M} \approx 2\frac{\Delta t}{t}.$$

The factor of 2 occurs because $m \propto t^2$. The ratio $M/\Delta M$ is called the *mass resolution* of the spectrometer.

$$t = L\left(\frac{M}{2V_{\mathrm{acc}}e}\right)^{1/2} = 50 \times 10^{-2}\left(\frac{36 \times 1.66 \times 10^{-27}}{2 \times 100 \times 1.6 \times 10^{-19}}\right)^{1/2} = 21.6\ \mu\mathrm{s},$$

where we have taken $M = (35 + 37)/2 = 36$. Hence,

$$\Delta M \approx 2M\frac{\Delta t}{t} = 2 \times 36 \times \frac{0.1}{21.6} = 0.33\ \mathrm{u}.$$

ΔM is much smaller than the mass difference (2 u) between the two isotopes, which could therefore be easily resolved by the spectrometer.

Avogadro's number and the mole

The number of atoms in 12 g (0.012 kg) of ^{12}C is

$$\frac{0.012\ \mathrm{kg}}{1.992647 \times 10^{-26}\ \mathrm{kg}} = 6.022 \times 10^{23}.$$

This is an important number, which is called *Avogadro's number*, and is given the symbol N_A. Again, it is convenient to have units that result in numerical values that are not too large. For the measure of an amount of substance, an appropriate unit is the *mole* (mol). It is defined as the amount of substance that contains as many elementary entities as there are atoms in 12 g (0.012 kg) of ^{12}C. Entities include atoms, molecules, and ions. Hence, 1 mol of helium contains 6.022×10^{23}atoms of helium atoms, 1 mol of oxygen gas contains 6.022×10^{23} molecules of O_2, and 1 mol of copper contains 6.022×10^{23} atoms of copper.

Suppose that we measure out a mass, *in grams*, of an element numerically equal to its relative atomic mass A_r. The number of atoms in this amount of the element is

$$\frac{A_r \times 10^{-3}\ \mathrm{kg}}{\text{Mass of atom (kg)}}, \tag{1.5}$$

which on substituting for A_r from Equation (1.4) is

$$\frac{12 \times 10^{-3}\ \mathrm{kg}}{\text{Mass of a }^{12}\text{C atom (kg)}} = N_A. \tag{1.6}$$

We see that a mass A_r g of an element, having relative atomic mass A_r, contains Avogadro's number of atoms. And the amount of substance that this number corresponds to is 1 mol. For example, 1.2 kg of lead ($A_r = 207.3$) contains

$$\frac{1.2}{207.3 \times 10^{-3}} \times 6.022 \times 10^{23} = 3.5 \times 10^{24} \text{ atoms or } 5.8 \text{ mol.}$$

1.2 The size of an atom

Atoms do not have a sharp edge, as we will discuss in Section 1.3. Nevertheless, from a practical point of view, atoms do possess what amounts to a definite 'size'. We can see this from the fact that solids are essentially incompressible, which means that we cannot push atoms into each other. A number of ingenious methods were initially used to measure the sizes of atoms and molecules. In one method, a drop of a liquid that did not mix with water was placed onto a smooth body of water. The liquid spread out over the water and by measuring the area of the dispersed liquid, and knowing the volume of the initial drop of liquid, an upper limit to the diameter of the molecules in the liquid could be deduced. Now, using X-ray crystallography, we can determine the spacing between lattice planes in a crystal (see Section 8.4). Since crystals resist being compressed, we can take this separation as a measure of the size of the atoms.

We can estimate the size of an atom in the following way. We imagine the atoms in a solid to be packed together as illustrated by Figure 1.4. If the atomic radius is R, each atom will occupy a volume $\approx(2R)^3$. We recall that a mass A_r of an element contains Avagadro's number of atoms. And this mass has a volume $V = A_r/\rho$, where ρ is the density of the element. Hence, we have

$$V = \frac{A_r}{\rho} \approx N_A(2R)^3, \tag{1.7}$$

giving

$$R \approx \frac{1}{2}\left(\frac{A_r}{\rho N_A}\right)^{1/3}. \tag{1.8}$$

Table 1.1 lists the estimated radius of a number of elements across the periodic table that have been obtained using Equation (1.8). Also shown, in the last column of the table, are more accurate values of

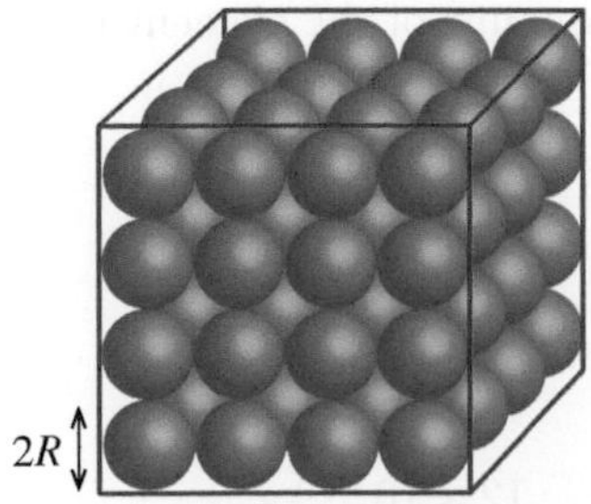

Figure 1.4 We can estimate the size of an atom by imagining the atoms in a solid to be packed together as in this figure. If the atomic radius is R, each atom will occupy a space of volume $\approx(2R)^3$.

Table 1.1 Estimated atomic radii obtained from Equation (1.8) for a number of elements across the periodic table. Also shown, in the final column, are accurate values of the radii obtained by sophisticated theoretical calculations. Note the small variation in atomic size with respect to atomic mass.

Element	ρ (g/cm^3)	A_r	Estimated atomic radius (nm)	Accurate atomic radius (nm)
Beryllium	1.85	9.01	0.10	0.11
Aluminium	2.70	26.98	0.13	0.12
Iron	7.87	55.85	0.16	0.16
Molybdenum	10.20	95.94	0.13	0.19
Barium	3.50	137.33	0.20	0.25
Platinum	21.50	195.08	0.13	0.18
Uranium	19.10	238.03	0.14	0.18

the atomic radii that have been obtained by sophisticated theoretical calculations. Despite the simplicity of our model, the two sets of values for the atomic radii are in reasonably good agreement. What is most revealing about the table is that light and heavy atoms have much the same size. As we see from Table 1.1, atomic radii lie typically within the range 0.1–0.3 nm. The essential reasons for this are as follows. As the mass of the elements increases, so also does the charge Ze on the nucleus, where Z is the *atomic number*. The electrons that are close to the nucleus feel the strong attraction of the nucleus and are drawn close to it. These electrons provide electrostatic shielding for electrons that are further away from the nucleus. Indeed the electrons that are furthest from the nucleus are almost completely shielded from the electrostatic attraction of the nucleus. And so they feel an attraction that is not so different from an electron in the singly charged ($Z = 1$) hydrogen atom. The size of an atom is determined solely by the electrons surrounding the nucleus, which has a size of $\sim 10^{-15}$m, while nearly all the mass of the atom is concentrated in the nucleus.

1.2.1 Scanning probe microscopy

We use a microscope to see small objects. However, the ultimate resolution of an optical microscope is limited by diffraction of the light that we use to observe the object. The diffraction limit, i.e. just how small an object we can resolve, is roughly half the wavelength λ of the light. Visible light spans the wavelength range from about 400 to 750 nm, and in practice, the best we can do with traditional optical microscopy is to resolve objects about 200 nm across. This is almost three orders of magnitude larger than the diameter of an atom, and clearly it is not possible to resolve atoms or molecules with an optical microscope. However, our ability to 'see' atoms changed dramatically in 1982 with the invention of *scanning probe microscopy* by Gerd Binnig and Heinrich Rohrer, an invention for which they were awarded the Nobel Prize in Physics in 1986. The principle of a scanning probe microscope is illustrated schematically in Figure 1.5. The microscope has a quartz crystal to which is attached an *extremely* sharp tip. This tip is positioned less than about 1 nm above the surface of the sample to be imaged. Crystalline quartz has a property called *piezoelectricity*. This means that when a potential difference is applied across the ends of a quartz crystal, the dimensions of the crystal change. Hence, by applying an appropriate voltage across the quartz crystal in the microscope, the tip can be moved across the surface of the sample. Indeed the tip can be moved in steps of less than 0.01 nm. In a similar way, the height of the tip above the surface can be controlled. Scanning probe microscopy exploits interactions that may exist between the tip and the atoms at the surface of the sample; there are no lenses or mirrors. Such tip-surface interactions produce a signal that can be measured. The tip is scanned in successive rows across the surface and as it does, the signal varies depending on the topology of the surface.

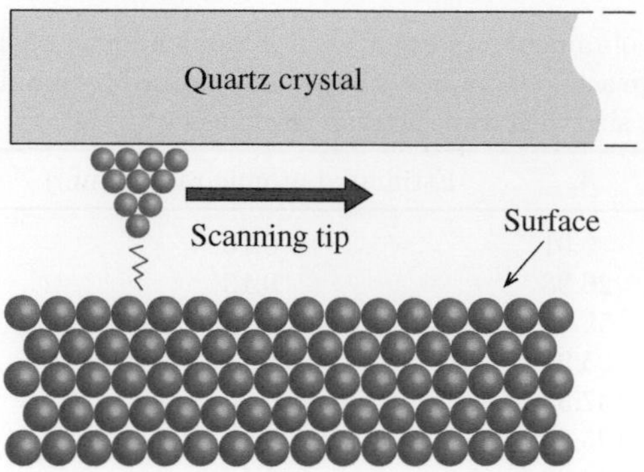

Figure 1.5 A schematic diagram to illustrate the principle of a scanning probe microscope. The microscope has a quartz crystal to which is attached an *extremely* sharp tip, which is positioned less than about 1 nm above the surface of the sample to be imaged. By applying a suitable voltage across the quartz crystal, which is piezoelectric, the tip can be moved in successive rows across the surface of the sample in steps of less than 0.01 nm. An interaction between the tip and the atoms at the surface of the sample produces a signal that can be measured. As the tip is scanned across the surface, the signal varies depending on the topology of the surface. The recorded signal is then processed to produce an image of the surface. Sub-atomic resolution is achieved because tip-surface interactions are very short range and because of the sharpness of the tip.

The recorded signal is then processed to produce an image of the surface. Sub-atomic resolution can be achieved, with details less than 0.1 nm being resolvable. This is because the tip-surface interactions have a very short range and because the tip can be made atomically sharp.

There are various scanning probe techniques that exploit the different possible interactions between the tip and the surface. *Scanning tunnelling microscopy* (STM) is based on the measurement of an electric current that flows between the tip and the surface. This current flow is due to *quantum mechanical tunnelling* of electrons across the narrow gap between the tip and surface. *Atomic force microscopy* (AFM) exploits various interatomic forces that the surface exerts on the tip. For example, this may be the van der Waals force that we will discuss in Section 2.3. The force can be detected as a deflection of the crystal to which the tip is attached or as a change in the vibrational frequency of the tip-crystal assembly.

The principle of one type of AFM, which exploits the change in the vibrational frequency of the tip-crystal assembly, is illustrated in Figure 1.6. This technique employs a quartz tuning fork, exactly like the kind that is found in a quartz watch. An atomically sharp tip is attached to one of the prongs of the tuning fork. The piezoelectric effect is exploited to drive the quartz tuning fork into oscillation by applying across it an AC voltage. Conversely, mechanical vibrations of the tuning fork produce an AC voltage signal, and this signal is used to determine the frequency of the vibrating tuning fork. The tuning fork is driven at its resonance frequency. The normal resonance frequency f_0 of such a quartz tuning fork is $2^{15} = 32.768$ kHz, but this is reduced somewhat by the addition of the sharp tip.

f_0 is the frequency of the tuning fork when it is far above a surface. When the tip is positioned just above a surface, the atoms on the surface exert a short-range force on the tip. This force causes the resonance frequency of the tuning fork to change to frequency f. The output frequency of the oscillator is adjusted to match the new resonance frequency f and the frequency difference $\Delta f = (f - f_0)$ is measured. The frequency difference Δf is recorded as the tip is scanned in rows across the surface. The recorded variation in Δf is then processed to produce an image of the surface.

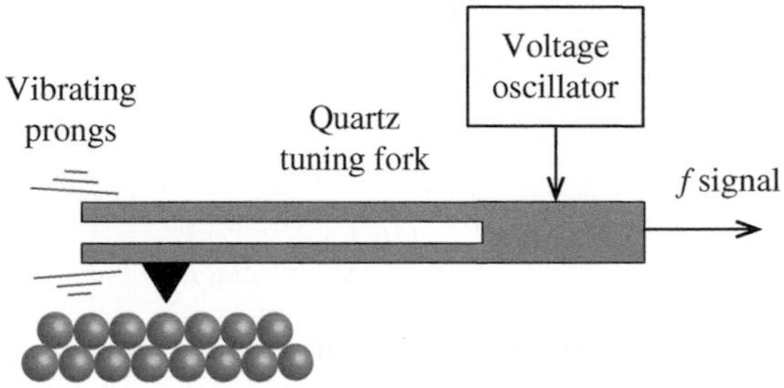

Figure 1.6 A schematic diagram to illustrate the principle of one type of atomic force microscope. An atomically sharp tip is attached to one of the prongs of a quartz tuning fork, exactly like the kind in a quartz watch. The piezoelectric effect is used to drive the quartz tuning fork into oscillation by the application of an AC voltage. The mechanical vibrations of the tuning fork produce an AC voltage signal, so that the vibrational frequency of the tuning fork can be measured. The tuning fork is driven at its resonance frequency. When the tip is positioned just above a surface, the atoms on the surface exert a short-range force on the tip, which causes the resonance frequency f_0 of the tuning fork to change to frequency f. The frequency difference $\Delta f = (f - f_0)$ is measured as the tip is scanned across successive rows of the surface. The recorded variation in Δf is processed to produce an image of the surface.

We can model the quartz tuning fork and its change in resonance frequency by the more familiar system of a mass m on the end of a spring of force constant k_0; see Figure 1.7a. The resonant frequency of vibration of the mass-spring system is

$$f_0 = \frac{1}{2\pi}\sqrt{\frac{k_0}{m}}. \tag{1.9}$$

If we add an additional spring of force constant k to the mass as in Figure 1.7b, the extra spring exerts an additional force on the mass and the resonant frequency changes to $f = (1/2\pi)\sqrt{(k_0 + k)/m}$. The frequency difference Δf is

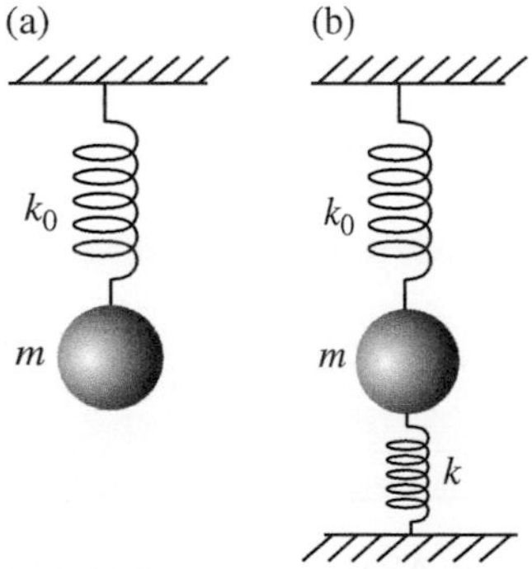

Figure 1.7 A model of the quartz tuning fork in an AFM instrument based on a mass-spring system. (a) This corresponds to the situation where the tip of the instrument is not close to the surface. The resonance frequency of the mass-spring system is $f_0 = (1/2\pi)\sqrt{k_0/m}$. (b) This corresponds to the situation where the tip of the instrument is very close to the surface. By attaching a second spring, the resonance frequency of the mass-spring system increases to $f = (1/2\pi)\sqrt{(k_0 + k)/m}$. To a good approximation $\Delta f = (f - f_0) = (f_0 - k_0)k$, when $k_0 \gg k$.

$$\Delta f = f - f_0 = \frac{1}{2\pi\, m^{1/2}}\left[(k_0 + k)^{1/2} - k_0^{1/2}\right]. \tag{1.10}$$

If $k_0 \gg k$, then to a good approximation

$$(k_0 + k)^{1/2} = k_0^{1/2}\left(1 + \frac{k}{2k_0}\right). \tag{1.11}$$

Substituting this result into Equation (1.10) and substituting for m from Equation (1.9), we obtain

$$\Delta f = \frac{f_0}{k_0} k. \tag{1.12}$$

The change in frequency Δf is a measure of the strength k of the additional force acting on the spring. Similarly, in the case of the quartz tuning fork, the difference Δf in its resonance frequency is a measure of the strength of the force acting on the tip due to the surface. Different parts of the surface will exert different amounts of force on the tip. Hence, the recorded variation in Δf can be processed to produce an image of the surface or indeed of a molecule attached to the surface.

The bottom half of Figure 1.8 shows an AFM image that was obtained of a pentacene molecule ($C_{22}H_{14}$), which was attached to an atomically flat surface. The image is essentially the measured variation of Δf for

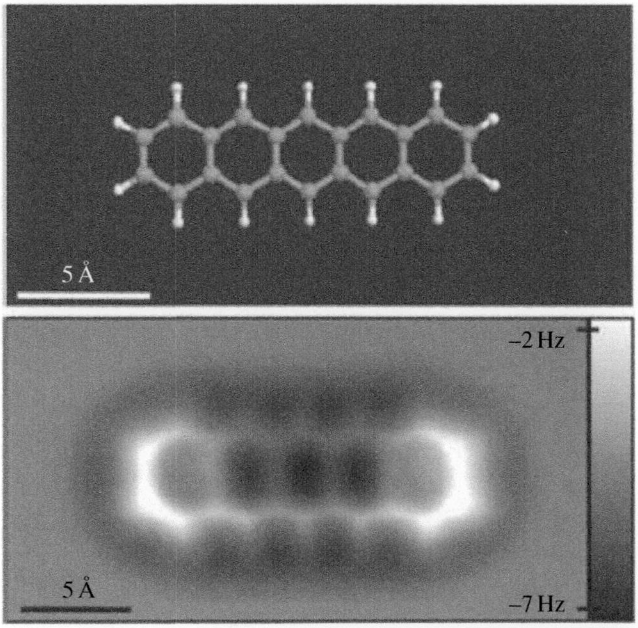

Figure 1.8 The bottom half of this figure shows an AFM image that was obtained of a pentacene molecule ($C_{22}H_{14}$). The image is essentially the measured variation of Δf for the tuning fork as the tip was scanned across the molecule. In this case, the end of the tip was terminated in a *single* atom of carbon monoxide that had been attached to the tip. The image clearly shows the structure of the pentacene molecule and its internal bonds. Indeed the AFM image closely resembles the ball and stick construction that is traditionally used to represent the molecule, which is shown in the top half of the figure. Source: Gross et al. (2009)/Reproduced from American Association for the Advancement of Science – AAAS.

the tuning fork as the tip was scanned across the molecule. In this case, the end of the tip was terminated in a *single* atom of carbon monoxide that had been attached to the tip. The image clearly shows the structure of the pentacene molecule and its internal bonds. Indeed the AFM image closely resembles the ball and stick construction that is traditionally used to represent the molecule, which is shown in the top half of Figure 1.8. In an extension of AFM, the tip can be used to manipulate the position of single atoms on a surface and hence to produce artificial structures on the atomic scale, i.e. *nanostructures*.

Worked example

Compare the number of water molecules in a teaspoon of water with the number of teaspoons of water in all the oceans of the world.

Solution

The molecular weight of water is 18 u and so 18 g of water contains Avagadro's number N_A of molecules. Taking the volume of a teaspoon to be 5 ml, it contains 5 g of water. The number of molecules in 5 g water

$$= 5 \times \frac{6.02 \times 10^{23}}{18} = 1.7 \times 10^{23}.$$

About 97% of Earth's water is found in its oceans. According to the United States Geological Survey, there are 1.3×10^{18} m^3 of water in the oceans. 1 m^3 = 10^6 ml.

Hence, the number of teaspoons of water in the oceans

$$= \frac{1.3 \times 10^{18} \times 10^6}{5} = 2.6 \times 10^{23}.$$

We see that the number of water molecules in a teaspoon of water is roughly the same as the number of teaspoons of water in all the oceans of the world.

1.3 Atomic structure

By the beginning of the twentieth century, the *emission spectra* of a variety of atoms had been measured. These were obtained by exciting the atoms in an electrical discharge and analysing the emitted radiation with either a prism or a diffraction grating in an *optical spectrometer*. The resulting emission spectra appear as a set of *discrete* spectral lines of different wavelengths (or colours) that are characteristic of the atom. Quite early on, it was realised that these spectra provide valuable information about the atoms that produce them. The scientist Anders Ångström measured the emission spectrum of atomic hydrogen and observed a *series* of four visible spectral lines. The spectrum of hydrogen is illustrated in Figure 1.9, showing the

Figure 1.9 The emission spectrum of hydrogen that lies in the visible region of the electromagnetic spectrum, showing the four lines of the Balmer series.

different colours of the lines: violet, blue, green, and red. Then in 1885, Johann Balmer, a school teacher, discovered that the wavelengths of the four hydrogen lines could be beautifully fitted by the simple formula

$$\lambda = 364.6\frac{n^2}{(n^2-4)}\ \text{nm}, \tag{1.13}$$

where $n = 3, 4$, or 5. Clearly, any successful theory of atomic structure would need to provide an explanation for this formula. In 1911, Rutherford proposed his model of the atom in which all the positive charge of the atom and essentially all its mass are concentrated in a tiny region called the nucleus. And around this nucleus, the atomic electrons circulate. Rutherford based his model on the observation that alpha particles could be scattered through a large angle by traversing a single atom, which could only happen if the nucleus was extremely small ($\sim 10^{-15}$ m). The physicist Niels Bohr worked for a brief period in Rutherford's laboratory in Manchester. And in 1913, adopting Rutherford's model, he devised a theory of the hydrogen atom. Bohr's theory was spectacularly successful in accounting for the observed hydrogen spectrum and for this work, Bohr was awarded the 1922 Nobel Prize in Physics.

1.3.1 The Bohr model of the hydrogen atom

In the Bohr model of the hydrogen atom, a single electron moves in a circular orbit around the nucleus, a proton; in a somewhat analogous way in which a planet orbits the Sun. Bohr's model is based on classical physics and herein lie two particular difficulties. Since the electron is continuously changing direction in its circular orbit, it is continuously accelerating. According to classical physics, an accelerating charge emits electromagnetic radiation. Therefore the electron should continuously lose energy and spiral into the nucleus. Cleary, it does not; atoms are observed to be stable. The other difficulty is that classical theory allows for all possible orbits and all orbital radii. But, the spectral lines observed in emission spectra are discrete, i.e. the emitted photons have a well-defined wavelength and energy. This indicates that electrons occupy energy levels with fixed energies so that when an electron makes a transition between them, the photon that is emitted also has a well-defined energy. To circumvent these difficulties Bohr made a number of postulates; specific quantitative predictions that could be tested by experiment. His postulates are:

1. An electron in an atom moves in a circular orbit about the nucleus under the influence of the electrostatic attraction between the electron and the nucleus, obeying the laws of classical mechanics.

2. Instead of the infinity of orbits that would be possible in classical mechanics, it is only possible for an electron to move in an orbit for which its orbital angular momentum L is an integral multiple of a fundamental unit of angular momentum, denoted as $\hbar$, which is Planck's constant h divided by 2π. Note that this postulate introduces the concept of *quantization*. Hence, we have

$$L = n\hbar, \quad n = 1, 2, 3, \ldots. \tag{1.14}$$

3. Despite the fact that the electron is accelerating, it moves in an allowed orbit without radiating electromagnetic energy and its total energy E remains constant.

4. Electromagnetic radiation is emitted if an electron discontinuously makes a transition from an orbit of total energy E_i to an orbit of energy E_f. The frequency ν of the radiation is given by

$$\nu = \frac{(E_i - E_f)}{h} \tag{1.15}$$

Note that this equation is just the Planck–Einstein relation that the energy of a photon is equal to $h\nu$. We see that Bohr retained the classical description of the electrostatic interaction between the nucleus and the electron. But he did not retain the classical behaviour for the electron to continuously lose energy; with the conclusion that this behaviour does not apply to electrons in atoms.

In the Bohr model, the equation of motion of the electron about the nucleus is

$$\frac{mv^2}{r} = \frac{1}{4\pi\varepsilon_0}\frac{Ze^2}{r^2}, \tag{1.16}$$

where r is the radius of the orbit, v; m and e are the velocity, mass, and charge of the electron, respectively; Ze is the charge on the nucleus ($Z = 1$ for hydrogen); and ε_0 is the permittivity of free space. (Here we have made the approximation that the nucleus is infinitely more massive than the electron, which is reasonable since the ratio of the masses of the proton and the electron is 1836:1.) From Equation (1.14), the orbital angular momentum of the electron is

$$L = mvr = n\hbar. \tag{1.17}$$

Solving for v and substituting into Equation (1.16), we obtain

$$r_n = 4\pi\varepsilon_0\frac{n^2\hbar^2}{mZe^2}. \tag{1.18}$$

We see that the quantization of angular momentum has restricted the possible circular orbits that the electron can occupy. The corresponding radii are proportional to n^2. The smallest orbit has $n = 1$ and taking this value of n, we have

$$r_1 = 4\pi\varepsilon_0\frac{\hbar^2}{me^2} = a_0, \tag{1.19}$$

where a_0 is called the *Bohr radius*. Using the values of the fundamental constants in Equation (1.19), we find that the smallest radius has the value 0.053 nm, a value that is consistent with the characteristic size of an atom, see Table 1.1.

The electron has both kinetic energy K and potential energy V. From Equation (1.16), the kinetic energy of the electron is given by

$$K = \frac{1}{2}mv^2 = \frac{1}{2}\frac{1}{4\pi\varepsilon_0}\frac{Ze^2}{r}. \tag{1.20}$$

The potential energy of the electron is equal to the work done in taking the electron from infinity to distance r from the nucleus:

$$V = \frac{1}{4\pi\varepsilon_0}\int_{\infty}^{r}\frac{Ze^2}{r^2}\,\mathrm{d}r = -\frac{1}{4\pi\varepsilon_0}\frac{Ze^2}{r}. \tag{1.21}$$

The potential energy is negative as the electron is bound to the nucleus, i.e. work must be done to remove the electron from its orbit to infinity. The total energy E of the electron, kinetic plus potential, is then

$$E = K + V = -\frac{1}{2}\frac{1}{4\pi\varepsilon_0}\frac{Ze^2}{r} = -\frac{1}{2}V. \tag{1.22}$$

Using Equations (1.21) and (1.22), we find that in the $n = 1$ orbit, the electron has a negative potential energy of -27.2 eV and the kinetic energy of $+13.6$ eV; the kinetic energy is positive of course. Hence, its total energy is -13.6 eV. Notice that the total energy E is equal to half the potential energy V, and that the kinetic energy K is numerically equal to half the potential energy. These relations hold for the motion of any classical (or quantum) system with a potential of the form $V \propto -(1/r)$, and are examples of the *virial theorem*. Substituting for r from Equation (1.19) into Equation (1.22), we obtain the energy E_n of the nth level:

$$E_n = -\frac{mZ^2e^4}{(4\pi\varepsilon_0)^2 2\hbar^2}\frac{1}{n^2}, \tag{1.23}$$

where n is the corresponding *quantum number* of the level. Evaluating Equation (1.23) with the known values of the fundamental constants and taking $Z = 1$, we obtain the possible energy levels of the hydrogen atom. These are shown in Figure 1.10, together with the corresponding values of n. The energies are given in electron volts (eV). This is a convenient unit of energy to use when dealing with atomic energy levels, and is defined as the energy an electron gains when it falls through a potential of 1 V. In terms of the joule,

$$1\,\text{eV} = 1.602 \times 10^{-19}\,\text{J}.$$

The $n = 1$ level corresponds to the *ground state* of hydrogen, which has the lowest energy, i.e. the most negative energy with a value of -13.6 eV. If we wanted to completely remove the electron from the $n = 1$ level to infinity, we would have to supply 13.6 eV of energy. This is the *binding energy* of the electron. It follows that we can write the energy E_n of the nth level as $E_n = -13.6(1/n^2)$ eV.

If the atom is excited by, for example, an electrical discharge, the electron is promoted to a higher lying level, with $n > 1$. After a short time, $\sim 10^{-8}$ seconds, the electron will return to a lower lying level. If the

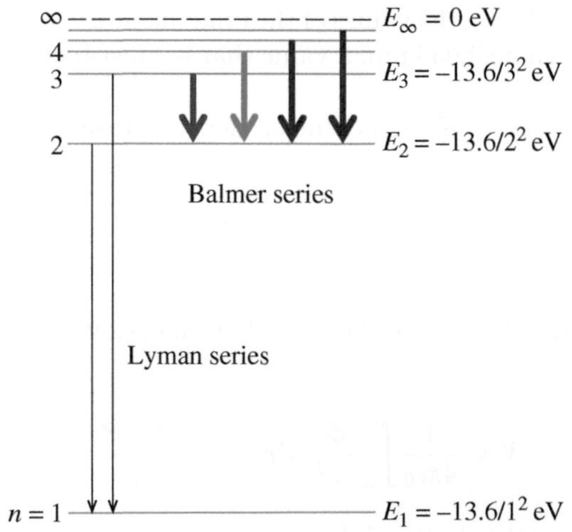

Figure 1.10 The allowed energy levels of the hydrogen atom according to the Bohr model. Also indicated are the values of the quantum number n and the corresponding binding energies for the levels, i.e. the amount of energy required to remove an electron from an orbit to infinity. The four visible emission lines of the Balmer series are also illustrated, as well as two members of the Lyman series.

electron makes a transition between energy levels with quantum number n_i and n_f, respectively, a photon is emitted with an energy given by

$$(E_i - E_f) = \frac{mZ^2 e^4}{(4\pi\varepsilon_0)^2 2\hbar^2}\left(\frac{1}{n_f^2} - \frac{1}{n_i^2}\right). \tag{1.24}$$

The wavelength of this photon is

$$\lambda = \frac{hc}{(E_i - E_f)}, \tag{1.25}$$

in accord with Bohr's fourth postulate. Then substituting for $(E_i - E_f)$ and rearranging Equation (1.25), we obtain

$$\lambda = \frac{4\pi\hbar^3 c(4\pi\varepsilon_0)^2}{mZ^2 e^4} \frac{n_f^2 n_i^2}{(n_i^2 - n_f^2)}. \tag{1.26}$$

Equations (1.25) and (1.26) are the essential predictions of the Bohr model.

When we take $n_f = 2$ and use the accepted values of the fundamental constants in Equation (1.26), we obtain

$$\begin{aligned}\lambda &= \frac{4\pi(1.056\times10^{-34})^3 \times (3\times10^8) \times (4\pi\times8.854\times10^{-12})^2}{9.109\times10^{-31}\times(1.602\times10^{-19})^4} \frac{4n_i^2}{(n_i^2-4)} \\ &= 366.4\frac{n_i^2}{(n_i^2-4)}\ \text{nm}.\end{aligned}$$

We see that Bohr's model agrees exactly with the form of Balmers' formula, Equation (1.13), and moreover, the value of the constant (366.4) is in agreement with that obtained by Balmer, within the accuracy of the experimental measurements available to him. The interpretation of the Balmer formula then is that it represents the wavelengths for the transitions of an electron from a higher lying energy level with $n > 2$, to the $n = 2$ level. Figure 1.10 shows the electron transitions corresponding to the Balmer series, where the colours of the arrows indicate the colour of the emitted radiation.

Equation (1.26) also predicts the wavelengths of the photons emitted when an electron makes a transition to levels with other values of n_f. The series of wavelengths for $n_f = 5$ had in fact already been observed and is called the Pfund series. However, the series for $n_f = 1$, 3 and 4 had not. But when they were looked for, they were indeed observed and their wavelengths were fitted accurately by Equation (1.26). Making predictions is an important test of any theory and these predictions were a triumph for the Bohr model. These series in hydrogen are named after the investigators who first observed them and are respectively the Lyman series, with $n_f = 1$, the Paschen series, with $n_f = 3$, the Brackett series with $n_f = 4$, and the Pfund series with $n_f = 5$.

Bohr's model also explains the sharp spectral lines in the *absorption* spectrum of an atom. As the electron can only be in one of the allowed energy levels, the atom can only absorb discrete amounts of electromagnetic energy, i.e. photons of a particular wavelength. Bohr's model also works well for hydrogen-like atoms such as singly ionized helium, He^+ ($Z = 2$), and doubly ionized lithium, Li^{++} ($Z = 3$).

Despite its spectacular success in explaining the hydrogen spectrum, the Bohr model has significant shortcomings. There is no justification for the postulates of fixed stable orbits, the absence of energy loss of the circulating electrons, or for the quantization of angular momentum except for the fact that the model accurately agrees with experimental data obtained for the hydrogen spectrum. Furthermore, the Bohr model does not explain the spectra of atoms that are more complicated than hydrogen; even the spectrum of helium that has just two electrons. A correct description of the atomic structure had to wait for the arrival of *quantum mechanics.* But despite its shortcomings, the Bohr model gives a useful pictorial representation of the atom that is easy to visualise and, moreover, the mathematics involved are easy to understand. And indeed, the Bohr model is often useful as a first step in explaining a variety of phenomena in matter.

Worked example

In one type of *extrinsic semiconductor*, a tiny amount of phosphorous is added to a crystal of silicon. The phosphorous atoms occupy sites in the crystal lattice normally occupied by silicon atoms. Silicon has four electrons that are involved in bonding with other silicon atoms. But phosphorous has five available electrons, only four of which can bond to the silicon atoms. The remaining electron is then only weakly bound to the phosphorous atom and can easily be removed from it. We can view this situation as a single electron bound to a positively and singly charged phosphorous atom, analogous to the hydrogen atom. Use the Bohr model to determine the radius of the electron's orbit in a phosphorous atom and the amount of energy required to remove it from the atom. Note that when an electron moves in a crystal, it has an effective mass m_e due to the periodic nature of the electrical potential it experiences in the crystal lattice. The relative permittivity ε_r of silicon is 11.6 and the effective mass m_e is $0.26m$, where m is the mass of an electron.

Solution

From Equation (1.19), the radius of the $n = 1$ orbit of the hydrogen atom is

$$r_H = 4\pi\varepsilon_0 \frac{\hbar^2}{me^2} = a_0.$$

Taking $n = 1$ again for the orbit of the electron in the case of the phosphorous atom, we have

$$r_P = 4\pi\varepsilon_r\varepsilon_0 \frac{\hbar^2}{m_e e^2} = 4\pi(11.6)\varepsilon_0 \frac{\hbar^2}{(0.26)me^2} = \frac{11.6}{0.26} \times a_0 = 2.4\,\text{nm}.$$

From Equation (1.23), we find that the energy of the $n = 1$ level of the hydrogen atom is

$$E_H = -\frac{me^4}{(4\pi\varepsilon_0)^2 2\hbar^2} = -13.6\,\text{eV}.$$

This means that the electron is bound by 13.6 eV, and this is the amount of energy that is required to remove it from the atom. For the case of a phosphorous atom,

$$E_P = -\frac{(0.26)me^4}{(4\pi\varepsilon_0)^2(11.6)^2 2\hbar^2} = -\frac{0.26}{(11.6)^2} \times 13.6 = -0.026\,\text{eV}.$$

Thus, the energy required to remove the electron from the phosphorous atom is 0.026 eV. This is about the same as the *thermal energy* a particle has at room temperature. This means that the spare electron in nearly all the phosphorous atoms is liberated and is able to move freely through the silicon crystal. This increases the electrical conductivity of the crystal enormously.

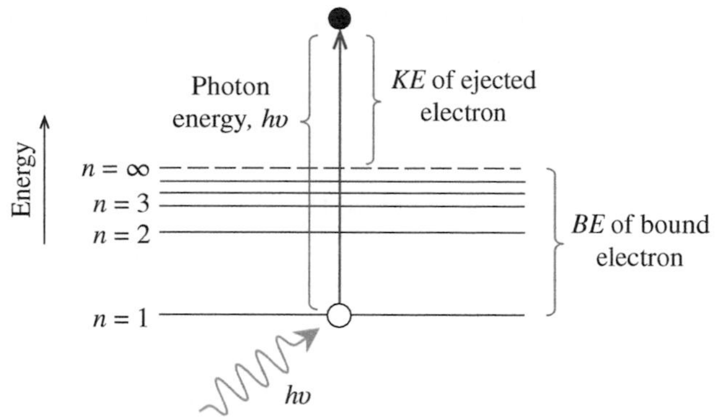

Figure 1.11 An energy level diagram for the situation where an energetic photon ejects the electron from the $n = 1$ orbit of hydrogen. The kinetic energy KE of the ejected electron is equal to the photon energy $h\nu$ minus the binding energy BE of the bound electron.

We have seen that we need to supply 13.6 eV of energy to completely remove the electron from a hydrogen atom. If we excite the atom with a photon that has energy greater than this, the ejected electron will have a non-zero kinetic energy and it will move away from the ionized atom. If we know the energy $h\nu$ of the incident photon and we measure the kinetic energy KE of the ejected electron, we can determine the energy required to remove the electron, i.e. its binding energy BE. The equation to describe this is just

$$BE = h\nu - KE. \tag{1.27}$$

An energy level diagram for this *ionization process* is illustrated in Figure 1.11. Of course, we know the binding energy for atomic hydrogen but we can use this principle to determine unknown binding energies in other atoms. And, in turn, this allows us to map out the energy levels of the atoms. This is the basis of *photoelectron spectroscopy* that is widely used to obtain this kind of information in gases, solids, and even liquids.

1.3.2 The Schrödinger equation

Following Bohr's model of the atom, key advances were made in the development of a quantum theory of the atom. It had already been observed that light has a particle-like nature. This was evident, for example, in the photoelectric effect. It was then natural to ask whether particles might have a wave-like nature. In 1924, Luis de Broglie put forward this idea in his doctoral dissertation, even though there was no experimental evidence to support it. de Broglie postulated that the wavelength λ associated with a particle is

$$\lambda = \frac{h}{p}, \tag{1.28}$$

where p is the momentum of the particle and h is Planck's constant. It was not long before experimental evidence arrived to support de Broglie's hypothesis. In 1927, the C. J. Davisson and L. H. Germer observed that a beam of electrons was diffracted by a crystal of nickel, just as X-rays are diffracted by a crystal. And shortly afterwards, G.P. Thompson performed diffraction experiments that also demonstrated the wave-like properties of electrons.

We can apply the hypothesis of de Broglie to Bohr's theory of the hydrogen atom. From Bohr's postulate for the quantization of angular momentum,

$$L = mvr = pr = n\hbar,$$

and substituting for p from the de Broglie relationship, Equation (1.28), we readily obtain

$$n\lambda = 2\pi r. \tag{1.29}$$

Thus, the allowed orbits are those for which the circumference of the orbit contains exactly an integral number of de Broglie wavelengths. We imagine an electron moving along its orbit at constant speed accompanied by its de Broglie wave. If the de Broglie wave satisfies the condition $n\lambda = 2\pi r$, constructive interference can occur and a standing wave will be produced. (In an analogous way, a standing wave is set up on a taut string that is fixed at both its ends if the waves on the string satisfy the condition $n\lambda/2 = L$, where L is the length of the string.) The standing wave for the case of the $n = 4$ orbit of hydrogen is illustrated in Figure 1.12. On the other hand, if the de Broglie wavelength does not satisfy this condition, destructive interference will result and a standing wave will not be formed; with the conclusion that an electron only exists in the orbit if Equation (1.29) is satisfied. We see that the quantum conditions of Bohr's theory can be interpreted on the basis of a *wave picture.*

Using a wave picture to describe a particle in a physical situation is the approach of *quantum mechanics.* And it is the key to understanding the behaviour of matter on the atomic, molecular, and nuclear scales. The wave equation to express this wave picture was developed by Erwin Schrödinger in 1926 and is the famous *Schrödinger equation.* It is the fundamental equation of quantum mechanics just as Newton's laws are fundamental to classical mechanics. And just like Newton's laws of motion, the Schrödinger equation cannot be derived. Its validity, like Newton's laws of motion, lies in its agreement with the experiment.

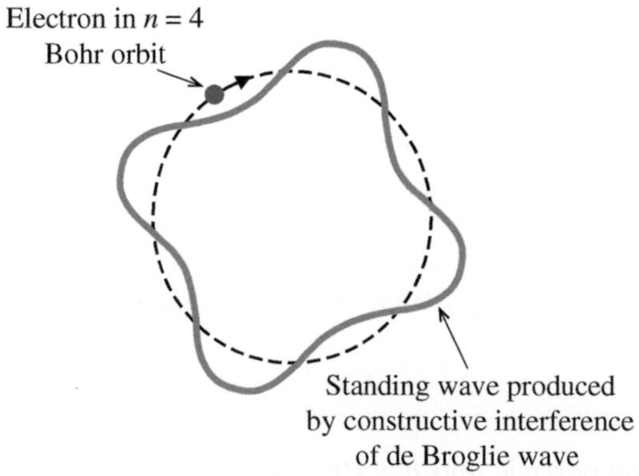

Figure 1.12 A picture that relates the Bohr orbit of an electron to a wave representation showing the formation of a standing wave. We imagine an electron moving along its orbit at constant speed accompanied by its de Broglie wave. If the de Broglie wave satisfies the condition $n\lambda = 2\pi r$, constructive interference can occur and a standing wave will be produced. This is illustrated for the case of the $n = 4$ orbit.

The Schrödinger equation

Schrödinger's equation, in Cartesian coordinates, is

$$\frac{\hbar^2}{2m}\left[\frac{\partial^2\psi}{\partial x^2}+\frac{\partial^2\psi}{\partial y^2}+\frac{\partial^2\psi}{\partial z^2}\right]+(E-V)\psi=0 \tag{1.30}$$

where E is the total energy of the particle, V is its potential energy and $\psi \equiv \psi(x, y, z)$ is the *wave function* that represents the wave nature of the particle. (Strictly speaking, Equation (1.30) is the *time-independent* Schrödinger equation, which can be used when the forces acting on the particle do not change with time t. This is the case for electrons in their bound states in an atom.) A physical interpretation of the wave function ψ was first stated as a postulate by Max Born in 1926. According to this postulate, the wave function ψ is a *complex* function of spatial coordinates and the *square of the magnitude* of ψ is a measure of the probability of the particle being at a particular place. More specifically, $|\psi|^2\,\mathrm{d}V$ *gives the probability of the particle being in the elemental volume* $\mathrm{d}V$. In a more compact notation, the Schrödinger equation can be written as

$$\frac{\hbar^2}{2m}\nabla^2\psi+(E-V)\psi=0, \tag{1.31}$$

where ∇^2 is the *Laplacian operator*:

$$\nabla^2=\frac{\partial^2}{\partial x^2}+\frac{\partial^2}{\partial y^2}+\frac{\partial^2}{\partial z^2}. \tag{1.32}$$

Schrödinger used his equation to solve the atomic structure of the hydrogen atom, obtaining the wave functions of the bound states of the electron and the associated energies. He found that the allowed energies are quantised just as Bohr had found. Indeed, Schrödinger's equation gives *exactly* the same expression for the allowed energy levels in the hydrogen atom as the Bohr model. But Schrödinger did not have to postulate quantization. Instead, as we shall see, the *boundary conditions* imposed on Schrödinger's equation naturally led to energy quantization.

We make again an analogy with the familiar situation of standing waves on a taut string. The one-dimensional wave equation for such waves is

$$\frac{\partial^2 y}{\partial t^2}=v^2\frac{\partial^2 y}{\partial x^2} \tag{1.33}$$

where y is the transverse displacement of the wave at position x along the length of the string and v is the wave velocity. The string is stretched between two fixed points, which we take to be at $x=0$ and $x=L$. Since Equation (1.33) is a linear differential equation with constant coefficients, we can separate the variables so that the solution can be written

$$y(x,t)=X(x)T(t), \tag{1.34}$$

where $X(x)$ is a function of x alone and $T(t)$ is a function of t alone. From Equations (1.33) and (1.34), we obtain

$$\frac{\partial^2 y}{\partial x^2}=T(t)\frac{\partial^2 X}{\partial x^2}, \tag{1.35}$$

$$\frac{\partial^2 y}{\partial t^2} = X(x)\frac{\partial^2 T}{\partial t^2}, \tag{1.36}$$

and, hence,

$$\frac{1}{X}\frac{\partial^2 X}{\partial x^2} = \frac{1}{v^2 T}\frac{\partial^2 T}{\partial t^2}. \tag{1.37}$$

The left-hand side of this equation depends only on x and the right-hand side depends only on t. This can only be the case if both sides are equal to the same constant, which we will call $-\omega^2/v^2$. Hence, we obtain

$$\frac{1}{X}\frac{\partial^2 X}{\partial x^2} = \frac{1}{v^2 T}\frac{\partial^2 T}{\partial t^2} = -\frac{\omega^2}{v^2}. \tag{1.38}$$

We now have two ordinary differential equations

$$\frac{\mathrm{d}^2 T}{\mathrm{d}t^2} = -\omega^2 T; \tag{1.39}$$

$$\frac{\mathrm{d}^2 X}{\mathrm{d}x^2} = -\frac{\omega^2}{v^2}X. \tag{1.40}$$

Equation (1.39) has the same form as that of simple harmonic motion (SHM) and has the general solution $T = A\cos(\omega t + \phi)$. ϕ is a phase angle that we can choose to be equal to zero for the sake of convenience. Equation (1.40) has the general solution $X = B\cos(\omega/v)x + C\sin(\omega/v)x$, where B and C are constants. From Equation (1.34), we therefore have

$$y(x,t) = \left(B\cos\frac{\omega}{v}x + C\sin\frac{\omega}{v}x\right)A\cos\omega t. \tag{1.41}$$

We now impose the boundary conditions $y = 0$ at $x = 0$ and at $x = L$ for all $t > 0$. The first condition gives $B = 0$. The second condition gives $C\sin([\omega/v]L) = 0$, which is satisfied if

$$\frac{\omega}{v}L = n\pi, \quad n = 1, 2, 3, \ldots. \tag{1.42}$$

(Since we are not interested in the trivial solution $y(x, t) \equiv 0$, we exclude the value $n = 0$). Hence, n must take one of the discrete values given by Equation (1.42), and so we write it as

$$\omega_n = \frac{n\pi v}{L}, \tag{1.43}$$

where for each n we have an associated ω_n. We have the familiar result that the standing waves on a taut string can only have certain frequencies of vibration; an example of a physical system where integral numbers occur naturally. In an analogous way, we will see that the Schrödinger equation *plus* the boundary conditions we impose on it leads naturally to quantized energy levels of the hydrogen atom.

A particle in an infinite potential well

We illustrate the application of the Schrödinger equation with the example of a particle trapped in a one-dimensional potential well.

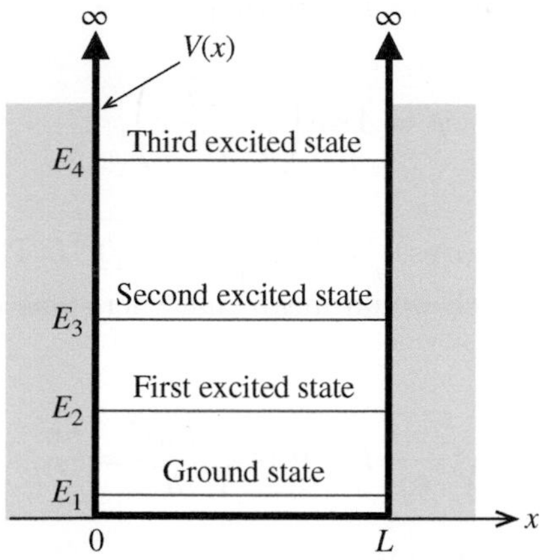

Figure 1.13 An infinite potential well. The walls of the well are at $x = 0$ and $x = L$. They are rigid and impenetrable so that a particle that is trapped in the well can never be outside the walls. This means that the potential energy $V(x)$ must be infinite beyond the walls of the well. Inside the well, $V(x) = 0$. The first four allowed energy levels, for $n = 1$, 2, 3 and 4, are superimposed on the potential well and scale as n^2. The particle has the lowest energy when it is in the energy level with $n = 1$, which is called the ground state for the particle. This energy is finite and is called *zero-point energy.*

The potential well is illustrated in Figure 1.13. The walls of the well are at $x = 0$ and $x = L$. They are rigid and impenetrable so that the particle can never be outside the walls. This means that the potential energy $V(x)$ must be infinite beyond the walls of the well. Inside the well we take $V(x)$ to be zero. Thus we have

$$\begin{aligned} V(x) &= \infty, \quad x < 0, x > L \\ &= 0, \quad 0 \leq x \leq L. \end{aligned} \tag{1.44}$$

The one-dimensional Schrödinger equation, see Equation (1.30), is

$$\frac{\hbar^2}{2m}\frac{\mathrm{d}^2\psi}{\mathrm{d}x^2} + (E - V)\psi = 0. \tag{1.45}$$

As the particle can never penetrate the walls, its wave function must be zero beyond the walls. Hence,

$$\psi = 0 \text{ for } x = \leq 0 \text{ and for } x = \geq L.$$

Inside the well, the potential energy $V(x) = 0$, and hence

$$\frac{\mathrm{d}^2\psi}{\mathrm{d}x^2} = -\frac{2m}{\hbar^2}E\psi, \quad 0 > x < L.$$

This equation has the general solution

$$\psi = A\sin\left(\sqrt{\frac{2mE}{\hbar^2}}x\right) + B\cos\left(\sqrt{\frac{2mE}{\hbar^2}}x\right).$$

The boundary condition $\psi = 0$ at $x = 0$ gives $B = 0$. Hence,

$$\psi = A \sin\left(\sqrt{\frac{2mE}{\hbar^2}}x\right). \tag{1.46}$$

The boundary condition $\psi = 0$ at $x = L$ gives $0 = A \sin \sqrt{(2mE)/\hbar^2}\,L$. The solution $A = 0$ is not acceptable as that would give $\psi = 0$ for all x, which would mean that there was no particle in the well. Hence, we must have

$$\sin\sqrt{\frac{2mE}{\hbar^2}}L = 0, \text{ or } \sqrt{\frac{2mE}{\hbar^2}} = \frac{n\pi}{L}.$$

Substituting this result in Equation (1.46) gives

$$\psi_n = A_n \sin\left(\frac{n\pi x}{L}\right), \quad n = 1, 2, 3, \ldots. \tag{1.47}$$

We have a family of solutions, each corresponding to a value of the integer n. Note that we exclude the $n = 0$ value since this would also give $\psi = 0$ for all x.

The condition $\sqrt{(2mE)/\hbar^2} = n\pi/L$ gives the allowed energies that the particle can have, which we write as

$$E_n = \frac{\pi^2\hbar^2}{2mL^2}n^2, \quad n = 1, 2, 3, \ldots. \tag{1.48}$$

The energies are quantized and this quantization arises because of the boundary conditions, Equation (1.44). The allowed levels for $n = 1$, 2, 3 and 4 are superimposed on the potential well in Figure 1.13. They scale as n^2. The particle has the lowest energy when it is in the energy level with $n = 1$, which is the ground state of the particle. This energy is finite; it is not zero. This is a consequence of the *Heisenberg uncertainty principle*, which we will discuss later. A particle bound in a potential well must always have a finite amount of energy, which is called the *zero-point energy*.

We determine the value of the constant A_n in Equation (1.47) by requiring the wave functions to be normalised. This is the mathematical statement that the particle must be *somewhere* between the two walls of the potential well. Hence,

$$\int_0^L |\psi_n|^2 \mathrm{d}x = A_n^2 \int_0^L \sin^2\left(\frac{n\pi x}{L}\right) \mathrm{d}x = 1. \tag{1.49}$$

Evaluating the integral gives $A_n = \sqrt{2/L}$, for any value of n. Hence, the normalized wave functions are

$$\psi_n = \sqrt{\frac{2}{L}}\sin\left(\frac{n\pi x}{L}\right), \quad n = 1, 2, 3, \ldots. \tag{1.50}$$

The wave functions for the $n = 1$, 2 and 3 states are shown in the top half of Figure 1.14. The close similarity between the wave functions and standing waves on a taut string is apparent.

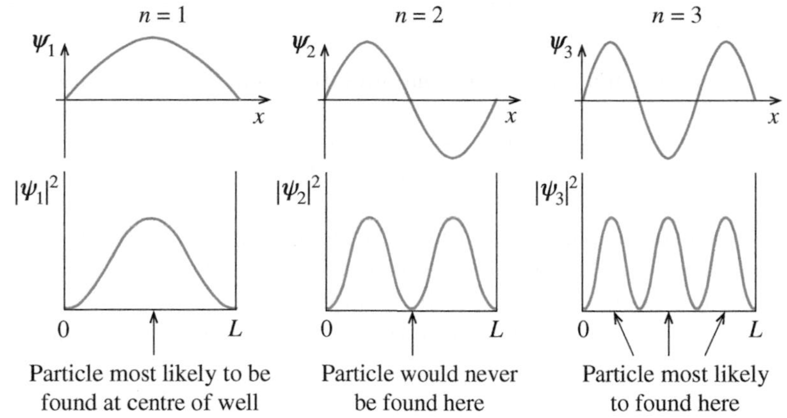

Figure 1.14 The top half of the figure shows the wave functions ψ_n for the $n = 1$, 2 and 3 states of a particle in an infinite potential well. The close similarity between the wave functions and standing waves on a taut string is evident. The bottom half of the figure shows the probability densities $|\psi_n|^2$ for the $n = 1$, 2 and 3 states. There are places where the particle is likely to be found and places where it is never to be found.

The connection between the wave function $\psi(x)$ and the behaviour of the associated particle is expressed in terms of the *probability density* $P(x)$, where

$$P(x) = |\psi(x)|^2. \tag{1.51}$$

The quantity $P(x)$ specifies the probability per unit length of the x-axis of finding the particle near the coordinate x. In other words, *if a measurement is made to locate a particle associated with the wave function* $\psi(x)$, *then the probability that the particle will be found between* x *and* $x + dx$ *is equal to* $P(x)\mathrm{d}x = |\psi(x)|^2\mathrm{d}x$. (This result is just the one-dimensional statement for the more general statement given previously that $|\psi|^2\mathrm{d}V$ gives the probability of the particle being in the elemental volume $\mathrm{d}V$.) It follows that the probability of the particle being in the range between x_1 and x_2 is

$$\int_{x_1}^{x_2} P(x)\mathrm{d}x = \int_{x_1}^{x_2} |\psi(x)|^2\mathrm{d}x. \tag{1.52}$$

In the present case, we have

$$\int_{x_1}^{x_2} |\psi_n|^2\mathrm{d}x = \frac{2}{L}\int_{x_1}^{x_2} \sin^2\left(\frac{\pi x}{L}\right)\,\mathrm{d}x.$$

The functions $|\psi_n|^2$ for $n = 1$, 2, and 3 are shown in the bottom half of Figure 1.14. We see that there are places where the particle is likely to be found and places where it is never to be found.

The function $P(x)$ is an example of a *probability distribution function.* We will encounter a number of other examples of probability distributions in our study of the physics of matter.

Worked example

A particle of mass m is confined in a one-dimensional infinite potential well of width L, where

$$V(x) = \infty, \quad x<0, x>L$$
$$= 0, \quad 0 \le x \le L.$$

For the ground state, ($n = 1$) of the particle, determine the following probabilities: (a) the probability that the particle will be between $x = 0$ and $x = L/2$, (b) the probability that the particle will be between $x = 0$ and $x = L/4$, (c) the probability that the particle will be in an interval of width $0.01L$ at $x = 0.5L$, and (d) if $L = 0.2$ nm, and the mass of the particle is equal to the electronic mass, what is the ground state energy of the particle?

Solution

For the ground state with $n = 1$, we have $|\psi_1|^2 = (2/L)\sin^2(\pi x/L)$.

(a) From the symmetry of the function $|\psi_1|^2$, we can say straight away that the probability that the particle will be between $x = 0$ and $x = L/2$ is 0.5.
(b) The probability that the particle will be between $x = 0$ and $x = L/4$ is equal to

$$\frac{2}{L}\int_0^{L/4} \sin^2\left(\frac{\pi x}{L}\right) dx.$$

We evaluate the integral using the standard integral

$$\int \sin^2(ax) = \frac{x}{2} - \frac{\sin(2ax)}{4a},$$

and find that it is equal to 0.091, i.e. the probability of the particle being within the first quarter of the well is 9.1%. See Figure 1.15.

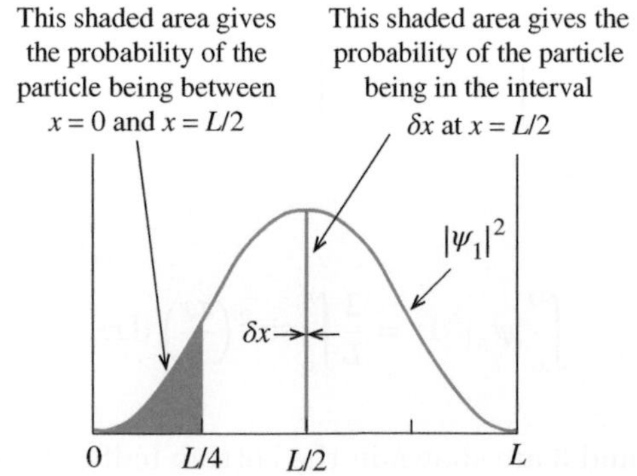

Figure 1.15 The probability density $|\psi_1|^2$ for the ground ($n = 1$) state of a particle in an infinite potential well. The area under the curve for a given range of x gives the probability of finding the particle within that range.

(c) The interval, $0.01L$, which we shall call δx, is much smaller than the length L of the well. Hence we can approximate the area under the curve to be $\psi_1^2 \delta x$, where ψ_1^2 is evaluated at the interval; in this case at $x = 0.5L$. Hence, the probability is

$$\frac{2}{L}\sin^2\left(\frac{\pi \times 0.5}{L}\right) \times 0.01L = 0.02 = 2\%.$$

(d) $$E_1 = \frac{\pi^2\hbar^2}{2mL^2} = \frac{\pi^2 \times (1.055 \times 10^{-34})^2}{2 \times 9.11 \times 10^{-31} \times (0.2 \times 10^{-9})^2} = 1.51 \times 10^{-18}\,\text{J} = 9.42\,\text{eV}.$$

This energy is similar to that of the electron in the ground state of hydrogen.

1.3.3 The Schrödinger equation and the hydrogen atom

The potential energy for the hydrogen atom is

$$V(r) = -\frac{1}{4\pi\varepsilon_0}\frac{Ze^2}{r},$$

with $Z = 1$. This form of this potential is illustrated in Figure 1.16. It is plotted twice as a function of r, with increasing r on both sides of the origin ($r = 0$), to emphasise the spherical nature of the potential well in which the electron finds itself. The blue horizontal lines represent the allowed energy levels in this potential well. As the potential depends only on radial distance r, it is more convenient to use the spherical coordinates, r, θ, and ϕ instead of the rectangular coordinates x, y and z. They are related by

$$\begin{aligned} z &= r\cos\theta \\ x &= r\sin\theta\cos\phi \\ y &= r\sin\theta\sin\phi \end{aligned} \tag{1.53}$$

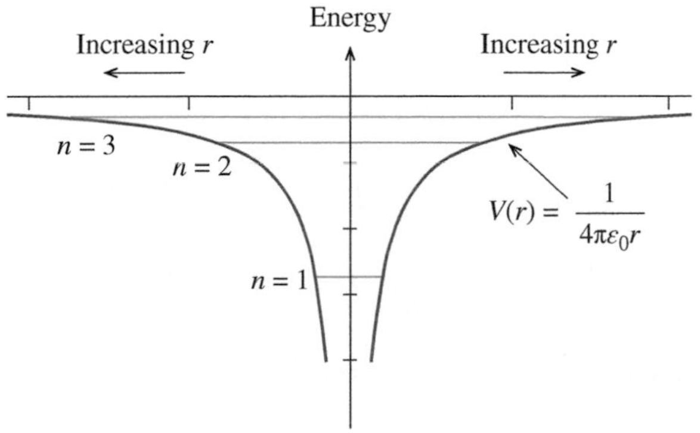

Figure 1.16 The form of the $1/r$ Coulomb potential. It is plotted twice as a function of r, with increasing r on both sides of the origin ($r = 0$), to emphasise the spherical nature of this potential. The blue horizontal lines represent the allowed energy levels in this potential well.

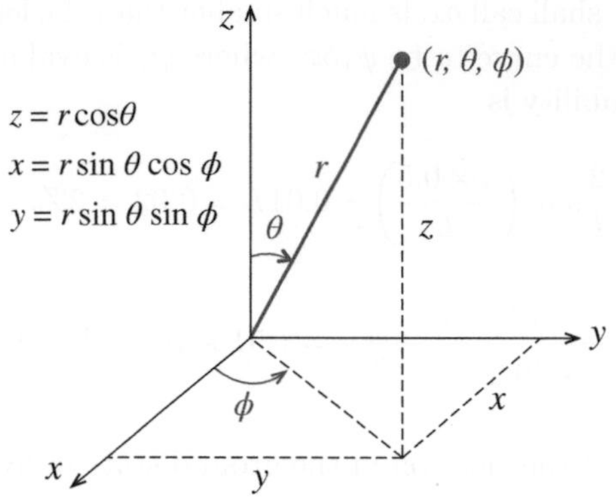

Figure 1.17 This figure shows the relationship between the rectangular and spherical coordinate systems.

as shown in Figure 1.17. The transformation of Schrödinger's from rectangular coordinates to spherical coordinates is straightforward but involves a good deal of algebraic manipulation. The resulting form of the equation is

$$-\frac{\hbar^2}{2mr^2}\frac{\partial}{\partial r}\left(r^2\frac{\partial\psi}{\partial r}\right)-\frac{\hbar^2}{2mr^2}\left[\frac{1}{\sin\theta}\frac{\partial}{\partial\theta}\left(\sin\theta\frac{\partial\psi}{\partial\theta}\right)+\frac{1}{\sin^2\theta}\frac{\partial^2\psi}{\partial\phi^2}\right]+V(r)\psi=E\psi \tag{1.54}$$

This equation looks formidable but there are standard methods to solve such equations, either analytically or by using *numerical methods*. In the case of the hydrogen atom, Schrödinger's equation can be solved exactly.

The first step in solving Equation (1.54) is to separate the variables by writing the wave function as a product of functions of every single variable; just as we did for standing waves on a taut string:

$$\psi(r,\theta,\phi)=R(r)\Theta(\theta)\Phi(\phi), \tag{1.55}$$

where $R(r)$ depends only on radial distance r, $\Theta(\theta)$ depends only on angle θ and $\Phi(\phi)$ depends only on angle ϕ. When this form of ψ is substituted into Equation (1.54) the equation can be transformed into three separate ordinary differential equations; one for $R(r)$, one for $\Theta(\theta)$ and one for $\Phi(\phi)$:

$$\frac{\mathrm{d}^2\Phi}{\mathrm{d}\phi^2}=-m_l^2\Phi \tag{1.56}$$

$$-\frac{1}{\sin\theta}\frac{\mathrm{d}}{\mathrm{d}\theta}\left(\sin\theta\frac{\mathrm{d}\Theta}{\mathrm{d}\theta}\right)+\frac{m_l^2\Theta}{\sin^2\theta}=l(l+1)\,\Theta \tag{1.57}$$

$$\frac{1}{r^2}\frac{\mathrm{d}}{\mathrm{d}r}\left(r^2\frac{\mathrm{d}R}{\mathrm{d}r}\right)+\frac{2m_l}{\hbar^2}(E-V)R=l(l+1)\frac{R}{r^2}, \tag{1.58}$$

where m_l and l are constants. The solution of Equation (1.56) is

$$\Phi(\phi)=A\exp(im_l\phi), \tag{1.59}$$

where A is a constant and $i=\sqrt{-1}$, as can be readily be shown by substituting this solution into Equation (1.56). One of the conditions on the wave function $\psi(r,\,\theta,\,\phi)$ is that it must be *single valued*, i.e. the probability of finding the particle at any point (r, θ, ϕ) must have only one value. For example, the value of $\psi(r, \theta, \phi)$ must be the same at angles ϕ and $(\phi + 2\pi)$, see Figure 1.17. It then follows that $\Phi(\phi) = \Phi(\phi + 2\pi)$, or

$$\Phi(\phi) = A\exp(im_l\phi) = A\exp(im_l[\phi + 2\pi]). \tag{1.60}$$

This can only be true when m_l is 0 or a positive or negative integer, $\pm1, \pm2, \pm3, \ldots$. For these values of m_l in the differential equation for variable θ, Equation (1.57), this equation has only acceptable solutions for certain values of $l(l+1)$, for which

$$l = |m_l|, |m_l| + 1, |m_l| + 2, |m_l| + 3, \ldots.$$

And, for these values of $l(l+1)$ in Equation (1.58), this equation is found to have acceptable values only for certain values of the total energy E. Hence, the imposition of the boundary conditions gives the result that the *energy E is quantized*. Further analysis of Equation (1.58) gives the following expression for the allowed energies:

$$E_n = -\frac{mZ^2e^4}{(4\pi\varepsilon_0)^2 2\hbar^2}\frac{1}{n^2}, \tag{1.61}$$

where $n = l+1, l+2, l+3, \ldots$. Strikingly, Equation (1.61) is exactly the same result as given by the Bohr model for the allowed energies of the hydrogen atom, Equation (1.23).

n is called the *principal quantum number*, l is called the *orbital quantum number* and m_l is called the *magnetic quantum number* and their allowed values are more conveniently expressed as

$$\begin{aligned} n &= 1, 2, 3, \ldots \\ l &= 0, 1, 2, \ldots (n-1) \\ m_l &= -l, (-l+1), \ldots (l+1), l. \end{aligned} \tag{1.62}$$

For any given value of n there are n values of l and for each value of l there are $2l+1$ values of m_l. The principal quantum number n is associated with the dependence of the wave function ψ_{nlm} on radial distance r. Since the potential energy $V(r)$ depends only on r and not on θ or ϕ, the energy of an allowed energy level in hydrogen depends only on n; see Equation (1.61). The quantum number l gives the angular momentum L of the electron according to

$$L = \sqrt{l(l+1)}\hbar. \tag{1.63}$$

That the energy of the electron does not depend on l is a peculiarity of the inverse-square force and holds only for an inverse $(1/r)$ potential. Notice that the quantum mechanical result for the angular momentum is different from the Bohr result $L = n\hbar$.

The quantum number m_l, which is associated with angle ϕ, is related to the angular momentum of the electron along a certain direction in space. For an isolated atom, all directions are equivalent. But when we place an atom in say an external magnetic field, we do introduce a particular direction, which is conventionally called the z-direction. Then the z-component of the angular momentum of the electron is given by

$$L_z = m_l\hbar. \tag{1.64}$$

In the absence of an external magnetic field, the energy does not depend on m_l.

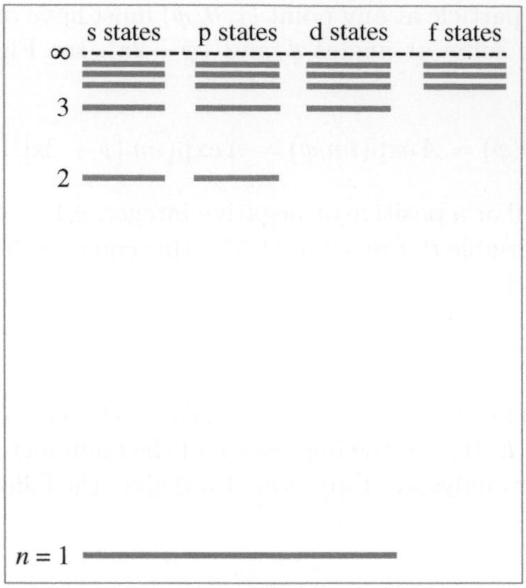

Figure 1.18 The energy levels of the hydrogen atom given by the Schrödinger equation. This figure is similar to Figure 1.10 from the Bohr model, but now, electronic states with the same value of n but different values of l are shown.

It is conventional to refer to the wave function ψ_{nlm} of an electron in an atom as an *orbital*, and an electron that is described by a particular wave function is said to occupy that orbital. The orbital gives all the information about the spatial position of the electron. It replaces the Bohr description of an electron moving in an orbit.

The energy levels for the hydrogen atom, as given by the solution of Schrödinger's equation, are shown in Figure 1.18. This figure is similar to Figure 1.10 from the Bohr model, but now, electronic states with the same value of n but different values of l are shown. By tradition, the various values of the orbital quantum number l are usually labelled with letters, according to the following scheme:

$l = 0$:	s states
$l = 1$:	p states
$l = 2$:	d states
$l = 3$:	f states
$l = 4$:	g states

The historical origin of this labelling system dates back to the early days of atomic physics when spectral lines were labelled as s for *sharp*, p for *principal*, d for *diffuse* and f for *fundamental*.

The ground state of the hydrogen atom

For the ground state of hydrogen, $n = 1$ and hence the angular momentum quantum number l must be equal to 0. Consequently, the electron is in the 1s orbital. Note, that as $l = 0$, the electron has zero angular momentum. This is a quantum effect and is clearly in contrast to the Bohr model in which an electron in the $n = 1$ orbit has angular momentum $L = \hbar$. The magnetic quantum number m_l is also equal to zero.

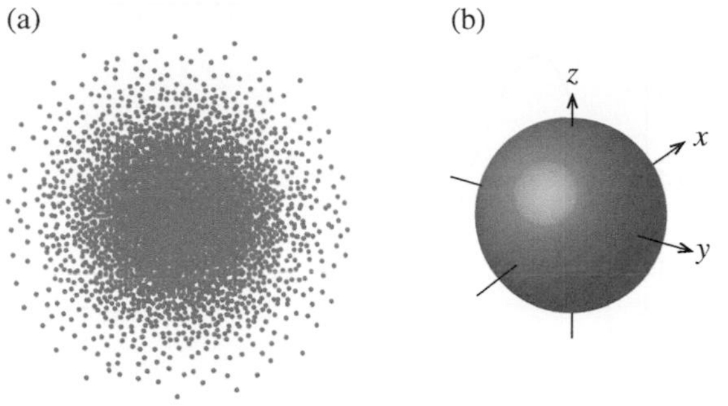

Figure 1.19 (a) A pictorial representation of the probability density for the ground state of hydrogen. The density of dots represents the probability density; the higher the dot density, the more likely the electron is to be found at that place. This gives a reasonably good impression of the charge distribution in the hydrogen ground state. (b) The boundary surface of the hydrogen ground state. There is a 90% probability that the electron will be within the volume enclosed by the surface; and a 10% probability that the electron will be outside that volume.

The solution of Schrödinger's equation for the hydrogen atom gives the following normalized wave function for its ground state:

$$\psi_{100} = \frac{1}{\sqrt{\pi a_0^3}} \exp\left(-\frac{r}{a_0}\right), \tag{1.65}$$

where $a_0 = 4\pi\varepsilon_0(\hbar^2/me^2) = 0.53$ nm, which we recognise as the Bohr radius. Inspection of Equation (1.65) shows that the wave function depends only on radius r and has no dependence on θ or ϕ, i.e. the wave function has the same amplitude at all points of the same radial distance from the nucleus, regardless of direction; the wave function is said to be *spherically symmetric*.

We recall that the physical significance of the wave function is that $|\psi|^2 \mathrm{d}V$ gives the probability of the particle being within the elemental volume d V; the quantity $|\psi|^2$ is the probability density. As the electron is obviously charged, its probability density then gives the charge distribution in an atom, which is central to the chemical reactivity of the atom, i.e. the way atoms interact with each other to form matter. Figure 1.19a is a pictorial representation of the probability density for the ground state of hydrogen. In this figure, the density of dots represents the probability density; the higher the density of the dots, the more likely the electron is to be found at that place. An alternative way to represent $|\psi|^2$ is as a closed surface called a *boundary surface*, such that there is a 90% probability that the electron will be within the volume enclosed by the surface. For the hydrogen ground state,

$$|\psi_{100}|^2 = \frac{1}{\pi a_0^3} \exp\left(-\frac{2r}{a_0}\right), \tag{1.66}$$

and the boundary surface for this state is shown in Figure 1.19b. Since $|\psi|^2$ is a function of r alone, the boundary surface also has a spherical shape.

We are often more interested in knowing the probability that an electron will be found at a given radial distance from the nucleus, regardless of its angular position. In that case, the elemental volume d V is the volume of the thin spherical shell between r and $r + \mathrm{d}r$, which is equal to $4\pi r^2 \mathrm{d}r$. Thus the probability of the electron being between r and $r + \mathrm{d}r$ is

$$P(r)\mathrm{d}r = 4\pi|\psi|^2 r^2 \mathrm{d}r. \tag{1.67}$$

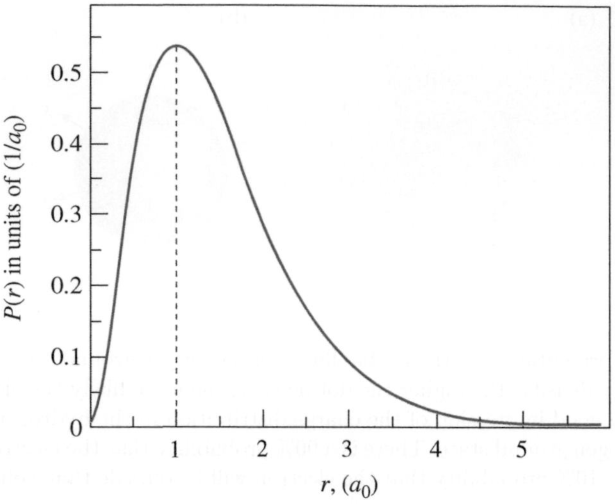

Figure 1.20 The radial probability density $P(r)$ as a function of radial distance r for the ground state of hydrogen. Note that $P(r)$ is plotted in units of $(1/a_0)$. We see that the radial position of an electron given by the Schrödinger equation is markedly different to the prediction of the Bohr model. In the Bohr model, the radius of an orbit corresponding to allowed energy is well defined. In quantum mechanics, we do not have a well-defined radius but a probability of finding the electron at a particular value of r.

$P(r) = 4\pi|\psi|^2 r^2$ is called the *radial probability density*. Substituting the ground state wave function into Equation (1.67) gives

$$P(r)\mathrm{d}r = 4\pi\frac{1}{\pi a_0^3}\exp\left(\frac{-2r}{a_0}\right)r^2\mathrm{d}r. \tag{1.68}$$

Figure 1.20 shows $P(r)$ plotted as a function of r. We see that the radial position of an electron given by the Schrödinger equation is markedly different to the prediction of the Bohr model. In that model, the electron moves in an orbit with a well defined radius as given by Equation (1.18). In quantum mechanics, we do not have a well-defined radius but a *probability* of finding the electron at a particular value of r. Since the probability of the electron being between r and $r + \mathrm{d}r$ is $P(r)\mathrm{d}r$, it follows that the most probable radius for the electron is that which gives the maximum value of $P(r)$. This is found by differentiating $P(r)$ with respect to r and equating the result to zero. From Equation (1.68) we have

$$\frac{\mathrm{d}}{\mathrm{d}r}P(r) = \frac{4}{a_0^3}\left[2r\exp\left(\frac{-2r}{a_0}\right) + r^2\exp\left(\frac{-2r}{a_0}\right)\left(-\frac{2}{a_0}\right)\right].$$

Putting $\frac{\mathrm{d}}{\mathrm{d}r}P(r) = 0$, cancelling the non-zero exponential term and simplifying, we obtain $r = a_0$. We find that the most probable radius occurs at $r = a_0$. Strikingly, this is the radius of the electron orbit that is predicted by the Bohr model.

Worked example

Determine the radius of the boundary surface for the ground state of hydrogen.

Solution

There is a 90% probability that the electron will be within the volume enclosed by a boundary surface, which for the hydrogen ground state has a spherical shape; see Figure 1.19. If the radius of the boundary surface is r_b, then the probability of the electron being within a sphere of radius r_b is

$$\int_0^{r_b} P(r)\mathrm{d}r = 4\pi\int_0^{r_b} |\psi|^2 r^2 \mathrm{d}r = 4\pi\frac{1}{\pi a_0^3}\int_0^{r_b} \exp\left(\frac{-2r}{a_0}\right) r^2 \mathrm{d}r,$$

where we have substituted for the ground state wave function. The integral can be solved by the integration of parts. Putting in the limits, the result is

$$4\pi\frac{1}{\pi a_0^3}\int_0^{r_b} \exp\left(\frac{-2r}{a_0}\right) r^2 \mathrm{d}r = \frac{4}{a_0^3}\left[\left(-\frac{a_0 r_b^2}{2} - \frac{a_0^2 r_b}{2} - \frac{a_0^3}{4}\right)\exp\left(\frac{-2r_b}{a_0}\right) + \frac{a_0^3}{4}\right].$$

This must be equal to 0.9, corresponding to a 90% probability. Expressing r in units of the Bohr radius a_0, we then obtain

$$4\left[\left(-\frac{r_b^2}{2} - \frac{r_b}{2} - \frac{1}{4}\right)\exp(-2r_b) + \frac{1}{4}\right] = 0.9.$$

Letting the function on the left-hand side of the equation be $y(r_b)$, we plot in Figure 1.21, $y(r_b)$ against r_b, in units of a_0. We see that the value of r_b that gives a value of 0.9 for $y(r_b)$ is $2.7a_0$, and hence this is the value of the corresponding radius for the boundary surface of the hydrogen ground state.

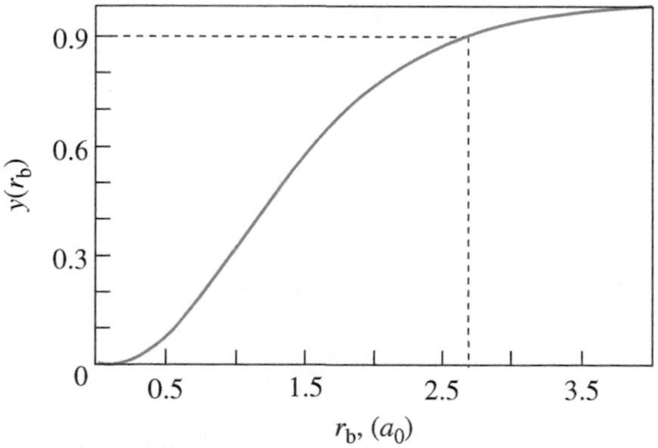

Figure 1.21 A plot of the function $y(r_b)$ against radial distance r_b, measured in units of the Bohr radius a_0.

Heisenberg's uncertainty principle

The fact that the radial position of the electron in a hydrogen atom does not have a well-defined orbit but is governed by a probability distribution is an illustration of Heisenberg's uncertainty principle. This principle is a statement about our knowledge of the properties of a particle. If we want to know where a particle is located, we measure its position, say x. That measurement will not be absolutely perfect but will have some uncertainty Δx. Similarly, if we want to know how fast a particle is going, we need to measure its velocity v_x or equivalently its momentum p_x. This measurement will also have some uncertainty Δp_x. Classical physics places no limits on how small the uncertainties Δx and Δp_x can be. It says that a particle at any instant of time has an exact position and an exact momentum and we can measure both x and p_x with arbitrary precision. The Heisenberg's uncertainty principle makes the bold statement that no matter how well we make the measurement, we cannot measure both x and p simultaneously with arbitrarily good precision. Any measurement we make is limited by the condition

$$\Delta x \Delta p_x \geq \frac{\hbar}{2}. \tag{1.69}$$

This is a statement of Heisenberg's uncertainty principle. The principle arises because of the wave-like nature of particles. In the case of the ground state of hydrogen, the electron is spread out in space and there is not a precise value of its radial position, see Figure 1.20. Similarly, the de Broglie relationship between momentum and wavelength implies that we cannot know the momentum of a particle more accurately than we know its wavelength.

The position-momentum uncertainty relationship, Equation (1.69), provides a powerful way of estimating the ground-state energy of a particle in a given potential $V(r)$. It is especially useful in cases where the form of the potential makes an exact analytical solution of the Schrödinger equation, difficult or impossible. The basis of the calculation is the assumption that the uncertainty Δp in the momentum of the particle is of the same order as the momentum p itself. As an example, consider a particle of mass m in a one-dimensional, infinite potential well of width L. For the value of Δx, we take the value of L itself. Using Equation (1.69), we find that the value of Δp is $\approx \hbar/L$. On the assumption that p is of the same order as Δp, we have

$$p \approx \frac{\hbar}{L}.$$

Taking the potential V to be zero at the bottom of the potential well, the energy E of the particle is equal to its kinetic energy $p^2/2m$. Hence, we have

$$E = \frac{p^2}{2m} \approx \frac{\hbar^2}{2mL^2}.$$

This approximate result is in reasonable agreement with the exact result from Equation (1.48):

$$E = \frac{\pi^2 \hbar^2}{2mL^2}.$$

If we apply the approximate result to the case of an electron confined to an atom with a typical dimension of 0.1 nm, we find

$$E \approx \frac{\hbar^2}{2mL^2} \approx \frac{\left(1 \times 10^{-34}\right)^2}{2 \times 9 \times 10^{-31} \times \left(1 \times 10^{-10}\right)^2} \approx 6 \times 10^{-19}\ \text{J} \approx 4\ \text{eV}.$$

This is roughly the kinetic energy of an electron in the ground state of an atom.

The first excited state of the hydrogen atom

The principal quantum number n is 2 for the first excited state of hydrogen. Hence, l can be either 0 or 1. With $l = 0$ and $m_l = 0$, the state is designated as the 2s state. The solution of Schrödinger's equation gives the following normalized wave function for the 2s state:

$$\psi_{200} = C_{200}\left(2 - \frac{r}{a_0}\right)\exp\left(-\frac{r}{2a_0}\right),$$

where C_{200} is the normalization constant, and we have taken the hydrogen nuclear charge $Z = 1$. As for the ground state of hydrogen, this wave function is spherically symmetric, with no dependence on θ or ϕ. Consequently, the probability density $|\psi_{200}|^2$ is also spherically symmetric. Its boundary surface looks just like that of the hydrogen ground state (see Figure 1.19b), but it is correspondingly larger.

The radial probability density $P(r)$ for the 2s state is shown in Figure 1.22. We see that $P(r)$ has two maxima. The largest maximum occurs close to $r = 5a_0$, and there is also a maximum at a much smaller radial distance; indicating that the electron spends a substantial amount of time close to the nucleus. The radial probability density is zero at $r = 2a_0$, as can also be deduced from the form of the wave function ψ_{200}. Hence, the electron is not to be found at this position.

For $n = 2$, $l = 1$, designated as 2p states, m_l can be $+1$, 0 or -1. The corresponding wave functions are

$$\psi_{210} = C_{210}\left(\frac{r}{a_0}\right)\exp\left(-\frac{r}{2a_0}\right)\cos\theta$$

and

$$\psi_{21\pm1} = C_{21\pm1}\left(\frac{r}{a_0}\right)\exp\left(-\frac{r}{2a_0}\right)\sin\theta\exp(\pm i\phi).$$

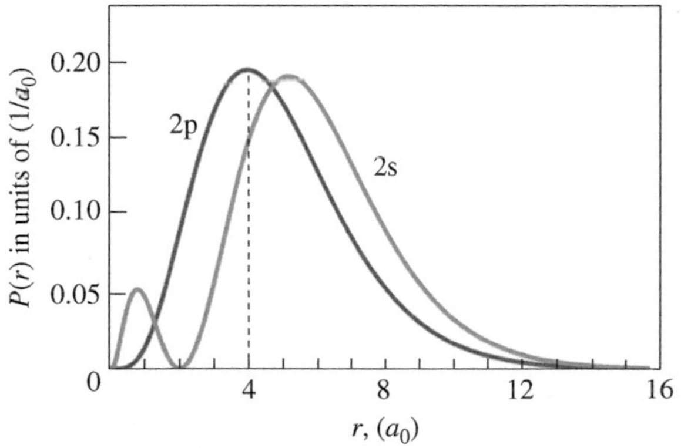

Figure 1.22 The radial probability densities $P(r)$ for the 2s and 2p states in hydrogen. There are two maxima in the 2s probability density. The largest maximum occurs close to $r = 5a_0$. However, there is also a maximum at a much smaller radial distance indicating that the electron spends a substantial amount of time close to the nucleus. The radial probability density is zero at $r = 2a_0$, and hence the electron is not found at this position. Note that the radial probability densities are different for the 2s and 2p states as $P(r)$ depends on quantum number l. However, $P(r)$ does not depend on angles θ and ϕ and so $P(r)$ is the same for all the three possible 2p wave functions; $l = 0, \pm 1$. Strikingly, the most probable radius for the 2p state has the same value as the prediction of the Bohr theory for the $n = 2$ orbit.

These wave functions *do* contain an angular dependence in addition to a radial dependence. The wave function ψ_{210} depends on θ and the $|\psi|^2_{21\pm1}$ wave functions depend on both θ and ϕ.

The radial probability density $P(r)$ is the same for all three possible 2p wave functions, with $l = 0, \pm 1$, as radial probability density depends only on radius r and not on the angular part of the wave equation. $P(r)$ for the 2p states is shown in Figure 1.22. Interestingly, the most probable radius for a 2p state has the same value as the Bohr radius for the $n = 2$ orbit. This is a general characteristic of the hydrogen atom; the most probable radius for the highest value of l for a given value of n is the same as the prediction of the Bohr model for that value of n.

The angular dependences of the 2p wave functions do, however, affect their probability densities $|\psi|^2$, which are

$$|\psi_{210}|^2 = C_{210}^2 \left(\frac{r}{a_0}\right)^2 \exp\left(-\frac{r}{a_0}\right) \cos^2\theta$$

and

$$|\psi|^2_{21\pm1} = C_{21\pm1}^2 \left(\frac{r}{a_0}\right)^2 \exp\left(-\frac{r}{a_0}\right) \sin^2\theta.$$

The probability densities have a radial part that is multiplied by an angular part. The probability density $|\psi_{210}|^2$ has a $\cos^2\theta$ term and the $|\psi|^2_{21\pm1}$ probability density has a $\sin^2\theta$ term. (The $\exp(\pm i\phi)$ term cancels out when the wave functions are squared.)

In Figure 1.23 we show the form of $\cos^2\theta$ and $\sin^2\theta$ in polar coordinates. These angular terms modulate the radial parts of the probability densities and we obtain the resultant probability densities for the 2p wave functions that are illustrated pictorially in Figure 1.24. Figure 1.24a illustrates the probability density for the 2p, $l = 0$ state and Figure 1.24b illustrates the probability density for the 2p, $l = \pm 1$ states. Note that since these probability densities do not depend on angle ϕ, they have rotational symmetry about the z-axis, i.e. the size and shape of the probability density do not change if it is rotated about the z-axis. Hence, in a three-dimensional view, the 2p, $l = \pm 1$ probability density looks like a fuzzy doughnut. In particular,

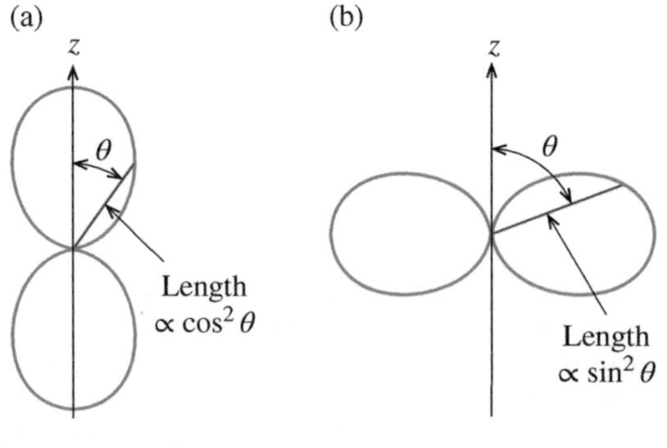

Figure 1.23 The form of (a) $\cos^2\theta$ and (b) $\sin^2\theta$ in polar coordinates.

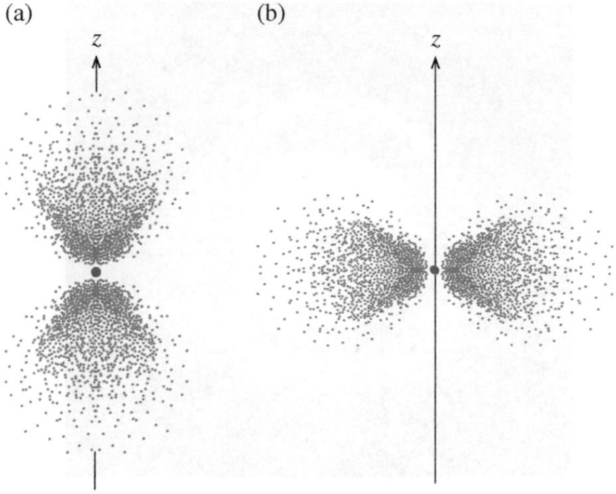

Figure 1.24 A pictorial representation of the probability density ψ^2 for (a) the 2p, $l = 0$ state and (b) the 2p, $l = \pm 1$ states in hydrogen. These probability densities do not depend on angle ϕ, and so have rotational symmetry about the z-axis. Hence, in a three-dimensional view, the 2p, $l = \pm 1$ probability density looks like a fuzzy doughnut. The figure illustrates the directionality of the probability densities and hence the electronic charge densities. This directionality can be of great importance when atoms combine to form molecules.

Figure 1.24 illustrates the directionality of the probability densities and hence the electronic charge densities. This directionality can be of great importance when atoms combine to form molecules.

Wave functions and probability densities may seem to be mathematical constructions. However, an international team of researchers have used a *quantum microscope* to directly observe the probability density for an excited state in the hydrogen atom.[1] In this elegant experiment, hydrogen atoms are ionized in a two-step process by laser radiation. The hydrogen atoms are in a static electric field. This field projects the ejected photoelectrons towards a two-dimensional electron detector, via an electrostatic lens. The detected photoelectrons produce a pattern on the detector that gives an image of the probability density. An example of the images obtained by the researchers is shown in Figure 1.25. It corresponds to the probability density for an excited state of hydrogen with $n = 30$, where the electron is far from the nucleus. In this experiment, the physical reality of a probability density is truly being observed.

Electron spin

To complete our description of an electronic state of hydrogen, we need to introduce the concept of *electron spin*. Spin is an intrinsic angular momentum that every electron possesses. Although it is tempting to think of an electron spinning about its axis like a top, that classical picture is not correct just as the Bohr model of classical orbits is not correct. Instead, it is best to think of the spin as a measurable intrinsic property of the electron, just like the electron charge or mass. The quantum number associated with electron spin is given the symbol s. Like orbital angular momentum, the spin angular momentum of an electron is quantized, but s

[1] A. S. Stodolna, A. Rouzée, F. Lépine, S. Cohen, F. Robicheaux, A. Gijsbertsen, J. H. Jungmann, C. Bordas, and M. J. J. Vrakking, Physical Review Letters, Vol. 110, 213001 (2013).

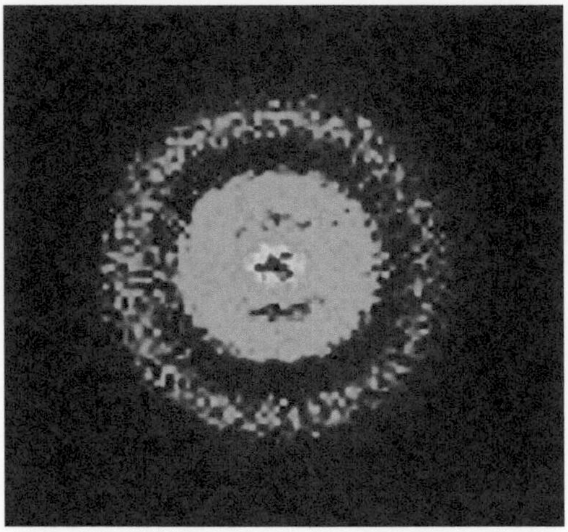

Figure 1.25 A direct visualization of the probability density for an excited state of hydrogen with $n = 30$, obtained using a quantum microscope. Source: A. S. Stodolna et al. (2013) / with permission of American Physical Society.

has the single positive value of ½. The z-component of spin is described by the magnetic quantum number m_s that can take just two values: $+½\hbar$ and $-½\hbar$, which is reminiscent of the expression $L_z = m_l\hbar$ for the z-component of orbital angular momentum. To completely specify the quantum state of an electron, we thus need four quantum numbers: n, l, m_l and m_s. For example, in the ground state of hydrogen, $n = 1$, $l = 0$, $m_l = 0$ and m_s can be either $+½$ or $-½$.

The existence of electron spin was first postulated in 1925 by two graduate students George Uhlenbeck and Samuel Goudsmit. They were trying to understand why certain spectral lines in the optical spectra of hydrogen and alkali metals are composed of closely spaced *pairs* of lines. The familiar yellow light of sodium, for example, is a *doublet* with wavelengths 588.995 nm and 589.592 nm, respectively. The quantum basis of electron spin was established later, in 1929, by P. A. M. Dirac who developed a relativistic theory of quantum mechanics. Dirac's theory showed that the electron *must* have an intrinsic angular momentum $s = ½$. Experimental evidence for electron spin also came from the *Stern Gerlach experiment*. In this experiment, a beam of silver atoms is passed through an inhomogeneous magnetic field. It is observed that the beam splits into two separate beams. The conclusion of the experiment is that the beam of silver atoms is displaced one way or the other according to the two spin states of the electron.

1.3.4 Multi-electron atoms

As we proceed through the periodic table, hydrogen ($Z = 1$), helium ($Z = 2$), lithium ($Z = 3$), etc., the charge on the nucleus increases by one unit from one element to the next, as does the number of atomic electrons. Each of the Z electrons will interact not only with the nucleus but also with every other electron. This introduces a great deal of complexity and, consequently, it is not possible to solve exactly the Schrödinger equation for a multi-electron atom; even for the helium atom, which has just two electrons. We can make a comparison here with planetary motion about the Sun. The mass of each planet is tiny compared to the mass of the Sun; the mass of the Sun is ten thousand times bigger than the mass of the largest planet, Jupiter. Consequently, the Sun's gravitational pull on a particular planet is very much greater than the gravitational pull of the other planets. Hence, when computing the orbit of a particular planet we can ignore the gravitational attraction of all the other planets, except for the most detailed calculations. In the case of a multi-electron atom, however, the mutual Coulomb repulsion of the electrons cannot be ignored; even though it will, in general, be smaller that the Coulomb attraction between an electron and the nucleus of charge $+Ze$. Fortunately, there are powerful *approximation methods* that do allow us to determine the wave functions and energy levels of electrons in a multi-electron atom to a high degree of accuracy. One approximation is called the *central field approximation.* In this approximation, each electron moves *independently* in a *net spherical potential* $V_{\text{net}}(r)$, which is due to the Coulomb attraction of the nucleus and the *average* effect of the Coulomb repulsion of all the other electrons. Hence, when solving the Schrödinger equation for a multi-electron atom, the potential energy $-e^2/4\pi\varepsilon_0 r$, that we use for the hydrogen atom, is replaced by the net potential $V_{\text{net}}(r)$. Solving the Schrödinger equation then gives the possible wave functions of an electron in that net potential. A crucially important advantage of the central field approximation is that we are now dealing with *single-electron* wave functions; the 'motion' of an individual electron is decoupled from the individual 'motion' of the other $Z-1$ electrons, in analogy to the case of planetary motion above.

In order to find the approximate form of the net potential $V_{\text{net}}(r)$, we start with the following considerations. When an electron is far from the nucleus, it is *screened* from the nucleus by the other $(Z-1)$ electrons. Effectively it sees a charge $+e$. We therefore expect, as r tends to infinity, a potential energy of the form

$$V_{\text{net}}(r) = -\frac{e^2}{4\pi\varepsilon_0 r}, \qquad r \to \infty.$$

But when r is small and the electron is close to the nucleus the electron sees a bare, unscreened nucleus and we expect a potential energy of the form

$$V_{\text{net}}(r) = -\frac{Ze^2}{4\pi\varepsilon_0 r}, \qquad r \to 0.$$

A simple model for $V_{\text{net}}(r)$ that reproduces this behaviour is

$$V_{\text{net}}(r) = -\frac{e^2}{4\pi\varepsilon_0 r}\left[(Z-1)e^{-r/a} + 1\right],$$

where a is the *screening radius* for a particular atom and is the order of the Bohr radius a_0. Then to find the possible wave functions of an electron, we solve the Schrödinger equation, Equation (1.54) using the net potential $V_{\text{net}}(r)$:

$$-\frac{\hbar^2}{2mr^2}\frac{\partial}{\partial r}\left(r^2\frac{\partial\psi}{\partial r}\right) - \frac{\hbar^2}{2mr^2}\left[\frac{1}{\sin\theta}\frac{\partial}{\partial\theta}\left(\sin\theta\frac{\partial\psi}{\partial\theta}\right) + \frac{1}{\sin^2\theta}\frac{\partial^2\psi}{\partial\phi^2}\right] + V_{\text{net}}(r)\psi = E\psi. \qquad (1.70)$$

As noted above, this differs from the Schrödinger equation for hydrogen only in that we have replaced the $-e^2/4\pi\varepsilon_0 r$ potential energy for the hydrogen atom with the net potential $V_{\text{net}}(r)$. Hence, the Schrödinger equation can be readily solved, although because of the more complicated form of the potential, the equation must be solved numerically.

The resulting probability distributions for the Z individual electrons give the distribution of electronic charge. We can then use Gauss' law of electrostatics to calculate the electric field $E(r)$ that this distribution produces. And from this, we can calculate an improved estimate of the net potential $V_{\text{net}}(r)$ that an electron experiences. In general, this refined or improved form of the potential will differ from the initial estimate of $V_{\text{net}}(r)$. If it is appreciably different, the above procedure is repeated using the improved form of $V_{\text{net}}(r)$ in the Schrödinger equation. This iterative procedure may be repeated over a number of cycles until the $V_{\text{net}}(r)$ obtained at the end of a cycle is essentially the same as that used at the beginning of the cycle. That is, there is *self consistency* between the form of $V_{\text{net}}(r)$ that is put into the Schrödinger equation and the form of it that is obtained from its solution.

We are dealing with single-electron wave functions in the central field approximation. This means that the wave functions for the electrons in a multi-electron atom can be labelled by the same quantum numbers used to label the wave functions of the hydrogen atom. These are the principal quantum number n, the orbital angular quantum number l, the magnetic quantum number m, and the electron spin quantum number m_s. Hence the wave functions are labelled 1s, 2s, 2p, 3s, 3p, 3d, etc., just as they are for hydrogen. The Z electrons fill the allowed energy levels in the sequence 1s, 2s, 2p, 3s, 3p, 3d, etc., in accord with the *Pauli exclusion principle* that we describe in the next section.

A major difference between the hydrogen atom and a multi-electron atom is that the energy of an electron in a multi-electron atom depends on both n and l. This arises because the form of the potential does not now have a $1/r$ dependence; $V_{\text{net}}(r)$ is more complicated than that. Whereas, for example, the 2s and 2p states of hydrogen have the same energy, their energies are different in a multi-electron atom.

This is illustrated in Figure 1.26, which shows the energy levels of the lithium atom ($Z = 3$). Also shown in the figure are the higher energy levels of hydrogen. While it takes 13.6 eV to ionize hydrogen, it takes just 5.4 eV to ionize lithium. Notice that the higher l levels in lithium, for a particular value of n, have higher

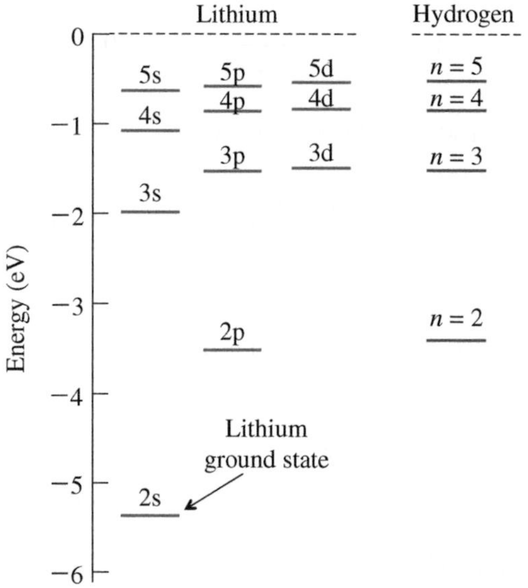

Figure 1.26 The energy levels of the lithium atom. Also shown in the figure are the higher energy levels of the hydrogen atom. Notice that the higher l levels in lithium, for a particular value of n have higher energy than the lower l levels. The electron configuration for the ground state of lithium ($Z = 3$) is $1s^2 2s$; the atom has two electrons in the 1s state and a third electron in the 2s state. The first excitation of lithium is obtained by raising the third electron to the 2p state.

energy than the lower l levels. The reason is that electrons in higher l states spend less time close to the nucleus and are therefore less tightly bound. Hence, it takes less energy to remove them from the atom and so their energy levels are closer to the ionization limit. Despite these differences, however, there is a good deal of similarity between the arrangement of the energy levels in lithium and hydrogen. This similarity supports the single-electron approach for the lithium atom.

The periodic table and the Pauli exclusion principle

Atoms of different atomic number Z have different physical and chemical properties. Strikingly, these properties vary with Z in a periodic way. For example, the alkali metals lithium, sodium, and potassium ($Z = 3$, 11, and 19, respectively) have similar chemical properties, while the halogens fluorine, chlorine, and bromine ($Z = 9$, 17, and 35, respectively) again exhibit similar properties to each other. Quantum mechanics explains why the properties of the elements vary in this periodic fashion. Two basic concepts are involved. First, each electron occupies a single-electron state with definite energy and second, the energy levels are filled with electrons according to the Pauli exclusion principle. This principle was formulated in 1925 by the physicist, Wolfgang Pauli. He developed it to explain the optical spectrum of helium. The Pauli exclusion principle states that *no two electrons can be in the same quantum state*, i.e. *that no two electrons can have exactly the same set of quantum numbers n, l, m_l, and m_s*. We emphasise that the properties of atoms and therefore of all matter depend crucially on this fundamental principle. Using the Pauli exclusion principle we list in Table 1.2 the possible sets of quantum numbers for electron states in an atom for the principal quantum number $n = 1$, 2, and 3. Electrons with the same value of n are described as being in the same *shell* as, roughly speaking, they all have similar values of radius. The 'number of states' column gives the maximum number of electrons that can be found in these states. For example, the maximum the number of electron states for $n = 3$ is 18. We thus obtain the following order in which the electrons fill the orbitals: $1s^2$, $2s^2$, $2p^6$, $3s^2$, $3p^6$, $3d^{10}$, $4s^2$, etc. where the superscripts indicate the number of electrons in a particular orbital. Table 1.3 shows how this works for the first 11 elements in the periodic table. The fourth column gives the *electron configuration* of the atomic ground states. We can see for example, that the ground state of lithium has two electrons in the 1s state and the third electron in the 2s state; see also Figure 1.26. The first excitation of lithium is obtained by raising the third electron to the 2p state. The table also lists the ionization potentials of the elements.

The physical and chemical properties of elements can be largely explained by their electron configurations. A striking example of this is provided by the variation in ionization potentials for the elements; the amount of energy required to remove the least-bound electron. Figure 1.27 shows this variation for elements up to $Z = 95$. The rare gas atoms helium, neon, argon, krypton, xenon, and radon have *closed shell*

Table 1.2 The possible sets of quantum numbers for electron states in an atom for principal quantum number $n = 1$, 2, and 3.

n	l	m_l	m_s	Spectroscopic notation	Number of states	Total number of states
1	0	0	±½	1s	2	2
2	0	0	±½	2s	2	8
2	1	−1, 0, +1	±½	2p	6	
3	0	0	±½	3s	2	18
3	1	−1, 0, +1	±½	3p	6	
3	2	−2, −1, 0, +1, +2	±½	3d	10	

Table 1.3 The electron configurations and ionization potentials for the first eleven elements in the periodic table.

Element	Symbol	Z	Electronic configuration	Ionization potential (eV)
Hydrogen	H	1	1s	13.6
Helium	He	2	$1s^2$	24.6
Lithium	Li	3	$1s^22s$	5.4
Beryllium	Be	4	$1s^22s^2$	9.3
Boron	B	5	$1s^22s^22p$	8.3
Carbon	C	6	$1s^22s^22p^2$	11.3
Nitrogen	N	7	$1s^22s^22p^3$	14.5
Oxygen	O	8	$1s^22s^22p^4$	13.6
Flourine	F	9	$1s^22s^22p^5$	17.4
Neon	Ne	10	$1s^22s^22p^6$	21.6
Sodium	Na	11	$1s^22s^22p^63s$	5.1

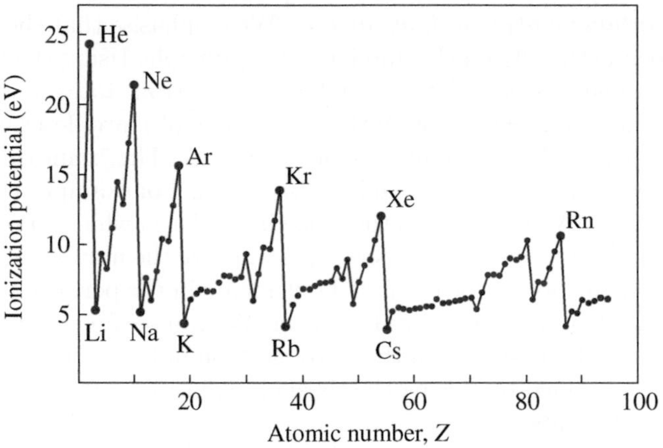

Figure 1.27 The variation of ionization potential with an atomic number for atoms up to $Z = 95$.

configurations. These are particularly stable structures and it takes a large amount of energy to excite one of its electrons. For example, neon has the closed-shell, electron configuration $1s^22s^22p^6$, and it takes 21.6 eV to remove a 2p electron from the atom.

This makes the rare gas atoms chemically inert, with little tendency to gain or lose an electron. On the other hand, the alkali metals lithium, sodium, potassium, rubidium, and caesium have just one electron in their outermost shell. This electron is far from the nucleus and is shielded by the inner electrons so that it effectively sees a charge of $+e$. Consequently, it is only loosely bound to the atom and the ionization potential is low. For example, it takes just 5.1 eV to remove the 3s electron from a sodium atom. Consequently, the alkali metals are extremely reactive. Note that because of the high stability of closed shell configurations most properties of atoms arise from those electrons that are outside the stable closed shells. These are called *valence* electrons.

Problems 1

1.1 (a) When a teaspoon of oil (a few cm^3) was placed on the surface of a calm lake, it was found that the oil covered an area of about 2000 m^2. Use this information to estimate the size of the oil molecules. (b) Suppose a car tyre lasts 30,000 km and that a layer of rubber 1 molecule thick is deposited on the road as the car moves along the road, estimate the size of the rubber molecules.

1.2 What is the time of flight for a Xe^{++} doubly-charged ion in a time of flight spectrometer with an acceleration voltage of 60 V and a 0.75 m flight tube. The atomic weight of xenon is 131 u.

1.3 Water can be converted into hydrogen and oxygen gases according to the reaction $H_2O \rightarrow H_2 + O$ by electrolysis. How many moles of these gases are produced from 5 l of water?

1.4 The density of potassium is 860 $\mathrm{kg/m^3}$. Estimate the diameter of a potassium atom. The atomic weight of potassium is 39 u.

1.5 One mole of any gas occupies 22.4 l at 0 °C and atmospheric pressure. Determine the mass of 1 m^3 of air at 0 °C. Assume a molecular weight of 30 u.

1.6 Obtain a value for the mass of the Earth's atmosphere. The radius of the Earth is 6380 km. Take atmospheric pressure to be 1.0×10^5 Pa.

1.7 Atoms with very high values of quantum number n can be produced. They are called *high Rydberg states* and are well described by the Bohr model. Calculate the radius of a hydrogen atom with $n = 150$. What is the separation in energy between the $n = 150$ and $n = 151$ states?

1.8 Calculate (a) the energy in eV required to ionize a hydrogen atom from its first excited state and (b) the energy required to fully ionize a Li^{++} doubly-charged ion.

1.9 It takes 24.6 eV to remove one electron from helium. What is the total amount of energy required to remove both the electrons from a helium atom? What is the doubly-charged ion of helium He^{++} usually called?

1.10 The mass of the muon particle is 207 times that of the electron, but can be treated to be similar in all other respects. Muonic hydrogen consists of a bound state of a proton and a negative muon. Calculate (a) the Bohr radius for muonic hydrogen, (b) the binding energy of the muon in muonic hydrogen and (c) the energy of the 1s to 2p transition.

1.11 Positronium is a short-lived atomic state consisting of an electron bound electrostatically to a positron. Calculate the ionisation energy of positronium and the wavelength of the light given off when positronium de-excites from $n = 2$ to $n = 1$. Note that in the case of hydrogen, we could assume that the mass of the nucleus was infinitely more massive than the electron. In the case of positronium, this is not so and the mass of the electron m must be replaced with the reduced mass of the positronium system, which is $m/2$, where m is the mass of an electron.

1.12 An electron microscope produces an electron beam in which the electrons are accelerated through a voltage of 2.5 keV. Estimate the theoretical spatial resolution of the microscope, i.e. the size of the smallest particle it can distinguish.

1.13 Use the uncertainty principle to estimate the energy of a proton in a nucleus. Take the diameter of the nucleus to be 2×10^{-15} m.

1.14 An estimate of the lowest energy of a particle in a potential well is obtained from the relationship $\Delta p \Delta x \sim \hbar$, where Δp and Δx are, respectively, the uncertainties in the momentum and position of the particle. (a) A particle of mass m moves in a vee-shaped potential of the form

$$V(x) = -bx \quad (x \le 0)$$
$$V(x) = +bx \quad (x \ge 0).$$

Use the above relationship to show that the energy of the lowest state is $(\hbar^2 b^2/m)^{1/3}$, within a numerical factor of order unity. Show that this result is correct dimensionally. (b) A particle undergoing SHM moves in a potential $V(x)$ that has the form $V(x) = \frac{1}{2}kx^2$, where k is the 'spring constant' and x is the displacement from equilibrium. If the mass of the particle is m, show that the energy of the lowest state of the particle is $\hbar\omega$ within a numerical factor of order unity, where ω is the angular frequency of the vibrational motion.

2

The forces that bind atoms together

At the beginning of Chapter 1, we made the statement that atoms attract each other when they are a little distance apart, but repel when they are squeezed together. Evidence that atoms are attracted to each other comes from the fact that atoms combine together to form solids and liquids. Evidence that atoms repel each other comes from the fact that solids and liquids resist being compressed. These physical properties arise from the forces that act between the atoms. These forces and the resulting bonding of the atoms are the subjects of the present chapter. We will describe the general characteristics of the forces that bind atoms together and the resulting potential energy of the atoms. And we will describe the principal kinds of interatomic bonding: van der Waals, ionic, covalent, and metallic bonding. We will see that all bonding is a consequence of the electrostatic interaction between nuclei and electrons. And we will see that the interatomic interactions that we will describe on the microscopic scale relate directly to the properties of matter that are observed in the laboratory.

2.1 General characteristics of interatomic forces

Before going into detail about particular types of interatomic force, we describe some general features of such forces. We are interested in whether the force acting between atoms is attractive or repulsive. We are also interested in the way the strength of the force varies with interatomic separation. To discuss these two aspects, we imagine an atom fixed in place at the origin ($r = 0$) of a coordinate system and a second atom a distance r away, where r is the distance between the *centres* of the two atoms. This arrangement is illustrated in Figure 2.1. We make the assumption that the force depends only on distance r. If the force is repulsive, the force acts to increase the separation of the two atoms, i.e. it points in the direction of increasing r. Hence, a repulsive force has a positive sign. Conversely, an attractive force acts to decrease r and so has a negative sign.

Physics of Matter, First Edition. George C. King.

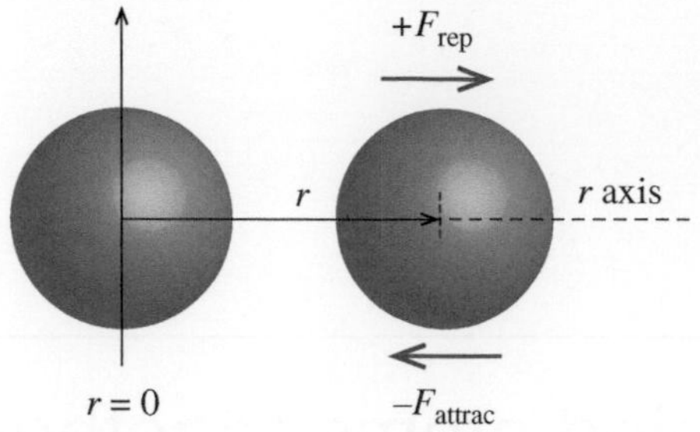

Figure 2.1 An arrangement of two interacting atoms. We imagine one atom is fixed in place at the origin ($r = 0$) of a coordinate system. The other atom is at distance r, where r is the distance between the centres of the two atoms. If the interatomic force is repulsive, the force acts to increase the separation of the two atoms, i.e. it points in the direction of increasing r. Hence, a repulsive force has a positive sign. Conversely, an attractive force acts to decrease r and so has a negative sign.

2.1.1 The range of a force

The way in which the strength of the force varies with separation is determined by its dependence on r. And this dependence determines the *range* of a force, i.e. the distance over which it has a significant influence. Some forces are described as long-range and others as short-range. For example, the gravitational force, which is proportional to $1/r^2$, is described as a *long-range* force because the strength of the force falls off relatively slowly with r. We know that gravitational forces extend over huge distances in a galaxy, and beyond.

To illustrate the difference between short- and long-range forces, we compare two forces, one that varies as $1/r^2$, and one that varies as $1/r^7$; we will encounter a force that is proportional to $1/r^7$ in our discussion of van der Waals bonding. It is often more convenient to measure distance in terms of a *standard length* that is characteristic of the physical situation than in terms of say metres or nanometres. This usually provides more physical insight. For example, when discussing interatomic bonding, the equilibrium separation of the atoms is a convenient standard length. We therefore write the long-range force as $F_l = -A(a/r)^2$, and the short-range force as $F_s = -A(a/r)^7$, where A is a constant and a is the standard length. Note that the forces have a negative sign as they are attractive. The two forces are plotted as a function of r in Figure 2.2, and have the same strength at $r = a$. Clearly, the influence of the $1/r^2$ force extends over a much larger distance than the $1/r^7$ force and consequently has the longer range. On the other hand, we see that the short-range force is dominant at short distances. The greater the exponent n in the r dependence of the force $1/r^n$, the shorter the range of the force.

2.1.2 Repulsive and attractive forces

Since atoms repel each other at small separations and attract at large separations, it follows that they are acted upon by both repulsive and attractive forces. We will discuss the physical nature of various repulsive and attractive forces in Section 2.3. For the moment, we note that in the different kinds of bonding that we will consider, the repulsive force has a much shorter range than the attractive force. The repulsive force is dominant at distances below about 0.2 nm and the attractive force is dominant at distances greater than this. This situation is illustrated in Figure 2.3. It shows a short-range repulsive force and a longer-range attractive force. It also shows the sum of the two forces, which is the net interatomic force acting between the two atoms. At some particular separation, the two forces must be equal and opposite so that the net

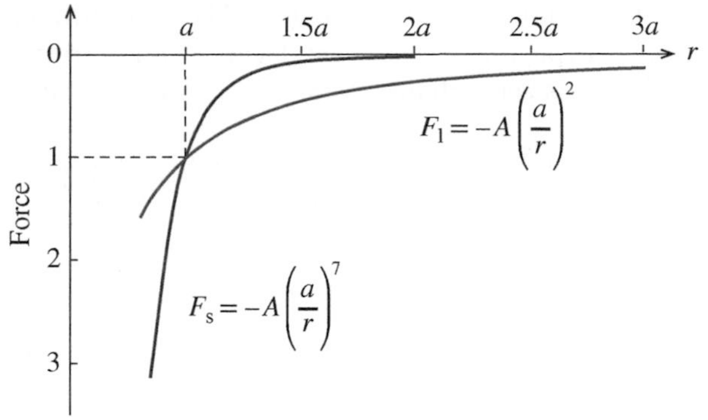

Figure 2.2 An illustration of the difference between short- and long-range forces. The long-range force is $F_l = -A(a/r)^2$, and the short-range force is $F_s = -A(a/r)^7$, where A is a constant and a is a standard length. The forces have a negative sign as they are attractive, and have the same strength at $r = a$. Clearly, the influence of the $1/r^2$ force extends over a much larger distance than the $1/r^7$ force. On the other hand, the short-range force is dominant at short distances.

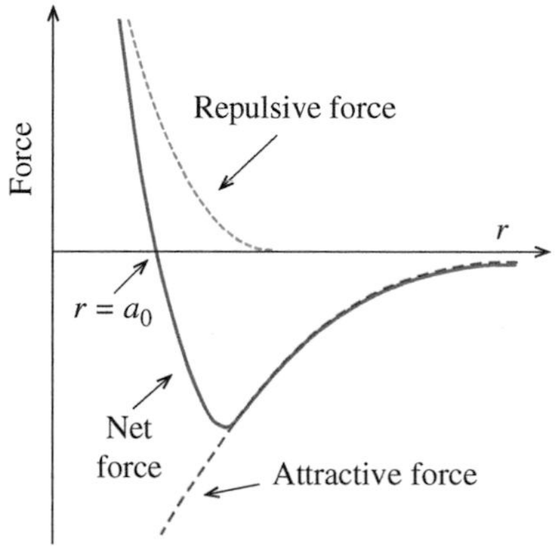

Figure 2.3 The figure shows a short-range repulsive force and a longer-range attractive force that act between two atoms. It also shows the sum of the two forces, which is the net interatomic force acting between the atoms. At some particular separation, the two forces must be equal and opposite so that the net interatomic force is zero. At this point, the atoms are at their equilibrium separation, for which $r = a_0$. The figure also illustrates that all interatomic forces reduce to zero as r goes to infinity, i.e. when the atoms are far apart.

interatomic force is zero. At this point, the atoms are at their *equilibrium separation*, for which $r = a_0$; see Figure 2.3. This figure also illustrates that all interatomic forces reduce to zero as r goes to infinity, i.e. when the atoms are far apart.

It is useful to have an expression that adequately describes the actual interatomic force $F(r)$ between two atoms but which has a simple analytical form. A suitable expression that has these properties is

$$F(r) = \frac{A}{r^m} - \frac{B}{r^n}, \tag{2.1}$$

where A and B are constants and m and n are positive integers. The first term represents the repulsive force and the second the attractive force. Typical values for m and n are 13 and 7, respectively. At the equilibrium separation, $r = a_0$, the net force $F(r)$ is zero. Hence, $B = Aa_0^{n-m}$, and we can write

$$F(r) = A\left[\frac{1}{r^m} - \frac{a_0^{n-m}}{r^n}\right] = \frac{A}{a_0^m}\left[\left(\frac{a_0}{r}\right)^m - \left(\frac{a_0}{r}\right)^n\right]. \tag{2.2}$$

2.1.3 Oscillations about the equilibrium separation

Figure 2.4 shows a much-enlarged plot of $F(r)$ in the region of the equilibrium separation a_0. This figure illustrates that for a range of r sufficiently close to the equilibrium distance a_0, $F(r)$ can be assumed to depend linearly on r. Then if the 'moveable' atom is slightly displaced from the equilibrium separation by distance δr as in Figure 2.5, the force $F(r)$ will change by δF, where

$$\delta F = \left(\frac{\mathrm{d}F}{\mathrm{d}r}\right)\delta r. \tag{2.3}$$

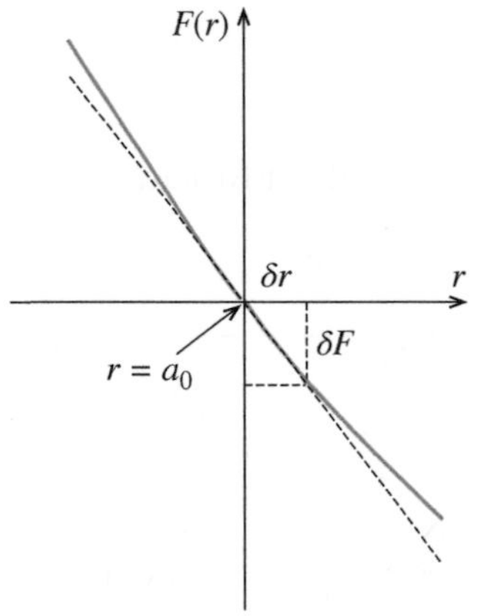

Figure 2.4 This figure shows a much-enlarged plot of the interatomic force $F(r)$ in the region of the equilibrium separation a_0. This figure illustrates that for a range of r sufficiently close to a_0, $F(r)$ can be assumed to depend linearly on r.

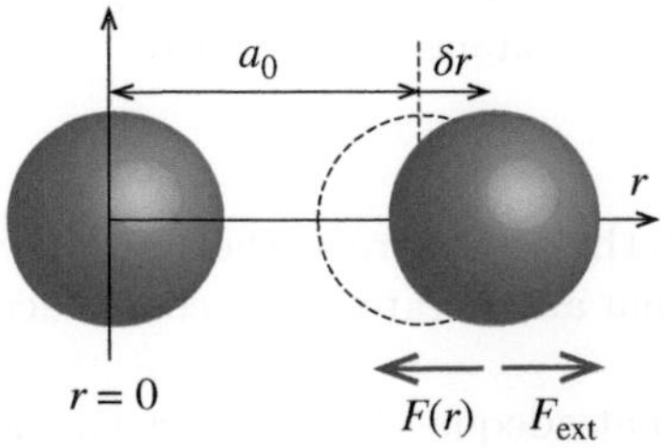

Figure 2.5 If the 'moveable' atom is displaced by a small amount δr, the atom will oscillate in simple harmonic motion about the equilibrium position. Moreover, when the atom is displaced, the potential energy $V(r)$ of the system increases. The relationship between the potential energy $V(r)$ and the interatomic force $F(r)$ is $F(r) = -\dfrac{\mathrm{d}V(r)}{\mathrm{d}r}$.

From Equation (2.2), we have

$$\frac{\mathrm{d}F}{\mathrm{d}r} = \frac{A}{a_0^m}\left[-m\frac{a_0^m}{r^{m+1}} + n\frac{a_0^n}{r^{n+1}}\right].$$

For r close to a_0, we take $r = a_0$, in evaluating $\mathrm{d}F/\mathrm{d}r$, which gives

$$\left.\frac{\mathrm{d}F}{\mathrm{d}r}\right|_{r=a_0} = -A\left[\frac{m-n}{a_0^{m+1}}\right].$$

Then substituting this result in Equation (2.3) gives

$$\delta F = -A\left[\frac{m-n}{a_0^{m+1}}\right]\delta r. \tag{2.4}$$

δF is a restoring force, which acts to return the atom to the equilibrium position. And this restoring force is directly proportional to displacement δr for small displacements. Equation (2.4) is exactly analogous to the familiar equation $F = -kx$, which is a signature of *simple harmonic motion* (*SHM*). Thus, if the atom is displaced by a small amount, it will oscillate in SHM about the equilibrium separation.

2.2 Interatomic potential energy

Considering the energy of a system is a powerful tool in solving physical problems. For one thing, scalar rather than vector quantities are involved, which usually simplify the analysis. In the present case, it is the potential energy of the interacting atoms that is relevant. This potential energy resides in the field of force between the two atoms in analogy with the potential energy stored in a stretched spring.

Consider Figure 2.5. In order to displace the 'moveable' atom by the amount δr, we must apply an external force F_{ext} to that atom. This external force must be equal to the force $F(r)$ that the interatomic interactions place on the atom, but it must act in the opposite direction. That is $F_{\text{ext}} = -F(r)$, where we take $F(r)$ to be given by Equation (2.2). The work done in moving the atom is $F_{\text{ext}} \times \delta r$. This is just equal to the increase δV in the potential energy $V(r)$. Hence, we obtain

$$\delta V = F_{\text{ext}}\delta r = -F(r)\delta r,$$

or

$$F(r) = -\frac{\delta V}{\delta r}.$$

Hence, in the limit $r \to 0$, we can write

$$F(r) = -\frac{\mathrm{d}V(r)}{\mathrm{d}r}. \tag{2.5}$$

We see that the force is equal to minus the gradient of the potential energy. Hence,

$$[V(r)]_{r_0}^{r} = -\int_{r_0}^{r} F(r)\,dr.$$

and

$$V(r) = V(r_0) - \int_{r_0}^{r} F(r)\,dr, \tag{2.6}$$

where r_0 is a standard reference point and $V(r_0)$ is the potential energy at that point. Usually, it is convenient to take $V(r_0) = 0$, as we are invariably interested in *changes* in potential energy; the absolute value of the potential energy is usually not relevant.

Equation (2.6) then reduces to

$$V(r) = -\int_{r_0}^{r} F(r)\mathrm{d}r.$$

Interatomic forces reduce to zero when the atoms are far apart, i.e. $r_0 \to \infty$. It follows that $r_0 = \infty$. Hence, the potential energy of the atom-atom system is

$$V(r) = -\int_{\infty}^{r} F(r)\mathrm{d}r. \tag{2.7}$$

We can now determine the potential energy arising from the repulsive and attractive forces described by Equation (2.2). The potential energy is

$$V(r) = -\frac{A}{a_0^m}\int_{\infty}^{r}\left[\left(\frac{a_0}{r}\right)^m - \left(\frac{a_0}{r}\right)^n\right]\mathrm{d}r = -\frac{A}{a_0^m}\left[-\frac{1}{(m-1)}\frac{a_0^m}{r^{m-1}} + \frac{1}{(n-1)}\frac{a_0^n}{r^{n-1}}\right].$$

To simplify this expression, we let $(m-1) = p$ and $(n-1) = q$ to obtain

$$V(r) = \frac{A}{a_0^p}\left[\frac{1}{p}\left(\frac{a_0}{r}\right)^p - \frac{1}{q}\left(\frac{a_0}{r}\right)^q\right]. \tag{2.8}$$

2.2.1 The Lennard-Jones 6–12 potential

One particular potential function that is widely used and which has the form of Equation (2.8) is the Lennard-Jones 6–12 potential, named after John Lennard-Jones. It has $p = 12$ and $q = 6$. This potential is used, in particular, for the *van der Waals force* that we shall describe in Section 2.2. Substituting for these values of p and q in Equation (2.8), we obtain

$$V(r) = \varepsilon\left[\left(\frac{a_0}{r}\right)^{12} - 2\left(\frac{a_0}{r}\right)^6\right], \tag{2.9}$$

where $\varepsilon = A/12a_0^{12}$ and a_0 are the equilibrium separation. A plot of the Lennard-Jones potential is shown in Figure 2.6. There are several important features of this potential. There is a minimum in the potential

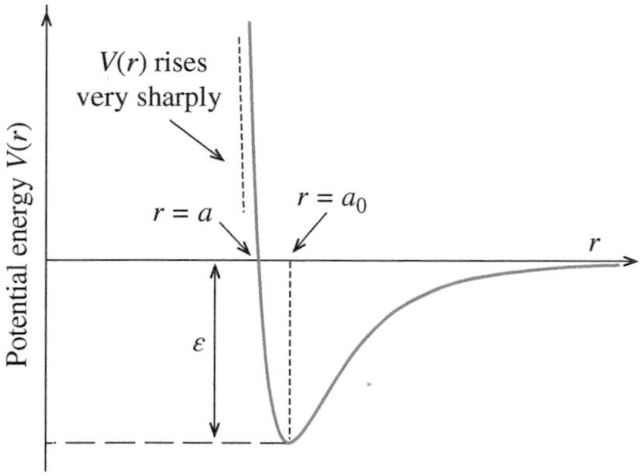

Figure 2.6 A plot of the Lennard-Jones 6–12 potential. There is a minimum in the potential energy, which occurs at the equilibrium separation $r = a_0$. At this point, the potential energy $V(r) = -\varepsilon$, where ε is the amount of energy required to separate the two atoms to infinity. The potential energy rises very steeply when the two atoms are pushed closer together. If we imagine the atoms to be hard spheres, as suggested by this behaviour of the potential energy, we can identify the value of a as the *diameter* of the atoms.

energy, which occurs at $r = a_0$. This result is readily obtained by differentiating $V(r)$ with respect to zero and equating the differential to zero. Any disturbance from this position would produce a force tending to return the system to the equilibrium separation, as we noted previously. At the equilibrium separation, $r = a_0$, the potential energy $V(r) = -\varepsilon$. Hence, ε is the *binding energy* of the two atoms, i.e. the amount of energy required to separate the two atoms to infinity. And because energy must be supplied to do this, the potential energy is negative. From Equation (2.9), we also find that $V(r) = 0$ when $r = a_0/\sqrt[6]{2} = a$, where the point a is marked on Figure 2.6. If the two atoms are pushed closer than point a, the potential rises very steeply, i.e. the force of repulsion increases greatly. If we imagine the atoms to be hard spheres, as suggested by this behaviour of the potential energy, then we can identify the value of a as the *diameter* of the atoms. As $a_0 = 1.12a$, both a and a_0 are good measures of the atomic diameter. If we make the substitution $a_0/\sqrt[6]{2}a$ in the Lennard-Jones potential, we obtain the following elegant form of the potential:

$$V(r) = 4\varepsilon\left[\left(\frac{a}{r}\right)^{12} - \left(\frac{a}{r}\right)^{6}\right]. \tag{2.10}$$

We can readily deduce the form of the force that gives arise to the Lennard-Jones potential by differentiating Equation (2.9). The result is

$$F(r) = -\frac{\mathrm{d}V(r)}{\mathrm{d}r} = \frac{12\varepsilon}{a_0}\left[\left(\frac{a_0}{r}\right)^{13} - \left(\frac{a_0}{r}\right)^{7}\right]. \tag{2.11}$$

The resultant force will be equal to zero when

$$\left(\frac{a_0}{r}\right)^{13} = \left(\frac{a_0}{r}\right)^{7},$$

i.e. when $r = a_0$, as expected.

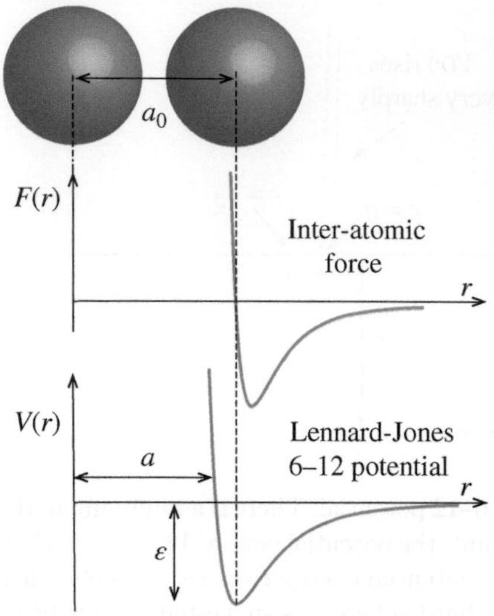

Figure 2.7 This figure shows the relationship between the Lennard-Jones 6–12 potential and the force that gives rise to it. The figure also shows the positions of the two interacting atoms at the equilibrium separation. Taking the diameter of the atoms to be a, we see that the two atoms are nearly touching when they are at the equilibrium separation.

In Figure 2.7, we show the relationship between the Lennard-Jones 6–12 potential and the force that gives rise to it. The figure also shows the positions of the two atoms at equilibrium. Taking the diameter of the atoms to be a, as earlier, we see that the two atoms are nearly touching when they are at their equilibrium separation.

Worked example

Obtain an expression for the potential energy of a pair of point charges, both of charge $+q$, as a function of their separation r.

Solution

The force on each charge is $F(r) = +\dfrac{1}{4\pi\varepsilon_0}\dfrac{q_1 q_2}{r^2}$, which is a positive force as it is a repulsive force and acts in the direction of increasing r. From Equation (2.5), $F(r) = -\dfrac{\mathrm{d}V(r)}{\mathrm{d}r}$.

Hence,

$$\int_{r_0}^{r} V(r) = V(r) - V(r_0) = -\int_{r_0}^{r} F(r)\mathrm{d}r$$

$$= -\frac{q_1 q_2}{4\pi\varepsilon_0}\int_{r_0}^{r}\frac{1}{r^2}\mathrm{d}r = +\frac{q_1 q_2}{4\pi\varepsilon_0}\left[\frac{1}{r}\right]_{r_0}^{r}.$$

The potential energy of the system is zero at $r = \infty$, and hence,

$$V(r) = \frac{1}{4\pi\varepsilon_0}\frac{q_1 q_2}{r}.$$

2.3 Types of interatomic bonding

The forces that act between atoms result in four principal types of bonding. These are: van der Waals, ionic, covalent, and metallic bonding. Here we describe these four types of bonding and how they arise from interatomic forces.

2.3.1 van der Waals bonding

Perhaps surprisingly, even atoms (and molecules) that are electrically neutral bind together. We know, for example, that inert-gas atoms such as neon first condense and then solidify at a low temperature and atmospheric pressure; the notable exception being helium, which does not. Similarly, nitrogen molecules condense to produce liquid nitrogen at low temperatures. The attractive forces that bind neutral atoms and molecules are called van der Waals forces, named after Johannes van der Waals. The most important of these forces is the *dispersive van der Waals force.* As we will see, this force arises from distortions that are induced in the electron charge distribution of the interacting atoms when they approach closely together. van der Waals bonding occurs in all atoms and molecules although as we shall see, its relatively low strength can be overshadowed when much stronger forces come into play. As usual, short-range repulsive forces ensure that the atoms or molecules do not coalesce.

To illustrate the origin of the van der Waals attractive force, we consider the interaction between two neon atoms. We recall from Section 1.3.4 that all the electronic shells of neon are full, which makes it an inert gas. Averaged over time, the electron charge distribution of the atom is spherically symmetric, i.e. the centre of the distribution coincides with the nucleus, as illustrated pictorially in Figure 2.8a. But as we saw in Section 1.3.3, atomic electrons are not fixed in space but can be found over a range of spatial positions. So, although, on average, the electron charge distribution is spherically symmetric, at any given instant,

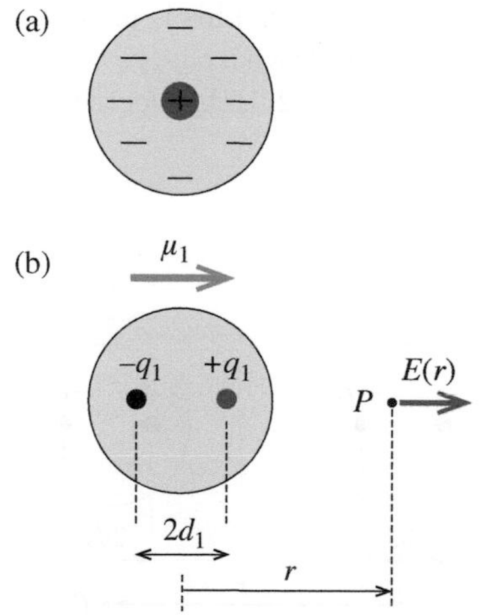

Figure 2.8 (a) Averaged over time, the electron charge distribution of an atom is spherically symmetric, i.e. the centre of the distribution coincides with the nucleus. (b) However, at any given instant, the effective centre of the distribution will not coincide with the nucleus. We can represent this instantaneous condition of the atom as two charges, $-q_1$ and $+q_1$, separated by distance $2d_1$. The resultant dipole electric field $E(r)$ at point P is the vector sum of the electric fields due to the two charges $-q_1$ and $+q_1$.

the effective centre of the distribution will not coincide with the nucleus. We can represent this *instantaneous* condition of the atom as two charges, $-q_1$ and $+q_1$, separated by distance $-2d_1$, as illustrated by Figure 2.8b. Such an arrangement of charges is called an *electric dipole*. The resultant electric field at any point is the vector sum of the electric fields due to the two charges $-q_1$ and $+q_1$. For the case shown in Figure 2.8b, the electric field $E(r)$ at point P, at distance r from the centre of the atom, is given by

$$E(r) = \frac{1}{4\pi\varepsilon_0}\left[\frac{q_1}{(r-d_1)^2} - \frac{q_1}{(r+d_1)^2}\right] = \frac{q_1}{4\pi\varepsilon_0 r^2}\left[\frac{1}{[1-(d_1/r)]^2} - \frac{1}{[1+(d_1/r)]^2}\right].$$

When $r \gg d_1$,

$$\frac{1}{[1-(d_1/r)]^2} \approx 1 + 2\frac{d_1}{r}.$$

Hence, within this approximation,

$$E(r) = \frac{2\times 2d_1 q_1}{4\pi\varepsilon_0 r^3} = \frac{2\mu_1}{4\pi\varepsilon_0 r^3}, \tag{2.12}$$

where $\mu = 2d_1q_1$ is the *dipole moment*. The dipole moment is a vector quantity that points from the negative to the positive charge. Note the $1/r^3$ dependence of the electric field from a dipole.

The effect of the electric field $E(r)$ on a second atom close by is to *polarize* that second atom, i.e. to distort its electronic charge distribution so that the negative electronic charge moves closer to the first atom and the positive nuclear charge moves further away. This is illustrated in Figure 2.9. Again, we represent this instantaneous condition of the atom as a dipole with charges $-q_2$ and $+q_2$ separated by distance d_2, with dipole moment $\mu_2 = 2d_2q_2$. At point P, see Figure 2.9, the electric field is $E(r)$. The electric field $E(r-d_2)$

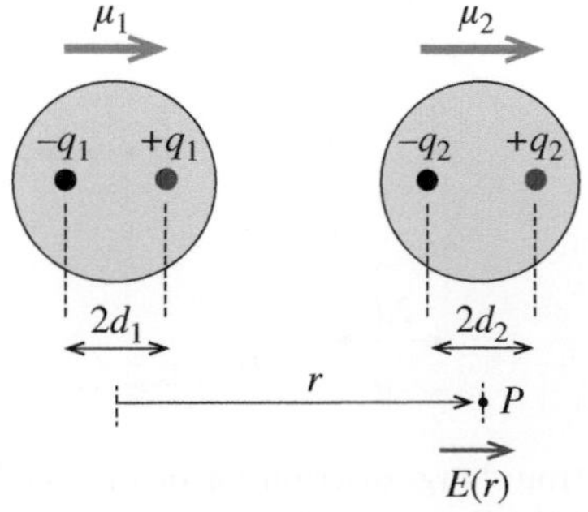

Figure 2.9 The dipole electric field $E(r)$ of the left-hand atom polarizes the right-hand atom so that its negative electron charge moves closer to the left-hand atom and its positive nuclear charge moves further away. The result is that the two atoms experience a mutual attraction.

acting on the negative charge $-q_2$ at $(r-d_2)$ produces a force $-E(r-d_2)q_2$, while the electric field $E(r+d_2)$ at the positive charge produces a force $+E(r+d_2)q_2$. Hence, the net force acting on the right-hand dipole is

$$F=[q_2E(r+d_2)-q_2E(r-d_2)]=q_2[E(r+d_2)-E(r-d_2)].$$

When $r \gg d_2$,

$$E(r+d_2)-E(r-d_2)\approx \frac{\mathrm{d}E}{\mathrm{d}r}2d_2.$$

Hence, we write

$$F=\mu_2\frac{\mathrm{d}E(r)}{\mathrm{d}r},$$

and substituting for $E(r)$ from Equation (2.12), we obtain

$$F(r)=-\frac{6\mu_1\mu_2}{4\pi\varepsilon_0 r^4}. \tag{2.13}$$

The magnitude of the induced dipole moment μ_2 is proportional to the strength of the electric field it experiences due to the first dipole, i.e. $\mu_2 \propto E(r)$, and hence $\mu_2 \propto 1/r^3$. Thus

$$F(r)\propto -\frac{1}{r^7},$$

and hence

$$V(r)\propto -\frac{1}{r^6}.$$

This is just the attractive term that we recognize from the Lennard-Jones potential.

Even though the first atom will go on to change the size and direction of its instantaneous dipole moment, the second atom will follow it. The two dipoles are *correlated*, with a positive charge on one atom appearing close to a negative charge on the other. Because of this correlation of the two relative positions of the positive and negative charges, the resulting attractive interaction between the two instantaneous dipoles does not average to zero. Instead, it gives a net attractive force. We have discussed the van der Waals interaction in terms of the first atom acting upon the second one, but, of course, the second atom must also influence the first in the same way.

The magnitude of the dipole moment induced in an atom depends on the ease with which the atom can be polarized; its *polarizability*. The polarizability of the rare gases increases with increasing atomic diameter. In a simple physical picture, the greater the atomic diameter, the less tightly bound are the outer electrons to the nucleus. The electron cloud is then more easily distorted by an external field, resulting in higher polarizability. With increasing polarizability comes increasing van der Waals attraction. One consequence of this is that the boiling points of the rare-gas atoms increase through the sequence – helium, neon, argon, krypton, and xenon as illustrated by Table 2.1.

Some molecules, such as hydrogen chloride (HCl) have a permanent dipole moment. This is because there is a *partial* movement of electron density between the constituent atoms; in this example, from the hydrogen atom to the chlorine atom. Not surprisingly, these molecules attract other molecules because

Table 2.1 The boiling points of the rare-gas atoms.

He	Ne	Ar	Kr	Xe
4.4 K	27.3 K	87.4 K	121.5 K	166.6 K

of their dipole electric field. They attract other molecules with a permanent dipole moment. However, they will also attract atoms and molecules that do not have a permanent dipole moment because they can polarize them in a similar manner to that discussed earlier.

2.3.2 Repulsive forces between atoms

The repulsive forces that atoms experience at small separations are associated with the overlap of the outer electron shells of the atoms. As the electron shells overlap, the positively charged nuclei become electrostatically less well-screened. In addition, repulsion arises because of the Pauli exclusion principle. This principle prevents two electrons from occupying the same quantum state; see Section 1.3.4. Overlap of the electron shells means that electrons must be promoted to higher atomic energy levels in order to comply with this requirement. The repulsive forces between atoms increase very rapidly with increasing overlap and this explains why inert gas atoms behave like hard spheres as suggested by Figure 2.6.

2.3.3 Binding energy and latent heat

The binding energy ε that characterizes the depth of the Lennard-Jones potential is closely allied to macroscopic quantities that we can measure in the laboratory; in particular, the latent heat of evaporation (liquid to gas) and latent heat of sublimation (solid to gas). So far, we have considered the situation where two atoms are bound together. But, of course, in a liquid or solid, an atom will have a number of nearest neighbours. Figure 2.10 is a two-dimensional picture to represent a particular atom, shaded in blue, which is surrounded by other atoms. In this two-dimensional picture, the atom has six nearest neighbours. In reality, for a liquid or solid, the number of nearest neighbours would be 10 or 12, respectively. The potential energy $V(r)$

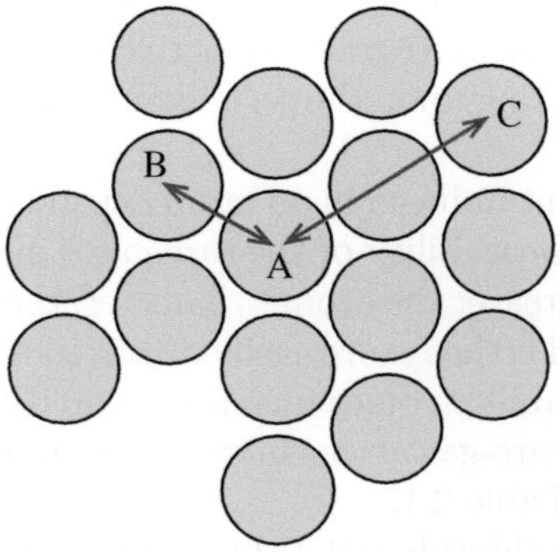

Figure 2.10 A two-dimensional picture to represent nearest and next-nearest neighbours. The potential energy of the pair of atoms A and B is $-\varepsilon$, while the potential energy of the pair of atoms A and C is approximately 30 times smaller.

of a particular pair of atoms with equilibrium separation $r = a_0$, say atoms A and B in Figure 2.10, is $-\varepsilon$, as we have already discussed. For the atoms A and C, we substitute $r = 2a_0$ in Equation (2.9) to calculate the potential energy due to their interaction. We find that this potential energy of the pair is $-\varepsilon/32$, i.e. much smaller than the potential due to the nearest neighbouring atoms. This large reduction reflects the short-range nature of the van der Waals force. Hence, to a good approximation, we need only count those atoms that are the nearest neighbours when calculating the potential energy of an assembly of atoms and we can ignore atoms that are further away.

With this picture in mind, we can obtain an approximate value of the binding energy ε from experimentally measured latent heat. Suppose that we have a quantity of liquid that contains N atoms and that each atom has n nearest neighbours. n is called the *coordination number*. If the binding energy for a pair of atoms is ε, then the amount of energy required to completely separate all the N atoms is $N \times 1/2n\varepsilon$. The factor of ½ arises because otherwise each atom–atom bond would be counted twice. If we are dealing with a mole of substance, then N is just Avogadro's number N_A, and we identify the total energy required to separate the atoms as the latent heat of vapourisation L_V. For liquid neon, L_V is 1.75 kJ/mol. Hence,

$$\varepsilon \approx \frac{2L_V}{N_A n} = \frac{2 \times 1.75 \times 10^3}{6.02 \times 10^{23} \times 10} = 0.58 \times 10^{-21} \approx 0.004\,\text{eV}.$$

This is a typical value for binding energy in van der Waals bonding. We note that it is relatively small, compared to, say, the value of kT at room temperature. We will return to the relevance of this important point in Section 2.4.

Although relatively weak, van der Waals forces play a dominant role in the bonding of a wide variety of atomic and molecular systems including, for example, organic molecules, colloidal systems, and pharmaceutical drugs. And, interestingly, it is thought that it is the van der Waals force that enables geckos to stick to smooth surfaces such as glass. We also saw, in Section 1.2, that atomic force microscopy exploits van der Waals forces between a surface and the AFM tip to image the surface.

2.3.4 Ionic bonding

In the case of van der Waals bonding, we saw that the electric dipole field of an atom distorts the charge distribution of a neighbouring atom. And this results in an attraction between the two atoms. By contrast, in ionic bonding, it is the *transfer* of an electron from one atom to another that produces the attraction. Consider the example of sodium chloride, (NaCl). An atom of sodium has a total of 11 electrons with the electron configuration $1s^2 2s^2 2p^6 3s$. It has one electron, the 3s electron, outside the closed-shell configuration of neon: $1s^2 2s^2 2p^6$. On the other hand, a chlorine atom has a total of 17 electrons and the electron configuration $1s^2 2s^2 2p^6 3s^2 3p^5$, which means that it is short of one electron compared to the closed-shell configuration of argon, which has a complete $3p^6$ shell. We recall from Section 1.3.4 that electronic configurations having full shells of electrons are particularly stable. Hence, by transferring an electron from a sodium atom to a chlorine atom, we obtain a combination of two particularly stable configurations. This results in a positively charged sodium ion, Na^+ and a negatively charged chlorine ion, Cl^-. These attract each other through the Coulomb force and form the molecule NaCl.

The potential energy between the sodium ion and the chlorine ion can be written as

$$V(r) = \frac{A}{r^p} - \frac{B}{r^6} - \frac{e^2}{4\pi\varepsilon_0 r}, \qquad (2.14)$$

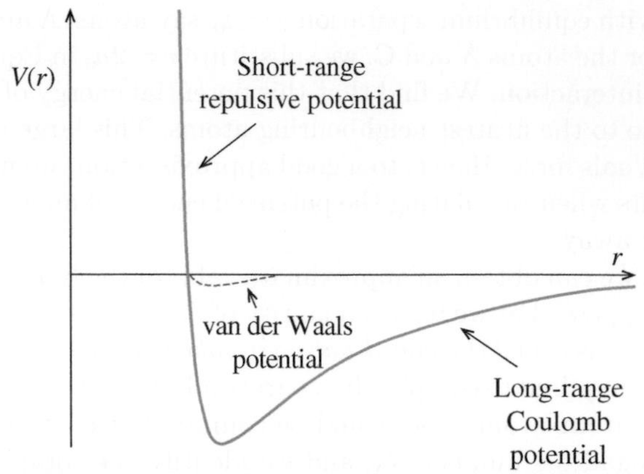

Figure 2.11 The potential energy of two ions of opposite polarity as a function of the distance r between their centres. The contribution to the total potential energy due to the Coulomb potential is much greater than the contribution due to the van der Waals potential.

where A and B are constants and p has a value close to 9. We recognize the first term on the right-hand side as due to the repulsion of the two ions as they are pushed together, due to the overlap of their electron charge clouds. We recognize the second term as the potential due to the van der Waals force of attraction. The third term is the Coulomb potential between two ions. Note that the Coulomb potential varies as $1/r$. This makes it a long-range potential and it is no longer justifiable to deal only with nearest-neighbour interactions. We must also take into account ions that are further away.

In Section 2.2, we saw that the depth of the potential well for atoms bound by the van der Waals force is ~0.01 eV. We can compare this value with that obtained from the Coulomb potential energy term in Equation (2.14). Taking a typical value of inter-ionic spacing of 0.2 nm for r, we have

$$\frac{e^2}{4\pi\varepsilon_0 r} = \frac{(1.602 \times 10^{-19})^2}{4\pi \times 8.8 \times 10^{-12} \times 0.2 \times 10^{-9}} \approx 7\,\text{eV}.$$

This is very much greater than the van der Waals potential energy and we get a picture of the resulting potential energy function between the two ions that is illustrated in Figure 2.11. We see that the Coulomb potential dominates and so we can ignore the van der Waals term in Equation (2.14) to obtain

$$V(r) = \frac{A}{r^p} - \frac{e^2}{4\pi\varepsilon_0 r}. \tag{2.15}$$

The relatively large value of the Coulomb potential energy means that sodium and chlorine atoms bind together to form solid NaCl at room temperature. If we want to produce gaseous NaCl molecules, we have to evaporate solid NaCl.

2.3.5 The Madelung constant and the Lattice energy

Solid NaCl exists in a cubic crystal structure, with the Na^+ and Cl^- ions situated at the alternate corners of a cube, as illustrated in Figure 2.12. Each Na^+ ion is surrounded by six equidistant Cl^- ions and each Cl^- ion is surrounded by six equidistant Na^+ ions. Clearly an ion in the crystal will feel the influence of many

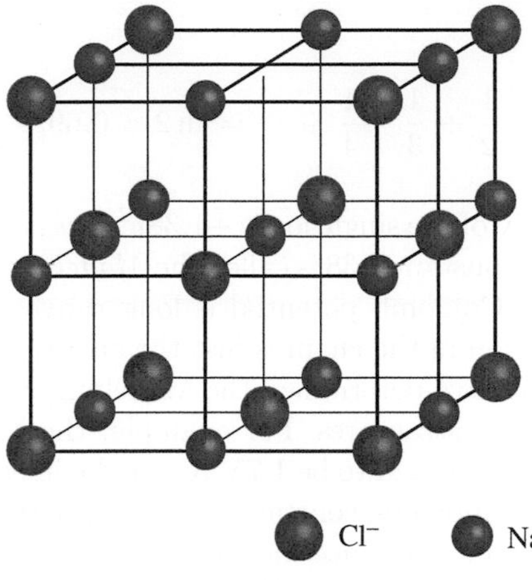

Figure 2.12 Crystalline NaCl exists in a cubic structure, with the Na^+ and Cl^- ions situated at the alternate corners of a cube. Each Na^+ ion is surrounded by six equidistant Cl^- ions and each Cl^- ion is surrounded by six equidistant Na^+ ions.

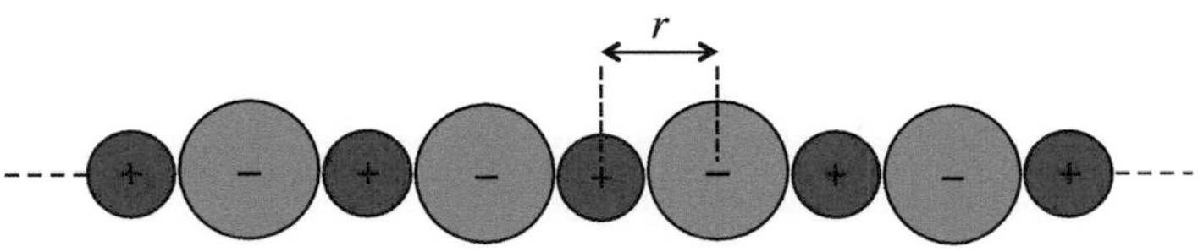

Figure 2.13 A one-dimensional model of an ionic crystal consisting of a linear chain of ions. Each ion has a charge e that is alternatively positive and negative. r is the distance between their centres.

neighbouring ions. In order to determine the potential energy of an ion in the crystal, we first consider a much simpler model. This consists of a linear chain of ions, as shown in Figure 2.13.

Moreover, we initially consider only the Coulomb interaction between the ions. The transfer of an electron from one atom to another in ionic bonding means that each ion has a charge e that is alternatively positive and negative. As noted previously, we cannot deal only with the nearest neighbours because of the long-range nature of the Coulomb force. Taking the inter-ionic distance to be r, the potential energy of a *single* ion due to its two neighbours, which have opposite signs to it, is $-2e^2/4\pi\varepsilon_0 r$. The next nearest neighbours, at distance $2r$ away and having the same sign as the ion under consideration, contribute potential energy $+2e^2/4\pi\varepsilon_0 2r$, and so on. Hence, the potential energy of a single ion in the infinitely long chain of ions is

$$\frac{-2e^2}{4\pi\varepsilon_0 r}\left(1-\frac{1}{2}+\frac{1}{3}-\frac{1}{4}+\cdots\right).$$

We recall that

$$\ln(1+x) = x - \frac{x^2}{2} + \frac{x^3}{3} - \frac{x^4}{4} + \cdots.$$

Hence,

$$1-\frac{1}{2}+\frac{1}{3}-\frac{1}{4}+\cdots=\ln 2=0.69.$$

Thus, the electrostatic potential energy of the single ion is $-1.38e^2/4\pi\varepsilon_0 r$, and this result applies to any ion, positive or negative, in the chain. The constant 1.38 is called the *Madelung constant* and is usually given the symbol α. We see that although the Coulomb potential is long range, we have the simplification that the Coulomb potential energy of any ion in the chain is just the energy of a nearest-neighbour pair times the Madelung constant. Although we have determined the Madelung constant for a linear chain of ions, it has been calculated for other crystal structures. For example, the Madelung constant for the cubic structure of crystalline sodium chloride is found to be 1.75. As for the linear chain, this Madelung constant is of the order of unity and indeed the Madelung constants for simple lattices of ions of opposite charge are all of the order of unity. This must be the case since the effect of, say, a positive charge is, to some extent, cancelled by the next nearest negative ion and so on. Hence, we can say that, in general, the electrostatic potential of an ion in a crystal is given by

$$V(r)_{\text{electrostatic}}=-\frac{\alpha e^2}{4\pi\varepsilon_0 r},\tag{2.16}$$

where α is the Madelung constant.

To obtain an expression for the total potential energy of a *single* ion in a crystal, we include a term to represent the repulsive interactions between the ions, giving

$$V(r)=\frac{A'}{r^p}-\frac{\alpha e^2}{4\pi\varepsilon_0 r}.\tag{2.17}$$

The first term on the right-hand side is the familiar term that represents the potential energy that the single ion experiences due to the surrounding ions, in this case, the nearest neighbours since the repulsive forces are short range. The constant p is called the *Born exponent.* Its value for a particular crystal can be determined experimentally from the *compressibility* of the crystal. For a sodium chloride crystal, its value is found to be 9.1.

The *lattice energy* is defined as the amount of energy required to separate one mole of a crystalline solid into a gas of its constituent *ions*. We can use Equation (2.16) to obtain a value for the lattice energy. Again, we use sodium chloride as an example. 1 mol of crystalline sodium chloride contains a total of $2N_A$ ions; N_A sodium ions and N_A chlorine ions, where N_A is Avagadro's number. Hence, the total potential energy of 1 mol of the crystal is

$$V(r)_{\text{mole}}=\frac{1}{2}2N_A\left(\frac{A'}{r^p}-\frac{\alpha e^2}{4\pi\varepsilon_0 r}\right).\tag{2.18}$$

The usual factor of ½ appears as we do not want to count the interaction between a given pair of ions twice. At equilibrium, the potential energy of the crystal will be minimized, for which

$$\frac{\mathrm{d}}{\mathrm{d}r}\left[V(r)_{\text{mole}}\right]=0.$$

Differentiating Equation (2.18), we obtain

$$\frac{1}{2}N_A\left[-p\frac{A'}{r^{p+1}}+\frac{\alpha e^2}{4\pi\varepsilon_0}\frac{1}{r^2}\right]=0,$$

which gives

$$A'=\frac{r^{p-1}}{p}\frac{\alpha e^2}{4\pi\varepsilon_0}.$$

Substituting for A' in Equation (2.17) and taking the inter-ion spacing to be $r = a_0$, we obtain the potential energy of 1 mol of the crystal:

$$V(r)_{\text{mole}}=\frac{N_A\alpha e^2}{4\pi\varepsilon_0 a_0}\left[\frac{1}{p}-1\right]. \tag{2.19}$$

This equation is known as the *Born–Landé equation*. The lattice energy is equal to

$$-V(r)_{\text{mole}}=\frac{N_A\alpha e^2}{4\pi\varepsilon_0 a_0}\left[1-\frac{1}{p}\right]. \tag{2.20}$$

The inter-ionic spacing a_0 for NaCl has been obtained by X-ray analysis to be 0.0282 nm and the value of p is 9.1. Taking $\alpha = 1.75$, we obtain the lattice energy of sodium chloride to be

$$\frac{6.02\times10^{23}\times1.75\times\left(1.602\times10^{-19}\right)^2}{4\pi\times8.85\times10^{12}\times0.282\times10^{-9}}\left[1-\frac{1}{9.1}\right]=765\ \text{kJ/mol}.$$

This value is within 3% of the measured value of the lattice energy, which is 786 kJ/mol. As the value of p is high, the lattice energy is dominated by the Coulomb potential energy. And to an accuracy of about 10%, the lattice energy is equal to $N_A\alpha e^2/4\pi\varepsilon_0 a_0$.

2.3.6 Covalent bonding

This type of bonding is the most prevalent in nature, and is intermediate between ionic and van der Waals bonding in the following sense. In ionic bonding, there is a transfer of an electron from one atom to another, whereas, in van der Waals bonding, there is no transfer of electrons. In covalent bonding, two atoms *share* a pair of electrons.

A pictorial representation of covalent bonding in the archetypal diatomic molecule H_2 is illustrated in Figure 2.14. In Figure 2.14a are shown the 1s wave functions of two isolated hydrogen atoms; see Section 1.3.3. When the two atoms are brought close together, as in Figure 2.14b, their wave functions overlap and combine to form the molecular wave function or *molecular orbital*. We recall that the square of a wave function gives the distribution of electron density and as can be seen in Figure 2.14c, there is a build-up of electronic charge between the two protons. Although the protons repel each other, each proton is attracted to the electron cloud that is situated between them. The net attraction of the electrons for each proton more than balances the repulsion of the two protons and the two electrons and a strong molecular bond is formed.

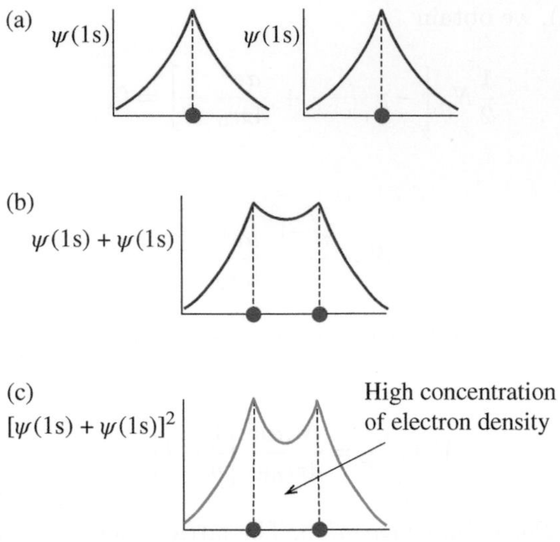

Figure 2.14 A pictorial representation of covalent bonding in the archetypal diatomic molecule H_2. (a) The 1s wave functions of two isolated hydrogen atoms. (b) When the two atoms are brought close together, their wave functions overlap and combine to form the molecular wave function. (c) The square of the wave function gives the distribution of electron charge density and as can be seen, there is a build-up of electron charge between the two protons. Although the protons repel each other, each proton is attracted to the electron cloud that is situated between them. The net attraction of the electrons for each proton more than balances the repulsion of the two protons and of the two electrons and a strong molecular bond is formed.

As we also saw in Section 1.3.4, the exclusion principle permits two electrons to be in the same *spatial* quantum state so long as they have opposite spins. Hence, opposite electron spins are an essential requirement for a covalent bond and it follows that no more than two electrons can occupy such a bond. The covalent bond is said to *saturate.* Hence, there is no tendency for a further atom to bond to the first two. For example, H_3 does not exist as a stable molecule, and two H_2 molecules will only attract each other by van der Waals forces.

In the case of ionic and van der Waals bonding, we were able to deduce how the potential energy due to the attractive force depends on interatomic separation r. We found the dependence $1/r^7$ for van der Waals bonding and $1/r$ for ionic bonding. However, detailed quantum mechanical calculations are required to determine how the potential energy depends on inter-nuclear separation in covalent bonding. Essentially, Schrödinger's equation is solved for the situation of the two, shared electrons in the presence of the two constituent atomic nuclei. The electrons are much lighter than the nuclei and this allows an important approximation to be made that makes this problem much more tractable. The approximation, called the *Born-Oppenheimer approximation,* is that the nuclei move much more slowly than the electrons. Then we can choose a particular, fixed separation for the nuclei and solve for the Schrödinger's equation for the electrons for that separation. Then we choose a different separation and repeat the calculation, and so on. In this way, we can determine how the potential energy of the molecule varies with separation and obtain the molecular potential energy curve.

A typical example of a molecular potential energy curve is shown in Figure 2.15. As before, we imagine one of the atoms to be fixed in position at the $r = 0$ axis and plot potential energy as a function of r, the separation of the two atoms. The curve is similar in shape to the potential energy curves we have seen before

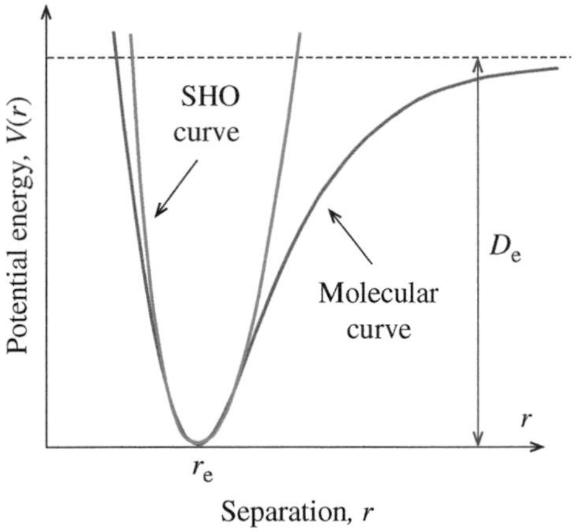

Figure 2.15 An example of a molecular potential energy curve, shown in blue. The potential energy $V(r) \rightarrow 0$ as $r \rightarrow \infty$, and increases sharply as the two atoms are pushed closer together. There is a minimum in the potential energy at $r = r_e$, which corresponds to the bond length of the molecule. The depth of the potential well is called the dissociation energy D_e. The green curve is the potential energy curve for a simple harmonic oscillator (SHO). Close to the equilibrium separation, the SHO overlaps the molecular potential energy curve.

and has the same features. The potential energy $V(r) \rightarrow 0$ as $r \rightarrow \infty$ and increases sharply as the two atoms are pushed closer together. And there is a minimum in the potential energy, corresponding to the equilibrium separation. In the case of diatomic molecules, the equilibrium separation is called the *bond length* r_e, and the depth of the potential well is called the *dissociation energy* D_e. Values of equilibrium separation are similar in magnitude to values for equilibrium separation that we have encountered for other types of bonding. For example, the bond length of the archetypal molecule H_2 is 0.074 nm. Consequently, diatomic molecules are not much larger than their constituent atoms. The depth of the well in covalent bonding is of a similar size to those encountered in ionic bonding, which means that the covalent bond is a strong bond. For the example of H_2, the depth of the well, i.e. the dissociation energy, is 4.52 eV. This means that it takes 4.52 eV to pull the molecule apart and form two isolated hydrogen atoms.

2.3.7 Vibrational motion of a diatomic molecule

A diatomic molecule, as far as its mass distribution is concerned, can be pictured as a dumbbell, as the electrons form only a tiny proportion of the mass distribution. However, it is not a rigid dumbbell as the molecule can vibrate about the equilibrium separation. Classically, we can picture a diatomic molecule as two masses m_1 and m_2, respectively, connected by a spring as shown in Figure 2.16. The spring represents the

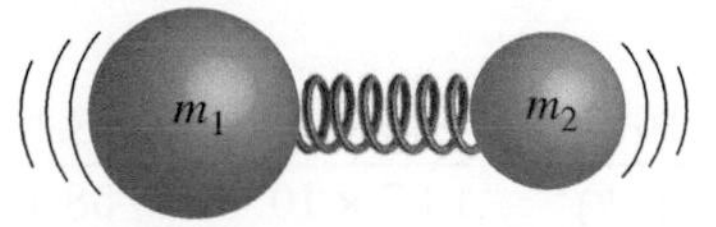

Figure 2.16 Classically, we can picture a diatomic molecule as two masses m_1 and m_2, respectively, connected by a spring. The spring represents the molecular bond.

molecular bond. Although this is a simple model, it provides a useful starting point for a discussion of molecular vibrations. The compression and extension of the molecular bond are described by Hooke's law, which we write as

$$F = -k(r - r_e). \tag{2.21}$$

F is the restoring force and k is the force constant, i.e. the strength of the molecular bond. This is the equation for a simple harmonic oscillator and when such an oscillator is disturbed from equilibrium, it oscillates with SHM at its resonance frequency

$$\nu_{osc} = \frac{1}{2\pi}\sqrt{\frac{k}{\mu}}\,\text{Hz}, \tag{2.22}$$

where μ is the *reduced mass* given by $\mu = \dfrac{m_1 m_2}{m_1 + m_2}$. Moreover, if this classical oscillator is driven by a sinusoidal force of the form $F_0 \sin 2\pi v t$, it will absorb maximum energy from the driving force at its resonance frequency, when $v = v_{osc}$. In an analogous way, a molecule will strongly absorb electromagnetic radiation when the frequency of the radiation matches the resonance frequency of the molecule. And in our classical model, this resonance frequency is given by Equation (2.22). Hence, by observing the frequency at which a molecule absorbs the radiation, the strength of its molecular bond can be deduced.

The potential energy curve for a simple harmonic oscillator, described by Hooke's law, Equation (2.19), is

$$V(r) = -\frac{1}{2}k(r - r_e)^2. \tag{2.23}$$

This function has a parabolic shape and such a parabolic curve is superimposed on the molecular potential energy curve in Figure 2.15. We see that close to the equilibrium separation, i.e. for small amplitude oscillations, the two curves overlap. Hence, under this condition, the classical model is a good approximation for molecular vibrations. Clearly, there is a strong similarity here to our discussion in Section 2.1 of oscillations about the equilibrium separation in interatomic bonding.

Worked example

The absorption spectrum of the molecule HCl shows very intense absorption at a wavelength of 3.35 μm. Use the classical model of molecular vibration to determine the strength of the molecular bond in HCl. The atomic masses of H and Cl are 1.67×10^{-27} kg and 58.1×10^{-27} kg, respectively.

Solution

In the classical model, the HCl molecule will absorb electromagnetic radiation whose frequency matches its resonance frequency v_{osc} given by Equation (2.22). The wavelength λ of the absorbed radiation $= c/v_{osc}$. Hence,

$$k = \frac{4\pi^2 c^2 \mu}{\lambda^2} = \frac{4\pi^2 (3 \times 10^8)^2}{(3.35 \times 10^{-6})^2} \times \frac{1.67 \times 10^{-27} \times 58.1 \times 10^{-27}}{1.67 \times 10^{-27} + 58.1 \times 10^{-27}} = 514\,\text{N/m}.$$

This value is in very good agreement with the accepted value of 516 N/m.

A mathematical expression that provides a good approximation to a molecular potential energy curve over an extended range of inter-nuclear separation is the *Morse function*:

$$V(r) = D_e[1 - \exp\{a(r_e - r)\}]^2, \tag{2.24}$$

This function is purely empirical. However, apart from being a good approximation to actual potential curves, it has an important advantage that it can be easily manipulated mathematically. Figure 2.17 shows a typical Morse function.

Classically, there is no restriction on the energy that an oscillator can have. But in quantum mechanics, the vibrational energy of an oscillator is quantized. And indeed, the vibrational levels of a molecule are quantized just as the energy levels are in an atom. This becomes apparent when the Morse function is used for the potential energy $V(r)$ in the Schrödinger's equation. Then the solution of the Schrödinger's equation gives a set of discrete vibrational energy levels. These are shown for the Morse potential in Figure 2.17. The vibrational levels are designated by the *vibrational quantum number* v, which can have values 0, 1, 2, 3, ... Because of the uncertainty principle, a molecule cannot sit at the bottom of the potential well. If it did, both its position and momentum would be completely defined. Instead, it sits in the energy level, with $v = 0$, just above the bottom of the well. The energy difference between the ($v = 0$) level and the bottom of the well is called the *zero point energy*. Because of the shape of a molecular potential well, the separation between successive vibrational levels steadily reduces and the levels eventually converge towards the top of the well.

From a quantum mechanical viewpoint, a molecule makes a transition from a particular vibrational level to a higher level when it absorbs a photon that has an energy equal to the energy difference between the two levels. As we will see in Section 3.8, molecules are most likely to be in the $v = 0$ level. So, in the case of the HCl molecule, for example, a photon of wavelength 2.35 μm can excite the molecule from the $v = 0$ level to the $v = 1$ level. This transition is indicated by the vertical arrow in Figure 2.17. A photon wavelength of 2.35 μm corresponds to a photon energy of 0.37 eV, which is therefore the energy separation of the two levels in HCl. We can compare this quantum viewpoint with the classical viewpoint in the worked example earlier.

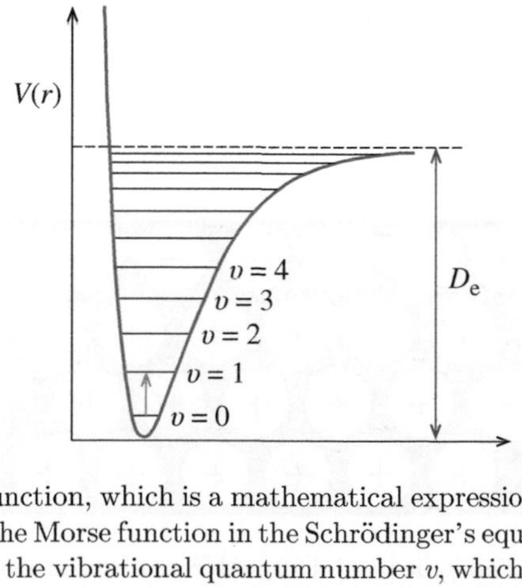

Figure 2.17 An example of a Morse function, which is a mathematical expression that provides a good approximation to a molecular potential curve. Using the Morse function in the Schrödinger's equation gives a set of discrete vibrational energy levels. These are designated by the vibrational quantum number v, which can have values 0, 1, 2, 3, ... Because of the uncertainty principle, a molecule cannot sit at the bottom of the potential well, but resides in the $v = 0$ energy level. A photon of appropriate wavelength can excite the molecule from the $v = 0$ level to the $v = 1$ level, as indicated by the vertical arrow.

The average temperature of planet earth is about +15 °C and, consequently, the blackbody radiation it emits lies in the infrared region of the electromagnetic spectrum. The vibrational frequencies of molecules also typically lie in the infrared region, as we have seen for the example of the HCl molecule. Consequently, atmospheric molecules such as water (H_2O) and carbon dioxide (CO_2) absorb the infrared radiation emitted by the earth; at wavelengths corresponding to their absorption frequencies. These excited molecules then re-radiate part of this excitation energy back to the earth when they de-excite. They thus act like a thermal blanket placed around the planet. By contrast sunlight, which encompasses the visible region of the electromagnetic spectrum, is not absorbed by the molecules in its passage to the Earth's surface. The overall result is a trapping of the solar heat and a net warming of the planet. Indeed, without this warming effect, the average temperature of the planet would be below 0 °C, the freezing point of water.

2.3.8 Metallic bonding

Metallic bonding can be thought of as a limiting case of covalent bonding in which electrons are shared by *all* the ions in a crystal. We take the example of metallic sodium. The sodium atom has the electronic configuration $1s^22s^23p^63s$. The 3s electron, in the outermost shell, is relatively weakly bound, i.e. it does not take a lot of energy to release it from the atom. When sodium atoms come together to form a crystal, energy is released when the atoms bond together. Recall the general form of an interatomic potential in which the bottom of the potential well lies below the potential energy of the separated atoms: see, for example, Figure 2.3. The energy released by this bonding process is sufficient to free the weakly bound 3s electrons from the sodium atoms. These freed electrons exert attractive forces on the positive ions that exceed the repulsion of the ions and so solid sodium is formed. The free electrons move in the combined potential of all the positive ions and are *shared* by all the atoms. Importantly, the free electrons are free to roam through the crystal from atom to atom and the electrons behave like an 'electron gas'. This situation is represented pictorially in Figure 2.18.

Because of the ability of the free electrons to move freely through the crystal, metallic solids are excellent conductors of electricity as the electrons readily move under the influence of an applied electric field. The free electrons also provide the principal mechanism for heat conduction by virtue of their kinetic energy, which they carry through the metal. Hence, metals are also excellent conductors of heat. For example, liquid sodium is used in some nuclear fission reactors to transport heat away from the nuclear core.

The problem of calculating the potential energy of an ion in the metallic lattice is a very complicated one because the free electrons can redistribute themselves if the separation of the ions is changed. However, we

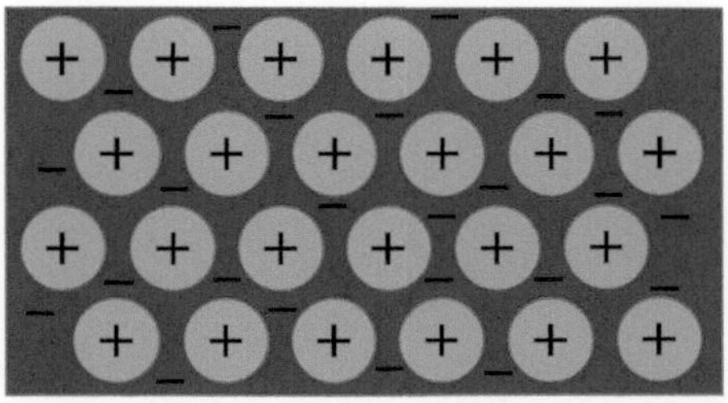

Figure 2.18 A pictorial representation of an ionic crystal formed from metallic atoms. Typically one or two electrons are donated by each atom in the crystal and these electrons move in the combined potential of all the positive ions. The electrons are free to roam through the crystal and behave like an 'electron gas'. Because of the ability of the free electrons to move freely through the crystal, metallic solids are good conductors of electricity and also of thermal energy.

can expect the form of the interatomic potential to have similar features to the potential curves we have seen before. Metals certainly form solids and they resist compression and there must be a minimum in the potential curve because it takes energy to evaporate the metal. The strength of a metallic bond depends on the number of electrons liberated from each atom. For example, the bond strength in magnesium is stronger than one in sodium because two electrons rather than one are freed in the case of magnesium. This is reflected in the latent heats of vaporization of sodium and magnesium, which are 97 kJ/mol and 128 kJ/mol, respectively.

2.4 Why gases, liquids, and solids

Matter may appear in various states or *phases*. A familiar example is provided by water. Depending on its temperature, it can be ice, water, or steam; i.e. in the solid, liquid, or gas phase, respectively. Following our discussion of interatomic forces and the resulting bonding, we are now in a position to understand why matter exists in a particular state. We will see that the three states of matter are the result of a competition between interatomic forces and thermal energy; or in other words, the competition between the potential energy and the kinetic energy of the atoms. The essential idea is this: if the kinetic energy of the atoms is much less than their binding energy, then the atoms will remain bound together as a solid. On the other hand, if their kinetic energy is much greater than their binding energy, the binding energy is not sufficient to hold the atoms together and they exist as a gas. And, roughly speaking, if the kinetic and potential energies are comparable, the atoms exist in the liquid state.

So far, in the present chapter, we have confined our attention to the potential energy of atoms. But now we take account of their kinetic energy also. As we will describe in Chapter 3, the kinetic energy of an atom is directly connected to its thermal energy. We will see that the mean translational kinetic energy of an atom at temperature T is equal to $3/2kT$ where k is the Boltzmann constant. *At room temperature,* kT *is equal to* 1/40 eV. This is a very important number to commit to memory as it turns up in many physical situations.

Figure 2.19a shows two atoms in a gas that have come together and become bound by the van der Waals force and another atom that is about to strike the bound pair. We recall that the binding energy involved in van der Waals bonding is ~0.01 eV. Since the kinetic energy of the incident atom at room temperature, ~0.025 eV, is much greater than the binding energy of the pair, the collision will cause the two bound atoms

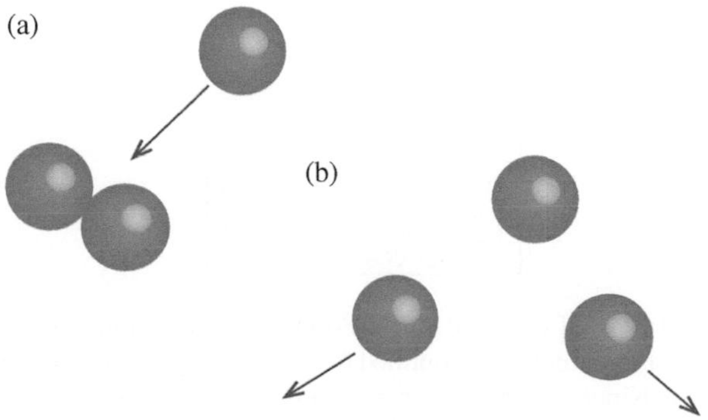

Figure 2.19 (a) This figure shows two atoms in a gas that have come together and become weakly bound by the van der Waals force and another atom that is about to strike the bound pair. (b) When the incident atom strikes the bound pair, the two atoms fly apart because their binding energy is much less than the kinetic energy of the incident atom.

to fly apart as in Figure 2.19b. And indeed any atoms that bond together will quickly, thereafter, fly apart in collisions with other atoms. Consequently, the collection of atoms remains in the gaseous state. We can reduce the kinetic energy of the atoms below their binding energy by reducing their temperature. This causes the atoms in the gas to condense into a liquid and eventually form a solid. Indeed, the implicit assumption in our previous discussion of van der Waals bonding was that the temperature of the atoms was sufficiently low that their kinetic energy was negligible compared to their binding energy.

At the other extreme, we have a situation where the kinetic energy of the atoms is very much less than the binding energy of the atoms. An example is an ionic crystal where the binding energy, ~5 eV, is certainly very much greater than kT at room temperature. And we may therefore correctly expect that the atoms stay at their fixed sites, i.e. the atoms remain in the solid state.

We see that we can make greatly simplified assumptions for the solid and gaseous states. In the solid state, potential energy dominates and so the atoms form rigid and well-ordered structures as illustrated by Figure 2.20a. The translation motion of the atoms is confined to vibration about their equilibrium positions. At the other extreme, in the gaseous state, kinetic energy dominates and we can describe the atoms as free particles that have random translational motion and negligible interaction with each other, as illustrated by Figure 2.20b. Liquids lie between these two extremes, where the kinetic energy of the atoms is comparable to their binding energy. Hence, we can make neither simplifying assumption and this makes the description of liquids much more complicated. We can, however, be guided by the macroscopic properties of liquids. We recall that they are essentially incompressible, but that a liquid can readily change its shape to fill its container. Thus, in a much simplified picture of what happens at the microscopic level, we may say the following. The interatomic attractive forces are sufficiently strong that they bind the atoms close together. However, they are not strong enough to prevent the translational movement of the atoms that arises from their thermal energy, i.e. to prevent atoms from sliding over each other. And this translational movement destroys any structural rigidity. The liquid state is illustrated in Figure 2.20c.

It follows that the liquid state of a substance only exists over a limited temperature range. Indeed, bromine and mercury are the only elements that are liquid at room temperature, although caesium, rubidium, and gallium are liquid just above it. Happily, water is liquid over the temperature range of 0–100 °C.

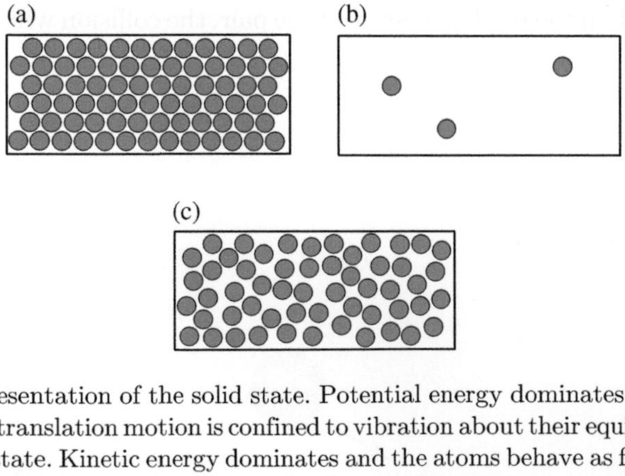

Figure 2.20 (a) Pictorial representation of the solid state. Potential energy dominates and the atoms form rigid and well-ordered structures. Their translation motion is confined to vibration about their equilibrium positions. (b) Pictorial representation of the gaseous state. Kinetic energy dominates and the atoms behave as free particles that have random translational motion and negligible interaction with each other. (c) Pictorial representation of the liquid state. The interatomic attractive forces are sufficiently strong that they bind the atoms close together. However, they are not strong enough to prevent the translational movement of the atoms that arises from their thermal energy, i.e. to prevent atoms from sliding over each other. This translational movement destroys any structural rigidity.

The *Goldilocks zone* refers to the habitable zone around a star where the temperature is just right, not too hot and not too cold, for liquid water to exist on a planet. Liquid water is essential for life as we know it. Consequently, much effort is going into the search for planets that lie in a Goldilocks zone and which could contain extra-terrestrial life.

Problems 2

2.1 The latent heat of evaporation of liquid nitrogen is 5.6 kJ/mol. Estimate the inter-molecular binding energy ε.

2.2 (a) Compare the relative strengths of the electrostatic force and the gravitational force between two electrons. (b) Compare the binding energy of the van der Waals bond between two helium atoms in He_2 (8.8×10^{-4} eV) with their energy due to gravitational attraction. The equilibrium separation of the atoms is 2.9×10^{-10} m.

2.3 Liquid argon has a latent heat of evaporation of 1.6×10^5 J/kg and a density ρ of 1400 kg/m^3. As we will see in Chapter 11, the surface tension of a liquid can also be related to the binding energy ε of a pair of atoms through the relation $\gamma = n\varepsilon/4a_0^2$, where a_0 is the diameter of an atom and n is the coordination number that can be taken to be equal to 10. The surface tension of argon is 0.013 N/m. Use these data to estimate (a) the diameter of an argon atom and (b) the binding energy ε.

2.4 The Lennard-Jones potential which describes the potential energy between two atoms may be written as

$$V(r) = 4\varepsilon\left[\left(\frac{a}{r}\right)^{12} - \left(\frac{a}{r}\right)^{6}\right],$$

where we identify a as the diameter of the atoms. Show that the minimum in $V(r)$ occurs at $r = \sqrt[6]{a}$. Show that the well depth is $-\varepsilon$.

2.5 An alternative representation of the interatomic potential, due to Morse, is

$$V(r) = D_e\{\exp[-2(r - r_e)/a] - 2\exp[-(r - r_e)/a]\},$$

where D_e, a and r_e are constants. What is the equilibrium separation r_e? What is the depth of the potential well?

2.6 Another representation of the interatomic potential between two atoms, called the *Mie potential*, is

$$V(r) = -\frac{A}{r^p} + \frac{B}{r^q}, \quad (0 < p < q),$$

where r is the separation of the atoms. If the equilibrium separation of the atoms is r_0, and the binding energy is ε, show that

$$A = \frac{q\varepsilon r_0^p}{q - p}, \quad B = \frac{p\varepsilon r_0^q}{q - p},$$

and hence, that $V(r)$ can be written as

$$V(r) = \frac{pq\varepsilon}{q - p}\left[-\frac{1}{p}\left(\frac{r_0}{r}\right)^p + \frac{1}{q}\left(\frac{r_0}{r}\right)^q\right].$$

2.7 Given that the interatomic potential between two hydrogen atoms is given by

$$V = \frac{pq\varepsilon}{q - p}\left[-\frac{1}{p}\left(\frac{r_0}{r}\right)^p + \frac{1}{q}\left(\frac{r_0}{r}\right)^q\right],$$

where r_0 is the equilibrium separation of the atoms, show that if the separation between the atoms is increased from r_0 to $r_0 + \delta$, there is a restoring force, $F(r)$ given by

$$F(r) = \frac{pq\varepsilon}{(q-p)r_0}\left[-\left(\frac{r_0}{r_0+\delta}\right)^{p+1} + \left(\frac{r_0}{r_0+\delta}\right)^{q+1}\right].$$

If the displacement is small, i.e. $\delta \ll r_0$, the denominator can be expanded by the binomial theorem. Show that to first order in δ,

$$F(r) = -\frac{pq\varepsilon\delta}{r_0^2}.$$

As the restoring force is proportional to the displacement, the molecule undergoes SHM. The frequency of vibration is 1.3×10^{14} Hz. Deduce a value for pq given that $\varepsilon = 4.4$ eV and $r_0 = 0.75 \times 10^{-10}$ m.

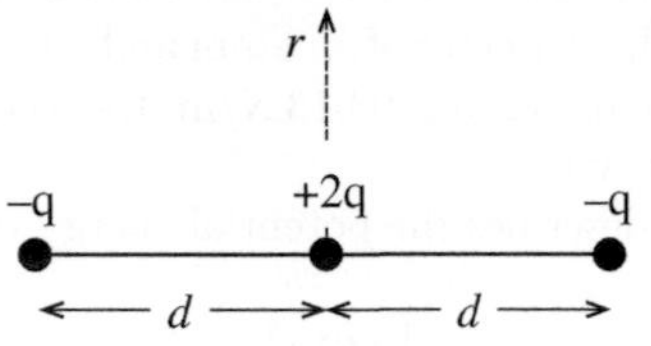

2.8 The figure shows an *electric quadrupole* that consists of three charges separated by distance d. We can think of the quadrupole as the superposition of two dipoles of opposite orientation so that their positive charges overlap. Show that the electric field at distance r along the direction *perpendicular* to the line of charges is given by

$$E(r) = \frac{3qd^2}{4\pi\varepsilon_0 r^4}.$$

2.9 The repulsive part of the potential energy experienced by an ion in a NaCl crystal can be expressed in the form of $A\exp(-r/\rho)$. (a) Obtain an expression for lattice energy, i.e. the total binding energy/mol. (b) Calculate the value of ρ given that the distance between nearest neighbours is 2.8×10^{-10} m, the Madelung constant is 1.75, and the binding energy is 763 kJ/mol.

2.10 As we will see in Chapter 3, the average kinetic energy of a molecule in a gas is $3/2kT$, where k is the Boltzmann constant and T is the temperature. (a) The binding energy ε of the Ne_2 molecule is 0.5×10^{-21} J. Would you expect Ne_2 molecules to exist in a gas of neon at room temperature? Explain. (b) The dissociation energy D_e of H_2 is 4.4 eV. At what value of T will the average kinetic energy be equal to D_e?

2.11 An incident photon with sufficient energy can dissociate a molecule. The dissociation energy D_e of H_2 is 4.4 eV. Calculate the maximum wavelength of light that can dissociate a hydrogen molecule. In what part of the electromagnetic spectrum does this wavelength lie?

3

Thermal energy of atoms and molecules

In Chapter 2, we described the forces that act between atoms, and the resulting potential energy of bound atoms. Now we turn our attention to the thermal energy of atoms and molecules. (For the sake of convenience, we will now talk in terms of molecules including atoms such as helium and argon as well as molecules such as molecular hydrogen and nitrogen.) This brings us naturally to the concept of temperature and its relationship to the kinetic energy of molecules. We touched upon the kinetic energy of molecules in Section 3.4, where we discussed why matter exists in the gaseous, liquid, or solid phase, and there we saw the connection between kinetic energy and temperature. This connection will become further evident in this chapter. We will extend our discussion to the rotational and vibrational motion that molecules may have in addition to their translational motion. We will relate temperature to the motions of molecules on the *microscopic scale*, just as we discussed interatomic forces and potentials between individual atoms.

Any laboratory sample of a gas contains a huge number of molecules; we recall that one mole of a gas contains $\sim 6 \times 10^{23}$ molecules. We cannot possibly know the positions and velocities of all the molecules to make detailed calculations of their individual motions. On the other hand, the extremely large number of molecules enables us to describe the state of a gas using statistical principles. For example, we can talk about the average or *mean* speed of the molecules and such mean values will enable us to describe the behaviour of a gas under various conditions. Our discussion of the thermal energy of molecules will involve the laws of Maxwell and Boltzmann and the equipartition of energy theorem, which are some of the most important results of classical physics. We will use classical physics to obtain the specific heat of a gas. And we will also see that the failure of classical physics to predict specific heat under certain conditions provided the first evidence of the breakdown of classical mechanics and the need for quantum mechanics.

3.1 Temperature and the translational kinetic energy of a molecule

We start by considering the pressure that a gas exerts on the walls of its container. We confine our attention to an ideal gas, which has the following properties: (i) the gas consists of a large number N of identical molecules of mass m, (ii) the molecules act as point particles of negligible size, (iii) the collisions of the molecules with other molecules and with the walls of the container are elastic, i.e. there is no loss of kinetic

Physics of Matter, First Edition. George C. King.

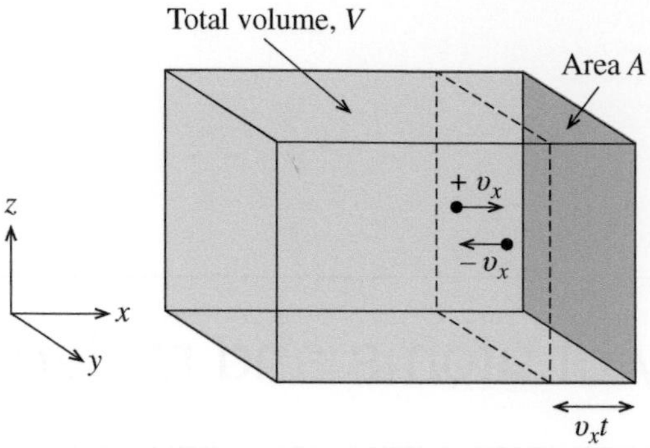

Figure 3.1 The figure shows a container of total volume V containing a total of N molecules, with a container wall of area A. The number of molecules that strike this wall in time t is $\frac{1}{2}(N/V)Av_x t$, where N/V is the number density and v_x is the component of velocity in the x-direction. Factor 1/2 arises because only half the molecules are travelling to the right.

energy in the collisions, and (iv) the motion of the molecules is random with no preferred positions or directions of travel.

Figure 3.1 shows a container of volume V containing a total of N molecules, and a wall of the container that has area A. Each molecule has velocity components v_x, v_y, and v_z. When a molecule makes a collision with the wall, the components of velocity that are parallel to the wall, v_y and v_z, are unchanged. Only the component, v_x, that is perpendicular is changed. That component reverses direction in the collision but, as the impact is elastic, the magnitude of the velocity component does not change. Hence, the momentum imparted to the wall by the molecule is $2mv_x$. For the moment, we suppose that all N molecules have the same value of the velocity component v_x. From Figure 3.1, we see that the number of molecules that strike the wall in time t is $\frac{1}{2} \times (N/V) \times A \times v_x t$, where N/V is the *number density* of the molecules. The factor $\frac{1}{2}$ arises because only half the molecules are travelling to the right. Therefore, the total momentum transfer in time t is $\frac{1}{2}(N/V)Av_x t \times 2mv_x$.

Then the rate of momentum transfer, i.e. the force acting upon the wall is $(N/V)Amv_x^2$, and the pressure P, i.e. force per unit area, exerted on the wall is

$$P = \frac{N}{V} m v_x^2. \tag{3.1}$$

Of course, all the molecules in a gas do not have the same velocity component. To allow for this, we replace v_x^2 with the mean value of v_x^2, for the N molecules, which we denote with a bar as $\overline{v_x^2}$. The speed c of a molecule is related to the velocity components v_x, v_y, and v_z by

$$c^2 = v_x^2 + v_y^2 + v_z^2. \tag{3.2}$$

For a very large number of molecules in random motion,

$$\overline{v^2} = \overline{v_x^2} + \overline{v_y^2} + \overline{v_z^2}, \tag{3.3}$$

and

$$\overline{v_x^2} = \overline{v_y^2} = \overline{v_z^2}. \tag{3.4}$$

Hence, from Equation (3.2)

$$\overline{v_x^2} = \frac{1}{3}\overline{c^2}, \tag{3.5}$$

where $\left(\overline{c^2}\right)^{1/2}$ is called the *root mean square speed* of the molecules. Dealing with speed rather than velocity has the advantage that we do not need to take into account the directions of travel of the molecules. Equations (3.3) and (3.4) seem to be entirely reasonable, but the proofs of these equations are actually quite involved. Hence, the pressure exerted by the molecules is given by

$$P = \frac{N}{3V} m\overline{c^2},$$

or

$$PV = \frac{2N}{3}\left(\frac{1}{2} m\overline{c^2}\right), \tag{3.6}$$

where $1/2\left(m\overline{c^2}\right)$ is the average translational kinetic energy of a molecule. We see that the product PV is equal to two-thirds of the total translational energy. And we see that we can think of pressure as energy per unit volume. Recalling that one mole of gas contains N_A molecules, the total translational energy of one mole of gas is

$$N_A\left(\frac{1}{2} m\overline{c^2}\right) = \frac{3}{2} PV. \tag{3.7}$$

3.1.1 The ideal gas equation

Experimental investigations of the behaviour of gases led to several conclusions: (i) the volume V of a fixed amount of gas varies inversely with pressure, if the temperature is constant, i.e. $PV = \text{constant}$, which is Boyle's law and (ii) the pressure is proportional to the absolute temperature, i.e. $P \propto T$, which is Charles' law. These relationships are combined in the ideal gas equation

$$PV = RT, \tag{3.8}$$

for one mole of gas. R is the gas constant, which is the same for all gases. An equation that expresses the relation between the three state variables pressure P, volume V, and temperature T is called an *equation of state*. If there are n moles of gas, the right-hand side of Equation (3.8) is multiplied by n. Strictly speaking, this equation is applicable to ideal gases. However, it also works for real gases within a few per cent up to several atmospheres and at temperatures well above those at which the gas liquefies.

We compare Equation (3.6) to Equation (3.8) to obtain

$$RT = \frac{2N_A}{3}\left(\frac{1}{2} m\overline{c^2}\right),$$

and hence,

$$\frac{1}{2}m\overline{c^2} = \frac{3}{2}\left(\frac{R}{N_A}\right)T. \tag{3.9}$$

The ratio R/N_A is the Boltzmann constant k. Its value is

$$k = \frac{R}{N_A} = \frac{8.314\ \text{J/mol·K}}{6.022 \times 10^{23}\ \text{molecules/mol}} = 1.381 \times 10^{-23}\ \text{J/K}.$$

Hence,

$$\frac{1}{2}m\overline{c^2} = \frac{3}{2}kT. \tag{3.10}$$

This is the key we have been looking for:

The temperature of a gas is directly related to the translational kinetic energy of the molecules.

This is a very important result of classical physics. Substituting for Equation (3.10) into Equation (3.6) we obtain

$$PV = NkT, \tag{3.11}$$

where N is the total number of molecules in volume V and P is the pressure.

Equation (3.10) gives the mean kinetic energy of a molecule. The root mean square speed of the molecules is then

$$\left(\overline{c^2}\right)^{1/2} = \sqrt{\frac{3kT}{m}}. \tag{3.12}$$

It follows from Equations (3.5) and (3.10) that

$$\frac{1}{2}m\overline{v_x^2} = \frac{1}{2}kT, \tag{3.13}$$

with corresponding equations for $\overline{v_y^2}$ and $\overline{v_z^2}$.

Notice that we have compared the result obtained from a discussion of the motion of molecules on a microscopic scale, Equation (3.6), with experimental measurements of volume, pressure, and temperature, obtained on the macroscopic scale, Equation (3.8). We have also introduced a statistical approach to the motion of a large collection, or assembly, of molecules. We pursue this statistical approach in the following section.

3.2 Probability distributions and mean values

We have obtained an expression for the root mean square speed of the molecules, (Equation 3.12), by considering the pressure exerted by a gas on the walls of its container. However, of course, not all the molecules will have the same speed. The molecules have a *distribution* of speeds. It is not possible to know the individual speeds of the $\sim 10^{23}$ molecules in a sample of gas. But this huge number is in fact a great advantage. It means we can use statistical methods to determine, with very high accuracy, the probability of a molecule having a speed within a certain range. We do not need to know which particular molecules have speeds in that range, only the fraction of molecules that do. Information about the distribution of speeds among the molecules is embodied in a so-called *probability distribution*, which we will present in Section 3.3. We will make much use of probability distributions in our discussion of the thermal properties of gases, and indeed probability distributions find application in a wide range of the physical sciences and beyond. So, we discuss them here in some detail.

To illustrate the concept of a probability distribution, we describe a quite different example, namely the distribution of heights in a group of people. Suppose that we want to know the distribution of heights in a particular group of people and the average or mean height of those people. We first note that height has a *continuous* range of values and because of this, no person would have exactly a height of, say, 174.3456 cm. This is in contrast to the score of a throw of dice, or the number of disintegrations of a radioactive nuclide in a given time period, which always give integer values. We take the continuous nature of height into account by dividing the range of heights into equal-sized intervals, Δh, and counting the number of people whose heights fall within that interval. This gives a definite number. We then construct a histogram where we plot the number of people n_i who have heights between h_i and $h_i + \Delta h$. An example is shown in Figure 3.2, where the width Δh of the intervals is 2 cm and the total number of people is 144. The histogram consists of adjacent rectangles, each of width Δh and height n_i. We see that there are some people who are taller than average and some that are shorter than average but most of the people have a height close to the mean height.

Rather than plot the number of people per height interval, we can plot the *fraction* of people having heights in a given interval. Moreover, it is advantageous to make the *area* of a given rectangle equal to the fraction of people having heights within that interval. If there are n_i people with heights within the height interval h_i to $h_i + \Delta h$, and the total number of people is $N = \sum_i n_i$, then the fraction of people in that height interval is n_i/N. We let the height of the corresponding rectangle to be $F_i(h)$. Then $F_i(h) \times \Delta h = \frac{n_i}{N}$, or

$$F_i(h) = \frac{n_i}{\Delta h N}. \tag{3.14}$$

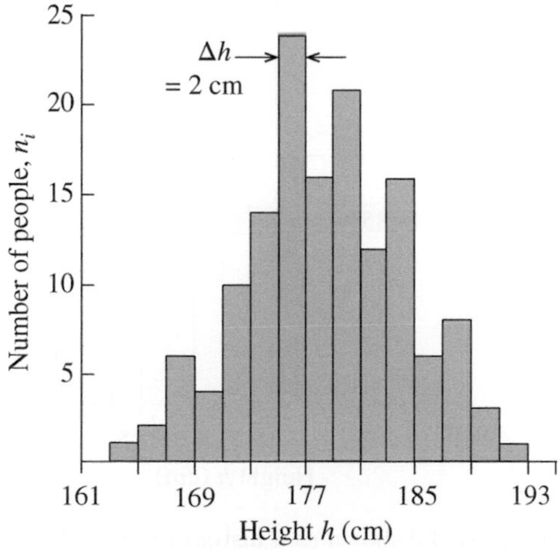

Figure 3.2 A histogram showing the height distribution in a group of people. The width of the height intervals is 2 cm and the total number of people is 144. The histogram shows that some people are taller than average and some are shorter than average, while most of the people have a height close to the average.

We see that $F_i(h)$ has the dimension of $[\text{length}]^{-1}$. Substituting for $\Delta h = 2$ cm in Equation (3.14), we obtain $F_i(h)$ as the *fraction per cm*, i.e. the *fraction per unit length*. It is useful to have the fraction per unit length because then $F_i(h)$ is independent of the width of the interval Δh. For example, in the histogram in Figure 3.2, the number of people with height within the interval 175–177 cm is 24. This gives the fraction $n_i/N = 24/144$. Then, for this rectangle,

$$F_i(h) = \frac{24}{2 \times 144} = 0.0833 \text{ cm}^{-1}.$$

In Figure 3.3, the height distribution of Figure 3.2 is shown as a plot of $F_i(h)$ against h, where the ordinate scale has been set so that $F_i(h) = 0.0833 \text{ cm}^{-1}$ for the interval 175–177 cm. Note that as $F_i(h)$ is the fraction per unit length, we must multiply $F_i(h)$ by a length to obtain the absolute fraction. For example, the fraction of people with heights between 173 and 175 cm, the heavily shaded rectangle in Figure 3.3, is $0.0485 \text{ cm}^{-1} \times 2 \text{ cm} = 0.097$, i.e. 9.7% of the people in the group have a height within this interval. In an entirely equivalent statement, we can say that the *probability* of an individual in the group having a height between 173 and 175 cm is 0.097. As each rectangle in the histogram of Figure 3.3 represents a fraction of the whole group, it follows that the sum of the areas of the rectangles must be numerically equal to 1:

$$\sum_i F_i(h)\Delta h = 1. \tag{3.15}$$

The histogram is said to be *normalised*.

To determine the mean height of the people in the group we proceed as follows. The fraction of people with heights between h_i and $h_i + \Delta h$ is the rectangular area $F_i(h) \times \Delta h$. This rectangle contributes

$$F_i(h)\Delta h \times (h_i + \Delta h/2)$$

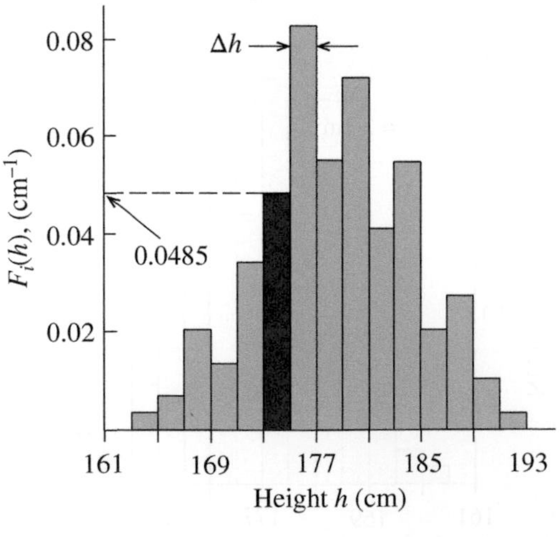

Figure 3.3 The height distribution of Figure 3.2 shown as a histogram of $F_i(h)$ against h, where the area $F_i(h)\Delta h$ of a given rectangle represents the fraction of people with heights between h_i and $h_i + \Delta h$. For example, the fraction of people with heights between 173 and 175 cm, the heavily shaded rectangle, is $0.0485 \text{ cm}^{-1} \times 2 \text{ cm} = 0.097$.

to the sum of all peoples' heights, where $\Delta h = 2$ cm and we take the value for the height at the mid-point of the rectangle. Hence, to find the mean height $\overline{h}$ for the group, we sum over all the rectangles:

$$\overline{h} = \sum_i F_i(h)(h_i + \Delta h/2)\Delta h. \tag{3.16}$$

The first four terms in this sum for the histogram in Figure 3.3 are shown in Table 3.1. The complete calculation gives a value for the mean height of 178.5 cm. Note that if, for some reason, we wanted to find the average of the square of the height $\overline{h^2}$, this is given by

$$\overline{h^2} = \sum_i F_i(h)(h_i + \Delta h/2)^2\Delta h. \tag{3.17}$$

If we take a larger group of people, say consisting of one million people, the histogram of $F_i(h)$ versus h becomes more smoothly stepped, as illustrated by Figure 3.4. This is because statistical fluctuations in the number of people in a particular height interval as a fraction of the total number of people in that interval reduce as the number of people in the group increases. In this figure, we again plot $F_i(h)$ against height h,

Table 3.1 Part of the calculation of the mean height $\overline{h}$ for the height distribution shown in Figure 3.3.

h_i (cm)	n_i	$F_i(h)$ (per cm)	$F_i(h) \times 2 \times (h_i + \Delta h/2)$ (cm)
164	1	0.00347	0.569
166	2	0.0694	2.306
168	6	0.0208	7.000
170	4	0.0139	4.722
…	…	…	…
			Total = 178.5

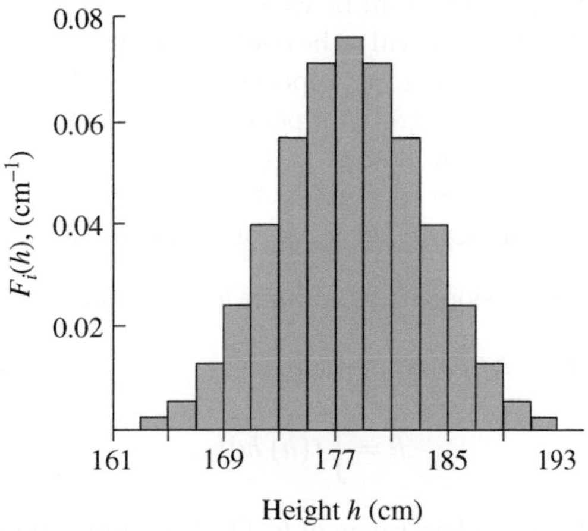

Figure 3.4 A histogram of $F_i(h)$ against h for a much larger group of people, than in Figures 3.2 and 3.3. The histogram is now more smoothly stepped. The scale of the vertical axis is the same as in Figure 3.3 because $F_i(h)$ is the fraction per unit length, not the fraction per interval.

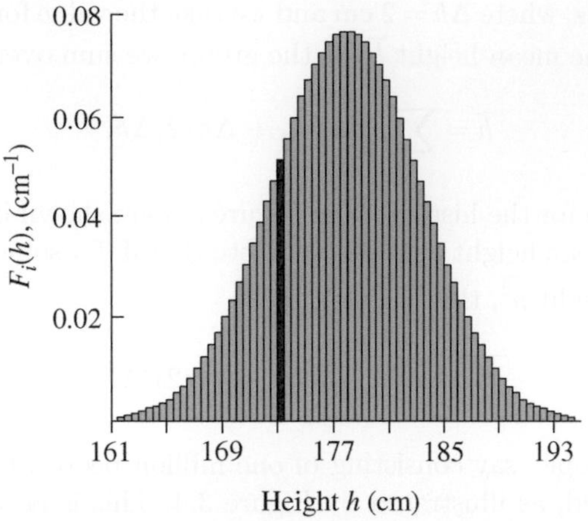

Figure 3.5 A histogram of $F_i(h)$ against h for an even larger group of people, in which it has been possible to decrease the size of the height interval, from 2 cm to 5 mm. As the interval Δh has reduced in width, the statistical information becomes more detailed. However, the fraction within each interval decreases, being proportional to the width of the interval. For example, the fraction of people with heights between 173 and 173.5 cm, the heavily shaded rectangle, is $0.051 \times 0.5 = 0.026$.

and again the total area of the histogram is equal to 1. Note that the ordinate scale for $F_i(h)$ is the same as in Figure 3.3 because $F_i(h)$ is the fraction per unit length, not the fraction per interval.

When we have a still larger group of people, say 10 million people, we can afford to decrease the size of the height interval, from 2 cm to say 5 mm. Then we obtain a histogram looking like that in Figure 3.5. The ordinate scale is the same as before, but as the interval Δh has reduced in width, the statistical information becomes more detailed. However, the fraction within each interval must decrease, being proportional to the width of the interval. For example, the fraction of people with heights between 173 and 173.5 cm, the heavily shaded rectangle in Figure 3.5, is $0.051 \times 0.5 = 0.026$, i.e. 2.6% of the people in the group have heights within this interval. The mean height can again be calculated in accordance with Equation (3.16).

We may imagine the width Δh of the interval to be continually reduced until the rectangles are very thin strips and the step-shaped contour of their tops appears to approach a smooth, continuous curve. This would, of course, require an extremely large group of people. Assuming that the height distribution follows an exact mathematical function as the number of people increases without limit and Δh is indefinitely reduced, then the set of $F_i(h)$s becomes the continuous function $f(h)$ and the interval Δh becomes the elemental width $\mathrm{d}h$. In this limit, the fraction $F_i(h) \times \Delta h$, of people in the range $h + \Delta h$ becomes:

$$\text{Fraction of people in the range } h \text{ to } h + \mathrm{d}h = f(h)\mathrm{d}h, \tag{3.18}$$

and mean height, $\overline{h} = \sum_i F_i(h)(h_i + \Delta h/2)\Delta h$, becomes

$$\overline{h} = \int f(h)\, h\mathrm{d}h, \tag{3.19}$$

where the integral extends over the complete range of h. The function $f(h)$ must be normalised, i.e. the area under the curve is made numerically equal to unity:

$$\int f(h)\mathrm{d}h = 1, \tag{3.20}$$

where again the integral extends over the complete range of h. More generally, the fraction of people having height between h_1 and h_2 is the area under the curve between $h = h_1$ and $h = h_2$, given by

$$\int_{h_1}^{h_2} f(h)\mathrm{d}h. \tag{3.21}$$

We can equate *fraction* of the whole to *probability* of being within that fraction. So, with respect to Figure 3.6, we can equivalently say:

$$\text{Probability of a person having a height in range } h \text{ to } h + \mathrm{d}h \text{ is } f(h)\mathrm{d}h, \tag{3.22}$$

and

$$\text{Probability that a person has a height between } h_1 \text{ and } h_2 \text{ is } \int_{h_1}^{h_2} f(h)\mathrm{d}h. \tag{3.23}$$

We can generalise Equation (3.19). If the probability of a variable x being between x and $x + \mathrm{d}x$ is $f(x)\mathrm{d}x$, the average value of an arbitrary function $g(x)$ of x is given by

$$\overline{g(x)} = \int f(x)g(x)\mathrm{d}x. \tag{3.24}$$

For example,

$$\overline{x^2} = \int f(x)x^2\mathrm{d}x.$$

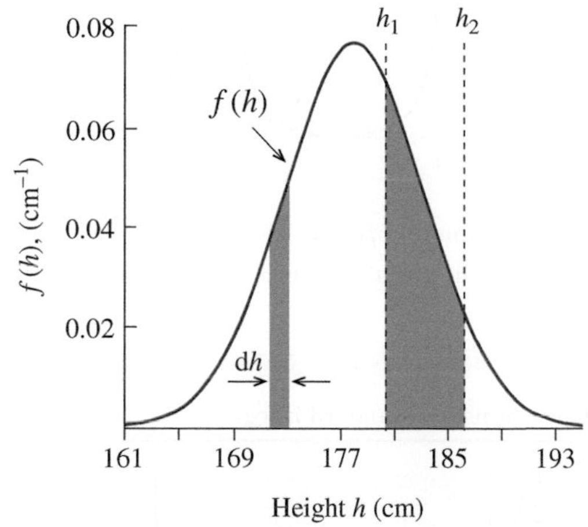

Figure 3.6 We may imagine the width Δh of the height intervals to be continually reduced until the rectangles are very thin strips and the step-shaped contour of their tops approaches a smooth, continuous curve. This requires an extremely large group of people. Assuming that the distribution of heights follows an exact mathematical function as the number of people increases without limit and Δh is indefinitely reduced, the set of $F_i(h)$s becomes the continuous function $f(h)$ and the interval Δh becomes the elemental width dh. The shaded area under the curve between h_1 and h_2 represents the fraction of people with heights within that range. The fraction of people with height between h to $h + \mathrm{d}h$ is $f(h)\mathrm{d}h$.

3.2.1 The normal or Gaussian distribution

The distribution of heights in a given population is usually taken to follow the normal or Gaussian distribution, which can be written as

$$f(h) = Ae^{-\left(h-\overline{h}\right)^2/2\sigma^2}, \tag{3.25}$$

where σ is a constant called the *standard deviation*, and is a measure of the width of the distribution. Hence, the probability distribution that a person has height between h and $h + dh$ is

$$f(h)dh = Ae^{-\left(h-\overline{h}\right)^2/2\sigma^2}dh, \tag{3.26}$$

Gaussian distributions are widely used in statistical analyses in the natural and social sciences to represent the distribution of variables. We will see them, for example, in our discussion of the velocity distribution of molecules in a gas. Some of the properties of the Gaussian function, often simply referred to as the Gaussian, are illustrated in the following worked example, and in Figure 3.7. Useful standard integrals for this example and for other derivations in this book are given in Table 3.2.

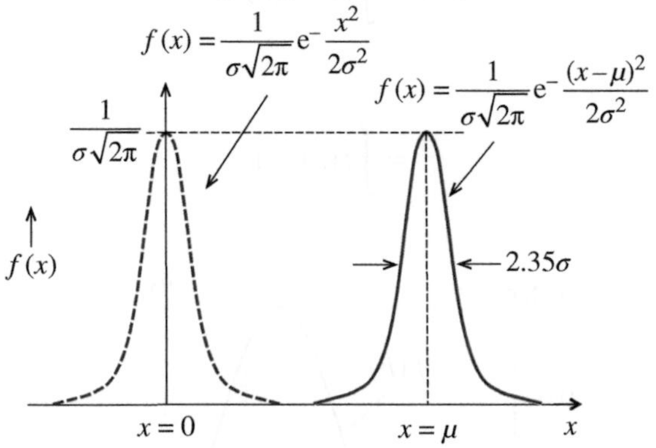

Figure 3.7 The solid curve is the Gaussian function $f(h) = Ae^{-(x-\mu)^2/2\sigma^2}$. The dashed curve is the Gaussian function $f(h) = Ae^{-x^2/2\sigma^2}$. We see that by changing the variable from x to $(x - \mu)$, the Gaussian has moved a distance to the right, but retains the same shape.

Table 3.2 Table of useful standard integrals.

$\int_0^{+\infty} e^{-\alpha x^2} dx = \frac{1}{2}\sqrt{\frac{\pi}{\alpha}}$	$\int_{-\infty}^{+\infty} e^{-\alpha x^2} dx = \sqrt{\frac{\pi}{\alpha}}$
$\int_0^{+\infty} x\, e^{-\alpha x^2} dx = \frac{1}{2\alpha}$	$\int_{-\infty}^{+\infty} x\, e^{-\alpha x^2} dx = 0$
$\int_0^{+\infty} x^2 e^{-\alpha x^2} dx = \frac{1}{4}\sqrt{\frac{\pi}{\alpha^3}}$	$\int_{-\infty}^{+\infty} x^2 e^{-\alpha x^2} dx = \frac{1}{2}\sqrt{\frac{\pi}{\alpha^3}}$
$\int_0^{+\infty} x^3 e^{-\alpha x^2} dx = \frac{1}{2\alpha^2}$	$\int_{-\infty}^{+\infty} x^3 e^{-\alpha x^2} dx = 0$

Worked example

Given the Gaussian distribution

$$f(x)\mathrm{d}x = A\mathrm{e}^{-(x-\mu)^2/2\sigma^2}\mathrm{d}x,$$

where the limits of x are $\pm\infty$,

(a) determine the constant A that normalises the function $f(x)$,
(b) show that $\mu = \overline{x}$, the mean value of x.
(c) show that $\overline{(x-\overline{x})^2} = \sigma^2$,
(d) what is the most probable value of x?
(e) show that the width of the Gaussian at half its height is 2.35σ.

Solution

(a) For $f(x)$ to be normalised we must have

$$\int_{-\infty}^{+\infty} f(x)\mathrm{d}x = A\int_{-\infty}^{+\infty} \mathrm{e}^{-(x-\mu)^2/2\sigma^2}\mathrm{d}x = 1.$$

We change the variable from x to $u = (x-\mu)$ to evaluate the integral. Then, as μ is a constant, $\mathrm{d}u = \mathrm{d}x$, and using a standard integral, we obtain $A\sqrt{2\sigma^2\pi} = 1$. Hence $A = \frac{1}{\sigma\sqrt{2\pi}}$.

(b) The mean value of x is given by

$$\overline{x} = \frac{1}{\sigma\sqrt{2\pi}}\int_{-\infty}^{+\infty} x\,\mathrm{e}^{-(x-\mu)^2/2\sigma^2}\mathrm{d}x.$$

Again changing the variable from x to $u = (x-\mu)$ to evaluate the integral, we obtain

$$\overline{x} = \frac{1}{\sigma\sqrt{2\pi}}\left[\int_{-\infty}^{+\infty} u\mathrm{e}^{-u^2/2\sigma^2}\mathrm{d}x + \int_{-\infty}^{+\infty} \mu\mathrm{e}^{-u^2/2\sigma^2}\mathrm{d}x\right].$$

Then using a standard integral, we find $\overline{x} = \frac{1}{\sigma\sqrt{2\pi}}\mu\sqrt{\pi 2\sigma^2} = \mu$.

(c) $\overline{(x-\overline{x})^2} = \frac{1}{\sigma\sqrt{2\pi}}\int_{-\infty}^{+\infty}(x-\overline{x})^2\mathrm{e}^{-\frac{(x-\mu)^2}{2\sigma^2}}\mathrm{d}x$. We change the variable from x to $u = (x-\overline{x})$. Then $\mathrm{d}u = \mathrm{d}x$, and we obtain

$$\overline{u^2} = \frac{1}{\sigma\sqrt{2\pi}}\int_{-\infty}^{+\infty} u^2\mathrm{e}^{-u^2/2\sigma^2}\mathrm{d}u.$$

Then using a standard integral, we obtain

$$\overline{u^2} = \frac{1}{\sigma\sqrt{2\pi}}\frac{1}{2}\sqrt{8\sigma^6\pi} = \sigma^2, \text{ and, hence, } \overline{(x-\overline{x})^2} = \sigma^2.$$

(d) The most probable value of x is when $f(x)$ is a maximum. By inspection of the function $f(x) = (1/\sigma\sqrt{2\pi})\mathrm{e}^{-(x-\mu)^2/2\sigma^2}$, we can see that the maximum of $f(x)$ occurs when $x = \overline{x}$. Alternatively, we can set $(\mathrm{d}/\mathrm{d}x)f(x) = 0$.

(e) The values of x at the half-height points of the Gaussian are given by the expression

$$\frac{1}{\sigma\sqrt{2\pi}}\mathrm{e}^{-(x-\mu)^2/2\sigma^2} = \frac{1}{2}\frac{1}{\sigma\sqrt{2\pi}}.$$

Simplifying this expression gives $x = \mu \pm \sigma\sqrt{2\ln 2}$.
Hence, width of Gaussian at half its height is $2\sigma\sqrt{2\ln 2} = 2.35\sigma$.
Some of these results may be familiar to the reader from the statistical treatment of experimental data.

3.3 The Maxwell–Boltzmann speed distribution

In Section 3.1, we considered the pressure exerted by a gas on the walls of its container and determined an expression for the root mean square speed of the molecules. However, this did not yield any details about the distribution of molecular speeds. The effect of the huge number of intermolecular collisions is *not* to even out the speeds of the molecules, but to distribute them over a broad range. Moreover, for a gas in thermal equilibrium, this distribution is stable and well defined.

It was James Clerk Maxwell who first solved the problem of finding the velocity distribution of molecules in a gas and their speed distribution. This is the same Maxwell who produced a unified theory of the electromagnetic field and used it to show that light is a type of electromagnetic wave. Maxwell showed that each component of molecular velocity in an ideal gas follows a Gaussian distribution, and that the molecular speeds follow what is now called the Maxwell–Boltzmann speed distribution. Maxwell published this work in 1860 and the passage in his paper describing this first derivation of the velocity distribution has been cited as 'one of the most important passages in physics'.

The Maxwell–Boltzmann speed distribution is

$$P[c]\mathrm{d}c = 4\pi\left(\frac{m}{2\pi kT}\right)^{3/2} c^2\mathrm{e}^{-mc^2/2kT}\mathrm{d}c, \tag{3.27}$$

where c is the molecular speed and the other symbols have their usual meanings. Although purely theoretical when proposed, the Maxwell–Boltzmann distribution has been confirmed by molecular beam experiments. In these experiments, the speeds of the molecules escaping through a small hole in a gas container are measured directly. We will derive the Maxwell–Boltzmann speed distribution in Section 3.6. In the meantime, we show in Figure 3.8 the speed distribution for molecular nitrogen at a temperature of 300 K. The probability function $P[c]$ has units of seconds per metre (s/m) as $P[c]\mathrm{d}c$ is dimensionless. Note that from now on, when dealing with probability distributions, it will be convenient to enclose the variable in square brackets to emphasise that it is a probability distribution.

The speed distribution in Figure 3.8 is normalised so that the total area under the curve is numerically equal to unity. As a quick check, we note that the distribution is roughly triangular with a height of about 2×10^{-3} s/m and a baseline width of 1000 m/s, which gives an area equal to 1. The shaded area under the curve represents the fraction of the total number of molecules that have speeds between 650 and 750 m/s.

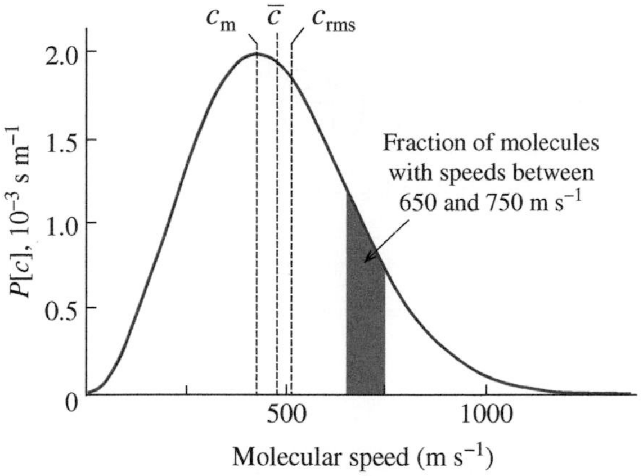

Figure 3.8 The Maxwell–Boltzmann speed distribution $P[c]\mathrm{d}c$ for molecular nitrogen at a temperature of 300 K. The distribution function $P[c]$ has units of seconds per metre as $P[c]\mathrm{d}c$ is dimensionless. The distribution is normalised so that the total area under the curve is numerically equal to unity. There are three characteristic speeds associated with the speed distribution, namely (i) the most probable speed c_{m}, (ii) the mean speed $\overline{c}$, and (iii) the root mean square speed $c_{\mathrm{rms}} = \left(\overline{c^2}\right)^{1/2}$, and these are marked on the figure. The shaded area gives the fraction of molecules with speeds between 650 and 750 m/s.

There are three characteristic speeds associated with the Maxwell–Boltzmann speed distribution, namely (i) the most probable speed c_{m}, (ii) the mean speed $\overline{c}$, and (iii) the root mean square speed $c_{\mathrm{rms}} = \left(\overline{c^2}\right)^{1/2}$. These are indicated in Figure 3.8.

(i) The most probable speed of a molecule is the speed for which the function $P[c]$ has its maximum value and is obtained by differentiating $P[c]$ with respect to c and equating the result to zero. This gives the most probable speed to be

$$c_{\mathrm{m}} = \left(\frac{2kT}{m}\right)^{1/2}. \tag{3.28}$$

(ii) The mean speed of the molecules is given by

$$\overline{c} = \int_0^\infty P[c]\,c\,\mathrm{d}c = \frac{4}{\sqrt{\pi}}\left(\frac{m}{2kT}\right)^{3/2}\int_0^\infty c^3 \mathrm{e}^{-mc^2/2kT}\,\mathrm{d}c.$$

Notice that the limits of integration are from 0 to ∞ since speed c cannot have a negative value. Using the standard integral

$$\int_0^\infty x^3 \mathrm{e}^{-ax^2}\,\mathrm{d}x = \frac{1}{2a^2},$$

we readily obtain

$$\overline{c} = \left(\frac{8kT}{\pi m}\right)^{1/2}. \tag{3.29}$$

(iii) The root mean square speed of the molecules is given by

$$\overline{c^2} = \int_0^\infty P[c]c^2 \mathrm{d}c.$$

Using the standard integral

$$\int_0^\infty x^4 \mathrm{e}^{-\alpha x^2} \mathrm{d}x = \frac{3}{8}\sqrt{\frac{\pi}{\alpha^5}},$$

we readily obtain

$$\overline{c^2} = \frac{3kT}{m},$$

and, hence,

$$c_{\mathrm{rms}} = \left(\frac{3kT}{m}\right)^{1/2}. \tag{3.30}$$

This gives a mean translational kinetic energy $\frac{1}{2}\left(m\overline{c^2}\right) = \frac{3}{2}(kT)$ in agreement with our previous result from Section 3.1, Equation (3.10). All these speeds have similar magnitudes, in the ratios $c_m : \overline{c} : c_{\mathrm{rms}} =$ 1:1.13:1.22. If we want to obtain an estimate of the speed of a molecule, probably the easiest relationship to remember is $\frac{1}{2}\left(m\overline{c^2}\right) = \frac{3}{2}(kT)$. For the example of molecular nitrogen at room temperature, this gives

$$\left(\overline{c^2}\right)^{1/2} = \left(\frac{3 \times 1.38 \times 10^{-23} \times 293}{28 \times 1.66 \times 10^{-27}}\right)^{1/2} = 511\,\mathrm{m/s}.$$

This is about the speed of a bullet fired from a rifle.

The speed distribution depends on the temperature T of the gas and the molecular mass m. Figure 3.9 shows the speed distribution for helium at two different temperatures, 273 K and 2000 K, respectively. Note that the areas under the curves are equal. As the temperature increases, the molecules move faster, and the width of the speed distribution increases. Figure 3.10 shows the speed distributions for molecular nitrogen and molecular hydrogen at 273 K. At the same temperature, the nitrogen and hydrogen molecules have the same average kinetic energy and consequently the less massive hydrogen molecules move much faster.

The Maxwell–Boltzmann speed distribution explains many physical observations. For example, although hydrogen is the most abundant element in the universe, it is very scarce in Earth's atmosphere. One reason for this is as follows. To escape the Earth's gravitational pull, a particle must have a speed greater than the escape velocity, which for Earth is 11.2 km/s. Perhaps surprisingly, the temperature at an altitude of ~500 km in the Earth's atmosphere is typically about 1000 K. For hydrogen molecules, this

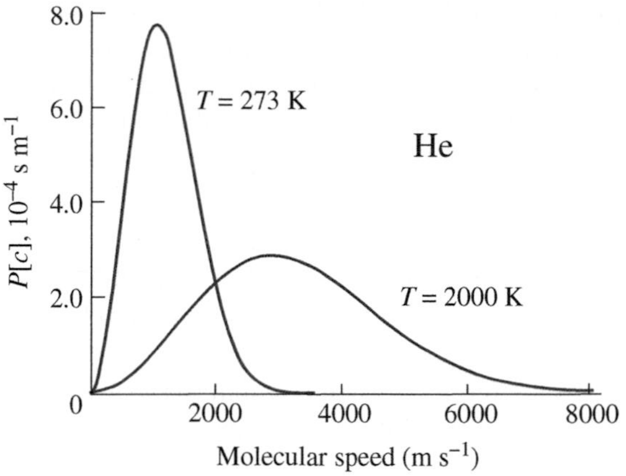

Figure 3.9 The Maxwell–Boltzmann speed distribution for helium at temperatures 273 K and 2000 K, respectively. Note that the areas under the curves are equal. As the temperature increases, the molecules move faster, and the width of the distribution increases.

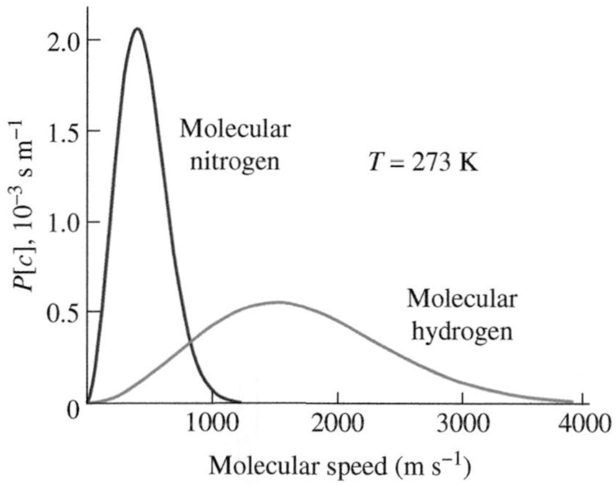

Figure 3.10 The Maxwell–Boltzmann speed distributions for molecular nitrogen and molecular hydrogen at 273 K. At the same temperature, the nitrogen and hydrogen molecules have the same average kinetic energy and consequently the much lighter hydrogen molecules move much faster.

corresponds to a mean speed of about 5 km/s. Although this value is less than the escape velocity, there will be a small but not negligible number of hydrogen molecules with speed in the high-energy tail of the speed distribution with a speed greater than the escape velocity. And these can fly away into space. Other atmospheric molecules, such as nitrogen are much heavier than hydrogen and their speeds are correspondingly lower. But more importantly, the exponential term $e^{-mc^2/2kT}$ in the Maxwell–Boltzmann speed distribution reduces far more rapidly with respect to speed c for, say, nitrogen than it does for hydrogen. Consequently, the probability of speeds in excess of the escape velocity is negligible. The speed distribution of molecules in a liquid can also be represented by a curve similar to the Maxwell–Boltzmann curve. Most of the molecules are not sufficiently energetic to escape from the liquid. However, again there will be those molecules with sufficiently high speed lying in the high-energy tail of the distribution, which can escape. It is these molecules, for example, in water, that evaporate from the liquid. As these high-speed molecules escape, carrying energy with them, the temperature of the remaining water reduces by evaporative cooling or is maintained by heat transfer from the surroundings. In the latter case, fast molecules, produced by favourable intermolecular collisions, take the place of those that have escaped, and the speed distribution is maintained.

3.3.1 The kinetic energy distribution

The Maxwell–Boltzmann speed distribution as given by Equation (3.27) can also be written as a translational kinetic energy distribution $P[K]\mathrm{d}K$, where K is the translational kinetic energy of a molecule. A particular value of speed c gives a definite kinetic energy K equal to $\frac{1}{2}(mc^2)$. Moreover, the fraction of molecules that have a speed between c and $c+\mathrm{d}c$ will be equal to the fraction of molecules that have energy between K and $K+\mathrm{d}K$, so long as the speed interval $\mathrm{d}c$ corresponds to the kinetic energy interval $\mathrm{d}K$:

$$P[c]\mathrm{d}c = P[K]\mathrm{d}K. \tag{3.31}$$

This result is illustrated by Figure 3.11, which shows (a) the distribution function $P[c]$ and (b) the distribution function $P[K]$. As $K = \frac{1}{2}(mc^2)$ and $\mathrm{d}K = mc\mathrm{d}c$, this means that

$$P[K] = P[c]\frac{\mathrm{d}c}{\mathrm{d}K} = P[c]\frac{1}{mc}. \tag{3.32}$$

Substituting for $P[c]$ from Equation (3.27), we obtain

$$P[K] = \frac{4\pi}{mc}\left(\frac{m}{2\pi kT}\right)^{3/2} c^2 \mathrm{e}^{-mc^2/2kT}.$$

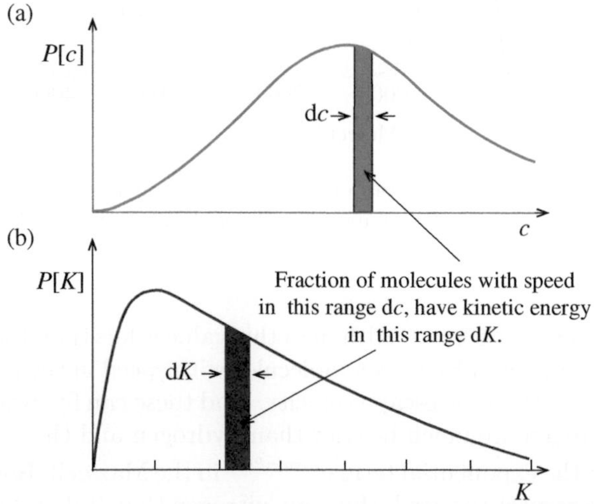

Figure 3.11 The Maxwell–Boltzmann speed distribution $P[c]\mathrm{d}c$, panel (a), can be converted into a kinetic energy distribution $P[K]\mathrm{d}K$, panel (b), where K is the kinetic energy of a molecule. The fraction of molecules that have a speed between c and $c+\mathrm{d}c$ will be equal to the fraction of molecules that have energy between K and $K+\mathrm{d}K$, so long as the speed interval $\mathrm{d}c$ corresponds to the energy interval $\mathrm{d}K$.

Simplifying and putting $K = \frac{1}{2}(mc^2)$, we obtain

$$P[K] = \left(\frac{4}{\pi k^3 T^3}\right)^{1/2} K^{1/2} e^{-K/kT},$$

and, hence,

$$P[K]dK = \left(\frac{4}{\pi k^3 T^3}\right)^{1/2} K^{1/2} e^{-K/kT} dK. \tag{3.33}$$

This is the probability distribution that gives the probability that a molecule has kinetic energy between K and $K + dK$. Using the same methods that we used for molecular speed, we readily find from this distribution that the most probable kinetic energy of a molecule is $kT/2$. This is much smaller than the mean kinetic energy of a molecule $(3kT/2)$ because of the shape of the energy distribution curve.

Statistical thermodynamics is the branch of physics that is concerned with applying probability concepts to interpret the thermal behaviour of matter. In the language of statistical thermodynamics, the energy distribution $P[K]dK$ is considered to be the product of two factors; one is called the *density of states*, which is proportional to $K^{1/2}$; the other is the probability of a state being occupied, which is proportional to $e^{-K/kT}$, and is called the *Boltzmann factor*.

Worked example

Energetic neutrons emanating from a nuclear reactor are slowed down by passing them through a *moderator*. The moderator absorbs kinetic energy from the neutrons converting them into 'thermal neutrons' at which point they are in thermal equilibrium at a temperature of 300 K. Determine the de Broglie wavelength of the thermal neutrons using the value of their most probable speed.

Solution

The most probable speed of the neutrons is given by $c_m = (2kT/m_n)^{1/2}$, where m_n is the mass of a neutron. Hence,

$$c_m = \left(\frac{2kT}{m_n}\right)^{1/2} = \left(\frac{2 \times 1.38 \times 10^{-23} \times 300}{1.67 \times 10^{-27}}\right)^{1/2} = 2.23\ \text{km/s}.$$

The de Broglie wavelength λ is given by $\lambda = h/p$, Equation (1.28), where p is the momentum of the particle and h is Planck's constant. Taking $p = c_m m_n$, we obtain

$$\lambda = \frac{h}{c_m m_n} = \frac{6.63 \times 10^{-34}}{2.23 \times 10^3 \times 1.67 \times 10^{-27}} = 1.79 \times 10^{-10}\ \text{m}.$$

This value is about the same size as the spacing of molecules in solid materials. This makes thermal neutrons a valuable probe to investigate the matter, especially since neutrons are uncharged and do not feel the electrostatic forces that exist in matter.

3.4 Boltzmann's law

Maxwell found the distribution of molecular velocities in a gas. And from this, the distribution of *kinetic energy* in a gas could be obtained, Equation (3.33), which shows that the probability of a molecule having kinetic energy, K is proportional to $e^{-K/kT}$. Later, Ludwig Boltzmann discovered a more general law that gives the probability of a particle having a *total energy* E, where E is the sum of its kinetic and potential energies. Boltzmann's law says that the probability of a particle, in thermal equilibrium at temperature T, having total energy E, is proportional to

$$e^{-K/kT}, \tag{3.34}$$

where $E = V + K$ and V is the total potential energy and K is the total kinetic energy. Boltzmann's law is one of the most important results of classical physics. It tells us how energy is shared between an assembly of particles in thermal equilibrium, in terms of the probability of a particle having a particular energy. Boltzmann's law works for classical systems and also for quantum systems. In fact, as we shall see in Section 3.8.5, because energies are discrete in quantum mechanical systems, Boltzmann's law is easier to implement.

3.4.1 General form of Boltzmann's law

In classical mechanics, we can completely specify the state of a particle by its coordinates of *position*, say x, y, and z, and by its components of *linear momentum*, say p_x, p_y, and p_z. The potential energy of a particle *always* depends on its position coordinates; for example, the potential energy of a particle due to its height in the gravitation field of the Earth. On the other hand, the momenta components *always* give rise to kinetic energy terms, e.g. $p_x^2/2m$. Note that we use momenta components instead of velocity components as the former are more widely applicable. The total energy of a particle is then the sum of contributions from all the position coordinates, which give rise to potential energy terms, and all the momenta components, which give rise to kinetic energy terms. We can then describe the state of a particle as a point in a six-dimensional space called *phase space*. This space has not only the familiar x, y, and z axes, but also p_x, p_y, and p_z axes corresponding to the three components of momenta. We may not be able to draw a six-dimensional space but we can certainly deal with it mathematically. The state of a particle is then specified by the point in phase space with coordinates (x, y, z, p_x, p_y, and p_z).

In classical physics, potential and kinetic energies have a continuous range of values, just as the height in a group of people has a continuous range of values. And just as we divided people's height into discrete intervals in Section 3.2, it is convenient to divide a particle's potential and kinetic energies into discrete intervals when we wish to determine their probability distributions. Hence, we consider the probability of a particle having spatial coordinates that lie within the ranges x to $x + dx$, y to $y + dy$, and z to $z + dz$, respectively, together with the probability of it having momenta coordinates that lie within the ranges p_x to $p_x + dp_x$, p_y to $p_y + dp_y$, and p_z to $p_z + dp_z$, respectively. In six-dimensional phase space, this gives an elemental volume $dx\,dy\,dz\,dp_x dp_y dp_z$. Boltzmann showed that the probability of finding a particle with a given spatial position and with a given momentum is proportional to $e^{-E/kT}$ per unit volume of phase space, where E is the total energy of the particle. Hence, Boltzmann's law says, *the probability of a particle having spatial and momenta coordinates between* x to $x + dx$, y to $y + dy$, ..., p_z to $p_z + dp_z$ is

$$P\left[x, y, z, p_x, p_y, p_z\right] dx\,dy\,dz\,dp_x dp_y dp_z = Ae^{-E/kT} dx\,dy\,dz\,dp_x dp_y dp_z, \tag{3.35}$$

where A is the normalisation constant and $\mathrm{d}x\,\mathrm{d}y\,\mathrm{d}z\,\mathrm{d}p_x\mathrm{d}p_y\mathrm{d}p_z$ is the elemental volume in phase space. This is our general formulation of Boltzmann's law. It is straightforward to extend the Boltzmann distribution to N particles. It is simply a product of N terms like that of Equation (3.35).

When an assembly of particles is in thermal equilibrium, the spatial coordinates and the momentum components of a particle are *all* independent of each other, i.e. wherever a particle is located, its momentum may be of any magnitude and any direction. And because of the exponential term in energy E, the Boltzmann factor, $\mathrm{e}^{-E/kT}$ can always be separated into the product of a potential energy term and a kinetic energy term. The right-hand side of Equation (3.35) can thus be written as

$$A\mathrm{e}^{-V(x,y,z)/kT}\mathrm{d}x\,\mathrm{d}y\,\mathrm{d}z\,\mathrm{e}^{-K(p_x,p_y,p_z)/kT}\mathrm{d}p_x\mathrm{d}p_y\mathrm{d}p_z, \tag{3.36}$$

where $V(x, y, z)$ is the total potential energy of the particle and $K(p_x, p_y, p_z) = p_x^2/2m + p_y^2/2m + p_z^2/2m$ is the total kinetic energy of the particle.

Equations (3.35) and (3.36) may seem rather complicated. However, in many physical situations, we are dealing with either just the potential energy or just the kinetic energy of a particle. And often, we are dealing with just one coordinate, say the spatial coordinate x or the momentum coordinate p_x. Then Boltzmann's law reduces to

$$P[x]\,\mathrm{d}x = A\mathrm{e}^{-V(x)/kT}\mathrm{d}x, \tag{3.37}$$

or

$$P[p_x]\,\mathrm{d}p_x = A\mathrm{e}^{-K(p_x)/kT}\mathrm{d}p_x, \tag{3.38}$$

respectively, where $V(x)$ and $K(p_x)$ are the potential energy and kinetic energy respectively. We can interpret Equation (3.37), for example, as follows. The probability of a particle having a value of position coordinate x between x and $x + \mathrm{d}x$ is proportional to $\mathrm{e}^{-[V(x)/kT]}$, where $V(x)$ is the potential energy at that value of x.

In Section 3.1, we introduced temperature T through the ideal gas law, $PV = RT$. Now we have introduced temperature through the Boltzmann law, and it is the latter that is the fundamental definition of temperature.

3.4.2 The probability distribution for a single component of molecular velocity

We consider an ideal gas so that the potential energy due to intermolecular interactions can be neglected. We also take a sufficiently small sample of gas so that any variation in the gravitational potential energy of the molecules can be neglected. As the potential energy of the molecules can be neglected, Boltzmann's law reduces to

$$P[p_x, p_y, p_z]\,\mathrm{d}p_x\mathrm{d}p_y\mathrm{d}p_z = A\,\mathrm{e}^{-K(p_x,p_y,p_z)/kT}\mathrm{d}p_x\mathrm{d}p_y\mathrm{d}p_z. \tag{3.39}$$

When an assembly of particles is in thermal equilibrium, the velocity components are independent of each other and so we can consider each component separately. Then from Equation (3.38), we can say that the probability of a molecule having momentum between p_x and $p_x + \mathrm{d}p_x$ is given by

$$P[p_x]\,\mathrm{d}p_x = A\mathrm{e}^{-p_x^2/2mkT}\mathrm{d}p_x. \tag{3.40}$$

Since the limits of the velocity component p_x are $\pm\infty$, we have

$$A\int_{-\infty}^{+\infty} e^{-p_x^2/2mkT}\,dp_x = 1.$$

Using a standard integral, we obtain

$$A = \left(\frac{1}{2\pi mkT}\right)^{1/2},$$

giving

$$P\,[p_x]\,dp_x = \left(\frac{1}{2\pi mkT}\right)^{1/2} e^{-p_x^2/2mkT}\,dp_x, \tag{3.41}$$

To convert from the probability function $P\,[p_x]$ to the probability function $P\,[v_x]$, we use the relationship

$$P\,[v_x] = P\,[p_x]\left(\frac{dp_x}{dv_x}\right), \tag{3.42}$$

cf. Equation (3.32). As $dp_x/dv_x = m$, we obtain

$$P\,[v_x] = \left(\frac{1}{2\pi mkT}\right)^{1/2} e^{-p_x^2/2mkT}\,m. \tag{3.43}$$

Substituting for $p_x = mv_x$, we finally obtain the probability distribution for the x-component of velocity

$$P\,[v_x]dv_x = \left(\frac{m}{2\pi kT}\right)^{1/2} e^{-mv_x^2/2kT}\,dv_x. \tag{3.44}$$

Note that the normalization constant $A = (m/2\pi kT)^{1/2}$ has units of probability per unit velocity, as can be checked from its dimensions.

The function $P\,[v_x] = (m/2\pi kT)^{1/2}e^{-mv_x^2/2kT}$ is plotted in Figure 3.12. We see that $P\,[v_x]$ is a Gaussian centred about the velocity component $v_x = 0$. This means that the most probable value of p_x is zero. Since the Gaussian is symmetric about $v_x = 0$, it is also the case that the mean value of v_x is zero, as for every value of $+\,v_x$, there is a corresponding value of $-\,v_x$.

Although $\overline{v_x}$ is equal to zero, $\overline{v_x^2}$ is not. We can see this as follows:

$$\overline{v_x^2} = \int_{-\infty}^{+\infty} v_x^2 P\;[v_x]dv_x = \int_{-\infty}^{+\infty} v_x^2\left(\frac{m}{2\pi kT}\right)^{1/2} e^{-mv_x^2/2kT}\,dv_x. \tag{3.45}$$

Using a standard integral, we obtain

$$\overline{v_x^2} = \frac{kT}{m}.$$

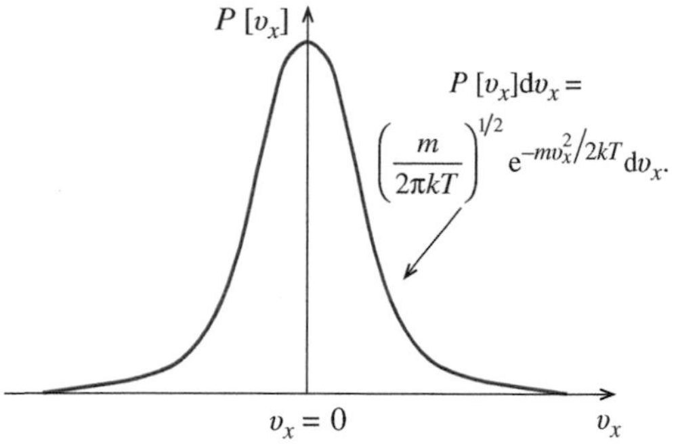

Figure 3.12 The probability function $P[v_x]$ for the x-component of molecular velocity as given by Boltzmann's law. We see that the function $P[v_x]$ is a Gaussian, which is symmetric and centred about $v_x = 0$.

Hence, the mean kinetic energy

$$\frac{1}{2}m\overline{v_x^2} = \frac{1}{2}kT, \tag{3.46}$$

which confirms the result we obtained in Section 3.1, Equation (3.13). However, now the result is obtained using the more fundamental definition of temperature from Boltzmann's law.

3.4.3 Doppler broadening of spectral lines

A physical effect that arises directly from the velocity distribution of molecules is the *Doppler broadening* of spectral lines. The Doppler effect is familiar in the case of sound waves, as, for example, in the apparent change in pitch of an emergency vehicle siren when the vehicle passes by an observer. As the vehicle approaches, the observer experiences a higher pitch than the actual pitch of the siren, but then experiences a lower pitch as the vehicle recedes into the distance. A molecule in an excited state will decay by emitting a photon of well-defined frequency ν_0. In a gas, the molecules have a distribution of velocities, as we have seen. Because of this motion, an observer viewing a photon emitted from a moving molecule will observe a frequency that is shifted away from ν_0. If the molecule is moving towards the observer, the frequency will increase compared to ν_0, and if the molecule is moving away from the observer, the observed frequency will decrease. The net effect is if the emitted radiation from a gas of excited molecules is viewed by an optical spectrometer, the observed spectral line is broadened. Moreover, the profile of the broadened spectral line follows the velocity distribution.

Figure 3.13 is a schematic diagram that shows a gas discharge lamp and an optical spectrometer. The spectrometer has a diffraction grating that disperses the radiation emitted by the lamp according to wavelength. The photon detector measures the intensity of the selected wavelength.

The Doppler equation is $\nu = \nu_0(v_w/v_w - v_s)$, where ν is the observed frequency, ν_0 is the emitted frequency when the source is stationary, v_w is the velocity of the wave, and v_s is the velocity of the moving source. In the present case, $v_w = c$, the velocity of light. When the radiation is viewed along the x-direction, it is the x-component of velocity v_x that gives rise to the Doppler broadening, no matter what is the direction of motion of the excited atom. Hence $v_s = v_x$, and we have

$$\nu = \nu_0\left(\frac{c}{c - v_x}\right) = \nu_0\left(\frac{1}{1 - v_x/c}\right) \cong \nu_0\left(1 + \frac{v_x}{c}\right), \tag{3.47}$$

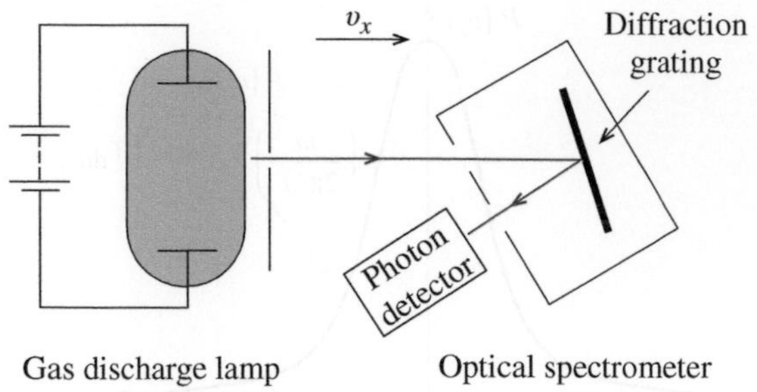

Figure 3.13 A schematic diagram showing a gas discharge lamp and an optical spectrometer. The spectrometer has a diffraction grating that disperses the radiation emitted by the lamp according to wavelength. The photon detector measures the intensity of the radiation selected by the spectrometer. Notice that the spectrometer views light emitted along the x-direction and so it is the distribution of the x-component of velocity that is relevant.

as $c \gg v_x$. Usually, an optical spectrometer measures wavelength λ rather than frequency, where $\nu\lambda = c$. Then, we obtain from Equation (3.47)

$$\frac{\lambda - \lambda_0}{\lambda_0} = \frac{v_x}{c} \tag{3.48}$$

where λ is the observed wavelength and λ_0 is the wavelength observed when the molecule is stationary.

The fraction of molecules that have velocity component v_x in the range v_x to $v_x + \mathrm{d}v_x$ is equal to $P\,[v_x]\mathrm{d}v_x$, which is given by

$$P\,[v_x]\mathrm{d}v_x = \left(\frac{m}{2\pi kT}\right)^{1/2} \mathrm{e}^{-\,mv_x^2/2kT}\mathrm{d}v_x.$$

Molecules that have velocity component v_x in the range v_x to $v_x + \mathrm{d}v_x$ give rise to wavelengths in the range λ to $\lambda + \mathrm{d}\lambda$. Hence, the fraction of molecules that emit radiation with wavelength in the range λ to $\lambda + \mathrm{d}\lambda$ is

$$P\,[\lambda]\mathrm{d}\lambda = P\,[v_x]\frac{\mathrm{d}v_x}{\mathrm{d}\lambda}\mathrm{d}\lambda, \tag{3.49}$$

where, from Equation (3.48), $\mathrm{d}v_x/\mathrm{d}\lambda = c/\lambda_0$. Then substituting for v_x, $\mathrm{d}v_x/\mathrm{d}\lambda$, and $P\,[v_x]$ in Equation (3.49), we find that the fraction of molecules that emit radiation with wavelength in the range λ to $\lambda + \mathrm{d}\lambda$ is

$$P\,[\lambda]\mathrm{d}\lambda = \left(\frac{mc^2}{2\pi kT\lambda_0^2}\right)^{1/2} \mathrm{e}^{-\,mc^2(\lambda-\lambda_0)^2/2kT\lambda_0^2}\mathrm{d}\lambda.$$

The function $P\,[v_x]$ has exactly the same form as the Gaussian function

$$f(x) = \frac{1}{\sigma\sqrt{2\pi}}\mathrm{e}^{-\,(x-\mu)^2/2\sigma^2},$$

which has a width at half height equal to $2\sigma\sqrt{2\ln 2}$; see Section 3.2. Comparing these two equations, we find that the width $\Delta\lambda$ of the spectral line at half height is given by

$$\Delta\lambda = 2\sqrt{2\ln 2}\left(\frac{kT\lambda_0^2}{mc^2}\right)^{1/2} = \left(\frac{8\,[\ln 2]kT}{m}\right)^{1/2}\frac{\lambda_0}{c}.$$

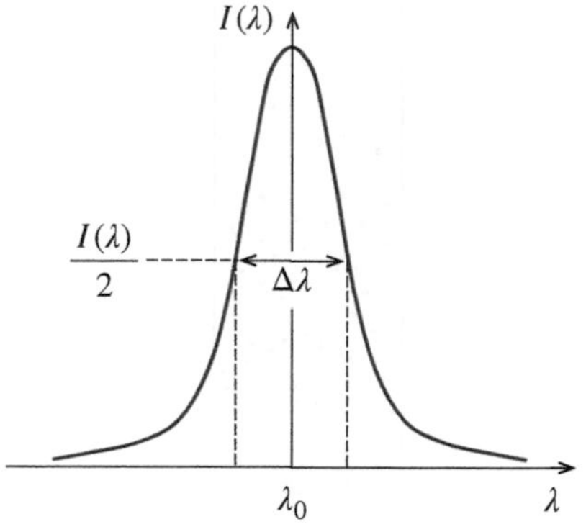

Figure 3.14 A spectral line that is Doppler broadened due to the thermal motion of molecules in a gas discharge lamp. The Gaussian shape of the line profile follows the shape of the distribution function of the v_x component of molecular velocity.

Hence, a spectral line that would be sharp in the absence of thermal motion of the atoms is broadened into a Gaussian shape whose width increases with temperature. The irradiance, or intensity, $I(\lambda)$ of the emitted radiation at a particular wavelength λ is directly proportional to the fraction of molecules that give rise to radiation at that wavelength. Hence, we can write

$$I(\lambda)\mathrm{d}\lambda = I(\lambda_0)\mathrm{e}^{-mc^2(\lambda-\lambda_0)^2/2kT\lambda_0^2}\mathrm{d}\lambda,$$

where $I(\lambda_0)$ is the intensity of radiation at $\lambda = \lambda_0$. Figure 3.14 illustrates the shape of a Doppler broadened spectral line.

Doppler broadening may be a hindrance to optical spectroscopy, as it limits the ability to resolve closely lying spectral lines. However, Doppler broadening has practical applications. In particular, it is used as a diagnostic tool to measure the temperature of a gas. For example, it is used in astronomy to determine the temperature of stars and to determine the temperature of interstellar gas clouds, although in the latter case, it is the spectral width in an absorption spectrum that determines the temperature. Doppler broadening is also used in plasma physics, for example, to determine the temperature of the plasma in a fusion reactor.

3.5 The isothermal atmosphere

In this section, we illustrate Boltzmann's law by considering the case of an *isothermal atmosphere*. This is a model approximating a real atmosphere like our own. The model consists of a tall column of gas at a uniform temperature that is acted upon by gravity. It is illustrated schematically in Figure 3.15. We experience the weight of our atmosphere as the pressure it exerts upon us. Atmospheric pressure is nominally 1.01×10^5 Pa or 1 Bar. As the height above sea level increases, the weight of the atmosphere above that height reduces and, consequently, so does the pressure. From this, we can obtain an expression for how the number density of the molecules varies with height. And we can relate this dependence to Boltzmann's law. Furthermore, we can use the variation in number density to determine the velocity distribution of the molecules.

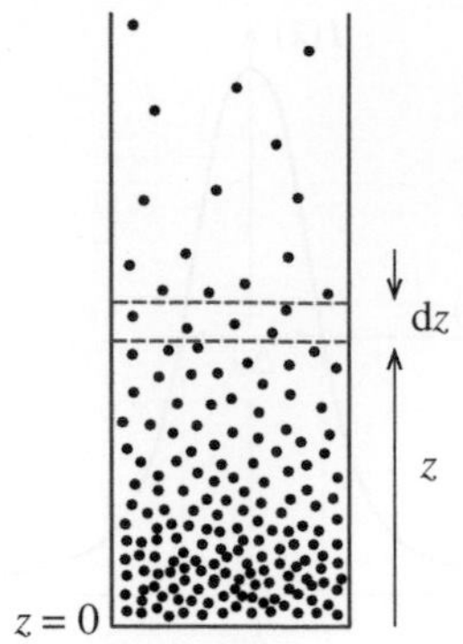

Figure 3.15 Schematic diagram of an isothermal atmosphere, which consists of a tall column of gas at a uniform temperature that is acted upon by gravity. It is an approximate model of our own atmosphere. Considering a horizontal slice of the column of unit area, the pressure at height z must exceed the pressure at height $(z + dz)$ by the weight of the intervening gas.

The model has a number of simplifying assumptions:

(i) The acceleration due to gravity is constant over the extent of the gas column. In the case of Earth's atmosphere, this is a good approximation as the height of the atmosphere is very small compared to the radius of the Earth.

(ii) The column of gas is in thermal equilibrium at constant temperature T. This is a rather crude approximation.

(iii) We neglect intermolecular interactions. This is also a good approximation of the Earth's atmosphere.

3.5.1 Potential energy distribution of the molecules

Neglecting intermolecular interactions, the potential energy of a molecule is due only to its gravitational potential. At height z, this potential energy is $V(z) = mgz$, where m is the mass of a molecule and g is the acceleration due to gravity. We have the ideal gas law, $PV = NkT$, Equation (3.11), where P is the pressure and N is the number of molecules in volume V. The number density $n = N/V$, and hence we can write

$$n(z) = \frac{P(z)}{kT}, \qquad (3.50)$$

where $n(z)$ and $P(z)$ are the number density and the pressure at height z, respectively. Considering a horizontal slice of the column of *unit area*, the pressure at height z must exceed the pressure at height $z + dz$ by the weight of the intervening gas. The number of molecules in the horizontal slice is $n(z)dz$. Hence, the pressure difference between the top and bottom of the slice is

$$P(z + dz) - P(z) = -mgn(z)dz.$$

As $P(z) = n(z)kT$, we can write

$$n(z + dz) - n(z) = -\frac{mg}{kT}n(z)dz.$$

Using

$$\frac{dn}{dz} = \frac{n(z + dz) - n(z)}{dz},$$

we obtain

$$\frac{dn(z)}{dz} = -\frac{mg}{kT}n(z), \tag{3.51}$$

which has the solution

$$n(z) = n_0 e^{-mgz/kT}, \tag{3.52}$$

where n_0 is the number density at $z = 0$. We see that the number density falls off exponentially with height z. We can also express this equation as

$$n(z) = n_0 e^{-z/z_0}, \tag{3.53}$$

where z_0 is called the *scale height* equal to mg/kT. Clearly, mg/kT must have the dimension of length, as the exponent of the exponential must be dimensionless. Taking $m = 30$ u and $T = 300$ K, we obtain $z_0 = 8.5$ km. Since pressure is directly proportional to number density, we can say straightaway that the pressure varies with height according to

$$P(z) = P_0 e^{-z/z_0}, \tag{3.54}$$

where P_0 is the pressure at ground level. Again, the pressure varies exponentially with height, and this is illustrated by Figure 3.16, where $P(z)/P_0$ is plotted against z/z_0.

As the number of molecules in the horizontal slice between z and $z + dz$ is $n(z)dz$, we have from Equation (3.52) that the *fraction* of molecules in this slice is

$$\frac{n(z)dz}{N_0} = \frac{n_0}{N_0} e^{-mgz/kT} dz, \tag{3.55}$$

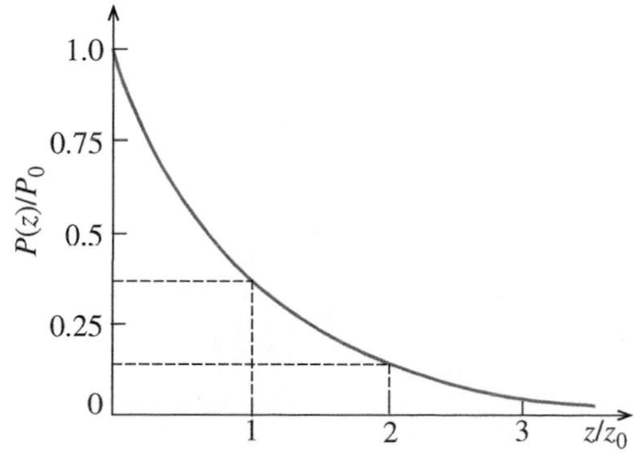

Figure 3.16 The exponential variation of pressure $P(z)$ with respect to height z in an isothermal atmosphere. $P(z)/P_0$ is plotted against z/z_0, where P_0 is the pressure at ground level, $z = 0$, and z_0 is the scale height.

where N_0 is the total number of molecules in the atmosphere. But this is just the probability that a molecule will be within the height range z to $z + \mathrm{d}z$. Hence, we have

$$P\,[z]\mathrm{d}z = \frac{n_0}{N}\mathrm{e}^{-\,mgz/kT}\mathrm{d}z, \tag{3.56}$$

which we recognise as a probability distribution function; where $P\,[z]$ is not to be confused with pressure $P(z)$. A molecule must be found somewhere in the column. Hence,

$$\int_0^\infty P\,[z]\,\mathrm{d}z = \frac{n_0}{N_0}\int_0^\infty \mathrm{e}^{-\,mgz/kT}\mathrm{d}z = 1,$$

which gives

$$\frac{n_0}{N_0} = \frac{1}{-\left[\frac{kT}{mg}\mathrm{e}^{-\,mgz/kT}\right]_0^\infty} = \frac{1}{kT/mg} = \frac{mg}{kT}.$$

Substituting for $n_0/N_0 = mg/kT$ in Equation (3.56), we find the probability distribution for a molecule being within the height range z to $z + \mathrm{d}z$:

$$P\,[z]\mathrm{d}z = \frac{mg}{kT}\mathrm{e}^{-\,mgz/kT}\mathrm{d}z, \tag{3.57}$$

which is in accord with Boltzmann's law.

3.5.2 Velocity distribution of the molecules

We may imagine a picture in which molecules rise from the ground to a maximum height, and then fall back to Earth, where they rebound elastically. This picture is illustrated in Figure 3.17. This figure illustrates that it is the thermal energy of the molecules that holds the atmosphere up. If they did not have thermal energy, the molecules would simply fall to the ground. If all the molecules had the same velocity as they left the ground, the atmosphere would end abruptly, which, of course, it does not. There is a variation in the number density with height and from this variation, given by Equation (3.52), we can deduce the velocity distribution of the molecules.

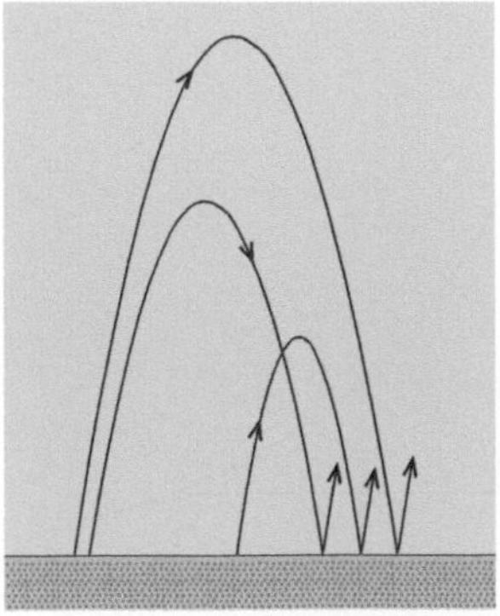

Figure 3.17 Pictorial representation of molecules rising from the ground to a maximum height, and then falling back to Earth, where they rebound elastically. This picture illustrates that it is the thermal energy of the molecules that holds the atmosphere up. If they did not have thermal energy, the molecules would simply fall to the ground.

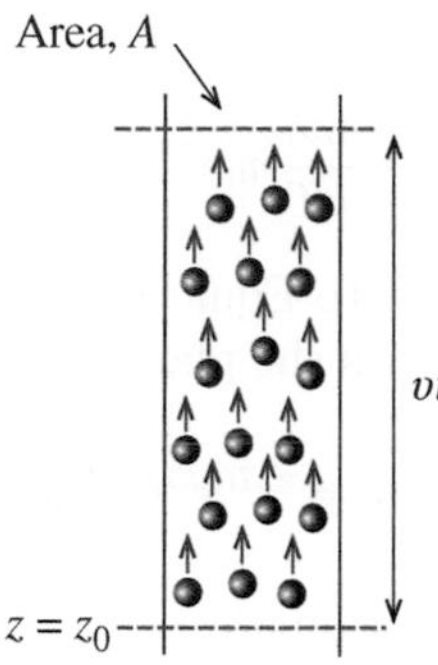

Figure 3.18 The flow of molecules through a tube of cross sectional area A. The number density of the molecules is n and they all have the same velocity v along the direction of flow. A volume vtA of gas will pass a given point on the tube, say $z = z_0$, in time t. This volume contains $vtAn$ molecules, and hence the number of molecules that pass a given point per unit time per unit area is vn.

In our one-dimensional model of the atmosphere, we consider molecules leaving the ground with velocities in the vertical direction. Consider for a moment the flow of molecules through a tube of cross sectional area A. The number density of the molecules is n and they all have the same velocity v along the direction of flow; see Figure 3.18. A volume vtA of gas will pass a given point on the tube, say $z = z_0$, in time t. This volume contains $vtAn$ molecules. Hence, the number of molecules that pass a given point per unit time per unit area is vn.

The molecules leaving the ground have a distribution of velocities described by a probability function $P[v]$, where $P[v]\mathrm{d}v$ is the probability of a molecule having vertical velocity between v and $v + \mathrm{d}v$. We recall that we can also think of $P[v]\mathrm{d}v$ as a fraction of the whole. Given that n_0 is the number density at ground level, $z = 0$, the number density for molecules that have velocities between v and $v + \mathrm{d}v$ is $n_0 P[v]\mathrm{d}v$. Therefore, the number of molecules leaving the ground per second per unit area with velocities between v and $v + \mathrm{d}v$ is

$$n_0 P[v]\mathrm{d}v \times v. \tag{3.58}$$

Molecules with velocity v will have a turning point at $z = v^2/2g$, where g is the acceleration due to gravity. Molecules with velocity $v + \mathrm{d}v$ will have a turning point at $z = (v + \mathrm{d}v)^2/2g$. We have

$$\frac{(v + \mathrm{d}v)^2}{2g} = \frac{v^2 + (\mathrm{d}v)^2 + 2v\mathrm{d}v}{2g} \cong \frac{v^2}{2g} + \frac{v\mathrm{d}v}{g} = z + \mathrm{d}z.$$

Hence, molecules with velocity $v + \mathrm{d}v$ will have a turning point at $z + \mathrm{d}z$.

The number of molecules passing in an upward direction through a horizontal plane at height z is $n(z)\overline{v}(z)$, where $\overline{v}(z)$ is the mean speed of the molecules at height z. The corresponding number of molecules passing through the plane at $z + \mathrm{d}z$ is $n(z + \mathrm{d}z)\overline{v}(z + \mathrm{d}z)$, where $\overline{v}(z + \mathrm{d}z)$ is the mean speed at $z + \mathrm{d}z$. The mean speed is determined by the temperature and since this is constant:

$$\overline{v}(z + \mathrm{d}z) = \overline{v}(z) = \text{constant} = \overline{v}.$$

Then, the number of molecules that cross the plane at height z but do not cross the plane at height $z + \mathrm{d}z$ is

$$\overline{v}[n(z) - n(z + \mathrm{d}z)] = -\overline{v}\left(\frac{\mathrm{d}n}{\mathrm{d}z}\right)\mathrm{d}z. \tag{3.59}$$

We have from Equation (3.51)

$$\frac{dn}{dz} = -\frac{mg}{kT} n_0 e^{-mgz/kT}.$$

Substituting for dn/dz from Equation (3.51) into Equation (3.59), we find that the number of molecules that cross the plane at height z but do not cross the plane at height $z + dz$ is

$$\frac{\overline{v}mg}{kT} n_0 e^{-mgz/kT} dz. \tag{3.60}$$

This is just the number of molecules given by Equation (3.58). Hence,

$$\overline{v}\frac{mg}{kT} n_0 e^{-mgz/kT} dz = n_0 \overline{v} P[\overline{v}] d\overline{v}. \tag{3.61}$$

Since $mgz = mv^2/2$ and $gdz = vdv$, we obtain

$$P[v]dv = \left(\frac{m\overline{v}}{kT}\right) e^{-mv^2/2kT} dv, \tag{3.62}$$

where the factor $(m\overline{v}/kT)$ is a constant. Equation (3.62) is the probability distribution for the molecules having velocity in the range v to $v + dv$, in accord with Boltzmann's law.

3.6 Derivation of the Maxwell–Boltzmann speed distribution

In Section 3.4, we obtained the probability distribution for a single component of molecular velocity. As noted there, it is usually more useful to know the speed distribution of the molecules. The difference is, of course, that speed is the magnitude of velocity and does not depend on the direction of travel. In this section, we derive the Maxwell–Boltzmann speed distribution. However, instead of going directly to the speed distribution of a three-dimensional gas, we first deal with the case of a *two-dimensional gas*, for which the analysis is easier to visualise.

The two-dimensional case may seem rather artificial, but, in fact, there are physical situations that can be described in this way. For example, it describes the behaviour of gases on the surface of a catalyst. We can imagine the molecules skating on and confined to the flat surface of the catalyst. And we note that *two-dimensional materials* can now be manufactured and are of much current interest for technological reasons. Moreover, some physical problems may be extremely challenging to solve in three dimensions. And then it is useful to first solve a two-dimensional version of the problem.

3.6.1 Two-dimensional gas

There is nothing special about the x-component of velocity. So, in analogy with Equation (3.44), we can immediately write down the probability distributions for the y and z velocity components:

$$P[v_y]dv_y = \left(\frac{m}{2\pi kT}\right)^{1/2} e^{-mv_y^2/2kT} dv_y;\ P[v_z]dv_z = \left(\frac{m}{2\pi kT}\right)^{1/2} e^{-mv_z^2/2kT} dv_z. \tag{3.63}$$

As noted previously, all three components of velocity are independent of each other. When we have independent probabilities, the probability of them all occurring is the product of their individual probabilities. A familiar example would be the probability of obtaining three consecutive heads when a coin is tossed, i.e.

$1/2 \times 1/2 \times 1/2 = 1/8$. Hence, the probability of a molecule having velocity components within the ranges v_x to $v_x + dv_x$, and v_y to $v_y + dv_y$, is the product of the individual probabilities:

$$P\left[v_x, v_y\right] dv_x dv_y = \left(\frac{m}{2\pi kT}\right) e^{-mv_x^2 + mv_y^2/2kT} dv_x dv_y. \tag{3.64}$$

In the case of a single component of velocity, the probability function $P\left[v_x\right]$ is a Gaussian function; see Equation (3.44). Now, the probability function $P\left[v_x, v_y\right]$ is a two-dimensional Gaussian function. The shape of $P\left[v_x, v_y\right]$ is illustrated in Figure 3.19, where the $P\left[v_x, v_y\right]$ axis is perpendicular to the plane containing the v_x and v_y axes. Figure 3.20 shows a different view of this two-dimensional Gaussian, now 'looking down' on

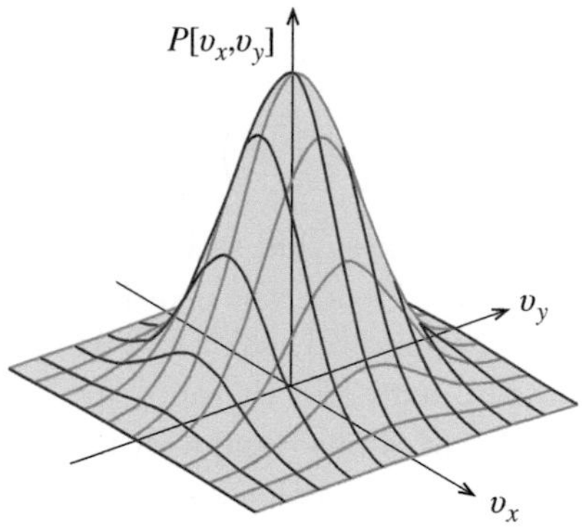

Figure 3.19 The shape of the function $P\left[v_x, v_y\right]$, where $P\left[v_x, v_y\right] dv_x dv_y$ is the probability distribution of a molecule having velocity components within the ranges v_x to $(v_x + dv_x)$, and v_y to $(v_y + dv_y)$. $P\left[v_x, v_y\right]$ is a two-dimensional Gaussian, where the $P\left[v_x, v_y\right]$ axis is perpendicular to the plane containing the v_x and v_y axes.

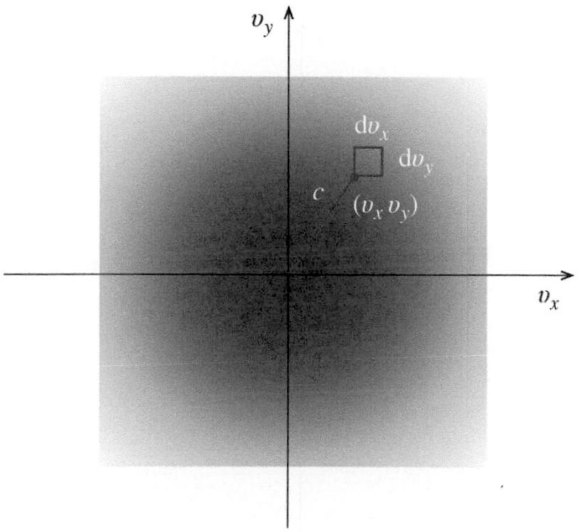

Figure 3.20 A view of the two-dimensional Gaussian function $P\left[v_x, v_y\right]$ 'looking down' on it from above. The Gaussian has its maximum value at the origin of the v_x, v_y coordinate system and in this view it has rotational symmetry about the $P\left[v_x, v_y\right]$ axis. The density of shading serves to represent the variation in the height of the Gaussian. The height varies smoothly with distance from the origin.

it from above. The Gaussian has its maximum value at the origin of the v_x, v_y coordinate system and in this view, it has rotational symmetry about the $P\left[v_x, v_y\right]$ axis. The density of shading serves to represent the variation in the height of the probability function.

For a two-dimensional gas, a molecular velocity with components v_x and v_y is represented by a point with coordinates $\left(v_x, v_y\right)$ in the $v_x - v_y$ plane, as indicated by the red dot in Figure 3.20. The associated velocity intervals $\mathrm{d}v_x$ and $\mathrm{d}v_y$ are represented by the rectangle with sides of length $\mathrm{d}v_x$ and $\mathrm{d}v_y$, respectively. The probability of a molecule having velocity components within the ranges v_x to $v_x + \mathrm{d}v_x$ and v_y to $v_y + \mathrm{d}v_y$ is equal to the value of the probability function at the point $\left(v_x, v_y\right)$, multiplied by the elemental area $\mathrm{d}v_x \times \mathrm{d}v_y$, that is $P\left[v_x, v_y\right]\mathrm{d}v_x\mathrm{d}v_y$. Since $\mathrm{d}v_x\mathrm{d}v_y$ is an area, we interpret $P\left[v_x, v_y\right]$ as the *probability per unit area.*

The length of the red, solid line in Figure 3.20, connecting the point $\left(v_x, v_y\right)$ to the origin is the magnitude of the velocity, i.e. the speed c of the molecule, where

$$c = \left(v_x^2 + v_y^2\right)^{1/2}. \tag{3.65}$$

Any molecular velocity that has $\left(v_x, v_y\right)$ coordinates that satisfy this condition will have the same speed, no matter in which direction the velocity points. Hence, the coordinates of these velocities lie on a circle of radius c in the $v_x - v_y$ plane. Moving to Figure 3.21, this circle is shown as the inner red, dotted circle. In this figure, we have the same v_x and v_y axes and the same probability function $P\left[v_x, v_y\right]$. But now the elemental area of interest is the annulus of radius c and width $\mathrm{d}c$. Any molecule that has velocity coordinates $\left(v_x, v_y\right)$ that lie within this elemental area has a speed between c and $c + \mathrm{d}c$. It follows that the probability of a molecule having a speed between c and $c + \mathrm{d}c$ is given by $P\left[v_x, v_y\right] \times 2\pi c\mathrm{d}c$, where $2\pi c\mathrm{d}c$ is the area of the annulus. This probability is $P\,[c]\mathrm{d}c$. Hence

$$P\,[c]\mathrm{d}c = 2\pi c\left(\frac{m}{2\pi kT}\right)\mathrm{e}^{-\,mv_x^2 + mv_y^2/2kT}\,\mathrm{d}c, \tag{3.66}$$

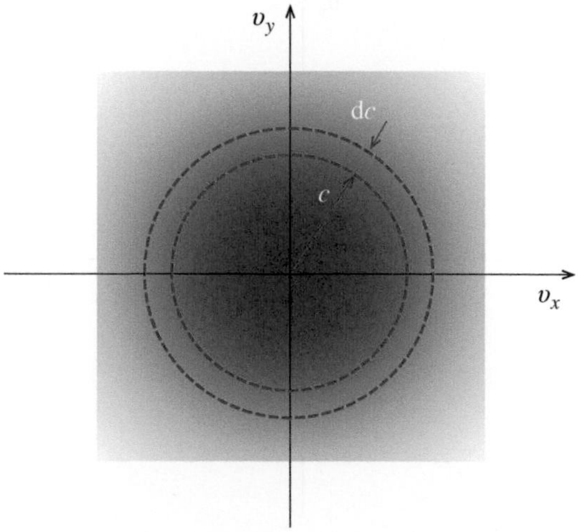

Figure 3.21 This figure reproduces Figure 3.20, but now the elemental area of interest is the annulus of radius c and width $\mathrm{d}c$. Any molecule that has velocity coordinates $\left(v_x, v_y\right)$ that lie within this elemental area has a speed between c and $c + \mathrm{d}c$.

and finally

$$P\,[c]\mathrm{d}c = \left(\frac{m}{kT}\right) c\,\mathrm{e}^{-\,mc^2/2kT}\;\mathrm{d}c, \tag{3.67}$$

where we have substituted $v_x^2 + v_y^2 = c^2$. Equation (3.67) is the speed distribution for a two-dimensional gas.

3.6.2 Three-dimensional gas

The analysis for the speed distribution of a three-dimensional gas is exactly analogous to that of the two-dimensional case. But now we are dealing with a three-dimensional coordinate system with axes corresponding to the velocity components v_x, v_y, and v_z. The probability of a molecule having velocity components within the ranges v_x to $v_x + \mathrm{d}v_x$, v_y to $v_y + \mathrm{d}v_y$, and v_z to $v_z + \mathrm{d}v_z$ is

$$P\left[v_x, v_y, v_z\right]\mathrm{d}v_x\mathrm{d}v_y\mathrm{d}v_z = \left(\frac{m}{2\pi kT}\right)^{3/2}\mathrm{e}^{-\,mv_x^2 + mv_y^2 + mv_z^2/2kT}\mathrm{d}v_x\mathrm{d}v_y\mathrm{d}v_z. \tag{3.68}$$

The probability function $P\left[v_x, v_y, v_z\right]$ now has the form of a three-dimensional Gaussian, which is positioned with its maximum at the origin of the v_x, v_y, v_z coordinate system, where $P\left[v_x, v_y, v_z\right]$ is the probability *per unit volume*. Interestingly, we saw an analogous probability distribution in our discussion of the ground state of hydrogen in Section 1.3.3.

The speed c of a molecule is related to its velocity components v_x, v_y, and v_z by

$$c = \left(v_x^2 + v_y^2 + v_z^2\right)^{1/2}, \tag{3.69}$$

which is the equation of a spherical surface of radius c. Hence, all molecules with speed c will have a combination of velocity components that satisfy this condition and their corresponding $\left(v_x, v_y, v_z\right)$ coordinates will lie on a spherical surface of radius c. Moreover, molecules with speed between c and $c + \mathrm{d}c$ will have coordinates that lie within a spherical shell of radius c and width $\mathrm{d}c$. The volume of this shell is $4\pi c^2\mathrm{d}c$. Hence, the probability $P[c]\mathrm{d}c$ that a molecule has a speed between c and $c + \mathrm{d}c$ is the product of $P\left[v_x, v_y, v_z\right]$ and the volume element $4\pi c^2\mathrm{d}c$:

$$P[c]\mathrm{d}c = P\left[v_x, v_y, v_z\right]4\pi c^2\mathrm{d}c = 4\pi\left(\frac{m}{2\pi kT}\right)^{3/2} c^2\,\mathrm{e}^{-\,mv_x^2 + mv_y^2 + mv_z^2/2kT}\mathrm{d}c. \tag{3.70}$$

Substituting for $v_x^2 + v_y^2 + v_z^2 = c^2$, we finally obtain

$$P[c]\mathrm{d}c = 4\pi\left(\frac{m}{2\pi kT}\right)^{3/2} c^2\,\mathrm{e}^{-\,mc^2/2kT}\mathrm{d}c.$$

This is the Maxwell–Boltzmann speed distribution that we first saw in Section 3.3.

The distribution function $P\,[c]$ is the product of two factors $4\pi(m/2\pi kT)^{3/2}c^2$ and $\mathrm{e}^{-\,mc^2/2kT}\mathrm{d}c$, respectively. The two factors are shown by the dashed curves in Figure 3.22, while the solid curve is their product, $P\,[c]$. With respect to the factor involving c^2, we note that

$$c^2 = 0 \text{ at } c = 0, \text{and } c^2 \rightarrow \infty \text{ as } c \rightarrow \infty.$$

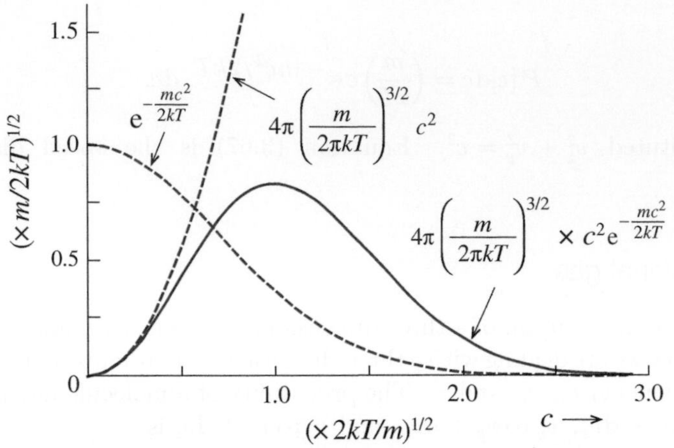

Figure 3.22 The speed distribution function $P[c]$ is the product of two factors $4\pi(m/2\pi kT)^{3/2}c^2$ and $e^{-mc^2/2kT}dc$, respectively. These two factors are shown by the dashed curves, while the solid curve is $P[c]$. $c^2 = 0$ at $c = 0$, and $c^2 \rightarrow \infty$ as $c \rightarrow \infty$. On the other hand, $e^{-mc^2/2kT} = 1$ at $c = 0$ and $e^{-mc^2/2kT} \rightarrow 0$ as $c \rightarrow \infty$. The exponential term reduces more rapidly than the c^2 term increases and hence dominates at large c. The overall result is that $P[c]$ reaches a maximum value, before decreasing at large c.

On the other hand

$$e^{-mc^2/2kT} = 1 \text{ at } c = 0, \text{and } e^{-mc^2/2kT} \rightarrow 0 \text{ as } c \rightarrow \infty.$$

The exponential term reduces more rapidly than the c^2 term increases and dominates at large c. The overall result is that the $P[c]$ reaches a maximum value, before decreasing at large c.

3.7 Equipartition of energy

We have seen that the mean kinetic energy associated with each component of molecular velocity is equal to $\frac{1}{2}kT$, and that the mean potential energy of a molecule in an isothermal atmosphere is also equal to $\frac{1}{2}kT$. We now consider other kinds of energy that a molecule might possess; in particular, energy contributions that arise from rotational and vibrational motion. We will see that the mean energy of each individual contribution to the total energy is always $\frac{1}{2}kT$, which is a statement of the *equipartition of energy* theorem.

3.7.1 Rotational motion of a diatomic molecule

A classical model for the rotation of a diatomic molecule consists of two masses connected by a rigid rod, as illustrated in Figure 3.23. For the sake of simplicity, we assume both masses to be equal, for example, the two atoms of hydrogen in molecular H_2. The diatomic molecule has two axes, a and b respectively, about which it can rotate. Both are perpendicular to the rod connecting the two masses. Rotation about the axis joining the two constituent atoms, shown as axis c in Figure 3.23, is not included because the atoms are considered to be point masses.

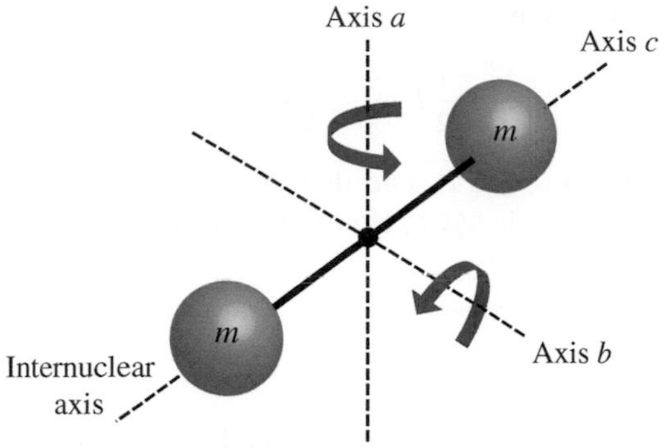

Figure 3.23 A classical model for the rotation of a diatomic molecule. It consists of two equal masses connected by a rigid rod, where the masses represent the constituent atoms. The molecule has two axes, a and b, about which it can rotate, both of which are perpendicular to the axis joining the two constituent atoms. Rotation about the axis joining the atoms, axis c, is not included because the atoms can be considered to be point masses.

The rotational kinetic energy of the molecule about axis a is

$$\frac{1}{2}I_a\omega_a^2, \tag{3.71}$$

where I_a is the moment of inertia about axis a and ω_a is the angular velocity about that axis. Rotational kinetic energy is analogous to translational kinetic energy, which we discussed in Section 3.4. There, using Boltzmann's law, we found that the probability distribution for the x-component of translational velocity is given by

$$P\left[v_x\right]\mathrm{d}v_x = \left(\frac{m}{2\pi kT}\right)^{1/2} \mathrm{e}^{-mv_x^2/2kT}\mathrm{d}v_x.$$

Similarly, Boltzmann's law says that the probability of a molecule having an angular velocity between ω_a and $\omega_a + \mathrm{d}\omega_a$ is

$$P\left[\omega_a\right]\mathrm{d}\omega_a = A\mathrm{e}^{-I_a\omega_a^2/2kT}\mathrm{d}\omega_a, \tag{3.72}$$

where A is the normalization constant, for which

$$\int_{-\infty}^{\infty} A\mathrm{e}^{-I_a\omega_a^2/2kT}\mathrm{d}\omega_a = 1,$$

and we take the limits of ω_a to be $\pm\infty$. Using a standard integral, we readily find that

$$A = \left(\frac{I_a}{2\pi kT}\right)^{1/2}. \tag{3.73}$$

Then the mean value $\overline{\omega_a^2}$ is

$$\left(\frac{I_a}{2\pi kT}\right)^{1/2}\int_{-\infty}^{\infty} \omega_a^2\,\mathrm{e}^{-I_a\omega_a^2/2kT}\mathrm{d}\omega_a = \left(\frac{I_a}{2\pi kT}\right)^{1/2}\frac{1}{2}\left[\pi\left(\frac{2kT}{I_a}\right)^3\right]^{1/2} = \frac{kT}{I_a},$$

and the mean rotational kinetic energy is

$$\frac{1}{2}I_a\overline{\omega_a^2} = \frac{1}{2}kT. \tag{3.74}$$

We see that the mean energy is equal to $\frac{1}{2}kT$, and does not depend on the moment of inertia. Similarly, the mean rotational energy about axis b is $\frac{1}{2}kT$. Hence, the mean total rotational energy of the diatomic molecule is kT.

3.7.2 Vibrational motion of a diatomic molecule

As we saw in Section 2.3.7, a diatomic molecule has vibrational motion. To describe this vibrational motion, we again use a model. This consists of two equal masses connected by a Hooke's law spring, as illustrated in Figure 3.24. The spring replaces the rigid rod of the simpler model we used for rotational motion. We might expect that the moment of inertia of the molecule would change due to the centrifugal force when the molecule rotates and that this would affect the rotational motion. It does, but this *centrifugal distortion* is relatively small and can be neglected for our purposes. For the sake of simplicity, we assume that one mass is fixed in position at the origin and the other mass vibrates along the x-axis, about its equilibrium position.

The vibrating mass has potential energy due to the stretching of the spring and kinetic energy due to its vibrational motion. The potential energy is equal to the work done in extending or compressing the spring. The work done on the spring in extending it from x' to $x' + \mathrm{d}x'$ is $\alpha x'\mathrm{d}x'$, where α is the *spring constant*, the restoring force per unit extension. Hence, the work done in extending the spring by distance x from its unstretched length x_0 is

$$\int_0^x \alpha x'\mathrm{d}x' = \frac{1}{2}\alpha x^2. \tag{3.75}$$

This is the potential energy $V(x)$ of the mass. In Section 3.4, we had an expression for Boltzmann's law with respect to potential energy $V(x)$:

$$P\left[x\right]\mathrm{d}x = A\mathrm{e}^{-\,V(x)/kT}\mathrm{d}x, \tag{3.37}$$

Hence, according to Boltzmann's law, the probability of the mass having an extension between x and $x + \mathrm{d}x$ is

$$P\left[x\right]\mathrm{d}x = A\mathrm{e}^{-\,\alpha x^2/2kT}\mathrm{d}x. \tag{3.76}$$

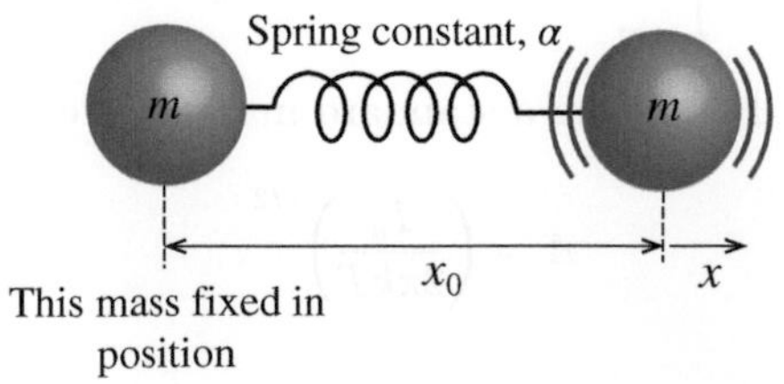

Figure 3.24 A classical model for the vibrational motion of a diatomic molecule. It consists of two masses connected by Hooke's law spring. The spring replaces the rigid rod of the simpler model shown in Figure 3.23. For the sake of simplicity, one mass is fixed in position at the origin and the other mass vibrates along the x-axis, about its equilibrium position. The equilibrium distance between the two masses is x_0.

This equation has exactly the same form as Equation (3.72). So, we can immediately write down

$$A = \left(\frac{\alpha}{2\pi kT}\right)^{1/2},$$

and

$$\overline{x^2} = \left(\frac{\alpha}{2\pi kT}\right)^{1/2}\left[\pi\left(\frac{2kT}{\alpha}\right)^3\right]^{1/2} = \frac{kT}{\alpha}.$$

Then the mean potential energy is

$$\frac{1}{2}\alpha\overline{x^2} = \frac{1}{2}kT. \tag{3.77}$$

We see that the mean potential energy of a one-dimensional harmonic oscillator is equal to $\frac{1}{2}kT$, and is independent of the spring constant. The translational kinetic energy of the mass is $\frac{1}{2}m\dot{x}^2$, where $\dot{x}$ is its velocity along the x-axis. In a completely analogous way to how we found the mean rotational energy and the mean potential energy, we readily find that the mean value of $\overline{\dot{x}^2}$ is kT/m and that the mean kinetic energy of the oscillator is again equal to $\frac{1}{2}kT$. Hence, modelling a diatomic molecule as a simple harmonic oscillator, we find that the molecule has potential energy and translational energy due to its vibrational motion. The total energy is

$$\frac{1}{2}\alpha x^2 + \frac{1}{2}m\dot{x}^2, \tag{3.78}$$

where both terms have a mean energy of $\frac{1}{2}kT$. Notice that both x and $\dot{x}$ appear as quadratic terms. Of course, a molecule does not have one of its constituent atoms held fixed in position. To take this into account, we simply need to use the reduced mass of the molecule, replacing mass m with the reduced mass μ where

$$\mu = \frac{m_1 \times m_2}{m_1 + m_2},$$

where m_1 and m_2 are the masses of the two constituent atoms that, in the present case, are equal to m.

In summary, a diatomic molecule has the following terms to express its total energy:

- Three kinetic energy terms corresponding to translational motion along its three orthogonal directions of motion, involving v_x^2, v_y^2 and v_z^2, respectively.
- Two rotational energy terms corresponding to rotational motion about two axes of rotation, involving ω_a^2 and ω_b^2, respectively.
- One potential energy term plus one kinetic energy term corresponding to vibrational motion along the axis connecting the two atoms, x^2 and $(\dot{x})^2$, respectively.

And each of these seven terms contributes $\frac{1}{2}kT$ to the total energy of the molecule, and hence the total mean energy of a diatomic molecule is $\frac{7}{2}kT$. This is an example of the equipartition of energy theorem, which can be stated as follows: *every independent quadratic term in the energy of a system has a mean value of ½kT, provided the system is in thermal equilibrium at temperature T.* Note that the terms are all of the same type; a constant times the square of a coordinate. Such terms are called *degrees of freedom.*

Worked example

Suppose that a particle, in thermal equilibrium with its surroundings at temperature T, has energy $a\xi^2$, where a is a constant and ξ may be either a position or a momentum coordinate. Show that the mean energy $a\xi^2$ is equal to $\frac{1}{2}kT$. Take the limits of ξ to be $\pm\infty$.

Solution

The particle has energy $a\xi^2$ that is a function of the coordinate ξ. According to Boltzmann's law, the probability of a particle having the value of that coordinate within the range ξ to $\xi + \mathrm{d}\xi$ is given by the probability distribution

$$P[\xi]\mathrm{d}\xi = A\mathrm{e}^{-a\xi^2/kT}\mathrm{d}\xi,$$

where A is the normalisation constant given by

$$A = \frac{1}{\int_{-\infty}^{\infty} \mathrm{e}^{-a\xi^2/kT}\mathrm{d}\xi}.$$

The mean value of ξ^2 is given by

$$\overline{\xi^2} = A\int_{-\infty}^{\infty} \xi^2\mathrm{e}^{-a\xi^2/kT}\mathrm{d}\xi = \frac{\int_{-\infty}^{\infty} \xi^2\mathrm{e}^{-a\xi^2/kT}\mathrm{d}\xi}{\int_{-\infty}^{\infty} \mathrm{e}^{-a\xi^2/kT}\mathrm{d}\xi}.$$

Using the appropriate standard integrals, we find

$$\overline{\xi^2} = \frac{1}{2}\left(\frac{\pi}{[a/kT]^3}\right)^{3/2}\left(\frac{[a/kT]}{\pi}\right)^{1/2} = \frac{kT}{2a}.$$

Hence, the mean energy $a\overline{\xi^2} = ½kT$, and is independent of the constant a.

3.7.3 The equipartition theorem applied to macroscopic bodies

The equipartition theorem also applies to macroscopic bodies. However, because the masses involved are large, the resultant thermal motions are relatively small. But these thermal motions can be observed and can have important consequences, especially with respect to the detection sensitivity of measuring apparatus, i.e. their ability to measure small signals. As an example, we consider the mirror galvanometer.

A schematic diagram of a mirror galvanometer is shown in Figure 3.25a. A coil of fine wire is suspended in a permanent magnet. When a current passes through the coil, a couple is set up, which causes the coil to

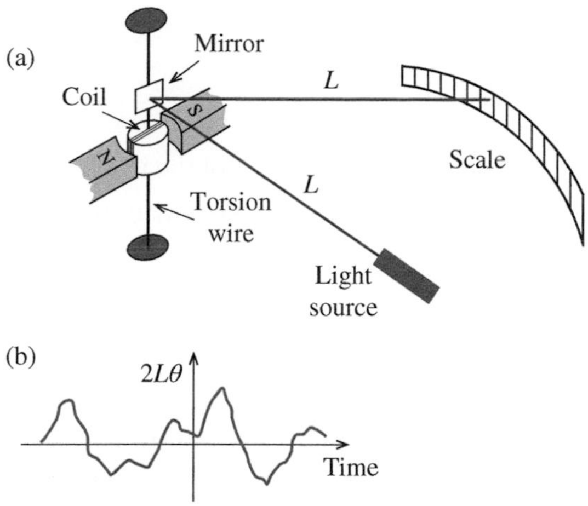

Figure 3.25 (a) Schematic diagram of a mirror galvanometer. A coil of fine wire is suspended in a permanent magnet. When a current passes through the coil a couple is set up which causes the coil/mirror assembly to deflect. An optical lever amplifies the angular deflection and the result is an extremely sensitive galvanometer, where the deflection of the light spot on the scale is a measure of the current passing through the coil. (b) It is found that the light spot of such a galvanometer does not stay put but continually jiggles, even when no current is passing through the coil. The light spot jiggles because of the thermal energy of the mirror/coil assembly. This jiggling represents a fundamental limit to the ability of the galvanometer to measure small currents, i.e. its sensitivity.

deflect. The coil is mounted on a torsional wire, to which is also attached a mirror. For angular displacement θ of the coil, the restoring couple produced by the wire is $-\alpha\theta$, where α is the torsional constant. The mirror reflects an incident light beam onto a scale. This arrangement acts as an optical lever and amplifies the angular deflection. The result is an extremely sensitive galvanometer, where the deflection of the light spot on the scale is a measure of the current passing through the coil. The deflection of the light spot is $2L\theta$, where L is the distance between the mirror and both the light source and the scale. It is found that the light spot of such a galvanometer does remain stationary but continually jiggles, even when no current is passing through the coil. This jiggling is illustrated in Figure 3.25b. It arises because of the thermal energy of the mirror/coil assembly, which behaves like a giant molecule. This jiggling represents a fundamental limit to the ability of the galvanometer to measure small currents, i.e. to its sensitivity.

Since the restoring couple produced by the wire is $-\alpha\theta$, we can describe the system as a simple harmonic oscillator. We recall that a harmonic oscillator has both potential energy and rotational kinetic energy. In the present case, there is rotational motion about just a single axis. Hence, the torsional potential energy is equal to $\frac{1}{2}\alpha\theta^2$, while the rotational kinetic energy is $\frac{1}{2}I\dot{\theta}^2$, where $\dot{\theta}$ is the angular velocity. Hence, the total energy is

$$E = \frac{1}{2}\alpha\theta^2 + \frac{1}{2}I\dot{\theta}^2.$$

According to the equipartition theorem:

$$\frac{1}{2}\alpha\overline{\theta^2} = \frac{1}{2}kT.$$

Hence,

$$\overline{\theta^2} = \frac{kT}{\alpha}.$$

The root mean square deflection of the mirror $\overline{\theta^2}$ is a measure of the jiggling and by finding the value of $\overline{\theta^2}$ we can determine the sensitivity of the galvanometer. An obvious way to reduce $\overline{\theta^2}$ and improve the sensitivity of the galvanometer is to cool it.

3.8 Specific heats of gases

When heat is added to a substance, the temperature of the substance usually rises. The amount of heat required to raise the temperature of the substance by one degree is called the *heat capacity* of the substance. The *specific heat* of a substance is the heat required to raise the temperature of a unit amount of the substance by 1°. The amount may be 1 kg of the substance or it may be 1 mol of the substance. We will use the symbol C for the specific heat per mole, which has units of J/K/mol.

We have seen the connection between temperature and the kinetic energy of molecules in a gas, and we can correctly expect, that a rise in temperature is accompanied by a rise in the kinetic energy of the molecules. We have also seen that a diatomic molecule can also have rotational and vibrational motion, and we can expect that the molecules will rotate faster and vibrate faster when heat is added to a gas. This is indeed the case if the temperature of the gas is sufficiently high, as we shall see. A picture of the possible motions of diatomic molecules in a gas is illustrated pictorially in Figure 3.26. We discuss the concept of *internal energy* in Section 6.4. For the moment, we can say that the internal energy U of a molecule is the sum of its translational kinetic energy and any rotational and vibrational energy that it may have:

$$U = E_{\text{trans}} + E_{\text{rot}} + E_{\text{vib}}. \tag{3.79}$$

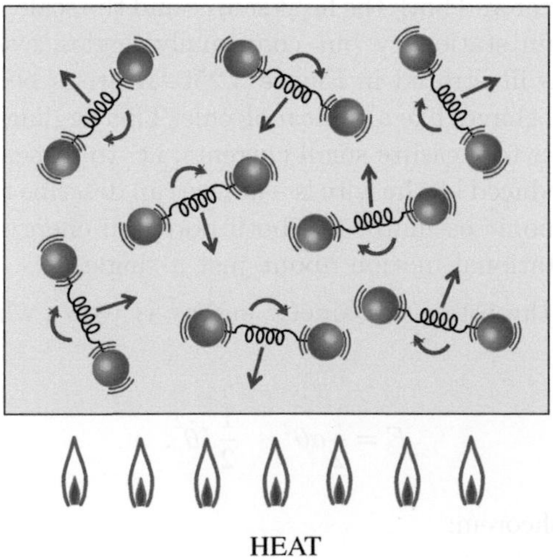

Figure 3.26 Pictorial representation of the possible motions of diatomic molecules in a gas, including translational, rotational, and vibrational motions. When heat is added to a gas at constant volume, all the heat is absorbed by the gas in increasing the energy of the molecules.

The specific heat of one mole of gas can be measured experimentally. On the other hand, we cannot measure the specific heat of an individual molecule. Nevertheless, we can obtain an understanding of the specific heat of gases from a molecular viewpoint, using the classical theory of molecular motions that we have described. Classical theory agrees with the experiment under certain conditions, but, importantly, it breaks down under other conditions. Historically, this was the first indication of the failure of classical physics and the need for a quantum theory of matter.

3.8.1 C_V, specific heat of one mole of an ideal gas at constant volume

We begin by considering a certain amount of gas, which we take to be one mole. We supply energy to the gas in the form of heat $\mathrm{d}Q$ and, keeping the volume of the gas constant, we measure the resulting rise in temperature $\mathrm{d}T$. Then the molar-specific heat C_V *at constant volume* is

$$C_V = \frac{\mathrm{d}Q}{\mathrm{d}T}. \tag{3.80}$$

Because the volume of the gas is kept constant, *all* the heat goes into increasing the internal energy of the molecules. If the gas were allowed to expand, some of this heat would have to go into doing work against the external pressure. When we heat the gas, we then have

$$\mathrm{d}U = \mathrm{d}Q, \tag{3.81}$$

where $\mathrm{d}U$ is the change in the internal energy and $\mathrm{d}Q$ is the amount of added heat. Hence,

$$\frac{\mathrm{d}Q}{\mathrm{d}T} = \frac{\mathrm{d}U}{\mathrm{d}T} = C_V. \tag{3.82}$$

A monatomic molecule such as helium or neon has only translational kinetic energy, with a mean value of $3/2kT$ per molecule or $3/2RT$ per mole, where $R = N_A k$ is the gas constant. Therefore,

$$C_V = \frac{\mathrm{d}}{\mathrm{d}T}\left(\frac{3}{2}RT\right) = \frac{3}{2}R. \tag{3.83}$$

Hence, we expect from the classical theory that monatomic gases will have a molar-specific heat equal to $3/2RT = 12.5$ J/mol·K. This is indeed the case. All the noble gases, He, Ne, Ar, Kr and Xe have values of C_V close to this, as can be seen in Table 3.3.

A diatomic molecule may have rotational and vibrational motion in addition to translational motion. As we described in our discussion of the equipartition theorem, translational motion gives rise to three quadratic energy terms, rotational motion gives rise to two quadratic energy terms and vibrational motion gives rise to another two quadratic energy terms. This gives

$$U = \left[\left(3 \times \frac{1}{2}\right)RT + \left(2 \times \frac{1}{2}\right)RT + \left(2 \times \frac{1}{2}\right)RT\right]RT = \frac{7}{2}RT,$$

Table 3.3 Molar-specific heat C_V for noble gases at 298 K.

Gas	He	Ne	Ar	Kr	Xe
C_V, J/mol·K	12.52	12.68	12.45	12.45	12.52

and

$$C_V = \frac{d}{dT}\left(\frac{7}{2}RT\right) = \frac{7}{2}R \text{ per mole.} \tag{3.84}$$

Thus, classical theory predicts a value of $7/2R$ for the molar-specific heat of a diatomic gas and that this value does not depend on temperature T.

3.8.2 C_P, specific heat of one mole of an ideal gas at constant pressure

When we heat a gas at constant pressure, the gas expands. Thus, part of the heat energy we supply goes into increasing the internal energy of the molecules and part goes into pushing back external forces, i.e. doing work. Then, according to the conservation of energy

$$dQ = dU + dW = dE_{int} + P dV,$$

where P is the external pressure. Hence

$$\left(\frac{\partial Q}{\partial T}\right)_P = \left(\frac{\partial U}{\partial T}\right)_P + P\left(\frac{\partial V}{\partial T}\right)_P = C_V + P\left(\frac{\partial V}{\partial T}\right)_P.$$

For 1 mol of ideal gas we have $PV = RT$, and hence $P(\partial V/\partial T)_P = R$, giving

$$\left(\frac{\partial Q}{\partial T}\right)_P = C_V + R,$$

or

$$C_P - C_V = R, \tag{3.85}$$

where C_P and C_V are the molar-specific heats at constant pressure and constant volume, respectively.

3.8.3 Ratio of specific heats γ

It is difficult to measure the specific heats of gases because they are small and the container containing the gas absorbs much more of the applied heat. It is much easier to measure the ratio of the specific heats defined by

$$\gamma = \frac{C_P}{C_V}. \tag{3.86}$$

Then C_P and C_V can be deduced from $C_P - C_V = R$.

3.8.4 The breakdown of the classical theory

Figure 3.27 shows the molar-specific heat C_V for molecular hydrogen H_2 as a function of temperature. Looking at this figure, we see that at low temperatures, C_V has the value of $3/2RT$. Thus, it appears that at low temperature, the H_2 molecule behaves as a monotonic gas with only translation motion. Then, as temperature increases, the value of C_V rises to $5/2RT$. This indicates that another kind of motion has started to

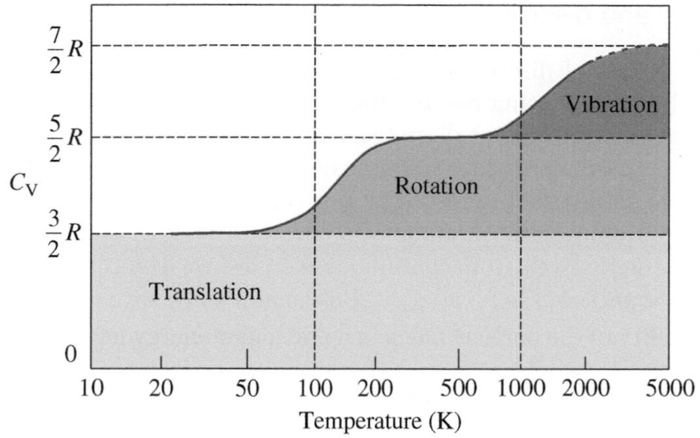

Figure 3.27 The molar-specific heat C_V for molecular hydrogen H_2 as a function of temperature. At low temperatures, C_V has the value of $3/2(RT)$, indicating that the H_2 molecule is behaving as a monotonic gas with only translation kinetic energy. As temperature increases, the value of C_V rises to $5/2(RT)$, and stays constant at this value over a limited temperature range. This indicates that another kind of motion has started to occur. At even high temperature, there is a second rise in the value of C_V, and eventually it approaches the predicted value of $7/2(RT)$ at a very high temperature. It does not quite reach the value of $7/2(RT)$ because at higher temperatures there is appreciable dissociation of the H_2 molecules into hydrogen atoms, as indicated by the dotted part of the C_V curve.

occur. At an even higher temperature, there is a second rise in the value of C_V, and eventually at very high temperature, it approaches the predicted value of $7/2RT$. It does not quite reach the value of $7/2RT$, as indicated by the dotted part of the C_V curve because at high temperatures, there is appreciable dissociation of the H_2 molecules into hydrogen atoms. Clearly, the experiment contradicts classical theory; C_V varies with temperature T and does not maintain the value of $7/2RT$.

The breakdown of the classical theory is due to the quantisation of rotational and vibration energy. According to quantum theory, only certain *discrete* energies of rotation and vibration are possible for a molecule, just as the electronic states of atoms and molecules are quantised. A certain amount of thermal energy is needed to start molecular rotation or vibration, i.e. to excite a molecule into a rotationally or vibrationally excited energy level. We have the physical picture of the molecules continually colliding. If the translational energy of a molecule is sufficiently high, it can excite another molecule into rotational or vibrational motion. Otherwise, it cannot.

The vibrational energy spacing of a molecule is always greater than the rotational energy spacing. Hence, rotational excitation will occur before vibrational excitation. At temperatures so low that kT is much less than the rotational level spacing, the probability of rotational excitation is very small. And if the molecules cannot absorb added heat by becoming rotationally excited, rotational motion does not contribute to the specific heat. However, when kT is the same order as the rotational spacing, a significant fraction of the molecules can become rotationally excited and absorb some of the added heat. And the specific heat increases. Similarly, only when kT is comparable to the vibrational energy spacing do a significant fraction of the molecules become vibrationally excited. Then they can absorb some of the added heat and again the specific heat increases. Because the vibrational energy spacing of a molecule is always greater than the rotational energy spacing, the first rise in C_V from $3/2RT$ to $5/2RT$ is attributed to rotation and the second rise from $5/2RT$ to $7/2RT$ is attributed to vibration. Note that since translation motion is not quantised in quantum theory, the classical result gives the correct result for the mean translational energy, which gives the contribution $3/2RT$ at all temperatures.

3.8.5 Boltzmann's law and discrete energy levels

Boltzmann's law applies to sets of discrete energies as well as to continuous energies like translational kinetic energy. Indeed, when formulating his distribution law, Boltzmann actually began by considering the distribution of particles among different discrete energy levels that were accessible to them. In the context of quantum mechanics, these energy levels are the quantised energy levels of the particles, such as the vibrational energy levels of a molecule. Interestingly, Boltzmann's use of discrete energies came some decades before the discovery of quantum mechanics.

To distinguish discrete energy levels from continuous energies, we denote them with the symbol ε. If the accessible energy levels for a particle are ε_1, ε_2, ε_3..., Boltzmann's law says that at thermal equilibrium at temperature T, the probability of the particle being in a particular energy level ε_i is proportional to $\mathrm{e}^{-\varepsilon_i/kT}$:

$$P[\varepsilon_i] \propto \mathrm{e}^{-\varepsilon_i/kT},$$

or

$$P[\varepsilon_i] = \alpha \mathrm{e}^{-\varepsilon_i/kT}, \tag{3.87}$$

where α is the normalization constant. Usually, quantum mechanical expressions are more complicated than the corresponding ones in classical mechanics. But in this case, the quantum mechanical result has a beautiful simplicity. If ε_0 is the lowest energy level for the particle, then it follows that

$$\frac{P[\varepsilon_i]}{P[\varepsilon_0]} = \frac{\mathrm{e}^{-\varepsilon_i/kT}}{\mathrm{e}^{-\varepsilon_0/kT}} = \mathrm{e}^{-\varepsilon_i - \varepsilon_0/kT}. \tag{3.88}$$

This means that the probability of the particle being in the excited state ε_i compared to its ground state ε_0 is $\mathrm{e}^{-\Delta\varepsilon/kT}$, where $\Delta\varepsilon = (\varepsilon_i - \varepsilon_0)$. Alternatively, we can say that the number of particles n_i in the excited level compared to the number n_0 in the ground state level is

$$\frac{n_i}{n_0} = \mathrm{e}^{-\Delta\varepsilon/kT}. \tag{3.89}$$

We see that it is less likely for the particle to be in a higher energy level than in a lower one.

The distinction between the quantum and classical regimes is as follows. If kT is comparable to or indeed less than the spacing of the quantised energy levels for a particle, as illustrated by Figure 3.28a, the

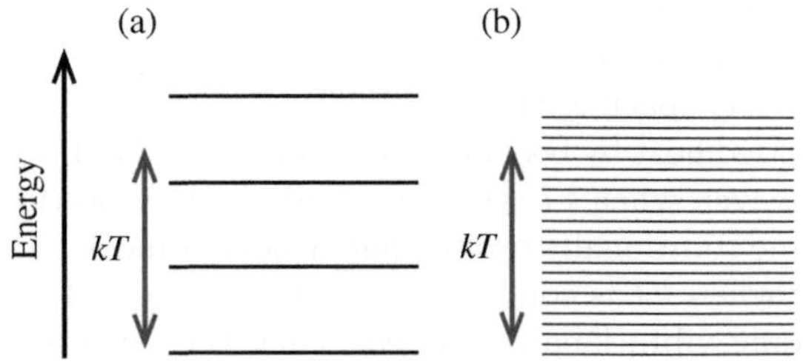

Figure 3.28 This figure illustrates the distinction between (a) the quantum and (b) the classical regimes. If the thermal energy kT of a particle is comparable to or indeed less than the spacing of the quantised energy levels for the particle, the quantisation must be taken into account. If on the other hand, kT is much larger than the level spacing, the quantised levels become indistinguishable from a continuum of energies, and a particle can have any energy. Then the quantisation of the levels can be ignored and classical theory can be used.

quantisation must be taken into account. If, on the other hand, kT is much larger than the level spacing, as illustrated by Figure 3.28b, the quantised levels become indistinguishable from a continuum of energies, and a particle can have any energy. Then the quantisation of the levels can be ignored and classical theory can be used. Hence, the equipartition theorem is always valid provided one uses degrees of freedom associated with fully excited energy states.

Worked example

Suppose that a particle in thermal equilibrium at temperature T has three accessible energy levels: $\varepsilon_1 = 0$, $\varepsilon_2 = \varepsilon$, and $\varepsilon_3 = 2\varepsilon$; see Figure 3.29. Determine the probabilities for the particle occupying each of these three levels. If the temperature is such that $kT = \varepsilon$, determine the relative populations of the three levels.

Solution

From Equation (3.87), we have

$$P[\varepsilon_1] = \alpha \mathrm{e}^{-\varepsilon_1/kT},$$
$$P[\varepsilon_2] = \alpha \mathrm{e}^{-\varepsilon_2/kT},$$
$$P[\varepsilon_3] = \alpha \mathrm{e}^{-\varepsilon_3/kT}.$$

We also have: $P[\varepsilon_1] + P[\varepsilon_2] + P[\varepsilon_3] = 1$.

Hence,

$$\alpha \mathrm{e}^{-\varepsilon_1/kT} + \alpha \mathrm{e}^{-\varepsilon_2/kT} + \alpha \mathrm{e}^{-\varepsilon_3/kT} = 1,$$

giving

$$\alpha = \frac{1}{\mathrm{e}^{-\varepsilon_1/kT} + \mathrm{e}^{-\varepsilon_2/kT} + \mathrm{e}^{-\varepsilon_3/kT}}.$$

Then, $P[\varepsilon_1] = \left(\mathrm{e}^{-\varepsilon_1/kT}\right)/\left(\mathrm{e}^{-\varepsilon_1/kT} + \mathrm{e}^{-\varepsilon_2/kT} + \mathrm{e}^{-\varepsilon_3/kT}\right)$, etc., or, in a more compact form,

$$P[\varepsilon_i] = \frac{\mathrm{e}^{-\varepsilon_i/kT}}{\sum_i \mathrm{e}^{-\varepsilon_i/kT}},$$

where the sum $\sum_i \mathrm{e}^{-\varepsilon_i/kT}$ is an example of a *partition function.*

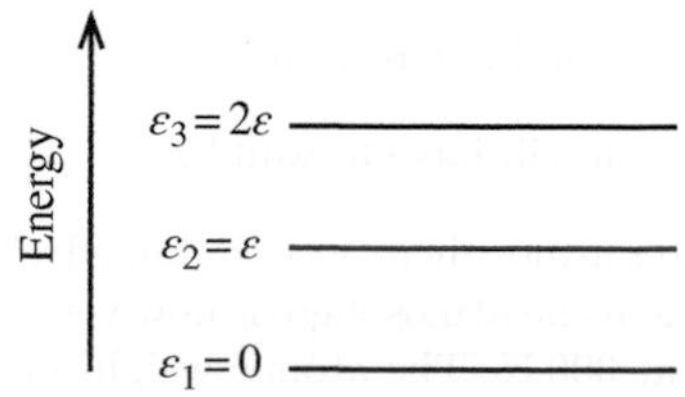

Figure 3.29 Three accessible energy levels of a particular particle in thermal equilibrium at temperature T.

Substituting $\varepsilon_1 = 0$, $\varepsilon_2 = \varepsilon$, and $\varepsilon_3 = 2\varepsilon$, and simplifying, we obtain:

$$P[\varepsilon_1] = \frac{1}{1 + e^{-\varepsilon/kT} + e^{-2\varepsilon/kT}},$$

$$P[\varepsilon_2] = \frac{1}{e^{\varepsilon/kT} + 1 + e^{-\varepsilon/kT}},$$

$$P[\varepsilon_3] = \frac{1}{e^{2\varepsilon/kT} + e^{\varepsilon/kT} + 1}.$$

For $T = \varepsilon/k$, we obtain

$$P[\varepsilon_1] = \frac{1}{1 + e^{-1} + e^{-2}} = 0.665,$$

and similarly $P[\varepsilon_2] = 0.245$, and $P[\varepsilon_3] = 0.090$. As a check, we see that these three numbers add up to 1.0, as required. These observations are in accord with our previous result that there is an appreciable probability of a particle being promoted to an excited energy level when its energy $\sim kT$ is comparable to the energy spacing.

Problems 3

3.1 In a low-pressure mercury discharge tube, the vapour is effectively at a temperature of 100 K. If the lamp emits a green spectral line at the wavelength of 564 nm, what will be the width of the line due to Doppler broadening? The atomic mass of mercury is 200.6 u.

3.2 You arrive at a bus stop to catch a bus that arrives at regular 20 minute intervals. Because your arrival time is random, the probability that you will have to wait, for example, less than 1 minute is the same as the probability that you will have to wait more than 19 minutes. The probability that you will have to wait a time between t and $t + dt$ is then given by

$$P[t]dt = A dt,$$

where A is the normalisation constant. Deduce:

(a) The value of the normalisation constant.

(b) Sketch the probability function $P[t]$.

(c) What is the probability that you will have to wait exactly 15 minutes?

(d) What is the probability that you will have to wait (i) less than 5 minutes, (ii) longer than 15 minutes?

(e) What is the average time that you will have to wait?

(f) What is the most probable time you will have to wait?

3.3 According to some theories, the Earth's atmosphere once contained a large percentage of helium gas. Explain the fact that the percentage of helium gas in the atmosphere is now very small. Assume that the temperature of the atmosphere has remained constant at 300 K. The atomic weight of helium is 4 u. Hint: Calculate the escape velocity from the Earth's gravitational field. The gravitational constant G is $6.67 \times 10^{-11}\ \mathrm{m^3/kg \cdot s^2}$, the mass and radius of the Earth are 6×10^{24} kg, and 6.4×10^6 m, respectively.

3.4 When a nucleus of deuterium ^{2}D combines with a nucleus of tritium ^{3}T in a fusion reaction, 17.6 MeV of energy is released, making nuclear fusion a potential energy source. For the fusion reaction to proceed, the two nuclei must come into contact, at which point the nuclear attractive force binds them together. However, to do this the nuclei must overcome the Coulomb force that mutually repels them; the so-called *Coulomb barrier*. Calculate the Coulomb potential when the two nuclei just touch. Hence, obtain an estimate of the temperature required to produce fusion reactions between deuterium and tritium nuclei in a hot plasma of deuterium and tritium ions. Note that in the hot plasma, the ions are fully stripped of their electrons and are therefore bare nuclei. The radius of a nucleus is given by $r = 1.2(A^{1/3}) \times 10^{-15}$ m, where A is the atomic mass.

3.5 The probability distribution for a molecule being within the height range z to $z + \mathrm{d}z$ in an isothermal atmosphere at temperature T is $P[z]\mathrm{d}z = A\mathrm{e}^{-mgz/kT}\mathrm{d}z$. (a) What is the mean gravitational potential energy of a molecule? (b) At what height is the mean thermal energy of a molecule equal to its potential energy? Assume a molecular mass of 30 u and a uniform temperature of 300 K. (c) What is the probability of a molecule being exactly at this height?

$$\text{Note,} \quad \int_0^\infty x\mathrm{e}^{-\alpha x}\mathrm{d}x = \frac{1}{\alpha^2}.$$

3.6 What holds the atmosphere up?

3.7 A centrifuge consists of a test tube rotating at an angular frequency ω in the horizontal plane. Particles of mass m and density ρ are in suspension in a liquid of density ρ_0 in the test tube. (a) Write down the force $F(r)$ on a particle when it is at radial distance r. (b) Deduce the potential energy $V(r)$ of the particle, defining $V(r) = 0$ at $r = 0$. (c) Hence, show that the particle density distribution $n(r)$ is given by

$$n(r) = A\exp\left(\frac{m' r^2\omega^2}{2kT}\right), \text{ where } A \text{ is a constant and } m' = \left(\frac{\rho - \rho_0}{\rho}\right)m.$$

(d) Small titanium particles of mass 1.0×10^{-19} kg and density 4200 kg/m^3 in suspension in water at a temperature of 300 K are placed in a centrifuge rotating at 10 rev/s. Show that at a distance of 100 mm from the axis of rotation, the concentration of particles drops by a factor of 1000 over a distance of less than 1 mm.

3.8 Given that the probability that a molecule of mass m at temperature T has a speed in the range between c and $c + \mathrm{d}c$ is given by

$$P[c]\mathrm{d}c = Ac^2\mathrm{e}^{-mc^2/2kT}\mathrm{d}c.$$

Calculate the most probable speed of an atom of xenon in thermal equilibrium at 600 K. The atomic weight of xenon is 131 u.

3.9 A gas of nitrogen molecules is in thermal equilibrium at a temperature $T = 300$ K.

(a) Calculate the mean speed $\overline{c}$, and the root mean square speed c_{rms} of the nitrogen molecules.

(b) Plot the Maxwell–Boltzmann speed distribution for the molecules.

(c) Verify from your graph that the most probable speed of the molecules is $\sqrt{2kT/m}$.

(d) What fraction of the molecules has a speed between (i) 600 m/s and 602 m/s and (ii) between 600 m/s and 900 m/s? For part (ii), you will need to estimate in some way the area under a graph between the two limits.

3.10 The probability that a molecule of mass m at temperature T has a speed in the range between c and $c + \mathrm{d}c$ is given by

$$P[c]\mathrm{d}c = 4\pi\left(\frac{m}{2\pi kT}\right)^{3/2} c^2\mathrm{e}^{-mc^2/2kT}\mathrm{d}c.$$

(a) Show that the probability that a molecule has a kinetic energy K in the range between K and $K+\mathrm{d}K$ is given by

$$P[K]\mathrm{d}K = \left(\frac{4}{\pi k^3 T^3}\right)^{1/2} K^{1/2}\mathrm{e}^{-K/kT}\mathrm{d}K.$$

(b) Obtain an expression for the most probable kinetic energy of a molecule.

(c) Obtain an expression for the mean kinetic energy of a molecule.

$$\text{Note,}\quad \int_0^\infty x^4\mathrm{e}^{-ax^2}\mathrm{d}x = \frac{3}{8}\left(\frac{\pi}{a^5}\right)^{1/2}.$$

3.11 A gas at low pressure and at temperature T is contained in a vessel from which it effuses through a small hole whose dimensions are small compared to the mean free path of the molecules. The speed distribution of the molecules that emerge from the hole is given by

$$P[c]\mathrm{d}c = Ac^3\mathrm{e}^{-mc^2/2kT}\mathrm{d}c,$$

where A is a constant. The term c^3 rather than the c^2 term that occurs in the Maxwell–Boltzmann speed distribution arises because molecules that pass through the hole per second will come from within a cone of height c and so faster molecules have a greater probability of passing through the hole in a given time. (a) Show that the most probable speed c_m of the molecules leaving the hole is $2kT$. (b) Show that the most probable kinetic energy K_m of the molecules leaving the hole is kT. (c) Show that the mean kinetic energy $\overline{K}$ of the molecules leaving the hole is $2kT$.

$$\text{Note}\quad \int_0^\infty x^3\mathrm{e}^{-ax^2}\mathrm{d}x = \frac{1}{2a^2},\quad \int_0^\infty x^5\mathrm{e}^{-ax^2}\mathrm{d}x = \frac{1}{a^3}.$$

3.12 The probability distribution for the kinetic energy of molecules hitting the wall of their container is given by

$$P[K]\mathrm{d}K = \left(\frac{1}{kT}\right)^2 K\mathrm{e}^{-K/kT}\mathrm{d}K.$$

(a) Show that the fraction of molecules that have a kinetic energy greater than ε is $(1+[\varepsilon/kT])\mathrm{e}^{-\varepsilon/kT}$.

(b) A molecule AB dissociates if it hits the surface of a catalyst if its kinetic energy is greater than 0.7 eV. Show that the rate of the reaction $AB \rightarrow A + B$ roughly doubles by raising the temperature by 10 °C from room temperature.

$$\text{Note}\quad \int_0^\infty x\mathrm{e}^{ax}\mathrm{d}x = \mathrm{e}^{ax}(ax-1).$$

3.13 A simple pendulum that consists of a 1 kg mass suspended at the end of a light string of length 1.2 m is nominally at rest. Is the pendulum exactly at rest or does it jiggle? If it does jiggle, what is the amount of jiggle? If the pendulum were suspended in an evacuated enclosure, would the situation be any different? Explain.

3.14 Consider a system of particles in thermal equilibrium at temperature T. A particle may be in one of two energy levels that are separated in energy by an amount ε. Write down an expression for the relative populations n_1/n_0 in the two energy levels. What will be the relative populations in the limits: (i) $T \rightarrow 0$, and (ii) $T \rightarrow \infty$.

3.15 (a) The quantised energy states associated with molecular rotation are of the order of magnitude $\hbar^2/I$, where $\hbar$ is Planck's constant over 2π and I is the moment of inertia of the molecule. Estimate the temperature at which you would expect the rotational motion to contribute to the specific heat of molecular hydrogen. For a diatomic molecule $I = [(m_1 m_2)/(m_1 + m_2)]d^2$, where m_1 and m_2 are the respective masses of the constituent atoms and d is their inter-nuclear separation. For molecular hydrogen, $d = 0.074$ nm. (b) The quantised energy states associated with molecular vibration are of the order of magnitude $h\nu$ where ν is the oscillation frequency of the vibrating molecule. The frequency can be deduced from the wavelength $\lambda = c/\nu$ at which the molecule absorbs light in its *Raman spectrum*. For hydrogen, λ is found to be 2.4 μm. Estimate the temperature at which you would expect the vibrational motion to contribute to the specific heat of molecular hydrogen.

4

Kinetic theory of gases: transport processes

In our discussion of gases so far, we have considered situations where the gas is in equilibrium. Now we turn our attention to situations where a gas is nearly but not quite in equilibrium; where there is a local non-uniformity in, say, temperature or the concentration of a particular molecular species. Under these circumstances, there is a tendency for the non-uniformity to die away. This is brought about by the movement of molecules in the gas; movements that are called *transport processes.* This chapter deals with three transport processes, namely molecular diffusion, heat conduction, and viscosity, which we will see are closely related. The empirical laws that describe these transport processes are presented. And, in each case, the process is explained in terms of the *kinetic theory of gases*, which relates the processes to microscopic quantities such as mass, speed, and *mean free path* of the molecules. And we will derive the important *diffusion equation* and *continuity equation.* Although the systems we will discuss are not quite in equilibrium in some way, we will assume that the departure from equilibrium is only small so that we can still use the Maxwell–Boltzmann distribution of molecular speeds. In this chapter, we also describe the so-called *random walk* and relate it to molecular diffusion.

4.1 Kinetic theory of gases

We have already made use of the kinetic theory of gases to relate the temperature of a gas to the translation energy of its molecules; see Section 3.1. The main features of an ideal gas as described by the kinetic theory are:

- A gas contains a huge number of individual molecules.
- The motion of the molecules is random with no preferred positions or directions of travel.
- The molecules continually make collisions with one another and these collisions are perfectly elastic, i.e. the total kinetic energy before and after a collision is the same.
- The molecules exert no forces of attraction or repulsion on one another.
- The average distance between molecules is much greater than the molecular size.

Physics of Matter, First Edition. George C. King.

In the simplest form of the theory, the molecules are point particles of negligible size. As a first improvement to the theory, the molecules are taken to be rigid spheres of finite size; although still widely separated from one another and with no attractive or repulsive interactions between them. This rigid-sphere model is the one we will use in our discussion of transport processes. The kinetic theory also has the two following hypotheses: (i) the molecular collisions obey the laws of classical mechanics and (ii) because of the huge number of molecules involved, the behaviour of a gas is expected to be the same as the average behaviour of the molecules, i.e. statistical methods can be used.

Although kinetic theory describes ideal gases, it is the case that real gases approximate to ideal gases if they are sufficiently dilute and if the potential energy due to their mutual attraction is negligible compared to their thermal energy. In practice, this means that gases such as argon and nitrogen behave as ideal gases under ordinary conditions of pressure and temperature. We discuss real gases in Chapter 5.

We will see that despite the simplicity of the kinetic theory outlined earlier, it provides a satisfactory explanation of the processes of molecular diffusion, thermal conduction, and viscosity. It gives the correct dependence of these processes on all the significant parameters, like temperature and pressure. And it enables us to make theoretical predictions of numerical values that agree within a factor of two or three with those obtained experimentally. Of course, as described in Chapter 2, molecules do exert attractive and repulsive forces on one another when they are very close together. And we will see that the theory does break down under certain conditions, such as when a gas is under a pressure of many atmospheres.

4.2 Molecular collisions and the mean free path

The picture we have is of molecules in a gas moving in random motion continuously making collisions with each other. The motion of a molecule is then something like that shown in Figure 4.1. The direction of travel of a molecule after a collision is entirely independent of its direction before the collision. The distance a molecule travels between collisions is called the *free path*. Of course, as indicated by

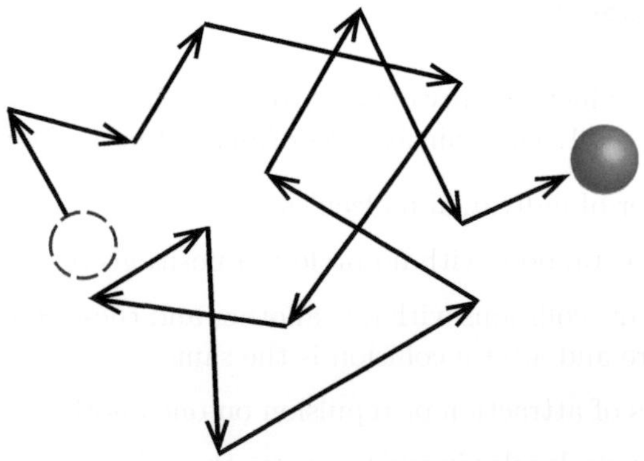

Figure 4.1 The molecules in a gas move in random motion continuously making collisions with each other. The direction of travel of a molecule after a collision is entirely independent of its direction before the collision. The distance a molecule travels between collisions is called the free path. As indicated, not all free paths are the same.

Figure 4.1, not all free paths are the same. However, we can talk about the *mean free path*, i.e. the average distance the molecules travel between collisions. The mean free path, denoted by the symbol λ, is one of the most useful concepts of the kinetic theory. We will see that it plays an important role in the kinetic theory of transport processes.

To obtain an expression for the mean free path, we first assume that all the molecules in the gas are stationary, except for one. This molecule travels through a column of gas making collisions with the stationary molecules, as depicted in Figure 4.2. The view that the moving molecule would see as it moves through the gas is shown pictorially in Figure 4.3, where the opaque circles represent the cross sections of the stationary molecules. The number density of the molecules and the length of the gas column are such that there is a negligible probability that the cross sections of two stationary molecules overlap. Clearly, the mean free path λ will depend on the number density n of the molecules, i.e. the number of molecules per unit volume. The larger the value of n, the smaller λ will be. Moreover, λ will depend on

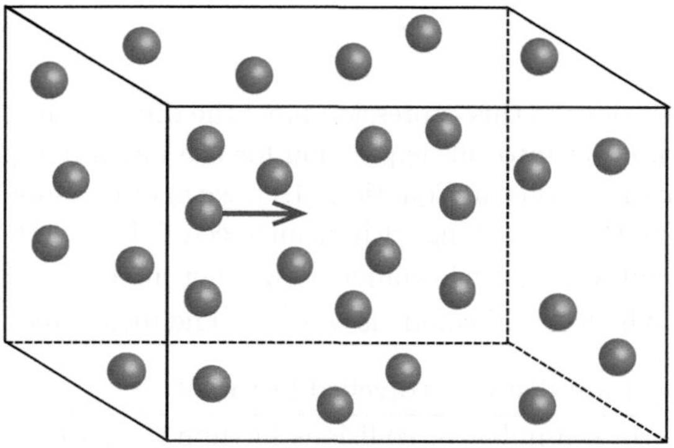

Figure 4.2 A pictorial illustration of a molecule as it moves through a gas of molecules. At atmospheric pressure, the average distance between molecules is about 10 molecular diameters.

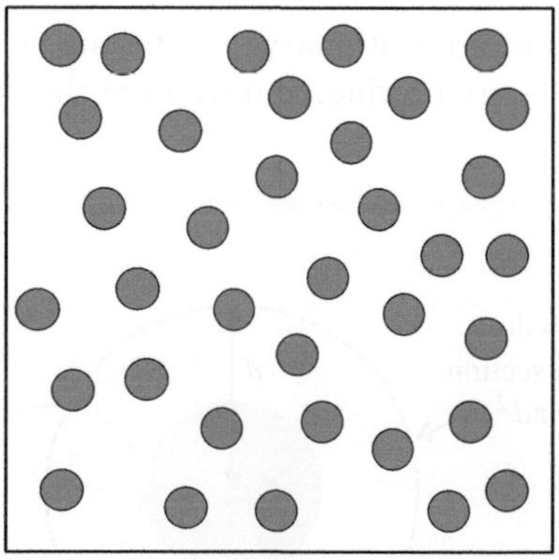

Figure 4.3 The view that a moving molecule would see as it moves through a gas. The opaque circles represent the cross sections of the other molecules that are assumed here to be stationary. The number density of the molecules and the length of the gas column are such that there is a negligible probability that the cross sectional areas of two stationary molecules overlap. The mean free path of the molecule depends on the number density and the collision cross section of the molecules.

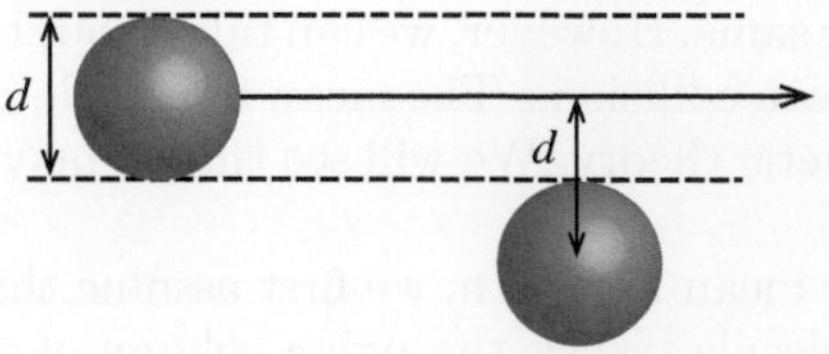

Figure 4.4 There will be a collision between two molecules if the distance between their centres is less than d, where d is the diameter of the molecules, which are modelled as rigid spheres.

the *collision cross section* of the molecules. Looking at Figure 4.4, we see that there will be a collision if the distance between the centres of the two molecules is less than d, where d is the molecular diameter. We thus define the collision cross section, denoted by the symbol σ, to be πd^2; see Figure 4.5. Again the larger the collision cross section, the smaller λ will be. We thus have

$$\lambda \propto \frac{1}{n\sigma}, \tag{4.1}$$

where $\sigma = \pi d^2$. Notice that both sides of this expression have the dimension of length. And notice that the speed of the molecules does not enter into this expression for the mean free path.

To find the constant of proportionality in Equation (4.1), we again assume that all the molecules in the gas are stationary except for one that is moving with mean speed $\overline{c}$. In time t, this molecule will sweep out a cylindrical volume $\pi d^2 \overline{c}\, t$ within which it will collide with other molecules. Hence, for number density n, the number of molecules with which it will collide is $n\pi d^2 \overline{c}\, t$. The mean free path is given by

$$\lambda = \frac{\text{total distance travelled in time } t}{\text{total number of collisions in time } t} = \frac{\overline{c}\, t}{n\pi d^2 \overline{c}\, t}. \tag{4.2}$$

Hence, we obtain

$$\lambda = \frac{1}{n\pi d^2}. \tag{4.3}$$

We have assumed that all the molecules are stationary, apart from the moving ones. When an account is taken of the fact that all the molecules are moving, and we take the relative speeds of the molecules into account, the expression becomes

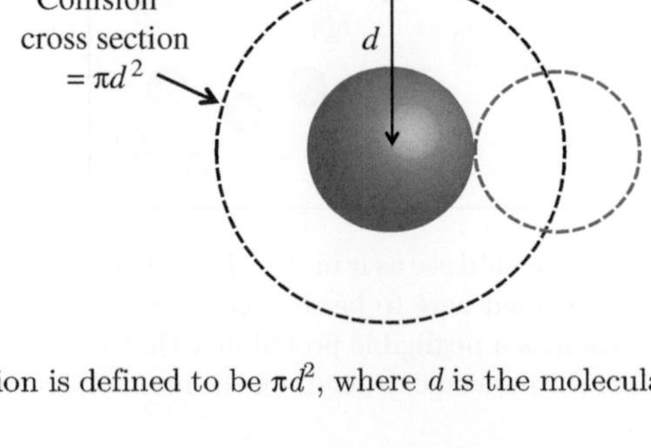

Figure 4.5 The collision cross section is defined to be πd^2, where d is the molecular diameter.

$$\lambda = \frac{1}{\sqrt{2}\, n\pi d^2}. \tag{4.4}$$

We take the example of air molecules at atmospheric pressure and room temperature. Taking pressure P to be 1.01×10^5 Pa and the diameter d to be 0.35 nm, and using the form of the ideal gas equation $PV = NkT$, where $n = N/V$, we have

$$n = \frac{N}{V} = \frac{P}{kT}, \tag{4.5}$$

and hence,

$$\begin{aligned} \lambda &= \frac{1}{\sqrt{2}\, n\pi d^2} = \frac{kT}{\sqrt{2} P\pi d^2} \\ &= \frac{1.38 \times 10^{-23} \times 293}{\sqrt{2} \times 1.01 \times 10^5 \times \pi \times (0.35 \times 10^{-9})^2} = 7.4 \times 10^{-8}\text{m}. \end{aligned} \tag{4.6}$$

This is about 200 molecular diameters. We can compare this value with the average separation of the molecules in the gas. The number density

$$n = \frac{P}{kT} = \frac{1.01 \times 10^5}{1.38 \times 10^{-23} \times 293} = 2.5 \times 10^{25}\ \text{molecules/m}^3.$$

If we imagine the molecules to be the same distance L apart in a cube of volume V, the number of molecules along each edge of the cube will be $(N)^{1/3}$. Hence, $(N)^{1/3} L = (V)^{1/3}$, or $L = (V/N)^{1/3}$. This gives

$$L = \left(\frac{1\ \text{m}^3}{2.5 \times 10^{25}}\right)^{1/3} = 3.5 \times 10^{-9}\ \text{m},$$

which is 10 molecular diameters. Hence, we see that the molecular diameter, the average distance apart and the mean free path are in the approximate ratio of 1 : 10 : 200. The mean free path is clearly much larger than the average molecular separation. We will be able to understand this difference when we discuss the distribution of mean free paths, in Section 4.3.

In the rigid-sphere model we are using, the mean free path of the molecules does not involve the mean speed $\bar{c}$. However, the number of collisions a molecule makes per second does involve $\bar{c}$. On average the number of collisions a *single* molecule makes per second is just $\bar{c}/\lambda$. As there are n molecules per unit volume and each of them makes $\bar{c}/\lambda$ collisions per unit time, the total number of collisions occurring per unit volume per unit time is $1/2(n\bar{c}/\lambda)$, where the factor of ½ is introduced to prevent each collision from being counted twice. This is called the *collision frequency*. From Equation (3.29), we have

$$\bar{c} = \left(\frac{8kT}{\pi m}\right)^{1/2},$$

and substituting for λ from Equation (4.4), we find that the collision frequency is

$$n^2 \pi d^2 \left(\frac{4kT}{\pi m}\right)^{1/2}.$$

Using appropriate values for air at atmospheric pressure and room temperature, we obtain a collision frequency

$$\left(2.6\times10^{25}\right)^2\times\pi\times\left(0.35\times10^{-9}\right)^2\left(\frac{4\times1.38\times10^{-23}\times293}{\pi\times30\times1.66\times10^{-27}}\right)^{1/2}$$
$$=8\times10^{34}\ \text{collisions/s·m}^3,$$

which is indeed a huge number.

4.3 The distribution of free paths

The distance a molecule travels between successive collisions varies; there is a distribution of free paths. In this section, we obtain this distribution, and from this, we will be able to determine the probability that a molecule will travel a certain distance *without* a collision.

Let the probability that a molecule travels distance l without collision be $P_S(l)$, where we emphasise that $P_S(l)$ is an *absolute* probability. And let the probability that a molecule travels distance $l+\mathrm{d}l$ without a collision be $P_S(l+\mathrm{d}l)$. If a molecule travels distance l without a collision, it may then have a collision within a further distance $\mathrm{d}l$. If we assume that the probability of a collision in length $\mathrm{d}l$ is proportional to $\mathrm{d}l$, we can say that the probability of a collision in length $\mathrm{d}l$ is equal to $\alpha\mathrm{d}l$, where α is a constant. This constant depends on number density n and collision cross section σ. Then, the probability of *no* collision in length $\mathrm{d}l$ is equal to $(1-\alpha\mathrm{d}l)$. The probability that a molecule will have a collision in distance $\mathrm{d}l$ is independent of the distance l from its previous collision. To make an analogy, if we toss a coin five times, the chance of getting heads on the sixth toss is independent of the results of the previous five tosses. We thus have

$$\begin{pmatrix}\text{Probability of}\\ \textit{no}\ \text{collision in}\ l\end{pmatrix}\times\begin{pmatrix}\text{Probability of}\\ \textit{no}\ \text{collision in}\ \mathrm{d}l\end{pmatrix}=P_S(l)(1-\alpha\mathrm{d}l). \tag{4.7}$$

But this is just equal to $P_S(l+\mathrm{d}l)$. Hence,

$$P_S(l+\mathrm{d}l)=P_S(l)(1-\alpha\mathrm{d}l). \tag{4.8}$$

Rearranging this expression, we obtain

$$\frac{P_S(l+\mathrm{d}l)-P_S(l)}{\mathrm{d}l}=-\alpha P_S(l).$$

By definition

$$\frac{\mathrm{d}}{\mathrm{d}l}[P_S(l)]=\frac{P_S(l+\mathrm{d}l)-P_S(l)}{\mathrm{d}l}.$$

Hence,

$$\frac{\mathrm{d}}{\mathrm{d}l}[P_S(l)]=-\alpha P_S(l).$$

As $P_S = 1$ for $l = 0$, we have

$$\int_1^{P_S} \frac{dP_S}{P_S} = -\alpha \int_0^l dl,$$

and hence

$$P_S(l) = e^{-\alpha l}. \tag{4.9}$$

$P_S(l)$ is the absolute probability of a molecule *surviving* a distance l without collision. We now find the distribution function $P[l]$, where $P[l]dl$ gives the probability of a molecule having a free path between l and $l + dl$. As usual, we use square brackets to indicate a distribution function.

The probability $P[l]dl$ of a free path being within the range l to $l + dl$ is equal to

$$\begin{pmatrix}\text{The probability of } no \\ \text{collision in distance } l\end{pmatrix} \times \begin{pmatrix}\text{The probability of a} \\ \text{collision in distance } dl\end{pmatrix} = P_S(l)\alpha dl. \tag{4.10}$$

Hence, $P[l]dl = P_S(l)\alpha dl$, and $P[l] = \alpha P_S(l)$. The mean free path is equal to

$$\int_0^\infty l\,P[l]dl = \alpha \int_0^\infty l P_S(l) dl = \alpha \int_0^\infty l\,e^{-\alpha l} dl = \frac{1}{\alpha},$$

using a standard integral. Hence, $\alpha = 1/\lambda$, and

$$P[l] = \frac{P_S(l)}{\lambda},$$

giving

$$P[l]dl = \frac{e^{-l/\lambda}}{\lambda} dl. \tag{4.11}$$

Note that $P[l]$ has the dimensions of probability per unit length, as it must. The function $P[l]$ is shown graphically in Figure 4.6, and we see that it reduces exponentially with distance l. In this figure, l is

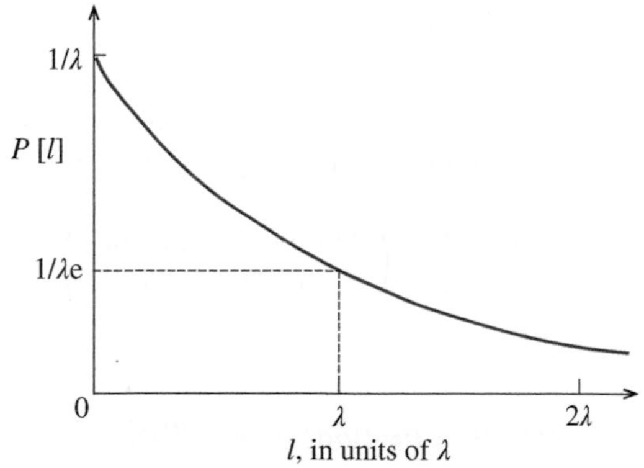

Figure 4.6 The exponential shape of the distribution function $P[l]dl$, which gives the probability of a molecule having a free path between l and $l + dl$. In the figure, l is measured in units of mean free path λ. The most probable free path is zero. However, because of the exponential shape of the distribution, there is the probability of large values of free path, which contribute to the mean value.

measured in units of λ. Perhaps surprisingly, the most probable free path is zero. Figure 4.6 also explains the difference between the value of the mean free path and the average distance between the molecules. Because of the exponential shape of the distribution, there is the probability of large values of the free path, which contribute to the mean free path. From Equation (4.9), with $\alpha = 1/\lambda$, we have

$$P_S(l) = e^{-l/\lambda}. \tag{4.12}$$

Hence, the probability of a molecule travelling a distance of at least one free path without a collision is $e^{-1} = 0.37$. Or equivalently, 63% of the molecules will have a collision within one mean free path length.

Worked example

The probability of a radioactive nucleus decaying in time interval dt is αdt, where α is a constant that is characteristic of the particular nucleus. Obtain expressions for the distribution of decay times and the mean decay time.

Solution

The solution is exactly analogous to that for the mean free path of a molecule. We let the probability of a nucleus *not* decaying in time t be $P_S(t)$. Then the probability of a nucleus not decaying in time $t + dt$ is $P_S(t + dt)$. We thus have

$$\begin{pmatrix}\text{Probability of}\\ \text{no decay in } t\end{pmatrix} \times \begin{pmatrix}\text{Probability of}\\ \text{no decay in } dt\end{pmatrix} = P_S(t)(1 - \alpha dt),$$

which is equal to $P_S(t + dt)$. Hence,

$$P_S(t + dt) = P_S(t)(1 - \alpha dt).$$

From which,

$$\frac{d}{dt}[P_S(t)] = -\alpha P_S(t).$$

The solution to this differential equation is $P_S(t) = e^{-\alpha t}$, cf. Equation (4.9).

The probability $P[t]dt$ of a decay being within the range t to $t + dt$ is equal to

$$\begin{pmatrix}\text{The probability of no}\\ \text{decay in time } t\end{pmatrix} \times \begin{pmatrix}\text{The probability of}\\ \text{decay in time } dt\end{pmatrix} = P_S(t)\alpha dt.$$

Hence,

$$P[t]dt = P_S(t)\alpha dt = \alpha e^{-\alpha t}dt.$$

The mean lifetime is

$$\alpha\int_0^\infty t e^{-\alpha t}dt = \frac{1}{\alpha}.$$

$P_S(t)$ is the probability that a nucleus does not decay in time t. Suppose that we have N_0 nuclei at time $t = 0$. It follows that the number of nuclei that have not decayed after time t is $N(t) = N_0 e^{-at}$, which is the familiar law of radioactive decay. We see that quite different physical situations can be described by analogous differential equations.

4.4 Diffusion

Diffusion is the transport of particles due to a difference in particle concentration. Diffusion can take place in gases, liquids, and also solids. A pictorial representation of gaseous diffusion is presented in Figure 4.7. Figure 4.7a shows a gas container with a partition separating it into two equal parts. Molecular nitrogen is contained to the right of the partition and carbon monoxide to the left. Importantly, both gases are at the same pressure, i.e. the number density is constant throughout the container. Imagine that the partition is suddenly removed. The nitrogen gas would start to diffuse into the region containing the carbon monoxide gas and vice versa. Note that since the two gases are at the same pressure, the movement of molecules is not due to a pressure gradient. Moreover, it is not due to any other bulk movement of the gas such as by convection. The movement is entirely due to the diffusion of the molecules. As time progresses, the two gases would diffuse further into the opposite sides of the container as illustrated by Figure 4.7b. Eventually, after a very long time, the two gases would become completely mixed as illustrated by Figure 4.7c. The concentrations of the two gases at the respective times are illustrated in Figure 4.8.

4.4.1 Fink's law of diffusion

The number of molecules crossing the unit area per second is called the flux J. It is found experimentally that the flux is directly proportional to the concentration gradient $\partial n/\partial x$:

$$J \propto -\frac{\partial n}{\partial x}. \tag{4.13}$$

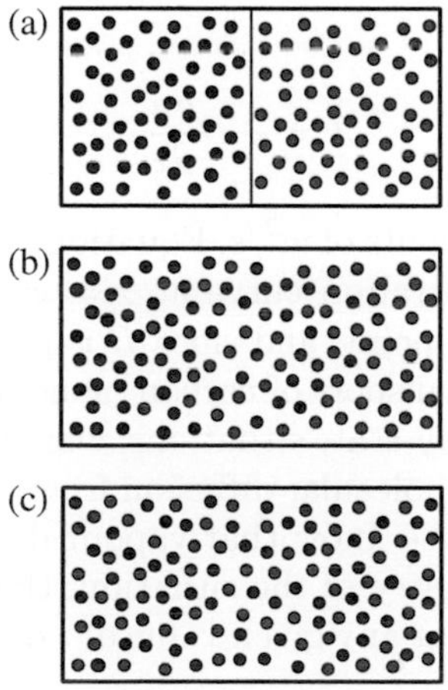

Figure 4.7 (a) Shows a gas container with a partition that separates it into two equal parts. Molecular nitrogen is contained to the right of the partition and carbon monoxide to the left. Importantly, both gases are at the same pressure. When the partition is removed, the nitrogen starts to diffuse into the region containing the carbon monoxide and vice versa, as illustrated by figure (b). Eventually, the two gases become completely mixed as in figure (c).

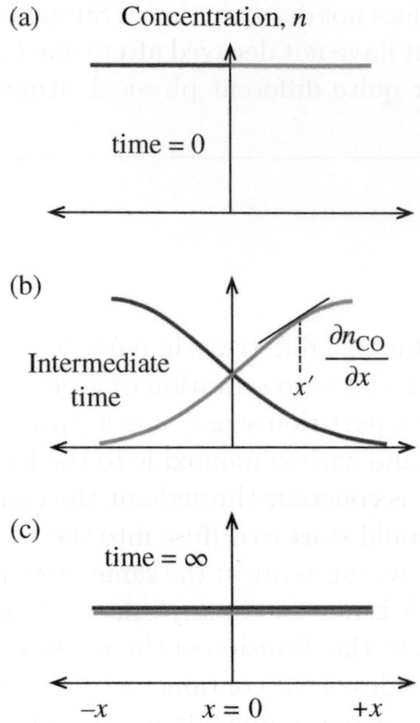

Figure 4.8 (a) Concentrations of nitrogen and carbon monoxide gases at the time when the partition separating the two parts of the container is suddenly removed. Note that since the two gases are at the same pressure, the movement of molecules is not due to a pressure gradient. (b) At some intermediate instant of time, the concentrations of the two gases varies with distance as illustrated. (c) Eventually, after a very long time, the two gases become completely mixed and there is uniform concentration of both gases throughout the container.

This equation describes the simplest case where J varies with respect to one coordinate only, which we will call the x-axis. The minus sign ensures that the direction of the flux is from high to low concentration. Hence, we can write

$$J = -D\frac{\partial n}{\partial x}, \tag{4.14}$$

where D is the *diffusion coefficient*, with units of m^2/s. Equation (4.14) is known as *Fink's law of diffusion.* In Figure 4.8b, the slope of the concentration curve at $x = x$ for the carbon monoxide molecules is shown. Since $\partial n_{CO}/\partial x$ is positive at this point, and since D is positive, it follows that the minus sign in Equation (4.14) indicates that the net transport of carbon monoxide molecules is in the $-x$ direction.

Equation (4.14) is applicable to steady-state situations, which do not change with time. As an example of a steady-state situation, consider molecular diffusion from a mothball (naphthalene) in a stagnant cupboard, as illustrated in Figure 4.9. We know that vapour diffuses from a mothball because we can smell it. Because of the spherical symmetry of the situation, the flux depends on just one coordinate; the radial distance r. Moreover, the flux at any given value of r does not change with time. Hence, Fink's law becomes

$$J(r) = -D\frac{dn(r)}{dr}. \tag{4.15}$$

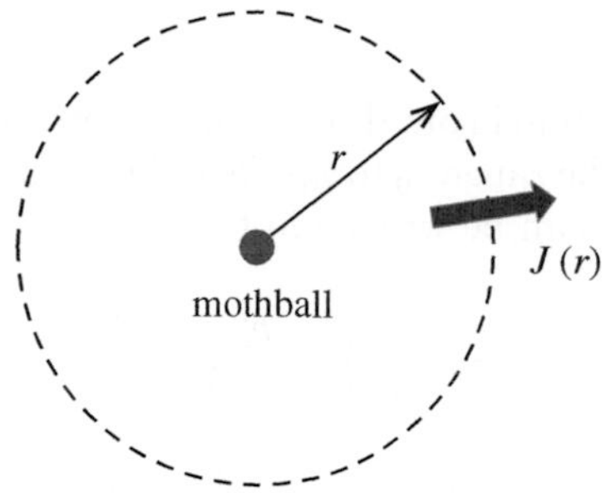

Figure 4.9 An example of steady-state molecular diffusion, where gas diffuses from a mothball in a stagnant cupboard. Because of the spherical symmetry of the situation, the flux of molecules depends on just one coordinate; the radial distance r. Moreover, the flux at any given value of r does not change with time.

where $J(r)$ is the flux and $n(r)$ is the molecular concentration and we have used a full differential as the flux depends on only the single variable r. The total flux out of a spherical surface of radius r is $4\pi r^2 J(r)$, which is equal to the total number N of molecules per second passing through that surface. This gives $J(r) = \frac{N}{4\pi r^2}$.

N must have the same value for all values of r, and hence $J(r)$ varies as $1/r^2$.

Substituting for $J(r)$ in Fink's law we obtain

$$N = -4\pi r^2 D \frac{dn}{dr}, \tag{4.16}$$

giving

$$dn = -\frac{N}{4\pi D r^2} dr,$$

which has the solution

$$n(r) = \frac{N}{4\pi D r} + \text{constant}.$$

Since $n = 0$ at $r = \infty$, the constant $= 0$ and

$$n(r) = \frac{N}{4\pi D r}. \tag{4.17}$$

If we assume that the number density $n(R)$ at the radius R of the mothball is given by the saturated vapour pressure P_{VP} of the gas emitted by the mothball, $n(R) = P_{VP}/kT$. Then,

$$N = 4\pi DR \frac{P_{VP}}{kT}.$$

Appropriate values of D, R, and P_{VP} are 1.0×10^{-5} m^2/s, 0.01 m, and 140 Pa, respectively, giving

$$N = \frac{4\pi \times 1.0 \times 10^{-5} \times 0.01 \times 140}{1.38 \times 10^{-23} \times 293} = 4.4 \times 10^{16} \text{ molecules/s}.$$

4.4.2 Taylor's theorem

We will make much use of Taylor's theorem in our discussion of transport properties. Taylor's theorem says that if $f(x)$ is a continuous function in the range x_0 to $x_0 + h$ and possesses derivatives of all orders at $x = x_0$, then Taylor's theorem states that $f(x)$ can be written in the form

$$f(x_0 + h) = f(x_0) + \frac{h}{1!}\left(\frac{\mathrm{d}f}{\mathrm{d}x}\right) + \frac{h^2}{2!}\left(\frac{\mathrm{d}^2 f}{\mathrm{d}x^2}\right) + \text{higher terms,} \tag{4.18}$$

where the derivatives $\mathrm{d}f/\mathrm{d}x$, etc., are evaluated at $x = x_0$. All of the functions we will be dealing with can be expanded in this way. Moreover, in our discussions of the three transport properties we describe, the relevant value of h is sufficiently small that we will be able to make the approximation

$$f(x_0 + h) = f(x_0) + h\left(\frac{\mathrm{d}f}{\mathrm{d}x}\right). \tag{4.19}$$

This result is illustrated graphically in Figure 4.10.

4.4.3 The diffusion equation

We now consider the more general case where the flux is a function of both position and time and so we write it as $J(x, t)$. To illustrate the general case, Figure 4.11 shows a thin slice in a volume of gas. The slice has area A, and lies between x and $x + \mathrm{d}x$, where the slice thickness $\mathrm{d}x$ is very much smaller than the dimensions of the gas container. The flux slowly varies in the x direction. If $J(x)$ is the flux at x, then using Taylor's theorem, we can say that the flux at $x + \mathrm{d}x$ is

$$J(x + \mathrm{d}x) = J(x) + \mathrm{d}x\left(\frac{\partial J}{\partial x}\right).$$

Then, for area A, we have

$$AJ(x) - AJ(x + \mathrm{d}x) = -A\mathrm{d}x\left(\frac{\partial J}{\partial x}\right).$$

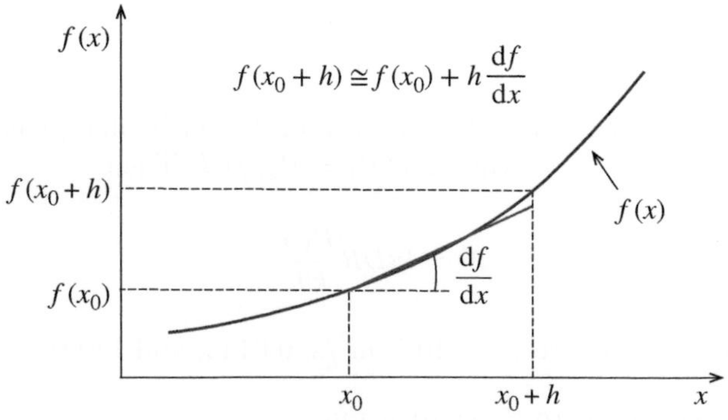

Figure 4.10 A graphical illustration of Taylor's theorem.

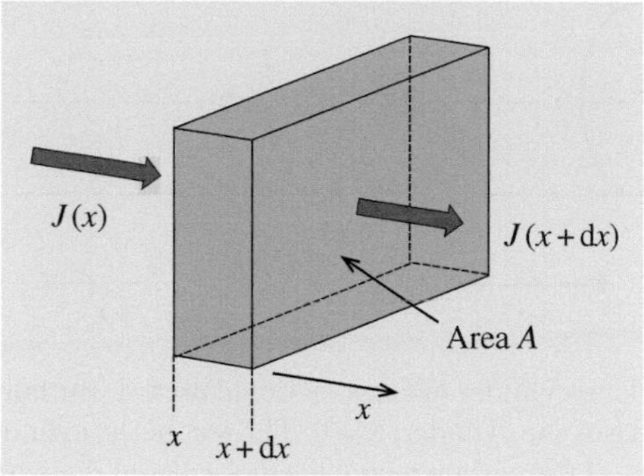

Figure 4.11 The figure shows a thin slice of gas of area A between x and $x+\mathrm{d}x$, where the slice thickness $\mathrm{d}x$ is very much smaller than the dimensions of the gas container.

This is just the rate of increase in the number of molecules in the slice, which is $\frac{\partial}{\partial t}(nA\mathrm{d}x)$. Therefore,

$$-A\mathrm{d}x\left(\frac{\partial J}{\partial x}\right)=\frac{\partial n}{\partial t}A\mathrm{d}x,$$

giving

$$-\frac{\partial J}{\partial x}=\frac{\partial n}{\partial t}. \tag{4.20}$$

This important equation is called the *continuity equation.* We also have Fick's law,

$$J=-D\frac{\partial n}{\partial x}.$$

Substituting Fick's law in the continuity equation gives

$$\frac{\partial n}{\partial t}=-\frac{\partial}{\partial x}\left(-D\frac{\partial n}{\partial x}\right)=D\frac{\partial^2 n}{\partial x^2},$$

or

$$\frac{\partial n}{\partial t}=D\frac{\partial^2 n}{\mathrm{d}x^2}. \tag{4.21}$$

This is the important one-dimensional *diffusion equation.*

As an example of the application of the one-dimensional diffusion equation, we imagine a very long cylinder of cross sectional area A. Initially, N molecules of carbon monoxide are situated in a thin slice in the middle of the cylinder at $x=0$; see Figure 4.12. The rest of the cylinder contains nitrogen molecules. Again we emphasise that the pressure, i.e. the total number of molecules per unit volume remains constant throughout the diffusion process. We are interested in how the concentration of the carbon monoxide molecules varies with respect to distance x as a function of time t.

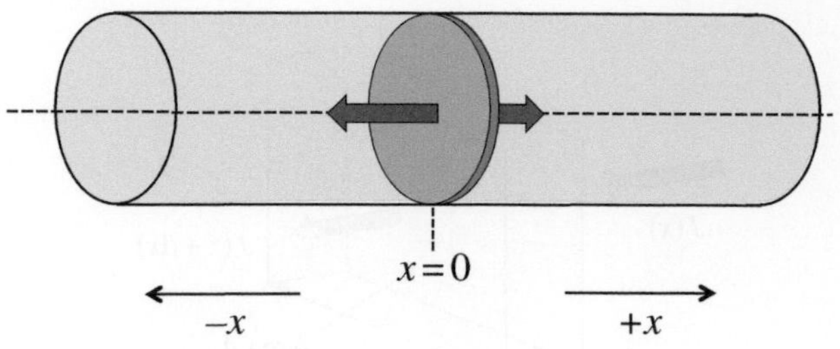

Figure 4.12 The figure shows a very long cylinder of cross sectional area A. Initially, N molecules of carbon monoxide are situated in a thin slice in the middle of the cylinder; $x = 0$. The rest of the cylinder contains nitrogen molecules. We are interested in how the concentration of the carbon monoxide molecules varies with respect to distance x as a function of time t. Importantly, the total number of molecules per unit volume remains constant throughout the diffusion process.

The appropriate solution to the diffusion equation for this example has the form $n(x, t) = (B/t^{1/2})\mathrm{e}^{-x^2/bt}$, where $n(x, t)$, is the concentration of the carbon monoxide molecules, and B and b are constants. We can verify this by substituting the solution into the diffusion equation. Thus

$$\frac{\partial n}{\partial t} = -\frac{B}{2}\left(\frac{\mathrm{e}^{-x^2/bt}}{t^{3/2}}\right) + \frac{B}{b}\left(\frac{x^2\mathrm{e}^{-x^2/bt}}{t^{5/2}}\right);$$

$$\frac{\partial^2 n}{\partial x^2} = \frac{4B}{b^2}\left(\frac{x^2\mathrm{e}^{-x^2/bt}}{t^{5/2}}\right) - \frac{2B}{b}\left(\frac{\mathrm{e}^{-x^2/bt}}{t^{3/2}}\right).$$

Substituting $\partial n/\partial t$ and $\partial^2 n/\partial x^2$ in the diffusion equation and equating coefficients of $\left(\left[x^2\mathrm{e}^{-x^2/bt}\right]/t^{5/2}\right)$ and $\left(\left[\mathrm{e}^{-x^2/bt}\right]/t^{3/2}\right)$ gives $b = 4D$ in both cases. The value of B depends on the total number of molecules present and is obtained from the normalisation condition

$$\int_{-\infty}^{+\infty} n(x, t)A\mathrm{d}x = N,$$

which must apply for any time t. Hence,

$$\frac{AB}{t^{1/2}}\int_{-\infty}^{+\infty} \mathrm{e}^{-x^2/4Dt}\mathrm{d}x = N.$$

Using a standard integral, we readily obtain $B = N/(A[4\pi D]^{1/2})$. Hence,

$$n(x, t) = \frac{N}{A(4\pi Dt)^{1/2}}\mathrm{e}^{-x^2/4Dt}. \tag{4.22}$$

We recognise this function as a Gaussian. And we see that the amplitude of the Gaussian reduces as t increases, and its width, determined by $4Dt$, reduces. This evolution of the concentration of the carbon monoxide molecules with time is illustrated in Figure 4.13.

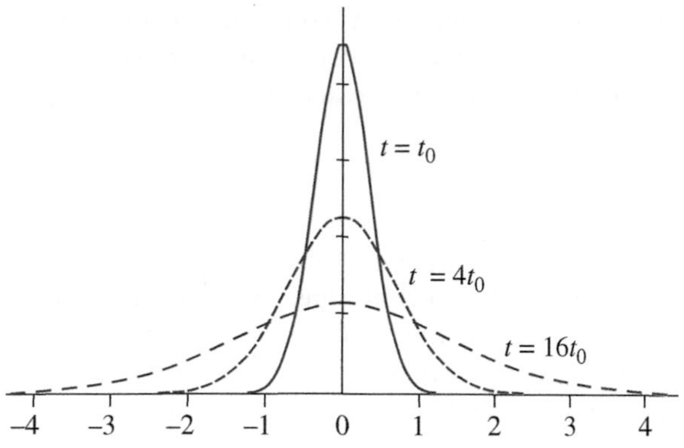

Figure 4.13 The evolution of the concentration of carbon monoxide molecules with time for the example shown in Figure 4.12.

The number of carbon monoxide molecules between x and $x + \mathrm{d}x$, at time t is

$$n(x,t)A\mathrm{d}x = \frac{N}{A(4\pi Dt)^{1/2}}\mathrm{e}^{-x^2/4Dt}A\mathrm{d}x.$$

Hence, the *fraction* of CO molecules between x and $x + \mathrm{d}x$, at time t is

$$\frac{n(x,t)A\mathrm{d}x}{N} = \left(\frac{1}{4\pi Dt}\right)^{1/2}\mathrm{e}^{-x^2/4Dt}\mathrm{d}x.$$

We can equally well interpret this as the probability distribution $P[x,t]\mathrm{d}x$ for a single CO molecule being between x and $x + \mathrm{d}x$ at time t. Hence,

$$P[x,t]\mathrm{d}x = \left(\frac{1}{4\pi Dt}\right)^{1/2}\mathrm{e}^{-x^2/4Dt}\mathrm{d}x. \tag{4.23}$$

Then the mean squared distance $\overline{x^2}$ travelled in time t is

$$\int_{-\infty}^{+\infty} x^2 P[x]\mathrm{d}x = \left(\frac{1}{4\pi Dt}\right)^{1/2}\int_{-\infty}^{+\infty} x^2\mathrm{e}^{-x^2/4Dt}\mathrm{d}x = 2Dt,$$

where we have made use of a standard integral. Hence, the root-mean-squared distance

$$\left(\overline{x^2}\right)^{1/2} = \sqrt{2Dt}. \tag{4.24}$$

We emphasise the dependence of the distance on the square root of t. This dependence on $\sqrt{t}$ is a general feature of diffusion processes.

The diffusion constant for the carbon monoxide gas in nitrogen gas has the value $D = 1.5 \times 10^{-5}\ \mathrm{m^2/s}$. Hence, in one hour, the carbon monoxide molecules 'spread' a distance of $\sim(1.5 \times 10^{-5} \times 60^2)^{1/2} \sim 0.25$ m. This is despite the fact that the molecules have a mean speed of several hundred metres per second.

The explanation is that the molecule makes numerous collisions with other molecules and after each collision, its direction of travel is completely random. Of course, in practice, there is the bulk movement of air due to wind and convection currents and these can cause much greater displacements of the molecules.

4.4.4 The kinetic theory of diffusion

We now consider diffusion from the microscopic viewpoint where we consider the movement of the molecules. We consider an imaginary plane X of area A in a gas, as shown in Figure 4.14. The plane is normal to the x direction. We consider the situation where the concentration of molecules to the right of the plane is greater than the concentration to the left of the plane. Molecules will diffuse through the plane in both directions. However, as there is a higher concentration of molecules to the right of the plane, there will be a net flow of molecules from right to left.

We make the assumption that every molecule that passes through plane X makes its last collision with another molecule at a distance of one mean path length λ from the plane, i.e. at imaginary planes P or Q; see Figure 4.14. Since λ is very much smaller than the size of a laboratory gas container, the molecular densities at P and Q can be taken to be $n - \lambda(\mathrm{d}n/\mathrm{d}x)$ and $n + \lambda(\mathrm{d}n/\mathrm{d}x)$, respectively, where $\mathrm{d}n/\mathrm{d}x$ is the concentration gradient.

In a gas, the molecules move in all directions. Since all directions of travel are equally likely, it follows that one-sixth of all molecules will be moving in any one direction, say the $+x$ direction. Then, the flux of molecules per unit area in a given direction is $n\bar{c}/6$, where n is the molecular concentration and $\bar{c}$ is the mean speed. This is called the one-sixth model. Using this model, we have the number of molecules from plane P crossing plane X per second is

$$\frac{1}{6}\bar{c}A\left[n - \lambda\frac{\mathrm{d}n}{\mathrm{d}x}\right].$$

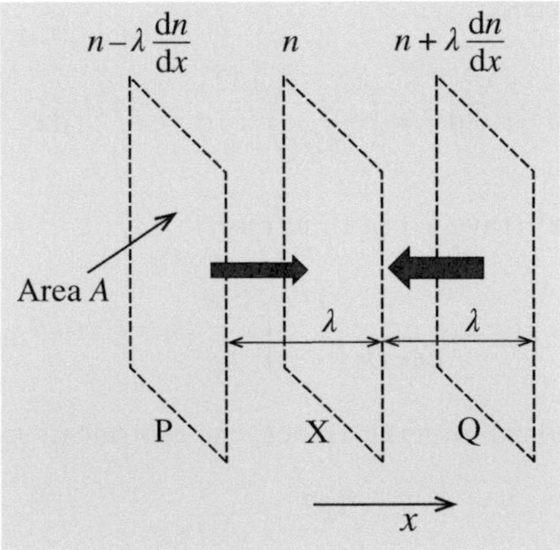

Figure 4.14 The figure shows an imaginary plane X in a gas. This plane, of area A, is normal to the x direction. Molecules diffuse through the plane in both directions. However, as there is a higher concentration of molecules to the right of the plane, there is a net flow of molecules from right to left. We make the assumption that every molecule that passes through plane X makes its last collision with another molecule at a distance of one mean path length λ from the plane, i.e. at imaginary planes P or Q.

Similarly, the number of molecules from plane Q crossing plane X per second is

$$\frac{1}{6}\overline{c}A\left[n+\lambda\frac{\mathrm{d}n}{\mathrm{d}x}\right].$$

Then, the net flux of molecules crossing through plane X in the $+x$ direction is

$$\frac{1}{6}\overline{c}A\left[n-\lambda\frac{\mathrm{d}n}{\mathrm{d}x}\right]-\frac{1}{6}\overline{c}A\left[n+\lambda\frac{\mathrm{d}n}{\mathrm{d}x}\right]=-\frac{1}{3}\lambda\overline{c}A\frac{\mathrm{d}n}{\mathrm{d}x}.$$

And the net flux of molecules per unit area is

$$-\frac{1}{3}\lambda\overline{c}\frac{\mathrm{d}n}{\mathrm{d}x}. \tag{4.25}$$

This result has the same form as the empirical law of Fink, $J=-D(\mathrm{d}n/\mathrm{d}x)$. Comparing the two expressions, we obtain

$$D=\frac{1}{3}\overline{c}\lambda. \tag{4.26}$$

Notice that this expression relates the macroscopic quantity D to the microscopic quantities $\overline{c}$ and λ of the gas.

The value of D depends on the particular diffusing molecules and on the gas molecules through which they diffuse; for example, carbon monoxide molecules diffusing through a gas of nitrogen. Of course, these can both be the same, in which case the process is called *self-diffusion*. For the example of self-diffusion in argon, at room temperature and atmospheric pressure, we have

$$\overline{c}=\left(\frac{8kT}{\pi m}\right)^{1/2}=\left(\frac{8\times 1.38\times 10^{-23}\times 293}{\pi\times 40\times 1.66\times 10^{-27}}\right)^{1/2}=394\,\mathrm{m/s}.$$

Taking $d=0.38\times 10^{-9}$ m for argon,

$$\lambda=\frac{1}{\sqrt{2}n\sigma}=\frac{kT}{\sqrt{2}P\pi d^2}=\frac{1.38\times 10^{-23}\times 293}{\sqrt{2}\times 1.05\times 10^5\times\pi\times\left(0.38\times 10^{-9}\right)^2}=6\times 10^{-8}\ \mathrm{m}.$$

Then,

$$D=\frac{1}{3}\times 394\times 6\times 10^{-8}=0.8\times 10^{-5}\,\mathrm{m^2/s}.$$

This value of D, obtained from the kinetic theory of gasses, is in reasonably good agreement with the experimentally measured value of 1.8×10^{-5} $\mathrm{m^2/s}$.

We have described the case of molecular diffusion. However, the diffusion equation and Fink's law have much wider applications. For example, they are used to model the diffusion of neutrons in the moderator of a nuclear reactor, and the diffusion of electrons and holes in the semiconductor materials of a solar cell.

4.5 Thermal conduction

Diffusion is the transport of particles due to a concentration gradient. Thermal conduction is also a diffusion process. But now it is the transport of thermal energy from a hotter to a colder region, due to a temperature gradient. We are familiar with the idea of heat flowing from a hot to a cold body as in, for example, a copper rod whose two ends are at different temperatures. For the one-dimensional case, the flow or flux of heat Q is $q = (\mathrm{d}Q/\mathrm{d}t)$ per unit area and has units of watts per square metre, $\mathrm{W/m^2}$. It is found empirically that the flux of heat per unit area is directly proportional to a temperature gradient. So, for the one-dimensional case, we can write

$$q = -\kappa \frac{\mathrm{d}T}{\mathrm{d}x}, \tag{4.27}$$

where T is temperature and κ is the *coefficient of thermal conductivity*, which has units $\mathrm{W/m \cdot K}$. The negative sign arises because heat flows from a higher to a lower temperature. This equation applies under steady-state conditions where the heat flow does not vary with time. It is found to be well obeyed in practically all gases, liquids, and isotropic solids, and is known as *Fourier's law.*

We note the similarity of Equation (4.27) with Fink's diffusion equation, Equation (4.14):

$$J = -D\frac{\mathrm{d}n}{\mathrm{d}x}.$$

And, as in the case of diffusion, we can use the kinetic theory of gases to deduce an expression for the thermal conductivity of a gas in terms of molecular parameters such as mean speed and mean free path.

We consider a gas across which there is a temperature gradient. Again we consider an imaginary plane X of area A in the gas through which molecules diffuse from left to right and from right to left; see Figure 4.15. The temperature at plane X is T. As before, we make the assumption that every molecule that

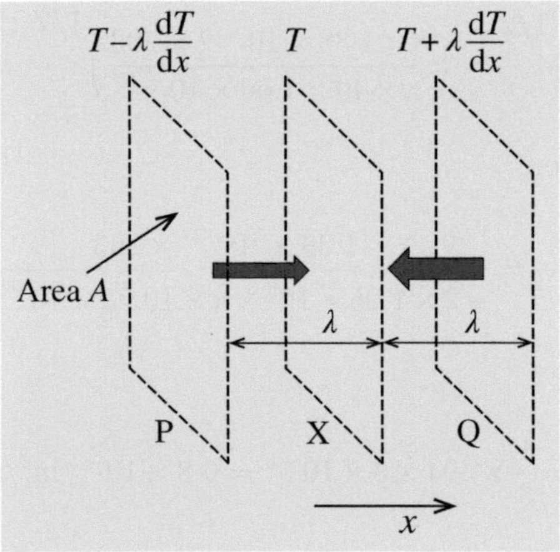

Figure 4.15 The figure shows an imaginary plane X of area A in a gas. This plane is normal to the x direction. Molecules diffuse through the plane in both directions, at equal rates. However, as the molecules to the right of the plane have larger thermal energy than those to the left there is a net flow of thermal energy from right to left. We make the assumption that every molecule that passes through plane X makes its last collision with another molecule at a distance of one mean path length λ from the plane, i.e. at imaginary planes P or Q.

passes through this plane makes its last collision at a distance of one mean free path λ from the plane, i.e. at plane P or Q. Since λ is very much smaller than the size of a typical gas container, we can take the temperatures at planes P and Q to be $T \pm \lambda(\mathrm{d}T/\mathrm{d}x)$, as shown in Figure 4.15.

Molecules will diffuse through plane X in both directions. Since we are assuming that the concentration n is uniform, the number diffusing in both directions will be the same. However, those molecules starting at plane P have a different translational energy than those starting at plane Q. The mean translational energy of molecules that collide in a region at temperature T is $3/2kT$. Therefore, molecules starting at plane P have mean kinetic energy $3k/2[T - \lambda(\mathrm{d}T/\mathrm{d}x)]$, while those starting at plane Q have mean kinetic energy $3k/2[T + \lambda(\mathrm{d}T/\mathrm{d}x)]$. Hence, there is a net transfer or flow of thermal energy across the plane. We assume that the temperature difference across the gas is sufficiently small that we can take the molecular mean speed $\overline{c} \propto T^{1/2}$, to be constant throughout the gas. The net flow of thermal energy, i.e. heat, across the plane in the $+x$ direction is

$$\frac{1}{6}n\overline{c}A\frac{3k}{2}\left[\left(T-\lambda\frac{\mathrm{d}T}{\mathrm{d}x}\right)-\left(T+\lambda\frac{\mathrm{d}T}{\mathrm{d}x}\right)\right] = -\frac{1}{2}n\overline{c}Ak\lambda\frac{\mathrm{d}T}{\mathrm{d}x}, \tag{4.28}$$

where n is the number density. The factor 1/6 arises from the one-sixth model that assumes one-sixth of the molecules will be travelling in a given direction. And the net flow of thermal energy per unit area is

$$-\frac{1}{2}n\overline{c}k\lambda\frac{\mathrm{d}T}{\mathrm{d}x}. \tag{4.29}$$

This has the same form as the empirical law of Fourier:

$$q = -\kappa\frac{\mathrm{d}T}{\mathrm{d}x},$$

and comparing the two expressions, we obtain

$$\kappa = \frac{1}{2}n\overline{c}k\lambda. \tag{4.30}$$

For argon at room temperature and atmospheric pressure, we take: $n = 2.6 \times 10^{25}/\mathrm{m}^3$, $\lambda = 6 \times 10^{-8}$ m, and $\overline{c} = 394$ m/s. With these values, Equation (4.30) predicts a value for κ of 4.2×10^{-3} W/m · K. This compares with the experimentally measured value of 17×10^{-3} W/m · K. We see that the theory gives a result that is of the correct order of magnitude.

Equation (4.30) is for a monotonic gas, like argon. For one mole of a monotonic gas, the molar-specific heat at constant volume is

$$C_\mathrm{V} = \frac{3}{2}N_\mathrm{A}k. \tag{4.31}$$

Substituting this expression in Equation (4.30) gives

$$\kappa = \frac{1}{3}\frac{C_\mathrm{V}}{N_\mathrm{A}}n\overline{c}\lambda. \tag{4.32}$$

We can readily extend this result to polyatomic molecules by using the measured value of the specific heat C_V of the gas. This takes into account other contributions to the total internal energy of the molecules due to rotational and vibrational motion.

The thermal conductivities for gases are much smaller than for, say, metals. For example, the thermal conductivity of air is 16,000 smaller than that of copper. Thermal insulating materials essentially consist of many small pockets of air and this explains their good thermal insulating properties.

4.5.1 Predictions for the thermal conductivity

From the results obtained from kinetic theory, we can make predictions about the behaviour of the thermal conductivity of a gas with respect to pressure and temperature. We recall that $\lambda = 1/(\sqrt{2}\, n\sigma)$ and $\bar{c} = ([8kT]/[\pi m])^{1/2}$. By substituting for λ and $\bar{c}$ in Equation (4.30), we find

$$\kappa \propto \frac{T^{1/2}}{\sigma}. \tag{4.33}$$

Hence, κ is predicted to be independent of n and hence pressure, and to be proportional to the square root of temperature. With respect to the pressure dependence, the explanation is that increasing n increases the number of molecules that transport the thermal energy but this also reduces the average distance a molecule travels before transferring energy to another molecule. And the two effects cancel out. For most gases, thermal conductivity is indeed found to be constant over a very wide range of pressure, from about 10^{-3} bar to about 50 bar. This constancy of thermal conductivity with pressure is one of the unexpected results of kinetic theory. The prediction of the theory at low pressures, below about 10^{-3} bar, does not hold because the mean free path λ of the molecules becomes comparable to the dimensions of the gas container. In that case, it is collisions between the molecules and the walls of the container that dominate, and the role of energy transfer between molecules diminishes. Indeed, in the limiting case of zero pressure, i.e. a vacuum, there is no heat conduction. This is the regime in which a thermos flask works. The prediction at high pressure breaks down because the picture of widely separated molecules no longer holds. However, under ordinary conditions, the prediction that the thermal conduction of a gas is independent of pressure is well obeyed.

The prediction of kinetic theory with respect to the dependence of thermal conductivity κ on $T^{1/2}$ holds less well. Experiment shows that the thermal conductivity increases more rapidly than $T^{1/2}$. This suggests, see Equation (4.33), that the collision cross section σ decreases as the molecular speed increases or that σ increases as the molecular speed decreases. This is evidence for the breakdown of our simple picture of molecules as non-interacting rigid spheres. As we saw in Chapter 2, there are in fact intermolecular forces that operate between molecules when they are close together. Suppose, for example, that two molecules come close together as illustrated pictorially in Figure 4.16. According to the rigid-sphere model, molecule A would simply pass by molecule B in a straight line trajectory, as they do not overlap. However, there is a force of attraction between the two molecules and molecule A is deflected from its straight-line trajectory toward molecule B, so that it actually collides with molecule B. The slower molecule A is travelling, the more time it will spend under the influence of the attractive force and the deflection will be correspondingly bigger, i.e. the *effective* collision cross section will increase. On the other hand, the faster molecule A is travelling, the less time it spends in the vicinity of molecule B and the cross section tends to our previous result πd^2. A useful empirical expression for the effective molecular diameter σ at temperature T in terms of the geometric diameter σ_0 is

$$\sigma^2 = \sigma_0^2\left(1 + \frac{K}{T}\right), \tag{4.34}$$

where $\sigma_0 = \pi d^2$, and K is a constant for the particular gas.

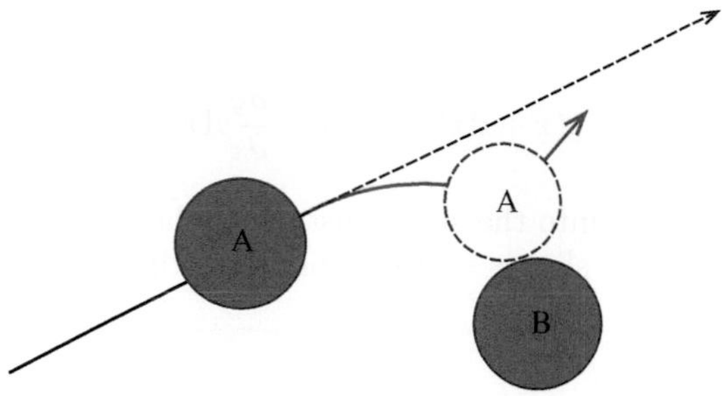

Figure 4.16 According to the rigid-sphere model, molecule A would simply pass by molecule B in a straight-line trajectory, as they do not overlap. However, there is a force of attraction between the two molecules and molecule A is deflected from its straight-line trajectory toward molecule B so that it actually collides with it. The slower molecule A is travelling, the more time it will spend under the influence of the attractive force and the deflection will be correspondingly bigger. On the other hand, the faster molecule A is travelling, the less time it spends in the vicinity of molecule B with the result that the collision cross section tends to πd^2.

4.5.2 The heat equation

Equation (4.27) applies to steady-state situations where the flow of heat does not change with respect to time. However, there are many practical situations where this is not the case. The heat equation is a differential equation that describes the general situation in which the distribution of heat varies with respect to both position and time.

To deduce the heat equation, we again consider a thin slice of gas of area A between x and $x + \mathrm{d}x$; see Figure 4.17. There is a temperature gradient across the slice and we can write

$$T(x + \mathrm{d}x) = T(x) + \frac{\partial T}{\partial x}\mathrm{d}x.$$

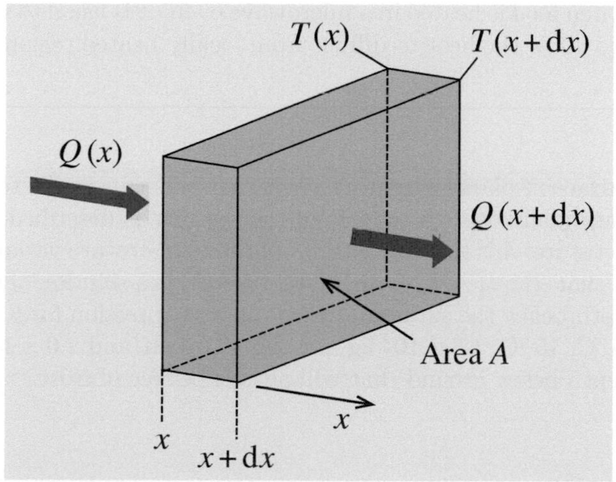

Figure 4.17 The figure shows a thin slice of gas of area A between x and $x + \mathrm{d}x$, across which there is a temperature gradient. The net flux of thermal energy going into the slice must equal the rate of increase of thermal energy in the slice.

and similarly,

$$q(x + \mathrm{d}x) = q(x) + \frac{\partial q}{\partial x}\mathrm{d}x.$$

The net flux of thermal energy going into the slice must equal the rate of increase of thermal energy in the slice. The thermal capacity of the slice is $c\rho A\mathrm{d}x$, where c is the specific heat per unit mass and ρ is the density. Hence, we have

$$A[q(x) - q(x + \mathrm{d}x)] = c\rho A\mathrm{d}x\frac{\partial T}{\partial t}, \tag{4.35}$$

from which we obtain

$$\frac{\partial q}{\partial x} = -\,c\rho\frac{\partial T}{\partial t}. \tag{4.36}$$

This is simply the conservation of energy and is another example of a continuity equation. We recall the empirical equation $q = -\,\kappa\frac{\mathrm{d}T}{\mathrm{d}x}$. Differentiating this equation with respect to x:

$$\frac{\partial q}{\partial x} = -\,\kappa\frac{\partial^2 T}{\partial x^2}.$$

Using this result in Equation (4.36), we obtain

$$\frac{\partial T}{\partial t} = \frac{\kappa}{c\rho}\frac{\partial^2 T}{\partial x^2}. \tag{4.37}$$

This is the *heat equation.*

The quantity κ/c is called the *thermal diffusivity.* It measures the *rate* of heat transfer from a hotter region to a colder region. When food is heated in a microwave oven, it is usual to wait a minute or so before the food is served. This is to allow the heat to diffuse from locally heated regions to the rest of the food.

Worked example

The Sun's rays heat the surface of planet Earth. Assume that this causes the temperature at a point on Earth's surface to vary sinusoidally over the course of the seasons as described by $T(t) = T_0 + A\sin\omega t$, where T_0 is the mean temperature, A is the "amplitude" of the temperature variation and $\omega = 2\pi/\tau$, where the period τ is 1 year. Show that $T(z, t) = T_0 + A\mathrm{e}^{-\beta z}\sin(\omega t - \beta z)$ is a solution of the one-dimensional heat equation, where z is the depth below the surface, and obtain an expression for β. Taking typical values of T_0, A, ρ, κ, and c to be 10 °C, 15 °C, 1.5×10^3 kg/m^3, 2.5 W/m · K and 2.0×10^3 J/kg · K, respectively, determine the minimum depth below ground that will always be free of frost.

Solution

Since the depths below ground that we will consider are tiny compared to Earth's radius, we can assume the ground to be flat and use the one-dimensional heat equation. Differentiating the given solution with respect to t once and with respect to z twice:

$$\frac{\partial T}{\partial t} = \omega A e^{-\beta z} \cos(\omega t - \beta z),$$

$$\frac{\partial^2 T}{\partial z^2} = 2\beta^2 A e^{-\beta z} \cos(\omega t - \beta z).$$

Substituting for $\partial T/\partial t$ and $\partial^2 T/\partial z^2$ in the heat equation, Equation (4.37), we obtain

$$\beta = \sqrt{\frac{\omega \rho c}{2\kappa}}.$$

For the given values of κ, ρ, and c,

$$\beta = \sqrt{\frac{2\pi \times 1.5 \times 10^3 \times 2.0 \times 10^3}{365 \times 24 \times 60 \times 60 \times 2 \times 2.5}} = 0.346 \text{ m}^{-1}.$$

The maximum and minimum values of $\sin(\omega t - \beta z)$ are ± 1. It follows that the minimum value of $T(z, t)$ at depth z is $T_0 - Ae^{-\beta z}$. For the ground to be frost-free at depth z, the minimum value of $T(z, t) = 0$. Hence, $0 = 10 - 15e^{-\beta z}$.

Then $\beta z = \ln\left(\frac{15}{10}\right) = 0.405$ and $z = 1.17$ m.

At this value of z, we have

$$T(t) = 10 + 15e^{-0.405} \sin(\omega t - 0.405),$$

where we have substituted the values of T_0 and A. The sinusoidal temperature variation at $z = 1.17$ m lags the sinusoidal variation at the surface by the phase angle 0.405 rad. In terms of the period τ, this is a time lag of $\frac{0.405}{2\pi}\tau = 0.065$ years $= 24$ days. Plots of (a) the temperature variation at the Earth's surface and (b) at a distance 1.17 m below the surface are shown in Figure 4.18 for these values.

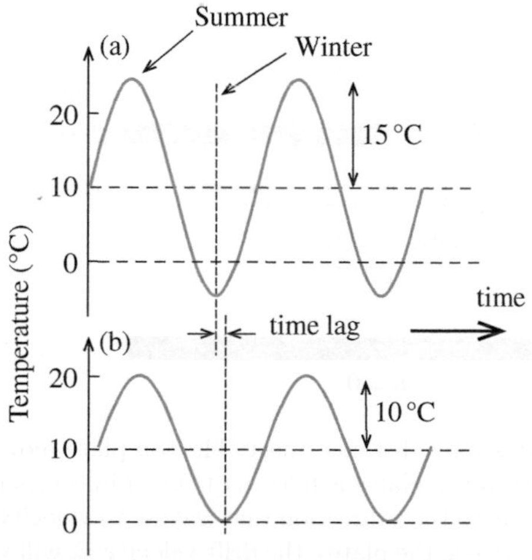

Figure 4.18 (a) Plot of the sinusoidal temperature variation at the Earth's surface over the course of the seasons. (b) Plot of the sinusoidal temperature variation at a depth of 1.17 m, which lags the variation at the Earth's surface by 24 days. At this depth, the temperature of the ground never goes below 0°.

4.6 Viscosity

Viscosity is the third transport property that we can understand on the basis of the kinetic theory of gases. In the case of viscosity, it is the transfer of momentum due to a gradient in molecular velocity. It is a more complicated process than diffusion or heat conduction and we will confine our attention to the steady-state case.

Consider two plates that are some distance apart, as shown in Figure 4.19. The top plate moves with velocity v, while the bottom plate is held stationary. The gap between the plates is filled with gas. Molecules near the top plate are dragged along and have a drift velocity u_x parallel to the x direction that is close to velocity v. But molecules near the bottom plate have a drift velocity close to zero. Between the plates, the drift velocity u_x varies between zero and v. In Figure 4.19, the lengths of the horizontal arrows indicate the magnitude of the drift velocity. For this particular case, the magnitude varies linearly with the distance between the plates. Molecules diffuse between the plates and since the number density n is assumed to be constant, the number of molecules diffusing in both the downward and upward directions is the same. As molecules moving in the direction from the top plate to the bottom plate have higher values of drift velocity u_x than those travelling in the opposite direction, there will be a net transport of momentum towards the bottom plate. The rate of change of momentum is equal to the force F_x that must be applied to the top plate to maintain its motion. The direction of this force is in the direction of the drift velocity. Experimentally, it is found that F_x is proportional to the area A of the plates and the velocity gradient $\mathrm{d}u_x/\mathrm{d}z$. Hence, we have

$$F_x = \eta A \frac{\mathrm{d}u_x}{\mathrm{d}z}, \tag{4.38}$$

where η is called the *coefficient of viscosity* with units Pa · s. (η may be more familiar from its appearance in Stoke's law $F = 6\pi a\eta v$ for a sphere of radius a travelling at velocity v through a fluid.) Equation (4.38) is known as Newton's law of viscosity.

Using the kinetic theory of gases, we can obtain an expression for η in an analogous way to that used to find expressions for the diffusion coefficient D and the thermal coefficient κ. We again consider the diffusion

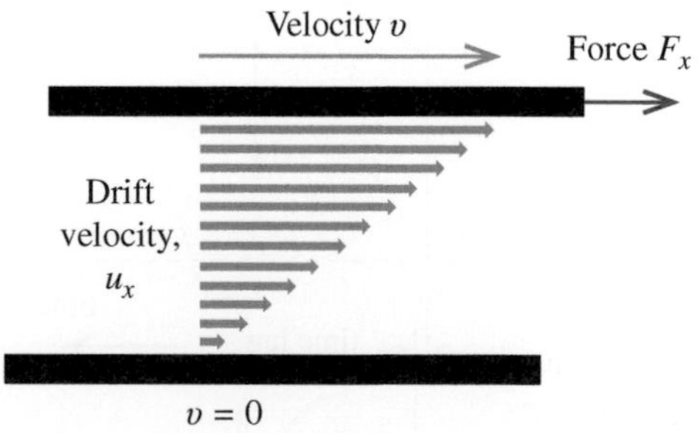

Figure 4.19 The figure shows two plates some distance apart. The top plate moves with velocity v, while the bottom plate is held stationary. The gap between the plates is filled with gas. Molecules near the top plate are dragged along and have a drift velocity u_x parallel to the x direction that is close to velocity v. Molecules near the bottom plate have a drift velocity close to zero. In between the plates, the drift velocity u_x will vary between zero and v. The length of a horizontal arrow indicates the magnitude of the drift velocity. For this particular case, the magnitude varies linearly between the plates.

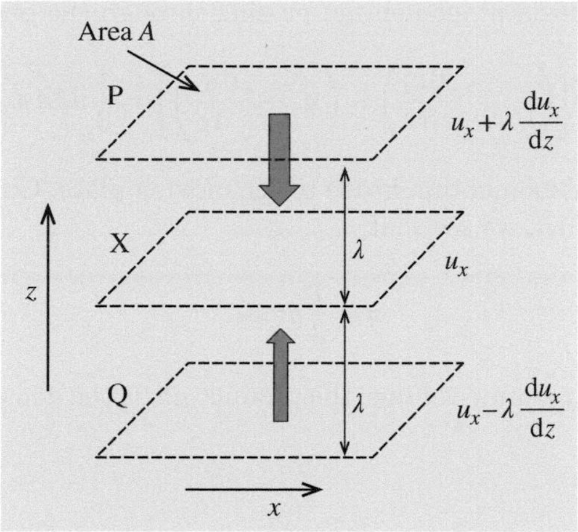

Figure 4.20 The figure shows an imaginary plane X of area A in a gas. This plane is normal to the z direction. Molecules diffuse through the plane in both directions. However, as the molecules coming from plane P have larger horizontal momentum than those coming from plane Q, there is a finite rate of change of horizontal momentum passing through plane X. This rate of change of horizontal momentum is due to the moving plate; see Figure 4.19.

of molecules across an imaginary plane of area A, in this case, a horizontal plane, as illustrated in Figure 4.20. Again, we make the assumption that every molecule that passes through plane X makes its last collision at a distance of one mean path length λ from the plane, i.e. at imaginary planes P or Q. Letting u_x be the drift velocity at plane X, we can say that the drift velocity at plane P, which is a distance λ above plane X is

$$u_x + \lambda \frac{\mathrm{d}u_x}{\mathrm{d}z},$$

and, similarly, the drift velocity at plane Q is

$$u_x - \lambda \frac{\mathrm{d}u_x}{\mathrm{d}z}.$$

The molecules passing through the imaginary plane in the downward direction have a horizontal component of momentum equal to $m(u_x + \lambda[\mathrm{d}u_x/\mathrm{d}z])$, where m is the mass of a molecule. The number of molecules crossing area A per second is $(n\bar{c}A)/6$, where the factor 1/6 comes from the usual one-sixth model. Therefore, the rate of change of horizontal momentum passing through the imaginary plane in the downward direction is

$$m\left(u_x + \lambda \frac{\mathrm{d}u_x}{\mathrm{d}z}\right)\frac{n\bar{c}A}{6}.$$

Similarly, the rate of change of horizontal momentum passing through the imaginary plane in the upward direction is

$$m\left(u_x - \lambda \frac{\mathrm{d}u_x}{\mathrm{d}z}\right)\frac{n\bar{c}A}{6}.$$

Hence, the rate of change of horizontal momentum passing through the plane is

$$\frac{1}{6}n\bar{c}Am\left[\left(u_x+\lambda\frac{\mathrm{d}u_x}{\mathrm{d}z}\right)-\left(u_x-\lambda\frac{\mathrm{d}u_x}{\mathrm{d}z}\right)\right]=\frac{1}{3}n\bar{c}Am\lambda\frac{\mathrm{d}u_x}{\mathrm{d}z}. \tag{4.39}$$

This rate of change of horizontal momentum is due to the moving plate. Comparing Equation (4.39) with the empirical law $F_x = \eta A(\mathrm{d}u_x/\mathrm{d}z)$, we see that

$$\eta = \frac{1}{3}nm\bar{c}\lambda. \tag{4.40}$$

As n is the number of molecules per unit volume, the product nm is the density ρ of the gas. Hence, we can also write Equation (4.40) as

$$\eta = \frac{1}{3}\rho\bar{c}\lambda. \tag{4.41}$$

4.6.1 Predictions for the coefficient of viscosity

From Equation (4.40), we can make various predictions about the behaviour of η as a function of pressure or temperature, just as we did for the coefficient of thermal conductivity κ. And indeed the conclusions are similar, and for similar reasons. Thus, as $\lambda \propto 1/n$, it follows that the viscosity of a gas should be independent of pressure. And indeed this is found to be the case over several orders of magnitude in pressure. Although the number density increases with increasing pressure, the distances over which the molecules transport their momentum decreases with pressure and the two effects cancel out. In the limit of zero pressure, there are no molecules to cause any viscous effects. This is exploited in modern flywheel storage systems. The flywheel assembly is mounted on magnetic bearings that prevent any contact between the stationary and moving parts and is contained in an evacuated chamber to eliminate any viscous effects. Perhaps the most surprising prediction of Equation (4.40) is that a gas becomes more viscous as it gets hotter since $\eta \propto \bar{c} \propto \sqrt{T}$. This is in stark contrast to the behaviour of liquids whose familiar behaviour is to become less viscous with rising temperature. Indeed gases do become more viscous with increasing temperature. However, the viscosity increases less rapidly than given by $\eta \propto \sqrt{T}$. This discrepancy between experiment and theory occurs because of the breakdown of the simple model of molecules as rigid spheres with no mutual interactions, as we described in the case of thermal conductivity.

4.7 Comparison of transport properties

Bringing together the results for the transport properties of a gas, we have

Diffusion:	$D = \frac{1}{3}\bar{c}\lambda.$
Thermal conductivity:	$\kappa = \frac{1}{3}\frac{C_V}{N_A}n\bar{c}\lambda.$
Viscosity:	$\eta = \frac{1}{3}nm\bar{c}\lambda = \frac{1}{3}\rho\bar{c}\lambda.$

Table 4.1 Transport coefficients for argon gas at atmospheric pressure and room temperature.

κ	1.7×10^{-2} W/m·K
ρ	$1.66\,\mathrm{kg/m^3}$
D	$1.8 \times 10^{-5}\,\mathrm{m^2/s}$
η	2.2×10^{-5} Pa·s
c_V	312 J/K·kg

There are clear similarities here and these exist because the underlying mechanism is common in all three cases; the diffusion of molecules and the collisions between them. Furthermore, the coefficients are related by

$$\frac{\rho D}{\eta} = 1; \tag{4.42}$$

$$\frac{\kappa}{\eta c_V} = 1, \tag{4.43}$$

where $c_V = C_V/N_A m$ and c_V is the specific heat at constant volume per unit mass of gas. Experimentally, it is found that for real gases, $(\rho D)/\eta$ lies between 1.3 and 1.5 instead of being unity, and $\kappa/(\eta c_V) = 1$, is found to lie between 1.4 and 2.5. For example, taking the parameters for argon gas shown in Table 4.1, we find $(\rho D)/\eta = 1.4$ and $\kappa/(\eta c_V) = 2.5$. In view of the simplicity of the molecular model, however, the agreement with theory can be considered to be very satisfactory.

4.7.1 Estimation of Avogadro's number

We can use the expression for viscosity to make an estimate of Avogadro's number. We have, Equation (4.40),

$$\eta = \frac{1}{3} nm\overline{c}\lambda.$$

We also have

$$\lambda = \frac{1}{\sqrt{2}n\pi d^2} \text{ and } \overline{c} = \left(\frac{8kT}{\pi m}\right)^{1/2}.$$

Substituting for λ and $\overline{c}$, we obtain

$$\eta = \frac{1}{3\sqrt{2}\pi d^2}\left(\frac{8kTm}{\pi}\right)^{1/2}.$$

We recall that $k = R/N_A$ and $m = (M \times 10^{-3})/N_A$ kg, where R is the gas constant, N_A is Avogadro's number and M is the molecular mass. Substituting for k and m, we obtain after rearrangement

$$N_A d^2 = \frac{1}{3\sqrt{2}\pi\eta}\left(\frac{8RT \times M \times 10^{-3}}{\pi}\right)^{1/2}. \tag{4.44}$$

From the discussion in Section 1.2, we can relate d to the molecular mass M and the density ρ_{solid} of the molecules in their *solid state* by

$$d \approx \left(\frac{M \times 10^{-3}}{\rho_{\text{solid}} N_{\text{A}}}\right)^{1/3},$$

And, hence,

$$N_{\text{A}} d^3 \approx \frac{M \times 10^{-3}}{\rho_{\text{solid}}}. \tag{4.45}$$

Dividing Equation (4.45) by Equation (4.44), we obtain

$$d \approx \frac{3\sqrt{2}\pi\eta}{\rho_{\text{solid}}} \left(\frac{\pi \times M \times 10^{-3}}{8RT}\right)^{1/2}. \tag{4.46}$$

Importantly, all the physical quantities on the right-hand side of this equation can be measured experimentally including ρ_{solid}, which can be determined by cooling the gas. Substituting the appropriate values for argon, we find

$$d \approx 3\sqrt{2}\pi \left(\frac{2.2 \times 10^{-5}}{1400}\right) \left(\frac{\pi \times 40 \times 10^{-3}}{8 \times 8.314 \times 293}\right)^{1/2} = 0.53 \times 10^{-9}\ \text{m}.$$

Then substituting this value of d in Equation (4.45), we find

$$N_{\text{A}} \approx \frac{40 \times 10^{-3}}{1400 \times (0.53 \times 10^{-9})^3} = 2 \times 10^{23}.$$

This value is within a factor of three of the accepted value. A very similar calculation was made by Loschmidt in 1865 to provide the first order of magnitude estimates of Avogadro's number and the sizes of molecules.

4.8 Effusion

Suppose we have a gas container that has a tiny hole through which molecules can escape. By tiny we mean that the dimensions of the hole are much smaller than the mean free path of the molecules. This means that there will be essentially no molecular collisions in the gas near the hole and the molecules escape individually. When this condition is obtained, the flow of gas is described as *effusion*. In this section, we obtain an expression for the rate at which molecules strike the walls of a container and hence determine the rate at which molecules escape through a tiny hole.

Figure 4.21 illustrates an elemental area $\mathrm{d}S$ in the wall of a gas container and an elemental volume $\mathrm{d}V$ in the gas. The spherical coordinate system is the natural one to represent this situation, where the spherical coordinates (r, θ, ϕ) are defined in Figure 4.21. To find the rate at which molecules strike area $\mathrm{d}S$, we consider: (i) the number of molecular collisions that occur in volume $\mathrm{d}V$ per second, (ii) the fraction of molecules

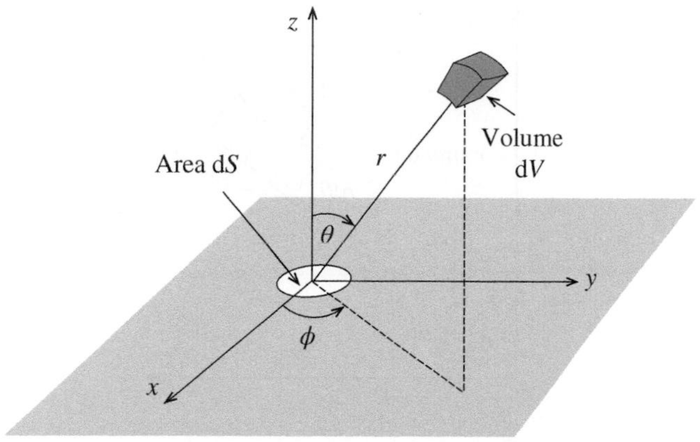

Figure 4.21 The figure shows an elemental area dS in the wall of a gas container, and an elemental volume dV in the gas. To find the rate at which molecules strike dS, we consider: (i) the number of molecular collisions that occur in dV per second, (ii) the fraction of molecules that are scattered towards dS after suffering a collision in dV, and (iii) the probability that those molecules will reach dS without having a further collision.

that are scattered towards dS after suffering a collision in dV, and (iii) the probability that those molecules will reach dS without having a further collision.

(i) The average number of collisions a *single* molecule makes per second is $\bar{c}/\lambda$, where $\bar{c}$ is the mean speed and λ is the mean free path. In dV there are ndV molecules where n is the molecular concentration. Then the total number of molecular collisions per second in dV is $(1/2)(n\bar{c}/\lambda)\mathrm{d}V$, where the factor of ½ is introduced so that each collision is not counted twice. Each collision results in two scattered molecules. Hence, the total number of molecules scattered out of dV per second is $n\bar{c}\mathrm{d}V/\lambda$.

(ii) The scattered molecules leave dV in all directions uniformly. The fraction of molecules that is scattered toward dS is given by the solid angle that dS subtends at dV divided by the total solid angle 4π. This is

$$\frac{\mathrm{d}S\cos\theta}{r^2}\frac{1}{4\pi}.$$

(iii) The probability that a molecule will travel distance r without having a collision is equal to $\mathrm{e}^{-r/\lambda}$, see Equation (4.12).

Hence, the number of molecules striking area dS per second from volume dV is

$$\frac{n\bar{c}\mathrm{d}V}{\lambda}\frac{\mathrm{d}S\cos\theta}{4\pi r^2}\mathrm{e}^{-r/\lambda}.$$

The elemental volume dV in terms of spherical coordinates is $r\sin\theta\,\mathrm{d}\phi\,r\mathrm{d}\theta\,\mathrm{d}r$; see Figure 4.22. Making this substitution, we obtain

$$\frac{n\bar{c}}{\lambda}\frac{\mathrm{d}S\cos\theta}{4\pi r^2}\mathrm{e}^{-r/\lambda}r\sin\theta\,\mathrm{d}\phi\,r\mathrm{d}\theta\,\mathrm{d}r.$$

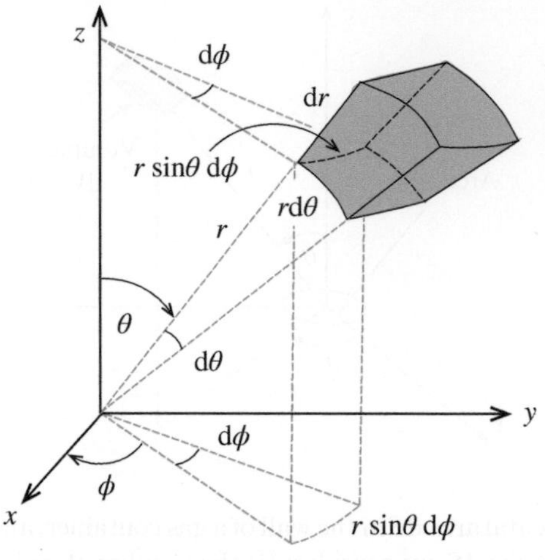

Figure 4.22 The elemental volume dV defined in terms of the spherical coordinates (r, θ, ϕ).

We integrate this result over the volume of space above the surface containing dS to obtain the total number of molecules striking dS per second:

$$\frac{n\bar{c}\mathrm{d}S}{4\pi\lambda}\int_0^{2\pi}\mathrm{d}\phi\int_0^{\pi/2}\sin\theta\cos\theta\,\mathrm{d}\theta\int_0^{\infty}\mathrm{e}^{-r/\lambda}\mathrm{d}r$$
$$=\frac{n\bar{c}\mathrm{d}S}{4\pi\lambda}[2\pi][1/2][\lambda]=\frac{n\bar{c}\mathrm{d}S}{4}.$$

And hence, the total number of molecules striking unit area per second is

$$\frac{1}{4}n\bar{c}. \tag{4.47}$$

If we cut a tiny hole of area A in the container wall, it follows from Equation (4.47) that the rate at which molecules pass through the hole is

$$\frac{1}{4}An\bar{c}. \tag{4.48}$$

4.8.1 Isotope separation

Figure 4.23 illustrates two gas containers that are connected by a tiny hole of area A. The gas pressure is higher in the left-hand container but this pressure is sufficiently low that the mean free path of the molecules is much greater than the dimensions of the hole. Then, according to Equation (4.48), the rate at which molecules in the left-hand chamber strike the hole is

$$\frac{1}{4}nA\bar{c}=\frac{1}{4}nA\left(\frac{8kT}{\pi m}\right)^{1/2}, \tag{4.49}$$

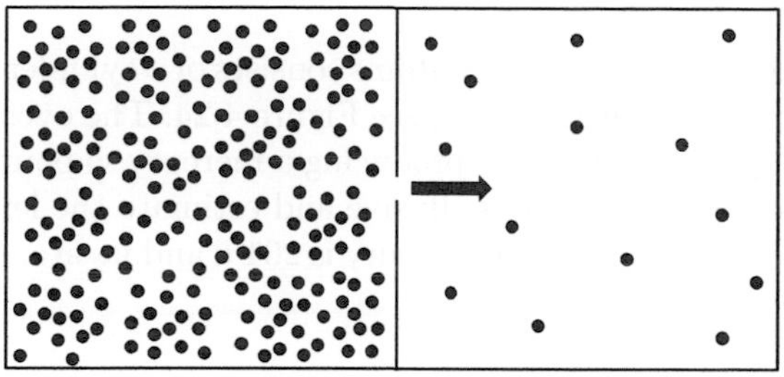

Figure 4.23 The figure illustrates two gas containers that are connected by a tiny hole. The gas pressure is higher in the left-hand container although this pressure is sufficiently low that the mean free path of the molecules is much greater than the dimensions of the hole. If the gas in the left-hand container consists of a mixture of two gases, the lighter gas will effuse through the hole at a faster rate than the heavier one. This enables the separation of different molecular isotopes.

where we have substituted for $\overline{c}$. We see that the rate at which molecules effuse through the hole is proportional to $n(T/m)^{1/2}$, i.e. the rate is directly proportional to the molecular concentration and inversely proportional to the square root of the mass.

Suppose that the gas in the left-hand container consists of a mixture of two gases with masses m_C and m_D, $(m_D > m_C)$ and concentrations n_C and n_D, respectively. The effusion rates of the two gases will be

$$\frac{1}{4}A\left(\frac{8kT}{\pi}\right)^{1/2} n_C \left(\frac{T}{m_C}\right)^{1/2},$$

and

$$\frac{1}{4}A\left(\frac{8kT}{\pi}\right)^{1/2} n_D \left(\frac{T}{m_D}\right)^{1/2},$$

respectively. As the two gases are at the same temperature, the lighter gas effuses through the hole at a faster rate than the heavier one. The ratio of the two effusion rates is then

$$\frac{n_C}{n_D}\left(\frac{m_D}{m_C}\right)^{1/2}. \tag{4.50}$$

This is also the ratio of the number densities of the two gases in the right-hand container and we see that compared to the left-hand container, this ratio has increased by the factor $(m_D/m_C)^{1/2}$. If this process is repeated in p stages of effusion, the ratio of molecular densities increases by the factor $(m_D/m_C)^{p/2}$. This is an important process because the various isotopes of a molecule have essentially the same chemical properties and so cannot be separated by chemical means. However, as effusion depends on a physical property, namely the mass of the particular species, it does allow different isotopes to be separated. It is the basis of the first technique that was able to produce, on an industrial scale, the enrichment of the uranium isotope ^{235}U from a naturally occurring sample of uranium. The molecule used in the process is uranium hexafluoride UF_6.

Worked example

An oven that is used to produce a beam of mercury atoms consists of a cylinder that is closed at one end and has a round orifice of diameter 0.5 mm at the other; see Figure 4.24. The oven contains a charge of 25 g of mercury and is heated to a temperature of 50 °C, producing a mercury vapour pressure of 3 Pa. Show that the flow of mercury atoms through the orifice is effusive and estimate the length of time before the oven needs refilling with mercury. The atomic mass of mercury is 201 u and its atomic radius is 0.15 nm. Assume a vacuum exists outside the oven.

Solution

The concentration of atoms in the mercury vapour

$$= \frac{P}{kT} = \frac{3}{1.38 \times 10^{-23} \times 323} = 6.73 \times 10^{20}\ \text{atoms/m}^3.$$

$$\text{Mean free path of the atoms} = \frac{1}{\sqrt{2}n\pi d^2} = \frac{1}{\sqrt{2} \times 6.73 \times 10^{20} \times \pi \times \left(0.3 \times 10^{-9}\right)^2} = 4\ \text{mm}.$$

This is an order of magnitude greater than the diameter of the orifice, and we can take the flow through the orifice to be effusive.

The mean speed of the atoms $= \left(\frac{8kT}{\pi m}\right)^{1/2} = \left(\frac{8 \times 1.38 \times 10^{-23} \times 323}{\pi \times 201 \times 1.66 \times 10^{-27}}\right)^{1/2} = 184\ \text{m/s}.$ Atoms pass through the orifice at the rate

$$\frac{1}{4}An\bar{c} = \frac{1}{4} \times \pi \times \left(0.25 \times 10^{-3}\right)^2 \times 6.73 \times 10^{20} \times 184 = 6.08 \times 10^{15}\ \text{atoms/s}.$$

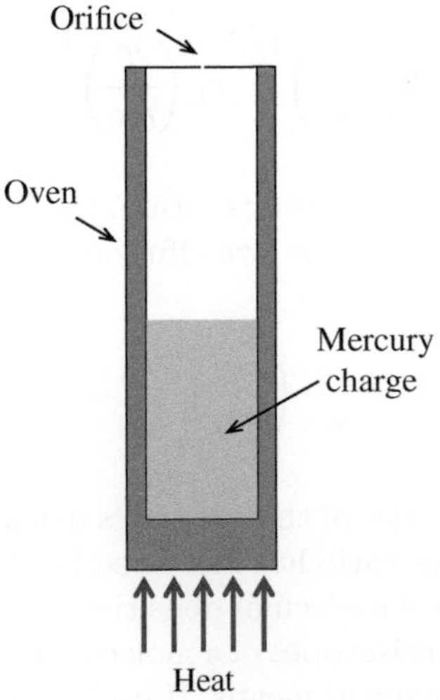

Figure 4.24 A schematic diagram of an oven that is used to produce a beam of mercury atoms. The oven consists of a cylinder that is closed at one end and has a round orifice at the other. The oven is heated to increase the vapour pressure of the mercury. The diameter of the hole is much smaller than the mean free path of the mercury atoms in the oven.

201 g of mercury contains Avogadro's number of atoms.

Therefore 25 g contains $\frac{25}{201} \times 6.02 \times 10^{23} = 7.5 \times 10^{22}$ atoms.

Hence, the charge of mercury will last $\frac{7.5 \times 10^{22}}{6.08 \times 10^{15}} = 1.23 \times 10^{7}\ \text{s} \sim 140$ days.

As noted above, a condition that is necessary for effusive flow is that the dimensions of the hole in the gas container must be small compared to the mean free path of the molecules. If, on the other hand, the mean free path is much smaller than the dimensions of the hole, the molecules cannot pass through the hole without making many collisions with other molecules in the vicinity of the hole. This results in the bulk flow of gas through the hole, which is called *viscous flow.*

4.9 The random walk[1]

Imagine that you are standing under a lamppost. You toss a coin. If it lands heads, you take one step to the right and if it lands tails, you take one step to the left. From your new position, you again toss a coin and if it lands heads you take one step to the right, and if it lands tails, you take one step to the left; and so on. Such an exercise is called a random walk since the direction of each step is independent of the previous one. Suppose you toss the coin, say 20 times. The question is: how far from the lamppost will you end up? One thing we can say straightaway is that because the chances of getting a head or a tail are the same, your most probable position is back where you started, under the lamppost.

Suppose now that 99 other people repeat this exercise. There would be a distribution in the distances reached from the lamppost. And we could plot a histogram of the number of people N that reach m steps from the lamppost. It would look something like that shown in Figure 4.25. We would find, see Equation (4.55), that the average absolute distance reached from the lamppost would be close to $\sqrt{20}$ steps.

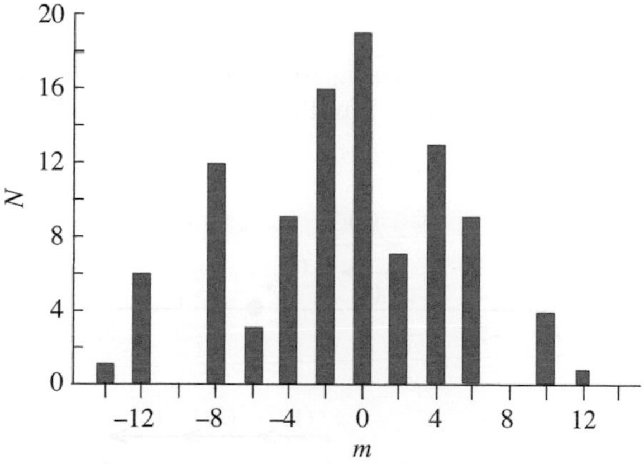

Figure 4.25 A possible distribution for the distance reached from the origin after a random walk of 20 steps.

[1] This section may be omitted at first reading of the book.

In this section we analyse the one-dimensional random walk. The one-dimensional case gives all the essential physics we need. Going to the three-dimension case only changes the constants in the final results and then only by a factor of the order of unity. At the end of the section, we relate the random walk to molecular diffusion.

In analogy with the exercise described above, we consider the random walk of a particle under the following conditions:

- The particle moves in one dimension only; the x-direction.
- The particle steps either to the left or right with a fixed step length l.
- The direction of each step, left or right along the x-axis, is independent of the previous one.

We want to find the probability of finding the particle at $x = ml$ after N steps, where m is an integer between $-N$ and $+N$.

We start with the particle at $x = 0$. In the first step, the particle may go left or right. Similarly, in the second step, it may go left or right, and so on. Figure 4.26 shows two possible paths the particle may take in three steps. In path a, indicated by the green arrows, it goes, left, right, and right again, arriving at $x = +l$. The probability for this path is $(1/2)^3 = 1/8$. In path b, indicated by the blue arrows, the particle goes right, right, and right again, arriving at $x = +3l$. The probability of this path is also $(1/2)^3 = 1/8$. Indeed the probability of the particle taking any *particular* path after three steps is $(1/2)^3 = 1/8$. However, there are three possible paths for the particle ending up at $x = +l$. These are indicated by the green, blue, and red arrows in Figure 4.27. Hence, the probability of the particle ending up at $x = +l$ after taking three steps *in any order* is $3 \times 1/8 = 3/8$. Notice that an odd number of steps, in this case three, gives rise to an odd value of displacement x, i.e. $x = \pm l$ or $\pm 3l$. Similarly, an even number of steps will give rise to even values of x.

More generally, we let n_1 be the number of steps taken to the right, and n_2 be the number of steps taken to the left. Then $n_1 + n_2 = N$, and $m = n_1 - n_2$, which gives $m = 2n_1 - N$. Then, let

p = probability that a step is to the right;

q = probability that a step is to the left.

The probability of a *particular* sequence of steps of n_1 steps to the right and n_2 steps to the left is $p^{n_1} q^{n_2}$. As we saw in the previous example, with $N = 3$, $n_1 = 2$ and $n_2 = 1$, the probability is $(1/2)^2(1/2)^1 = (1/8)$. The

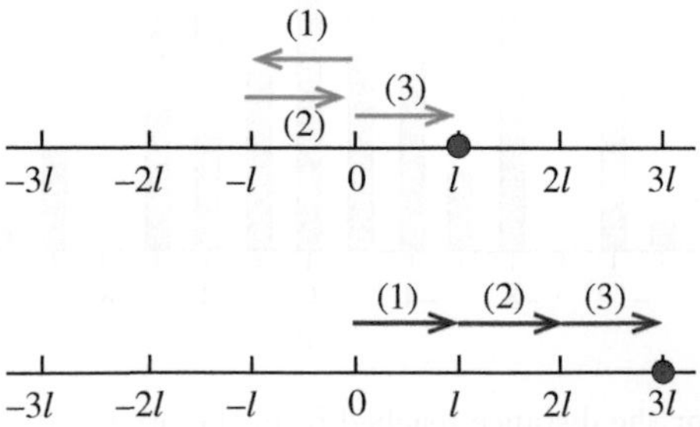

Figure 4.26 Two possible paths a molecule may take in three consecutive steps in a one-dimensional random walk, indicated by the green and blue arrows, respectively. The probability of the molecule taking any *particular* path after three steps is $(1/2)^3 = 1/8$.

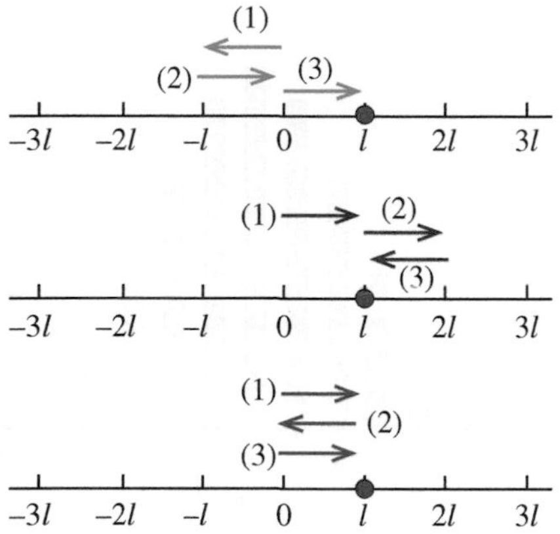

Figure 4.27 There are three possible paths for the molecule ending up at $x = +l$ in a one-dimensional random walk. These are indicated as the green, blue and red arrows, respectively. Hence, the probability of a molecule ending up at $x = +l$ after taking three steps in any order is $3 \times 1/8 = 3/8$. Notice that an odd number of steps, in this case three, gives rise to an odd value of net displacement x, i.e. $x = \pm l$ or $\pm 3l$. Similarly, an even number of steps will give rise to even values of x.

total number of different ways of taking n_1 steps to the right and n_2 steps to the left is $N!/(n_1!\,n_2!)$. For example, again with $N = 3$, $n_1 = 2$ and $n_2 = 1$, $N!/(n_1!\,n_2!) = (3 \times 2 \times 1)/(2 \times 1) \times 1 = 3$. Therefore, the probability of taking n_1 steps to the right and $n_2 = (N - n_1)$ steps to the left, *in any order*, is

$$\frac{N!}{n_1!n_2!}p^{n_1}q^{n_2}. \tag{4.51}$$

This is called *binomial distribution*. For $p = q = 1/2$, and substituting $n_1 = (N + m)/2$ and $n_2 = (N - m)/2$, we find that the probability $P(m, N)$ of a net displacement of m steps to the right, after N steps, is

$$\frac{N!}{\left(\frac{N+m}{2}\right)!\left(\frac{N-m}{2}\right)!}\left(\frac{1}{2}\right)^N. \tag{4.52}$$

By symmetry, this is also the probability of taking of m steps to the left after N steps. The binomial distribution $P(m, N)$ for $N = 10$ is shown in Figure 4.28. We see that after N steps, the most probable value of m is 0; the particle is most likely to be at its starting point, and that large values of $|m|$ are unlikely. We may also deduce that the root mean square of m is not zero.

For large N, Stirling's formula can be used for $N!$. It is

$$\ln N! = N \ln N - N. \tag{4.53}$$

Using Stirling's formula in Equation (4.52), we obtain, after some algebraic manipulation that we omit, the following expression for the probability distribution:

$$P(m, N) = \left(\frac{2}{\pi N}\right)^{1/2} e^{-m^2/2N}. \tag{4.54}$$

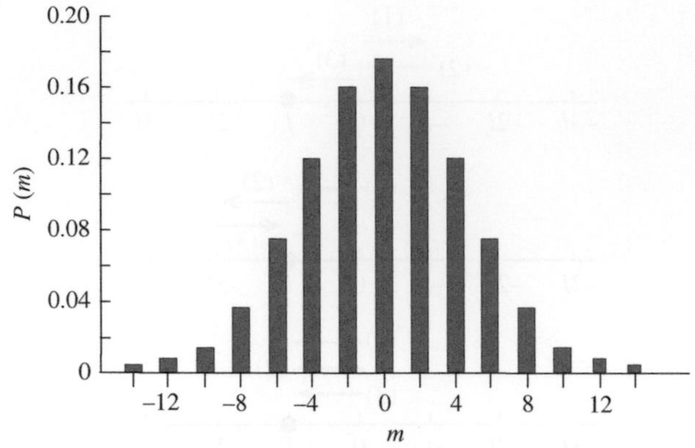

Figure 4.28 The binomial distribution $P(m, N)$ for $N = 10$, with $p = q = 1/2$. The figure illustrates that after N steps in a random walk, the most probable value of m is 0, that large values of $|m|$ are unlikely and that the root mean square of m is not zero.

This is the probability of the particle being m steps from its starting point. We recognise this function as a Gaussian. And we see that a binomial distribution acquires a Gaussian shape when N is large; in practice, this is the case for values of N larger than about 10. Equation (4.54) can be used to find the root mean square of m and it is found that

$$\left(\overline{m^2}\right)^{1/2} = \sqrt{N}. \tag{4.55}$$

This dependence on the square root of the number of steps is a general characteristic of a random walk.

4.9.1 Probability distribution $P[x]dx$ for displacement of the particle

We can express Equation (4.54) in terms of the displacement x of the particle, where $x = ml$ and l is the step length. We recall that $m = 2n_1 - N$, which means that m has integral values separated by $\Delta m = 2$. When l is small compared to the dimensions of the physical situation of interest, the fact that x can only take values in discrete increments of $2l$ is not important. Moreover, when N is large, $P(m, N)$ does not change appreciably from one possible value of m to the adjacent one. Therefore, $P(m, N)$ can be regarded as a smooth function. In this case, the discrete distribution of the type shown in Figure 4.28 assumes the form shown in Figure 4.29, where the vertical bars are very densely packed and their envelope forms a smooth curve. Under these circumstances, we can determine the probability that the particle is found between x and $x + \mathrm{d}x$ after N steps. The interval $\mathrm{d}x$ is chosen so that it is large compared to l but small compared to the dimensions of the physical situation. This is certainly the case, for example, in molecular diffusion where l is the order of the mean free path and $\mathrm{d}x$ is small compared to the size of the gas container. As m takes only integral values separated by $\Delta m = 2$ it follows that the range $\mathrm{d}x$ contains $\mathrm{d}x/2l$ possible values of m, which all occur with nearly the same probability $P(m, N)$; see Figure 4.29. Hence, the probability of finding the particle within the range x to $x + \mathrm{d}x$ is obtained by summing over all values of m lying in the interval $\mathrm{d}x$, i.e. by multiplying $P(m, N)$ by $\mathrm{d}x/2l$. This is just $P\,[x]\mathrm{d}x$ and so we have

$$P\,[x]\mathrm{d}x = P(m, N)\frac{\mathrm{d}x}{2l} = \left(\frac{2}{\pi N}\right)^{1/2} \mathrm{e}^{-m^2/2N}\frac{\mathrm{d}x}{2l}.$$

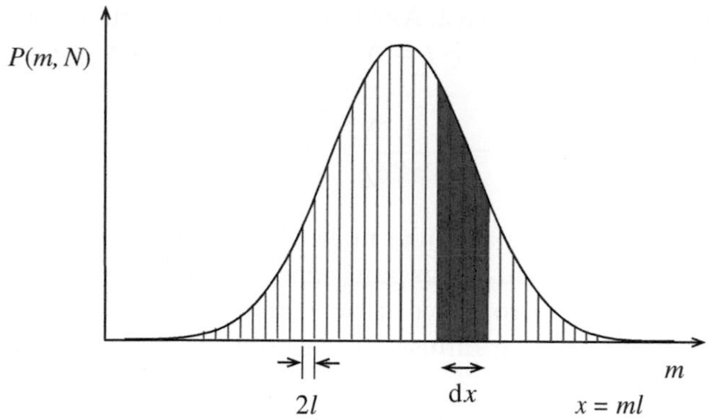

Figure 4.29 The probability distribution $P(m, N)$ for a net displacement of a particle in a random walk when the total number of steps N is very large and the step length l is very small. The interval $\mathrm{d}x$ is large compared to the separation $\Delta m = 2$ of adjacent m values but small compared to the dimensions of the physical situation, e.g. the size of a gas container. The interval $\mathrm{d}x$ contains $\mathrm{d}x/2l$ possible values of m.

Then, substituting $m = x/l$, we have that the probability of the particle ending up a distance between x and $x + \mathrm{d}x$ from its starting point is

$$P\,[x]\mathrm{d}x = \left(\frac{1}{2\pi N l^2}\right)^{1/2} \mathrm{e}^{-x^2/2Nl^2}\,\mathrm{d}x. \tag{4.56}$$

Although the most probable value of x is 0, the root mean squared value of x is not. We have

$$\begin{aligned}\overline{x^2} &= \int_{-\infty}^{+\infty} x^2 P\,[x]\mathrm{d}x = \left(\frac{1}{2\pi N l^2}\right)^{1/2} \int_{-\infty}^{+\infty} x^2 \mathrm{e}^{-x^2/2Nl^2}\,\mathrm{d}x \\ &= \left(\frac{1}{2\pi N l^2}\right)^{1/2} \frac{1}{2}\left(\pi\left[2Nl^2\right]^3\right)^{1/2} = Nl^2,\end{aligned}$$

where we have made use of a standard integral. Hence, we have

$$\left(\overline{x^2}\right)^{1/2} = \sqrt{N}l, \tag{4.57}$$

where again we note the dependence on the square root of the number of steps N.

4.9.2 The random walk and molecular diffusion

Consider a one-dimensional gas where a particular molecule, moving in the x direction, makes collisions with molecules that are at fixed positions, separated by distance l. And after each collision, the direction of the molecule, in the $+$ or $-x$ direction, is random. This one-dimensional model is described by Equation (4.56):

$$P\,[x]\mathrm{d}x = \left(\frac{1}{2\pi N l^2}\right)^{1/2} \mathrm{e}^{-x^2/2N\lambda^2}\,\mathrm{d}x, \tag{4.58}$$

where we have taken l to be the mean free path λ. Assuming the molecule moves with mean speed $\overline{c}$ between collisions, the time taken to make N collisions is $t = N\lambda/\overline{c}$. Substituting for N in Equation (4.58), we obtain

$$P\,[x]\mathrm{d}x = \left(\frac{1}{2\pi lt\overline{c}}\right)^{1/2} \mathrm{e}^{-x^2/2lt\overline{c}}\mathrm{d}x. \tag{4.59}$$

This expression gives the probability of a particular molecule being between x and $x + \mathrm{d}x$ at time t. Equivalently, it describes the spatial distribution of this particular type of molecule at time t, if there were a local concentration of these molecules in a narrow slice at $x = 0$ and $t = 0$. In Section 4.4, we used the diffusion equation to solve the problem of molecular diffusion in one dimension and obtained the result

$$P\,[x, t]\mathrm{d}x = \left(\frac{1}{4\pi Dt}\right)^{1/2} \mathrm{e}^{-x^2/4Dt}\mathrm{d}x.$$

Comparing this equation with Equation (4.59), we see that the essential features of the diffusion process are reproduced by a random walk, with the same dependence on time t. Moreover, this comparison gives $D = {}^1\!/_2\overline{c}\lambda$, which is consistent, within a factor of order unity, with the result $D = {}^1\!/_3\overline{c}\lambda$, obtained from the diffusion equation; see Equation (4.26).

We have related the random walk to molecular diffusion. However, the random walk finds application in a wide range of the physical sciences as well as in economics. For example, it is used to describe the diffusion of photons in the sun and the movement of stocks and shares in financial markets. The 'Quantum Cloud' is a contemporary sculpture by sculptor Antony Gormley and is shown in Figure 4.30. Interestingly, it is constructed from a collection of tetrahedral units whose steel sections were arranged using a computer model with a random walk algorithm.

Figure 4.30 Photograph of the sculpture 'Quantum Cloud' by sculptor Antony Gormley. The sculpture is constructed from a collection of tetrahedral units whose steel sections were arranged using a computer model with a random walk algorithm. Source: Andrew Bowden / Flickr / CC BY-SA 2.0.

Worked example

The net effect of thermonuclear fusion in the core of the Sun is the transformation of four protons into a helium nucleus:

$$4\text{p} \rightarrow {}^4_2\text{He} + 2\text{e}^+ + 2\nu + 2\gamma,$$

with the emission of two positrons, two neutrinos and two energetic photons. The underlying mechanism for the transport of radiant energy in the interior of the Sun is radiative diffusion. This is a random walk process in which the photons are continuously scattered as they make their way from the core to the surface of the Sun. They continuously lose energy in this process and appear as sunlight at the surface of the Sun. Assuming a step length of 1 mm between scattering events, estimate the length of time it takes the photons to reach the surface.

Solution

The net distance R the photons have to travel is $\sim l\sqrt{N}$, where R is the radius of the Sun, N is the number of scattering events and l is the step length. Hence, $N \sim (R/l)^2$. The time for each step is l/c, where c is the speed of light. Therefore, the total time to reach the surface is

$$\sim \frac{R^2}{l^2} \times \frac{l}{c} = \frac{R^2}{lc} \sim \frac{(7 \times 10^8)^2}{1 \times 10^{-3} \times 3 \times 10^8} \sim 52,000 \text{ years},$$

where we have taken R to be 7×10^8m. We see that it takes about 50,000 years for photons to travel from the Sun's core to its surface. By contrast, neutrinos, which hardly interact with matter, take $R/c = 7 \times 10^8/3 \times 10^8 \sim 2$ seconds to reach the surface.

Problems 4

4.1 What is the mean free path of the molecules in nitrogen gas at $T = 300$ K at (i) atmospheric pressure (1.01×10^5 Pa $= 1013$ mbar), (ii) 1×10^{-3} mbar, and (iii) 1×10^{-6} mbar. Take the diameter of the molecules to be 0.35×10^{-9} m.

4.2 In an undergraduate physics experiment, electrons are boiled off a heated filament by thermionic emission. The filament is surrounded by a cylindrical electrode called the anode, which collects electrons emitted from the filament. The current of electrons collected by the anode is measured as a function of the potential difference between the filament and anode. The apparatus is held in an evacuated chamber. Obtain a value for the maximum pressure that can be tolerated in the chamber if the radius of the anode is 25 mm. Take the diameter of the molecules to be 0.35×10^{-9} m and the molecular weight to be 30 u. The electron current emitted by the cathode is proportional to $T^2 \exp(-\phi/kT)$, where T is temperature k is the Boltzmann constant and ϕ is a constant called the *work function.* Which of the two terms T^2 and $\exp(-\phi/kT)$, would you expect to dominate as the temperature rises?

4.3 A vacuum chamber is maintained at a pressure of 1×10^{-9} mbar. A crystal is cleaved under vacuum in the chamber to yield a clean crystal surface. Estimate the length of time it takes for a monolayer of molecules to cover this clean surface. Assume the probability ε that a molecule sticks to the surface when it strikes it is 0.5. Take the diameter of a molecule to be 0.35×10^{-9} m and the molecular weight to be 30 u.

4.4 The molecules in a gas have a mean free path λ. Determine the most probable free path of the molecules.

4.5 One way to produce ultrahigh vacuum, i.e. ultralow pressure in a vacuum chamber is to include in the chamber a stainless steel plate that can be cooled to very low temperature, typically the boiling point of liquid nitrogen, 77 K. The cold temperature increases the probability that an incident molecule will stick to the plate. The chamber is first evacuated to $\sim 1 \times 10^{-6}$ mbar and then the plate is cooled so that residual molecules stick to the plate. Suppose that the plate has a total area of $0.15\ \mathrm{m}^2$, the volume of the chamber is $0.25\ \mathrm{m}^3$ and the probability ε of a molecule sticking to the plate is 1.0. If the plate is cooled very rapidly at time $t = 0$, how long will it take for the pressure to fall from 1×10^{-6} to 1×10^{-9} mbar? Assume a molecular weight of 30 u.

4.6 Obtain values for (i) the diffusion coefficient D, (ii) the viscosity η, and (iii) the thermal conductivity κ of nitrogen gas at atmospheric pressure and a temperature of 300 K. The molar specific heat of nitrogen gas is $5R/2$. Assume a molecular diameter of 0.35×10^{-9} m.

4.7 An apparatus to measure the thermal conductivity of a gas consists of a wire at the centre of a long cylinder of length l. The wire is heated by passing current through it and its temperature is deduced from its electrical resistance. The heat Q generated by the wire is equal to the product of the current and the voltage applied across the wire. If the wire and the cylinder are maintained at temperatures T_w and T_c, show that the thermal conductivity of the gas is given by

$$\kappa = \frac{Q \ln(r_\mathrm{c}/r_\mathrm{w})}{2\pi l (T_\mathrm{w} - T_\mathrm{c})},$$

where r_w and r_c are the radii of the wire and the cylinder respectively. Neglect any end effects due to the finite size of the cylinder and assume that heat losses due to radiation and convection are negligible. Given that the radii of the wire and cylinder are 5.0×10^{-5} m and 1.0×10^{-2} m, respectively, the length of the cylinder is 1.0×10^{-1} m, $T_\mathrm{w} = 150\,°\mathrm{C}$, and $T_\mathrm{c} = 20\,°\mathrm{C}$ and $Q = 0.25$ W, obtain a value for the thermal conductance of air.

4.8 An apparatus to measure the viscosity of a gas consists of two coaxial cylinders of radii r_a and $r_\mathrm{b} < r_\mathrm{a}$ and length l, with the gas contained between them. The outer cylinder is rotated at fixed angular velocity Ω rad/s. The inner cylinder is connected to a wire whose angle of twist measures the torque τ applied to the cylinder. Assume that $(r_\mathrm{a} - r_\mathrm{b})$ is small compared to both radii, so that a planar geometry may be assumed. Show that the viscosity of the gas is given by

$$\eta = \frac{\tau}{2\pi\Omega l} \frac{(r_\mathrm{a} - r_\mathrm{b})}{r_\mathrm{a} r_\mathrm{b}^2}.$$

In a particular apparatus, the radii of the outer and inner cylinders are 7.0×10^{-2} m and 6.8×10^{-2} m, respectively and are 25×10^{-2} m long. The outer cylinder rotates once every 50 seconds. Obtain a value for the viscosity of the gas if the measured torque on the inner cylinder is $7.2 \times 10^{-7}\ \mathrm{N \cdot m}$.

4.9 Repeat 4.8 but, taking the cylindrical geometry into account, show that

$$\eta = \frac{\tau}{4\pi\Omega l} \left(\frac{r_\mathrm{a}^2 - r_\mathrm{b}^2}{r_\mathrm{a}^2 r_\mathrm{b}^2} \right).$$

4.10 A test tube contains an amount of a volatile liquid. The surface of the liquid is at a distance h below the top of the test tube. Just above the liquid surface, the number density of vapour molecules is n_v, which is determined by the vapour pressure of the liquid. At the top of the test tube, the number density can be assumed to be equal to zero. The molecules have mass m. The molecules of the liquid diffuse slowly through the air above the liquid surface. Obtain the steady state equation for the number of molecules diffusing across any plane per second and hence, the mass per second crossing any plane. Relate this to the rate of loss of liquid to show that the distance h of the level below the open end is given by

$$h^2 = \frac{2Dn_\mathrm{v}m}{\rho} t,$$

where D is the diffusion constant for the liquid molecules diffusing through air and ρ is the density of the liquid. A test tube is filled to the top with water. How long will it take the water level to drop by 25 mm? The diffusion constant of water vapour through air is $0.3 \times 10^{-4}\ \mathrm{m}^2/\mathrm{s}$ and the saturated vapour pressure of water vapour is 3.2×10^3 Pa.

5

Real gases

So far, we have discussed ideal gases and described them in terms of kinetic theory. The main assumptions we made were that molecules have a negligible size and that there are no forces of attraction or repulsion operating between them. The molecules thus move independently of each other. Using kinetic theory, we derived the equation of state for an ideal gas; the equation that expresses the relation between the three state variables, pressure P, volume V, and absolute temperature T: $PV = RT$, for one mole of gas. We made the point that most gases, like helium and argon, obey this ideal gas law under ordinary conditions of pressure and temperature. However, they do deviate from the ideal gas law at high pressures and low temperatures. Perhaps the simplest way to see the breakdown of the ideal gas law for a real gas is to make a plot of PV/RT against increasing pressure P. Since the ideal gas law says that PV/RT is constant, a plot of PV/RT against P should give a straight horizontal line, as shown by the dashed line in Figure 5.1. However, for a real gas, the plot deviates from a straight line, as shown by the solid curves. At high pressure, the ideal gas law breaks down essentially because the average distance between the molecules becomes small, and so the finite size of the molecules and the intermolecular forces are no longer negligible. Further evidence that the ideal gas law breaks down under certain conditions is that gases can be liquefied and even solidified. It is the interactions between molecules that make matter condense into liquid and solid forms.

In this chapter, we discuss how the behaviour of real gases with respect to pressure, volume, and temperature can be described. The way this is done is to modify the ideal gas law $PV - RT$ in such a way that it can also deal with conditions where the finite size of the molecules and their mutual interactions cannot be neglected. There are two approaches to this. One approach adopts a molecular model in which an attempt is made to understand and account for the observed departures from the perfect gas law on the basis of simple assumptions about the intermolecular forces. Such an approach has produced a number of modified ideal-gas equations but the best known of these is the one proposed by Johannes van der Waals:

$$\left(P + \frac{a}{V^2}\right)(V - b) = RT. \tag{5.1}$$

Here the term a/V^2 takes account of intermolecular attraction and the b term accounts for the finite size of the molecules.

Physics of Matter, First Edition. George C. King.

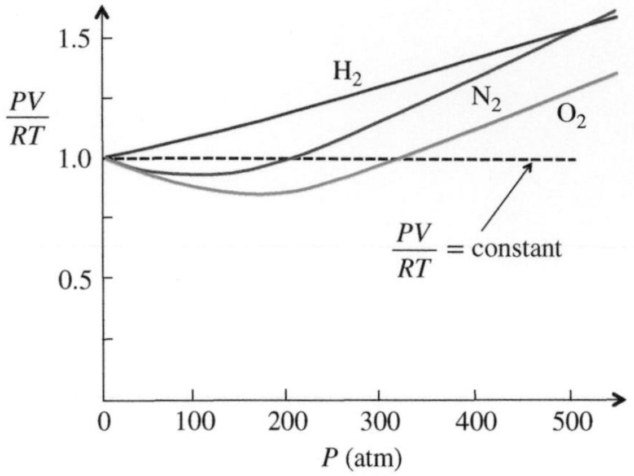

Figure 5.1 A plot of PV/RT against pressure P for various gases. As PV/RT is a constant for an ideal gas, a plot of PV/RT against P should give a straight horizontal line, as shown by the dashed line. However, the plot for a real gas deviates from a straight line, as shown by the solid curves. At high pressure, the ideal gas law breaks down essentially because the average distance between the molecules becomes small, and so the finite size of the molecules and the intermolecular forces are no longer negligible.

The second approach is empirical. Here we measure and plot the quantity PV/RT for a real gas as a function of either decreasing volume V or increasing pressure P. Such a plot is known as a *virial plot*. In the case of decreasing volume, we then fit the experimental data to an equation in terms of $1/V$ of the form

$$\frac{PV}{RT} = 1 + \frac{B}{V} + \frac{C}{V^2} + \frac{D}{V^3} + \cdots, \tag{5.2}$$

and adjust the values of the coefficients, B, C, D, ..., until a good fit to the experimental data is obtained. Using such a virial equation, it is possible to represent experimental data over a wide range of temperature and pressure with a small number of coefficients that diminish rapidly in size. And as the equation is based on actual experimental data, it makes accurate predictions.

The effects of intermolecular forces depend on the average separation of the molecules and therefore decrease with decreasing density. Similarly, the effects of the finite size of molecules also decrease with decreasing density. We can therefore expect the perfect gas law to hold within the limit of very low molecular densities, corresponding to large volumes. Indeed, both the van der Waals equation (5.1) and the virial equation (5.2) tend to be the ideal gas equation at large V.

5.1 The van der Waals equation

We recall the Lennard-Jones 6-12 potential that is used to describe the potential energy between two molecules; see also Section 2.2. This potential is shown in Figure 5.2. At distances of a few molecular diameters, the molecules attract one another. But at small separation, when the molecules come into close contact, they strongly repel each other and there is a steep increase in potential energy. It is this repulsion that gives the molecules a finite size.

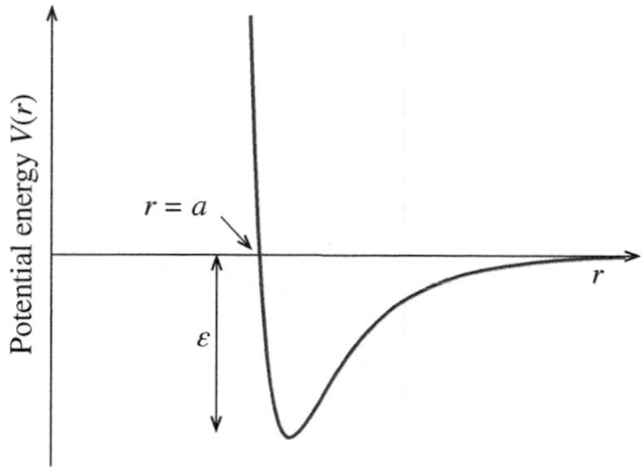

Figure 5.2 The Lennard-Jones 6–12 potential that describes the potential energy between two molecules. At distances of a few molecular diameters, the molecules attract one another. But at small separation, when the molecules come into close contact, they strongly repel each other and there is a steep increase in potential energy. It is this repulsion that gives the molecules a finite size. The diameter of the molecules is taken to be a, and ε is the depth of the potential well.

In the simplest form of kinetic theory, the finite size of the molecules and the interactions between the molecules are completely neglected. Despite its simplicity, the theory works for many gases under normal conditions of pressure and temperature because, under these conditions, the molecules are far apart.

In the first improvement of the theory, the assumption that molecules have zero size is abandoned and molecules are modelled as rigid spheres of finite size. The intermolecular potential energy for this rigid sphere model is shown in Figure 5.3. Here, the molecules have a definite size, while the molecules have no interaction with each other until they touch when the potential rises abruptly to infinity. This is the model we adopted in our discussion of the transport properties of gases in Chapter 4. In that discussion, we could still use the ideal gas equation for situations where the mean free path λ was much greater than the molecular diameter, i.e. where the average separation was so great that the molecules could be considered to move independently of one another.

The van der Waals equation follows from a model in which we treat the molecules as rigid spheres that interact weakly. This gives the more realistic form of the intermolecular potential that is shown in

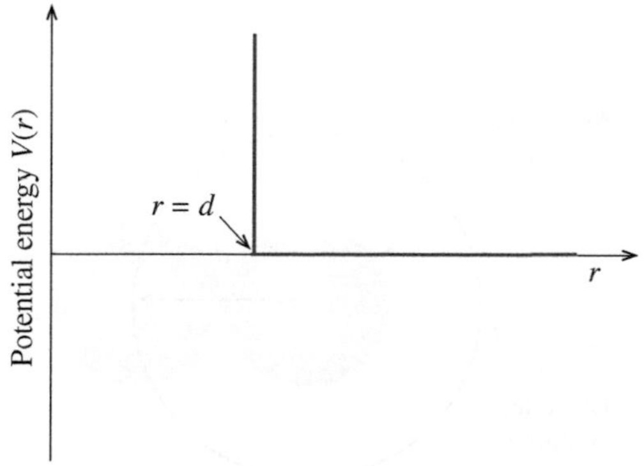

Figure 5.3 The intermolecular potential energy for two molecules that are modelled as rigid spheres of diameter d. In this model, the molecules have a definite size, while the molecules have no interaction with each other until they touch when the potential rises abruptly to infinity. This is the model we adopted in our discussion of the transport properties of gases in Chapter 4.

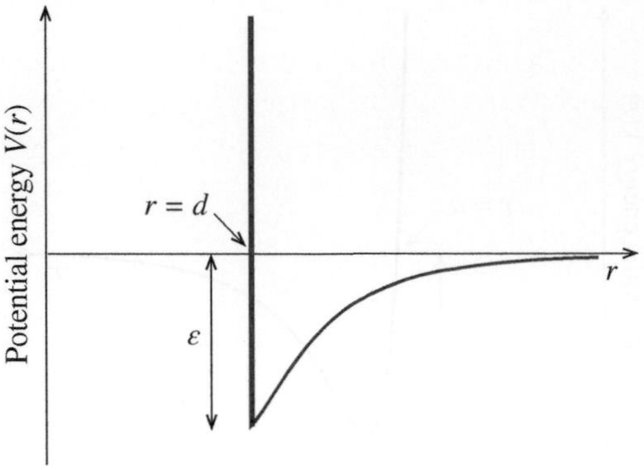

Figure 5.4 The intermolecular potential energy for two molecules that are modelled as weakly interacting rigid spheres of diameter d. This form of potential takes into account the effects of the finite size of the molecules and the intermolecular force of attraction. However, the repulsive part of the potential clearly rises too sharply compared to real molecules that are 'softer' than a rigid sphere. ε is the depth of the potential well.

Figure 5.4. This form of the potential takes into account the effects of (i) the finite size of the molecules, which reduces the volume available to the molecules and (ii) the intermolecular force of attraction. Importantly, the van der Waals equation decouples the two effects and this enables them to be dealt with separately. Note that the repulsive part of the intermolecular potential clearly rises too sharply compared to real molecules that are 'softer' than a rigid sphere.

5.1.1 The finite size of molecules

Molecules do have a finite size and hence a finite volume. Consequently, the volume of space that is available to the molecules is less than the volume V of the gas container. We can write the available volume as $(V - b)$ where b depends on the total volume of all the molecules, but is not the actual total volume, as we shall see. Taking d to be the diameter of a molecule in our rigid-sphere model, the molecular volume is $\frac{4}{3}\pi(d/2)^3$. If a molecule is approached by a second molecule, they will only come into contact when the distance between their centres is equal to d, see Figure 5.5. Thus, a molecule appears to carry an exclusion zone about itself of

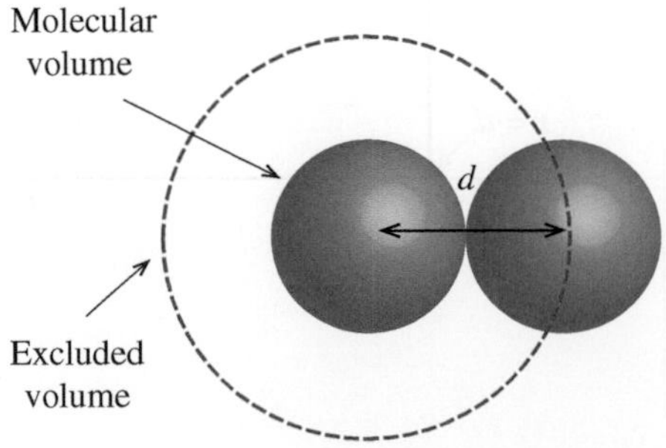

Figure 5.5 If a molecule is approached by a second molecule, they will not come into contact until the distance between their centres is equal to d, the diameter of a molecule. Thus, a molecule appears to carry an exclusion zone about itself of radius d. The volume of the exclusion zone or excluded volume is eight times the volume of a molecule.

radius d, i.e. of volume $\frac{4}{3}\pi d^3$, i.e. eight times the molecular volume. This is the *excluded volume* when a second molecule tries to approach it. Hence, the excluded volume *per molecule* is four times the molecular volume. For one mole of gas, there are Avogadro's number N_A of molecules. Hence, the total excluded volume for one mole of gas is $\frac{2}{3}N_A\pi d^3$. Therefore, in the van der Waals equation, the term V in the ideal gas equation is replaced by $(V - b)$, where

$$b = \frac{2}{3}N_A\pi d^3. \tag{5.3}$$

5.1.2 The intermolecular force of attraction

In the interior of a gas, a molecule will be attracted to some extent by all the other molecules around it. Over a period of time, these forces of attraction cancel out. However, molecules that strike the walls of the container experience only the attractive forces from the molecules behind them. This means that the resultant attractive force is not balanced; see Figure 5.6. Consequently, the kinetic energy of the molecules striking the wall is reduced. The result is that the pressure P exerted on the container walls, i.e. the pressure that is measured experimentally, is *lower* by a certain amount than the pressure in the interior of the gas.

van der Waals derived an expression for the pressure difference between the pressure in the interior of the gas and at the container wall. He considered that this difference was proportional to the number of molecules striking the wall per second and also the number of molecules behind them since this is a measure of the force of attraction. Both of these are proportional to the pressure P of the gas. Consequently, the pressure difference is proportional to P^2. For a given quantity of gas, the pressure is inversely proportional to volume V. Hence, the pressure difference can be represented by the term a/V^2, where a is a constant. Then the pressure in the interior of the gas is equal to P *plus* a/V^2, i.e. $P + a/V^2$.

The attraction of the walls on the incident molecules increases their velocity from say v to $v + \delta v$. After rebounding from the walls, however, this force of attraction reduces the speed of the molecules back to v

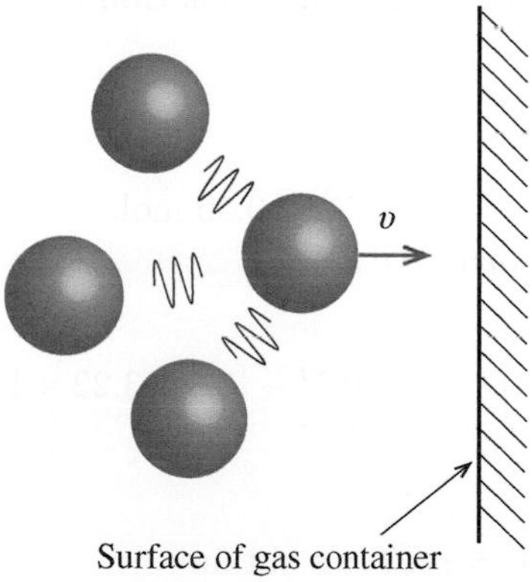

Figure 5.6 In the interior of a gas, a molecule will be attracted to some extent by all the other molecules around it. Over a period of time, these forces of attraction cancel out. However, molecules that strike the walls of the container experience only the attractive forces from the molecules behind them. This means that the resultant attractive force is not balanced. Consequently, the kinetic energy of the molecules striking the wall is reduced and hence the pressure exerted on the container walls is lower than the pressure in the interior of the gas.

again. Thus, the attraction of the walls has no net effect on the momentum change due to collisions with the wall, i.e. on the pressure exerted on the walls.

Combining the above two corrections to the ideal gas law, we obtain the van der Waals equation:

$$\left(P + \frac{a}{V^2}\right)(V - b) = RT$$

for one mole of gas. At high pressure, the number density of molecules is high, and then the volume factor b and the pressure defect a/V^2 become important. Conversely, at low pressure, implying large volume, the molecules are far apart on average and the gas behaves like an ideal gas obeying $PV = RT$.

For n moles of gas, the van der Waals equation becomes

$$\left(P + \frac{n^2 a}{V^2}\right)(V - nb) = nRT. \tag{5.4}$$

We see that the units of a are (volume)^2(pressure)/(mol)$^2 \equiv$ (m$^6 \cdot$ Pa)/mol^2, and the units of b are (volume)/(mol) $\equiv$ m^3/mol.

The van der Waals equation does not represent an exact law; it is only an approximate equation of state. The constants a and b are best regarded as empirical parameters for a particular molecule rather than as precisely defined molecular parameters. The van der Waals equation does, however, have the advantages that it is a relatively simple equation and it only introduces two extra parameters to the ideal gas equation. In addition, it provides physical insights into how intermolecular interactions contribute to the deviations of a real gas from the perfect gas law. We can judge the reliability of the van der Waals equation by comparing its predictions with experimental data, as we will do in Section 5.2.

Worked example

Using the van der Waals equation, calculate the pressure exerted by 2.0×10^{-2} kg of argon gas held at a temperature of 298 K in a vessel of volume 3.0×10^{-4} m^3. The van der Waals constants for argon are $a = 1.36 \times 10^{-1}$ m$^6 \cdot$ Pa/mol^2 and $b = 3.22 \times 10^{-5}$ m^3/mol. Compare your result with that obtained using the ideal gas equation.

Solution

2.0×10^{-2} kg of argon gas is $2.0 \times 10^{-2}/40 \times 10^{-3} = 0.50$ mol.

From the van der Waals equation (5.4), we have

$$\left(P + \frac{(0.5)^2 \times 1.36 \times 10^{-1}}{(3.0 \times 10^{-4})^2}\right)\left(3.0 \times 10^{-4} - \left[0.5 \times 3.22 \times 10^{-5}\right]\right) = 0.5 \times 8.31 \times 298.$$

This gives $P = 3.98 \times 10^6$ Pa.

If we use the ideal gas equation $PV = nRT$, we obtain

$$P = \frac{0.5 \times 8.31 \times 298}{3.0 \times 10^{-4}} = 4.13 \times 10^6 \text{ Pa}.$$

Notice that even at the relatively high pressure of 3.98×10^6 Pa, ~40 atm, the result from the van der Waals equation is only 4% different from that given by the ideal gas equation.

5.2 *P–V* isotherms for a real gas

An *isotherm* is a plot of the pressure of a gas against its volume at a constant temperature, i.e. P against V at constant T. A set of experimental isotherms for the particular example of carbon dioxide is shown in Figure 5.7 for various fixed temperatures. At the highest temperature of 50 °C, the isotherm resembles that for an ideal gas. Thus, as the gas is compressed, the pressure varies inversely with volume, $P \propto 1/V$, and the isotherm is part of a rectangular hyperbola. The isotherm at the temperature of 13 °C is very different. From A to B, the gas behaves as a gas. However, from B to C, the gas condenses as it is compressed and along this line, it becomes a mixture of gas and liquid. Eventually, at point C, it is completely liquid. Beyond point C, the isotherm rises very rapidly because a liquid is essentially incompressible. For the isotherm at the higher temperature of 22 °C, the region over which the mixture of gas and liquid coexists becomes smaller. Eventually, at the temperature of 31 °C, there is no condensation of the gas as it is compressed, no matter how high the pressure. Consequently, this is called the *critical temperature* T_C and the isotherm that passes through the critical temperature is called the *critical isotherm*. We can conclude that a gas cannot be liquefied by the application of pressure unless the temperature is below the critical temperature.

The critical temperature T_C can be readily understood in terms of intermolecular forces. The binding energy between two molecules corresponds to the depth of the Lennard-Jones potential well ε; see Figure 5.2. The thermal energy of a molecule $\sim kT$. If $kT > \varepsilon$, we can expect the molecules to be unbound, i.e. to have enough thermal energy to overcome the molecular force of attraction. This gives $T_C \sim \varepsilon/k$. In fact this prediction agrees quite well with experiment as illustrated by the data in Table 5.1.

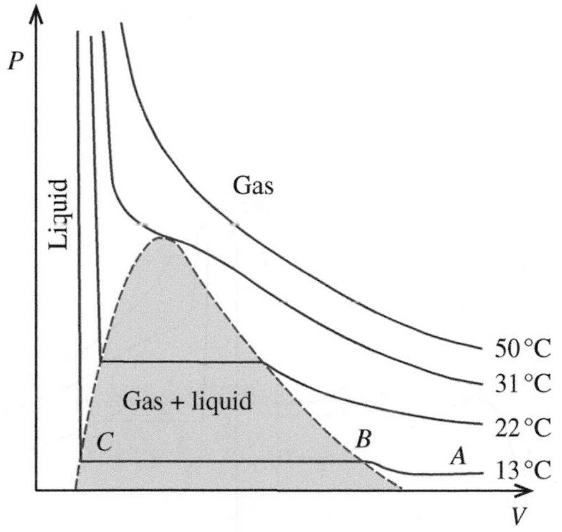

Figure 5.7 A set of experimental isotherms for carbon dioxide for various fixed temperatures. At the highest temperature of 50 °C, the isotherm resembles that for an ideal gas, i.e. part of a rectangular hyperbola. The isotherm at the temperature of 13 °C is very different. From A to B, the gas behaves as a gas. However, from B to C, the gas condenses as it is compressed and along this line, it becomes a mixture of gas and liquid. Eventually, at point C, it is completely liquid. Beyond point C, the isotherm rises very rapidly because a liquid is essentially incompressible. For the isotherm at the temperature of 22 °C, the region over which the mixture of gas and liquid coexists becomes smaller. Eventually, at the temperature of 31 °C, there is no condensation of the gas as it is compressed, no matter how high the pressure. Consequently, this is called the critical temperature T_C and the isotherm that passes through the critical temperature is called the critical isotherm.

Table 5.1 Critical temperatures for some gases.

Gas	ε (J)	ε/k	T_C (observed)
He	0.8×10^{-22}	6	5.2
H_2	4×10^{-22}	30	33
N_2	13×10^{-22}	100	127

The isotherms that the van der Waals equation predicts have the general shapes illustrated by the curves in Figure 5.8a. Above the critical temperature, van der Waals curves do resemble experimental isotherms quite well. However, van der Waals curves below the critical temperature have an 'S' shape, with two turning points. Such a curve is seen in Figure 5.8a, where the turning points occur at A and B. Such a curve predicts that between A and B the pressure increases as the volume increases. This is clearly unphysical behaviour.

We can see why there are two turning points in the van der Waals equation by recasting it in the form

$$PV^3 - (bP + RT)V^2 + aV - ab = 0. \tag{5.5}$$

This shows that the van der Waals equation is a cubic equation in V. Over a certain range of temperature and pressure, this cubic equation has three roots with two turning points. Figure 5.8b illustrates the way this unphysical behaviour of a van der Waals curve can be taken into account. A horizontal line is drawn between points C and D on the curve such that the two 'loops', one above and one below the horizontal line, have the same area. This may seem a rather arbitrary construction, but, in fact, it is based on sound physical reasoning and is called the *Maxwell construction.* The resultant curve including the horizontal line is then quite a good representation of an actual isotherm.

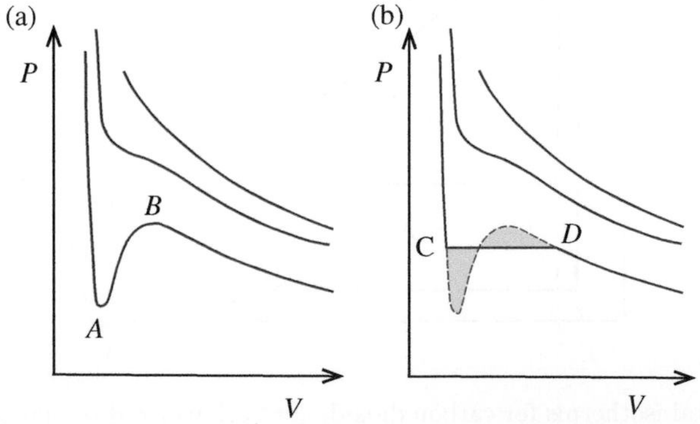

Figure 5.8 (a) The general shapes of the isotherms predicted by the van der Waals equation. Above the critical temperature, van der Waals curves do resemble experimental isotherms quite well. However, van der Waals curves below the critical temperature have an 'S' shape, with two turning points. The turning points occur at A and B. Such a curve predicts that between A and B, the pressure increases as the volume increases. This is clearly unphysical behaviour. (b) The unphysical behaviour of the van der Waals curve can be taken into account by drawing a horizontal line between points C and D such that the two 'loops', one above and one below the horizontal line, have the same area. This is called the Maxwell construction.

5.2.1 The critical points, T_C, P_C, and V_C

As the temperature increases, the two turning points in the van der Waals curve converge and at the critical temperature T_C, they merge to become a point of inflection; see Figure 5.9. Hence, at this temperature, both the first and second derivatives of the van der Waals equation with respect to V are equal to zero:

Multiplying out the van der Waals equation, we obtain

$$PV - \frac{ab}{V^2} + \frac{a}{V} - bP = RT. \tag{5.6}$$

The positions of the two turning points occur when $(\partial P/\partial V)_T$ of Equation (5.6) is zero.

Differentiating Equation (5.6) with respect to V, keeping T constant, we have

$$P + V\left(\frac{\partial P}{\partial V}\right)_T + \frac{2ab}{V^3} - \frac{a}{V^2} - b\left(\frac{\partial P}{\partial V}\right)_T = 0. \tag{5.7}$$

Setting $(\partial P/\partial V)_T = 0$ gives

$$P = \frac{a}{V^2} - \frac{2ab}{V^3}. \tag{5.8}$$

This is the equation of the red dashed curve in Figure 5.9, which traces out the positions of the turning points as the temperature is varied. The point of inflexion in the van der Waals curve occurs when $(\partial P/\partial V)_T$ of Equation (5.8) is zero. Then

$$\left(\frac{\partial P}{\partial V}\right)_T = -\frac{2a}{V^3} + \frac{6ab}{V^4} = 0, \tag{5.9}$$

which gives $V = 3b$. Taking the pressure and the volume at the critical temperature T_C to be P_C and V_C, respectively, we have

$$V_C = 3b. \tag{5.10}$$

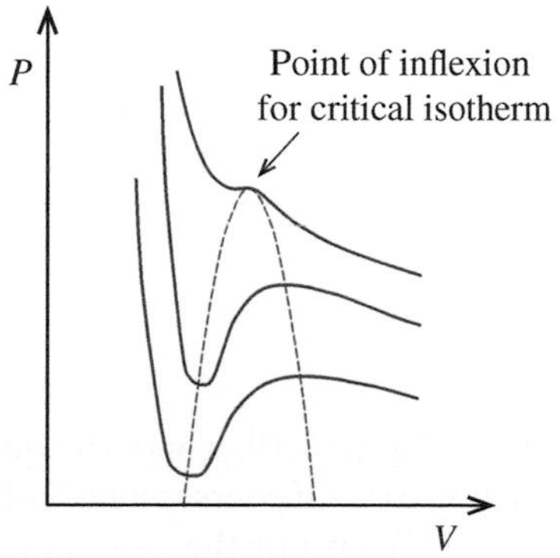

Figure 5.9 As the temperature increases, the two turning points in the van der Waals curves converge and at the critical temperature T_C, they merge to become a point of inflection. At this temperature, both the first and second derivatives of the van der Waals equation with respect to V are equal to zero. The red dashed curve traces out the positions of the two turning points as the temperature is varied.

Substituting for $V_C = 3b$ in Equation (5.8) gives $P_C = \dfrac{a}{27b^2}$. Then substituting for V_C and P_C in the van der Waals equation

$$\left(P + \frac{a}{V^2}\right)(V - b) = RT$$

gives $T_C = 8a/27bR$. The van der Waals equation thus predicts $P_C V_C/RT_C = 0.38$ for all gases. In fact, the experimental values generally lie between 0.2 and 0.3. For example, the ratio $P_C V_C/RT_C$ for argon is 0.29.

Worked example

The values of the critical temperature and critical pressure for nitrogen gas are measured to be 127 K and 3.43×10^6 Pa, respectively. Determine the values of the van der Waals constants a and b from these data. Hence, deduce a value for the diameter of a nitrogen molecule.

Solution

We have $P_C = a/27b^2$ and $T_C = 8a/27bR$.

Hence,

$$b = \frac{T_C R}{8P_C} = \frac{127 \times 8.31}{8 \times 3.43 \times 10^6} = 3.84 \times 10^{-5}\ \mathrm{m^3/mol},$$

and

$$a = \frac{27T_C^2R^2}{64P_C} = \frac{27 \times 127^2 \times 8.31^2}{64 \times 3.43 \times 10^6} = 0.137\ \mathrm{Pa{\cdot}m^6/mol^2}.$$

We have from Equation (5.3) $b = (2\pi/3)N_A d^3$, giving

$$d = \left(\frac{3b}{2\pi N_A}\right)^{\frac{1}{3}} = \left(\frac{3 \times 3.84 \times 10^{-5}}{2\pi \times 6.022 \times 10^{23}}\right)^{\frac{1}{3}} = 3.12 \times 10^{-10}\ \mathrm{m} = 0.312\ \mathrm{nm}.$$

This value is in good agreement with values of d obtained from other methods of determination; see also Section 1.2.

5.3 The virial equation

A series of virial plots for argon are shown in Figure 5.10, where the quantity PV/RT for 1 mol of the gas is plotted against $1/V$ for various fixed temperatures. Of course, such a plot would be a straight horizontal line for an ideal gas. But for a real gas like argon, this is not the case and the virial curves vary smoothly with inverse volume. To keep the scale of Figure 5.10 in perspective, a value of $1/V = 1 \times 10^4\ \mathrm{m^{-3}}$ corresponds to a pressure of approximately 360 atm at 400 K.

A smooth function $y = f(x)$ can be represented by a polynomial of the form $y = a + bx + cx^2 + dx^3 + \cdots$. Similarly, each virial plot in Figure 5.10 can be represented by an equation of the form

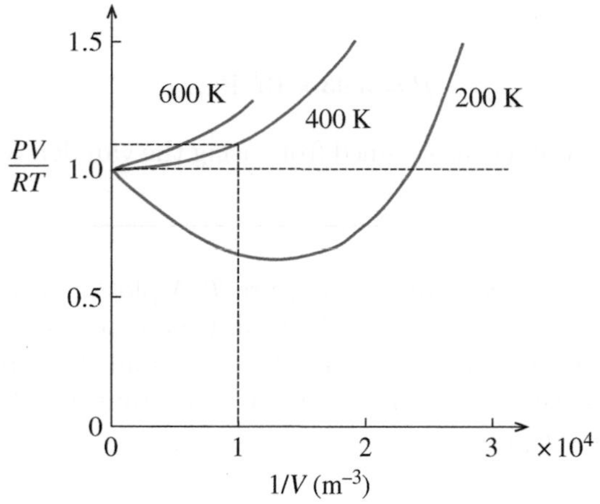

Figure 5.10 A series of virial plots for argon, where the quantity PV/RT is plotted against $1/V$ for various constant temperatures. Such a plot would be a straight horizontal line for an ideal gas, but for a real gas like argon, this is not the case and the virial curves vary smoothly with inverse volume. To keep the scale of the figure in perspective, a value of $1/V = 1 \times 10^4\ \text{m}^{-3}$ at $T = 400$ K corresponds to a pressure of approximately 360 atm.

$$\frac{PV}{RT} = 1 + \frac{B(T)}{V} + \frac{C(T)}{V^2} + \frac{D(T)}{V^3} + \cdots, \tag{5.11}$$

where the coefficients B, C, ... are called the second, third, and higher virial coefficients. As described earlier, the values of the coefficients are obtained by fitting the experimental data to Equation (5.11) and varying the coefficients until a good match is obtained. The third term, which depends on $1/V^2$, only becomes significant at high pressures, when V becomes small.

Worked example

Use the virial equation to calculate the pressure exerted by 2.0×10^{-2} kg of argon gas held at a temperature of 298 K in a vessel of volume $3.0 \times 10^{-4}\ \text{m}^3$. The virial coefficients at this temperature are $B = -18 \times 10^{-6}\ \text{m}^3/\text{mol}$ and $C = 1.1 \times 10^{-9}\ \text{m}^6/\text{mol}^2$.

Solution

For n moles of gas, the virial equation, Equation (5.11), becomes

$$\frac{PV}{nRT} = 1 + \frac{B(T)}{V} + \frac{C(T)}{V^2} + \frac{D(T)}{V^3} + \cdots, \tag{5.12}$$

2.0×10^{-2} kg of argon gas is $2.0 \times 10^{-2}/40 \times 10^{-3} = 0.50$ mol.

Taking $PV/(nRT) = 1 + (B/V) + (C/V^2)$, gives

$$\frac{P \times 3.0 \times 10^{-4}}{0.5 \times 8.31 \times 298} = 1 - \frac{18 \times 10^{-6}}{3.0 \times 10^{-4}} + \frac{1.1 \times 10^{-9}}{(3.0 \times 10^{-4})^2} = 1 - 0.06 + 0.012 = 0.952.$$

Hence,

$$P = 3.93 \times 10^6 \text{ Pa},$$

which is in good agreement with the result obtained from using the van der Waals equation in the previous worked example.

Note that the virial coefficients depend on temperature T. A plot of B against T for argon is shown in Figure 5.11. At low temperatures, B is negative, but it steadily increases as T increases. Notice also that the plot of B against T passes through zero at $T = 411$ K. Hence, around this temperature, and over a considerable range of pressure, the gas obeys the ideal gas law $PV =$ constant, i.e. Boyle's law. Consequently, this temperature is called the *Boyle temperature.*

5.3.1 Relationship between the van der Waals constants *a* and *b* and the virial coefficients *B* and *C*

We can expect there to be a relationship between the van der Waals constants a and b and the virial coefficients B and C. This is indeed the case. We write the van der Waals equation in the form

$$P = \frac{RT}{V-b} - \frac{a}{V^2}. \tag{5.13}$$

Hence,

$$\frac{PV}{RT} = \left(1 - \frac{b}{V}\right)^{-1} - \frac{a}{RTV}. \tag{5.14}$$

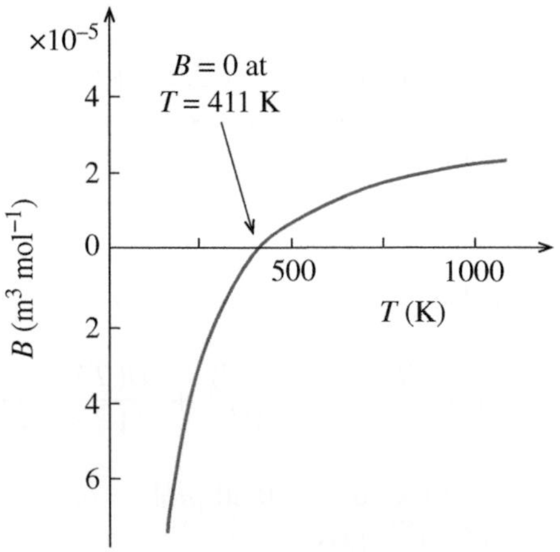

Figure 5.11 A graph of B against temperature T for argon. At low temperatures, B is negative but it steadily increases as T increases. Notice that the graph of B against T passes through zero at $T = 411$ K. Hence, around this temperature, and over a considerable range of pressure, the gas obeys the ideal gas law $PV =$ constant, i.e. Boyle's law. Consequently, this temperature is called the Boyle temperature.

For $b \ll V$, the binomial expansion gives

$$\left(1-\frac{b}{V}\right)^{-1}=1+\frac{b}{V}+\left(\frac{b}{V}\right)^{2}+\cdots.$$

Hence, we can write

$$\frac{PV}{RT}=1+\left(b-\frac{a}{RT}\right)\frac{1}{V}+\frac{b^2}{V^2}+\cdots. \tag{5.15}$$

Comparing this result with Equation (5.11) gives $B = b-(a/RT)$ and $C = b^2$. The van der Waals equation therefore predicts that the second virial coefficient B tends to $-\infty$ at very low temperatures and tends asymptotically to b at very high temperatures. These predications are in agreement with the argon data shown in Figure 5.11. Moreover, the Boyle temperature, when $B = 0$, is equal to a/Rb. Using the accepted values for argon: $a = 1.42 \times 10^{-1}\ \mathrm{Pa \cdot m^6/mol^2}$ and $b = 4.2 \times 10^{-5}\ \mathrm{m^3/mol}$, the resultant value of the Boyle temperature is

$$\frac{1.42\times 10^{-1}}{8.31\times 4.2\times 10^{-5}}=407\ \mathrm{K},$$

which compares well with the actual value of 411 K.

We can understand the temperature behaviour of B on the basis of the weakly interacting rigid spheres model. We recall that the van der Waals constant b relates to the molecular volume of the molecules, which in turn relates to the repulsive force between molecules. The constant a is determined by the attractive force between them. At temperatures so low that the kinetic energy of the molecules $\sim kT$ is small compared to the potential energy ε of two molecules when they are in close contact, the encounters between molecules are of relatively long duration. In that case, the influence of the attractive forces between the molecules is relatively large and so the term $-a/RT$ dominates. However, in the limit of high temperature, the encounters occur over a shorter time scale and the molecules are only slightly deflected by the attractive forces. Then the force of repulsion, represented by the b term, dominates.

5.4 Internal energy and specific heats of a van der Waals gas

We describe a gas that follows the van der Waals equation as a van der Waals gas. When the molecules of such a van der Waals gas are at near infinite distance apart, the attractive forces between them can be neglected. We can therefore take their potential energy to be zero. However, the molecules have kinetic energy, where each degree of freedom contributes $\frac{1}{2}RT$ of energy per mole. A monatomic gas like helium or argon has three degrees of freedom, and so the kinetic energy per mole is $\frac{3}{2}RT$. As the molecules come closer together, the attractive force between them leads to a lowering of their potential energy. From our discussion in Section 5.1.2, we conclude that the pressure change due to the attractive forces is $-a/V^2$. It follows that the potential energy of the gas is *reduced* by the amount

$$-\int_{\infty}^{V}\left(\frac{a}{V^2}\right)\mathrm{d}V=\frac{a}{V}. \tag{5.16}$$

Thus, the total internal energy per mole of a van der Waals gas is

$$U = \frac{3}{2}RT - \frac{a}{V}, \tag{5.17}$$

where the first term on the right-hand side is the kinetic energy and the second term is the potential energy. A negative value of potential energy may seem surprising. However, we recall, for example, that when a satellite is bound to a planet, the satellite has a positive kinetic energy but again its potential energy is negative.

5.4.1 The molar specific heats at constant volume and constant pressure

The molar specific heat at constant volume for a real gas is

$$C_{\mathrm{V}} = \left(\frac{\partial U}{\partial T}\right)_{\mathrm{V}} = \frac{\partial}{\partial T}\left(\frac{3}{2}RT - \frac{a}{V}\right)_{\mathrm{V}} = \frac{3R}{2}. \tag{5.18}$$

This is exactly the same result as for an ideal gas, and we can interpret it as follows. As the volume of the gas is constant, the average distance between the molecules remains the same and so there is no net change in their potential energy. In the case of the molar specific heat at constant pressure, the volume of the gas does change. It increases with increasing temperature. To calculate the molar specific heat at constant pressure, we note that a quantity of heat $C_{\mathrm{P}}\mathrm{d}T$ has to be supplied to raise the temperature by $\mathrm{d}T$. This heat has not only to increase the internal energy of the molecules but also to do work against the external pressure. Hence,

$$C_{\mathrm{P}}\mathrm{d}T = \mathrm{d}U + P\mathrm{d}V. \tag{5.19}$$

We recall that for a function $f(x)$, the differential $\mathrm{d}f$ is defined by

$$\mathrm{d}f = \frac{\mathrm{d}f}{\mathrm{d}x}\mathrm{d}x.$$

For the case of a function of two variables $f(x, y)$, the differential $\mathrm{d}f$ is

$$\mathrm{d}f = \left(\frac{\partial f}{\partial x}\right)_y \mathrm{d}x + \left(\frac{\partial f}{\partial y}\right)_x \mathrm{d}y.$$

Thus, from Equation (5.17), we have

$$\mathrm{d}U = \frac{3}{2}R\mathrm{d}T + \frac{a}{V^2}\mathrm{d}V. \tag{5.20}$$

Substituting for $\mathrm{d}U$ in Equation (5.19), we obtain

$$C_{\mathrm{P}}\mathrm{d}T = \frac{3}{2}R\mathrm{d}T + \frac{a}{V^2}\mathrm{d}V + P\mathrm{d}V = C_{\mathrm{V}}\mathrm{d}T + \left(\frac{a}{V^2} + P\right)\mathrm{d}V. \tag{5.21}$$

We can derive an expression for dV at constant pressure from the van der Waals equation, which we expand to give

$$PV - \frac{ab}{V^2} + \frac{a}{V} - bP = RT.$$

For a change in which P is constant, we must have

$$\frac{\mathrm{d}}{\mathrm{d}V}\left(PV - \frac{ab}{V^2} + \frac{a}{V} - bP\right)\mathrm{d}V = \frac{\mathrm{d}}{\mathrm{d}T}(RT)\mathrm{d}T. \tag{5.22}$$

Therefore

$$\left(P + \frac{ab}{V^3} - \frac{a}{V^2}\right)\mathrm{d}V = R\mathrm{d}T. \tag{5.23}$$

We can neglect the term in ab/V^3 and hence

$$\mathrm{d}V = \frac{R\mathrm{d}T}{(P - a/V^2)}. \tag{5.24}$$

Substituting for dV in Equation (5.21), we obtain after some rearrangement

$$C_{\mathrm{P}}\mathrm{d}T = C_{\mathrm{V}}\mathrm{d}T + \left(1 + \frac{a}{PV^2}\right)\left(1 - \frac{a}{PV^2}\right)^{-1} R\mathrm{d}T. \tag{5.25}$$

Applying the binomial theorem and ignoring second and higher order terms gives

$$C_{\mathrm{P}} - C_{\mathrm{V}} \approx R\left(1 + \frac{2a}{PV^2}\right) \approx R\left(1 + \frac{2a}{RTV}\right). \tag{5.26}$$

As $C_{\mathrm{V}} = \frac{3R}{2}$, we obtain

$$C_{\mathrm{P}} \approx \frac{5}{2}R + \frac{2a}{TV}. \tag{5.27}$$

for a monatomic gas obeying the van der Waals equation.

5.5 Phase diagrams

The particular *phase*, gaseous, liquid, or solid, that a substance is in depends on the state parameters: volume V, pressure P, and temperature T. For example, water at atmospheric pressure exists in the liquid phase over the temperature range 0–100 °C. A *phase diagram* is a convenient way to display the relationship between the gaseous, liquid, and solid phases of a substance and the state parameters. One kind of phase diagram plots the pressure of the substance as a function of volume. Figure 5.7 is an example of such a phase diagram. Another kind is to plot the pressure of the substance as a function of temperature at constant volume. Such a phase diagram is shown in Figure 5.13. And the relationship between pressure, temperature, and volume for the substance can be represented as a surface in a three-dimensional space with coordinates P, V, and T.

We can construct a phase diagram for a particular substance using an arrangement like that shown in Figure 5.12. This consists of a cylinder/piston arrangement contained in a thermostatically controlled enclosure. The cylinder contains a fixed amount of the substance. The pressure P and the volume V occupied by the substance can be varied by moving the piston and the temperature T of the arrangement can also be controlled. These three parameters P, V, and T are related by the equation of state. Hence, in general, if we set two of the parameters, the third is determined by that equation. In the case of an ideal gas, if we set, for example, the pressure P and temperature T, the volume V is determined by the ideal gas law: $PV = RT$.

Consider first the phase diagram shown in Figure 5.7, and, in particular, the 13 °C isotherm. Initially, the substance is in the gaseous phase. As the gas in the cylinder is compressed by pushing the piston into the cylinder, the pressure increases at first, broadly in agreement with Boyle's law until point B is reached. Above this point, the piston can be pushed in without any further increase in pressure until all the gas is converted into the liquid phase. As the piston is pushed in further, the liquid is compressed until point C is reached when the liquid is solidified. Obviously, it takes a large increase in pressure to make an appreciable change in the volume of the substance when in the liquid or solid phases.

Consider now the phase diagram shown in Figure 5.13 where the axes are pressure P and temperature T. This phase diagram is typical for many substances. There are various areas of the phase diagram corresponding to the gaseous, liquid, and solid phases, as indicated. In particular, only a single phase can exist

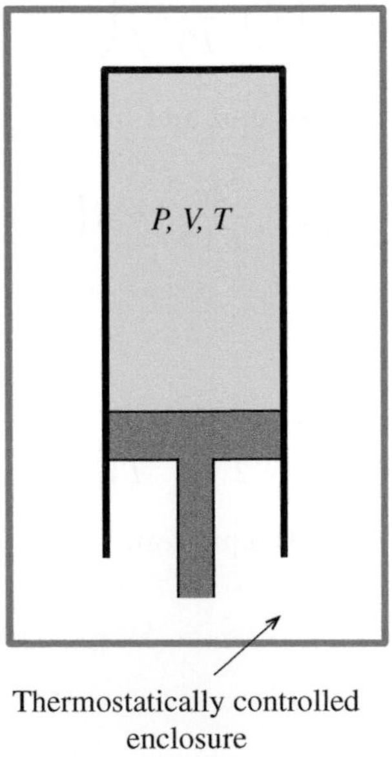

Figure 5.12 A system that can be used to construct the phase diagram of a substance. It consists of a cylinder/piston arrangement contained in a thermostatically controlled enclosure. The cylinder contains a fixed amount of the substance. The pressure P and the volume V occupied by the substance can be varied by moving the piston and the temperature T of the system can also be controlled.

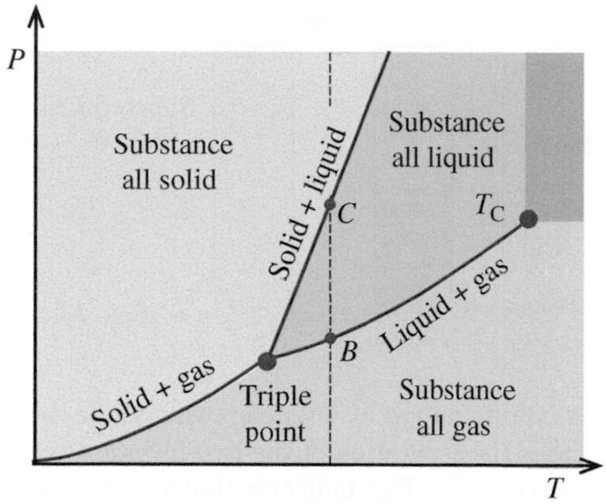

Figure 5.13 A typical phase diagram, where the axes are pressure P and temperature T. There are various areas of the phase diagram corresponding to the gaseous, liquid, and solid phases, as indicated. In particular, only a single phase can exist at each (T, P) point except for the points lying on the solid lines. These solid lines are called phase boundaries and along these lines, two phases can coexist in phase equilibrium. It is along the phase boundaries that transitions between two different phases take place. The three phase boundaries meet at the triple point. At this single point, and only at this point, can all three phases coexist.

at each (T, P) point except for the points lying on the solid lines. These solid lines are called *phase boundaries* and along these lines, two phases can coexist in *phase equilibrium*. The respective phases are indicated in Figure 5.13. It is along the phase boundaries that transitions between two different phases take place. The three phase boundaries meet at the *triple point*. At this single point, and only at this point can all three phases coexist. The dashed vertical line drawn on the phase diagram represents a line of constant temperature and can be related to the 13 °C isotherm in Figure 5.7.

The phase boundary between the liquid and gas phases does not continue indefinitely. Instead, it terminates at a point on the phase diagram corresponding to the *critical temperature* T_C; see Figure 5.13. This reflects the fact that, at extremely high temperatures and pressures, the liquid and gaseous phases become indistinguishable. This effect was discovered by Cagniard de la Tour in 1822. We can describe it as follows. Suppose we heat a liquid in a sealed vessel; see Figure 5.14. Initially, the temperature is such that the vessel contains liquid and gas and the meniscus at the top of the liquid indicates the separation of the two phases. As the temperature is raised, there comes a point, at the critical temperature, at which the meniscus disappears. This indicates that there is now no separation between the two phases. Instead, at the critical temperature and above, a single form of matter fills the vessel. This form of matter is called *a supercritical fluid*. It is not strictly a liquid because it never possesses a surface that separates it from the gaseous phase. However, it behaves like one in many respects. It has a similar density to a liquid and it can act as a solvent. An interpretation of the formation of a supercritical fluid is as follows. The density of the gas increases and the density of the liquid decreases as the temperature rises as illustrated schematically by Figure 5.15. Eventually, at the critical temperature, the densities of liquid and gas become the same and the two are identical; there is no separation between liquid and gas phases. The behaviour of a substance at its critical temperature is an example of *critical phenomena*, and such phenomena occur in many other branches of physics. For example, a ferromagnet near its Curie point behaves quite similarly to a liquid near its critical point.

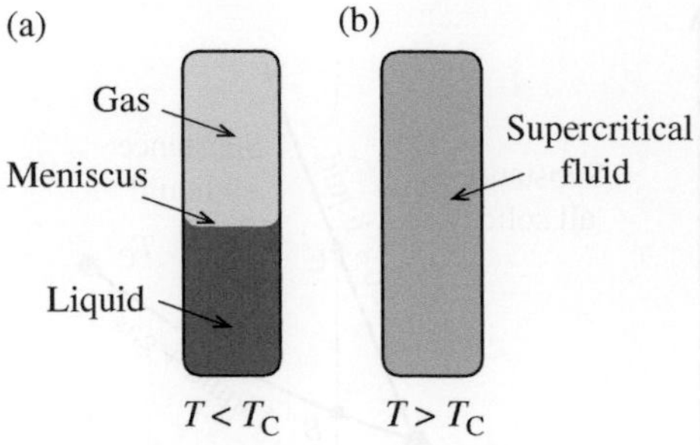

Figure 5.14 (a) A vessel that contains liquid and gas at a temperature below the critical temperature T_C. The meniscus at the top of the liquid indicates the separation of the two phases. (b) If the vessel is heated, there comes a point, at T_C, at which the meniscus disappears. This indicates that there is now no separation between the two phases. Instead, at the critical temperature and above, a single form of matter fills the vessel. This form of matter is called a supercritical fluid.

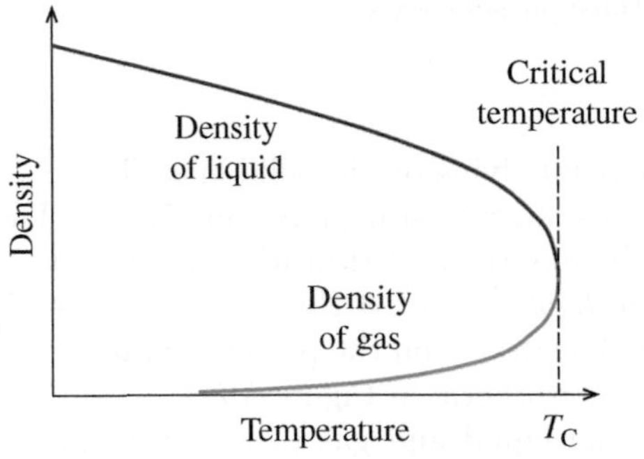

Figure 5.15 An interpretation of the formation of a supercritical fluid is as follows. The density of the gas increases and the density of the liquid decreases as the temperature rises. Eventually, at the critical temperature, the densities of liquid and gas become the same and the two are identical; there is no separation between liquid and gas phases.

As an example of the phase diagram of a real substance, the phase diagram of water is shown in Figure 5.16. For most substances, the solid–liquid phase boundary in the phase diagram has a positive slope so that the melting point increases with pressure, as in Figure 5.13. This is the case whenever the solid phase is denser than the liquid phase. Water is an exception to this. Ice is less dense than liquid water and, consequently, it has a solid–liquid boundary with a negative slope so that the melting point decreases with pressure. The fracturing of water pipes in extended periods of cold weather is a consequence of the fact that ice is less dense than water and so occupies a larger volume. The triple point of water occurs at a temperature of 0.01 °C (273.16 K) and a pressure of 0.0064 atm (612 Pa), and can be used as a calibration point to define the Kelvin temperature scale. The critical temperature of water is 374 °C. Note that for a given pressure, any two phases are in equilibrium at one specific temperature. For example, at atmospheric pressure, water and ice are in equilibrium at the fixed temperature of 0 °C.

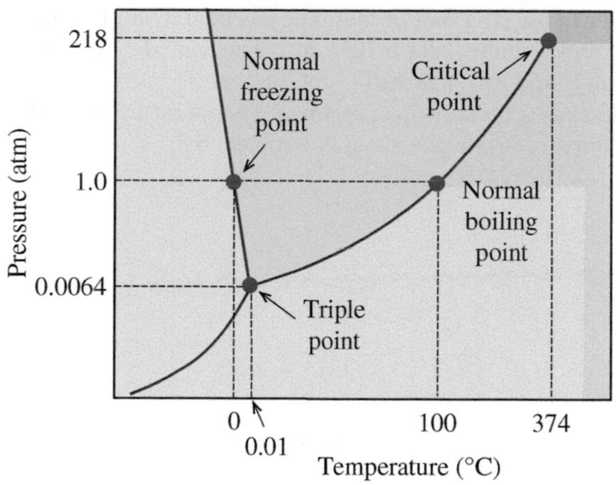

Figure 5.16 The phase diagram of water. The solid–liquid phase boundary of water has a negative slope, unlike most substances, because ice is less dense than liquid water. Hence, the melting point of ice decreases with pressure. The triple point of water occurs at a temperature of 0.01 °C and a pressure of 0.0064 atm (612 Pa). The critical temperature is 374 °C. Note that for a given pressure, any two phases are in equilibrium at one specific temperature. For example, at atmospheric pressure, water and ice are in equilibrium at the fixed temperature of 0 °C.

Problems 5

5.1 Show that the Boyle temperature at which a van der Waals gas behaves as an ideal gas is given by the expression $T_B = (a/bR)[1 - (b/V)]$, where a and b are the constants of the van der Waals equation of state. Assuming $b \ll V$, calculate the Boyle temperature of carbon dioxide, for which $a = 0.359\ \mathrm{m^2 \cdot Pa/mol}$ and $b = 4.27 \times 10^{-5}\ \mathrm{m^3/mol}$.

5.2 Calculate the temperature above which carbon dioxide cannot be liquefied. Use the values of the van der Waals constants given in Problem 5.1.

5.3 0.75 mol of argon is contained in a vessel of volume $5 \times 10^{-4}\ \mathrm{m^3}$ at a temperature of 600 K. Calculate the pressure of the gas using (i) the perfect gas equation, (ii) the van der Waals equation, and (iii) the virial equation. The van der Waals constants for argon are $a = 1.36 \times 10^{-1}\ \mathrm{m^6 \cdot Pa/mol^2}$ and $b = 3.22 \times 10^{-5}\ \mathrm{m^3/mol}$. The virial coefficients for argon at 600 K are $B = 1.10 \times 10^{-5}\ \mathrm{m^3/mol}$ and $C = 0.75 \times 10^{-9}\ \mathrm{m^6/mol^2}$.

5.4 Using the van der Waals data in Problem 5.3, estimate the diameter of an argon atom.

5.5 Starting from van der Waals equation of state, show that the second virial coefficient B for a gas of rigid spheres of diameter d without attractive forces is given by $B = \frac{2}{3} N_A \pi d^3$.

5.6 Rewrite the van der Waals equation in terms of the so-called *reduced variables* $P_r = P/P_C$, $V_r = V/V_C$ and $T_r = T/T_C$, where P_C, V_C and T_C are the critical pressure, volume, and temperature, respectively.

5.7 There exist other expressions for the equation of state of a real gas. Dieterici's equation of state is $P(V - b) = RTe^{-a/RVT}$. Show that this equation of state has the following critical constants: $V_C = 2b$, $T_C = a/4bR$ and $P_C = a/4e^2b^2$.

5.8 Measurements of the second and third virial coefficients of hydrogen gas at 298 K gave $B = 14.1 \times 10^{-6}\ \mathrm{m^3/mol}$ and $C = 350 \times 10^{-12}\ \mathrm{m^6/mol^2}$, respectively. Use these data to calculate the van der Waals constants a and b for hydrogen gas.

5.9 Show that the fractional difference between the pressure P_{vdw} predicted by the van der Waals equation of state and the pressure P_i predicted by the ideal gas equation is given by

$$\frac{\Delta P}{P_i} = \frac{P_{vdw} - P_i}{P_i} = \left[\frac{1}{V}\left(b - \frac{a}{RT}\right) + \frac{b^2}{V^2}\right].$$

Find the fractional difference for (i) 1 mol of methane gas at 0 °C and volume 2.24×10^{-2} m^3 and (ii) 1 mol of methane gas at 0 °C and volume 2.24×10^{-4} m^3. The van der Waals constants for methane gas are $a = 2.28 \times 10^{-1}\,\text{m}^6 \cdot \text{Pa/mol}^2$ and $b = 4.28 \times 10^{-5}\,\text{m}^3/\text{mol}$.

5.10 Calculate the molar specific heats C_V and C_P of argon at a temperature of 160 K and a volume of 6.0×10^{-4} m^3. The van der Waals constants for argon are given in Problem 5.3.

6

The First Law of Thermodynamics

So far, we have described matter in terms of the forces acting between molecules and molecular motions. This is a description of matter on the *microscopic* scale. One aspect of this approach is that the amount of matter encountered in any practical situation contains a huge number of molecules; of the order of Avogadro's number, 6×10^{23}. On the one hand, it is an impossible task to know the positions and momenta of such a huge number of molecules. On the other hand, this huge number enables us to use statistical methods to find average values of microscopic quantities that we can relate to *macroscopic* properties of gases. For example, in Section 3.1, we considered the impacts of the molecules of a gas on the walls of its container to obtain an expression for the pressure exerted on those walls. Moreover, we were also able to deduce the equation of the state of an ideal gas, the relationship between pressure P, temperature T, and volume V:

$$PV = RT, \tag{6.1}$$

where R is the gas constant. And in Chapter 4, we saw how the microscopic approach enabled us to obtain expressions for the transport properties of gases.

A quite different approach to the properties of matter is that of classical thermodynamics. This is based on a small number of basic principles that have been deduced from a large body of experimental observation. These basic principles are the laws of thermodynamics. These are phenomenological laws that are justified by their success in describing macroscopic phenomena. They avoid all molecular concepts, and instead deal exclusively with macroscopic variables such as pressure, temperature, and volume. This avoidance of molecular concepts limits the information that thermodynamics can provide about a system. For example, the equation of the state of a system cannot be derived from the laws of thermodynamics; it must be deduced from an experiment. Nevertheless, thermodynamics is very powerful and finds wide application throughout the physical sciences, including physics, chemistry, and biology.

This chapter deals with the first law of thermodynamics. This is essentially the conservation of energy when heat is taken into account. Much of thermodynamics deals with the closely related concepts of temperature, heat, and work, and we begin with a discussion of these basic concepts. We then consider the ways in which energy can be put into or taken out of a system, which may involve the transfer of heat and the performance of work.

Physics of Matter, First Edition. George C. King.

Much of our discussion will concentrate on the ideal gas, the fundamental thermodynamic system. However, gases under normal conditions of temperature and pressure behave as ideal gases and so the results we present have practical application. Moreover, the general principles that we will discuss apply to a much wider range of physical systems.

6.1 Thermodynamic equilibrium

To describe thermodynamic equilibrium, we consider a gas contained in a cylinder with a frictionless piston, as shown in Figure 6.1. The piston is free to move and is at rest at a position where the force exerted by the gas on the piston is equal to the external force exerted on the other side of the piston. The system is then said to be in *mechanical equilibrium* with its surroundings. The walls of the cylinder are *diathermic*, which means that they allow the flow of heat between the gas and the surroundings. This ensures that they are at the same temperature in *thermal equilibrium*. And as there are no changes in the chemical state of the gas, for example, no changes in chemical composition, the gas is said to be in *chemical equilibrium*. When the conditions for all three types of equilibrium are obtained, the system is said to be in a state of *thermodynamic equilibrium*. In that case, there will be no tendency for the system or its surroundings to move away from its state of equilibrium. If, on the other hand, any of the three conditions are not satisfied, the system is said to be in a non-equilibrium state.

The gas itself will then be in a state of equilibrium, in which the pressure and temperature of the gas are uniform throughout the cylinder. Of course, this is not a static equilibrium. At the microscopic level, we know that the molecules in a gas are continually moving in random motion. Clearly, therefore, the number density of molecules in a sufficiently small volume of the gas is not constant. However, over time, the *average* number of molecules in that volume will be constant and the macroscopic pressure of the gas that we measure will be constant. Similarly, the molecules have a range of thermal energy as described by the Boltzmann distribution. But again, the average thermal energy of the molecules is well defined and the macroscopic

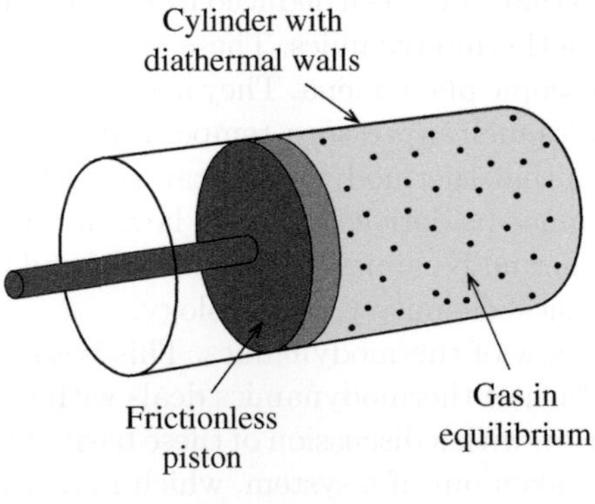

Figure 6.1 The figure illustrates a cylinder of gas in thermodynamic equilibrium. The piston is free to move and is at a position where the force exerted by the gas on the piston is equal to the external force exerted on the piston. This means that the system is in mechanical equilibrium with its surroundings. The diathermal walls of the cylinder allow heat to flow between the gas and the surroundings so that the system is in thermal equilibrium with its surroundings. And as there are no chemical changes in the state of the gas, the system is in chemical equilibrium. The gas itself is in equilibrium, with constant pressure and temperature throughout the cylinder.

temperature of the gas that we measure will be constant. Hence, we may say that over time, the system in thermodynamic equilibrium undergoes no macroscopically observable changes.

A crucially important property of an equilibrium state is that it can be specified by just a few macroscopic variables, and the description of a gas is particularly simple. A given amount of gas is fully determined by its volume V, temperature T, and pressure P. For an equilibrium state, these variables are called *functions of state.*

6.1.1 The equation of state

Suppose that we have a gas contained in a cylinder by a piston but now the cylinder is in contact with a heat reservoir at temperature T, as illustrated by Figure 6.2. By heat reservoir, we mean a body whose heat capacity is very large compared to that of the system to which it is connected. This means that the system will always be at a constant temperature, no matter what process it undergoes. We fix the volume of the gas at some arbitrary value V by moving the piston. And we chose a particular value of temperature T. Then we will find that we cannot vary the pressure P for those fixed values of T and P. Similarly, if P and T are chosen arbitrarily, we would not have a choice for the value of V. That is, of the three functions of state, P, T and V, only two are independent variables. This means that there must be an equation that connects these three variables. Such an equation is called the *equation of state* of the system. In the case of an ideal gas, the equation of state is particularly simple. It is Equation (6.1):

$$PV = RT,$$

for one mole of gas. The state of the gas is fully determined by just two of the three variables. If the pressure and volume are fixed, then

$$T(P, V) = \frac{PV}{R}.$$

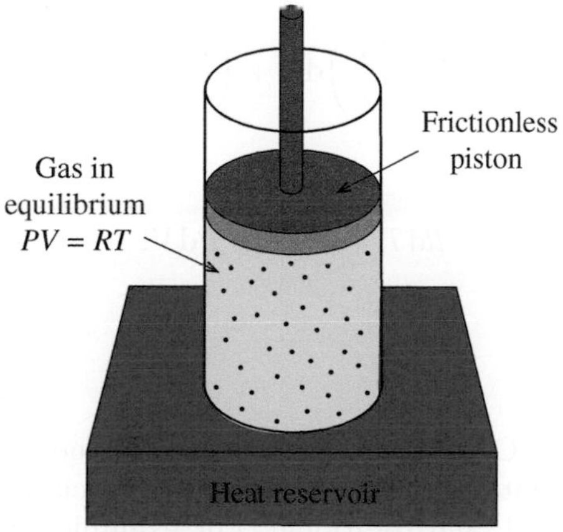

Figure 6.2 An arrangement that allows the volume and temperature of a gas to be varied. The state of the gas is determined by its equation of state $PV = RT$ in terms of the variables P, V, and T. If we choose particular values of two of the variables, the third one is fixed by the equation of state.

Similarly,

$$P = \frac{RT}{V}; \quad V = \frac{RT}{P}.$$

Equation (6.1), also called the *ideal gas law*, applies to an ideal gas, but most gases under normal conditions of pressure and temperature behave like an ideal gas and can be described by the ideal gas law. This makes it one of the most important equations in thermodynamics. And we emphasize that *the ideal gas law can always be applied when a gas is in thermal equilibrium*. In all of our discussions, we will assume the gas behaves like an ideal gas unless otherwise stated.

Every thermodynamic system has an equation of state, although it may be so complicated that it does not allow a simple analytic expression and must be found by experiment. For example, we saw in Section 5.1 that for real gases at high pressures, the equation of state is

$$\left(P + \frac{a}{V^2}\right)(V - b) = RT,$$

where a and b are best regarded as empirical parameters for a particular gas rather than as precisely defined molecular parameters.

An important property of a function of state is that it depends only on the state of the system and not on how that state was obtained. Thus, if we have one mole of gas in an equilibrium state, with pressure P and volume V, then the temperature T is uniquely defined through Equation (6.1). It does not matter how that pressure P and volume V were obtained; with those values of P and V the temperature will be $T = PV/R$.

As, say, the temperature T of a gas can be related to its pressure P and volume V by a mathematical equation, we can readily obtain an expression for the differential $\mathrm{d}T$; the change in temperature for infinitesimal changes in P and V. It is

$$\mathrm{d}T = \left(\frac{\partial T}{\partial P}\right)_{\mathrm{V}} \mathrm{d}P + \left(\frac{\partial T}{\partial V}\right)_{\mathrm{P}} \mathrm{d}V,$$

and hence,

$$\mathrm{d}T = \left(\frac{V}{R}\right)\mathrm{d}P + \left(\frac{P}{R}\right)\mathrm{d}V,$$

which gives

$$R\mathrm{d}T = V\mathrm{d}P + P\mathrm{d}V. \tag{6.2}$$

6.2 Temperature

The concept of temperature arose in Chapter 3 when we discussed the thermal energy of molecules. There we saw that temperature is an absolute measure of the energies associated with the *random* motions of the molecules: translational, vibrational, and rotational. For example, the average translational kinetic energy of a molecule in a gas in thermal equilibrium is equal to $3/2kT$.

Temperature also plays a central role in classical thermodynamics. In thermodynamics, we can consider temperature in the following way. Suppose we have two bodies that are initially at different temperatures.

If we place the two bodies in thermal contact, it is a familiar observation that the hotter body cools down and the colder body warms up until both bodies are at the same temperature. This equalization process is due to a net flow of thermal energy from the hot to the cold body, which we call heat. When heat no longer flows between the two bodies, i.e. when they are in equilibrium, they are at the same temperature. This tendency to equalization is one of the most remarkable features of temperature and gives us a meaning of what we mean by temperature, i.e. two bodies have the same temperature if they are in thermal equilibrium with each other.

6.2.1 The zeroth law of thermodynamics

Suppose that two bodies are in thermal equilibrium with each other. If a third body is in thermal equilibrium with one of the two bodies, then it follows that it must also be in thermal equilibrium with the other one. This is the *zeroth law of thermodynamics*. The zeroth law was not formulated until after the first and second laws had been formulated and numbered. However, the other laws depend on it. Moreover, the zeroth law has an important application. It enables the technique of thermometry. It follows from this law that if a third body, such as a mercury thermometer, is in thermal equilibrium with two separate bodies, then the two external bodies are in equilibrium with each other, i.e. have the same temperature. We do not need to bring the two bodies into intimate contact to see if they are in thermal equilibrium.

6.2.2 The measurement of temperature

To measure the temperature of a body we use a thermometer that has a physical property that varies with temperature in a known way. A mercury thermometer is a familiar example where the length of the column of mercury gives a measure of the temperature. Another example is the platinum resistance thermometer, where the resistance of the platinum gives a measure of the temperature. Ideally, the thermometric property of the thermometer would vary linearly with temperature. However, this is usually not the case. For example, the way the resistance of a platinum resistance thermometer varies with temperature over the range 0 °C and 100 °C can be described by

$$R_T = R_0\left(1 + AT + BT^2\right), \tag{6.3}$$

where R_0 is the resistance at 0 °C, R_T is the resistance at temperature T where T is measured with respect to 0 °C and A and B are constants that are determined by the experiment. The term in T^2 results in the non-linearity of the response, although for platinum $B \ll A$ and so, in practice, the error in a measured temperature is small.

For each of these thermometers, one can measure the thermometric property at two fixed points, say the *ice point*, i.e. the normal freezing point of water, and the *steam point*, i.e. the normal boiling point of water. Other temperatures are then found by interpolating between the two fixed points. For example, the Celsius or centigrade scale is defined by taking the temperatures of the ice and steam points to be 0 °C and 100 °C, respectively. Defining the thermometric property as X, the measured temperature T is given by

$$T = \frac{(X_{100} - X_0)}{100} X_T, \tag{6.4}$$

where X_{100} and X_0 are the values measured at 100 °C and 0 °C, respectively, and X_T is the value at temperature T.

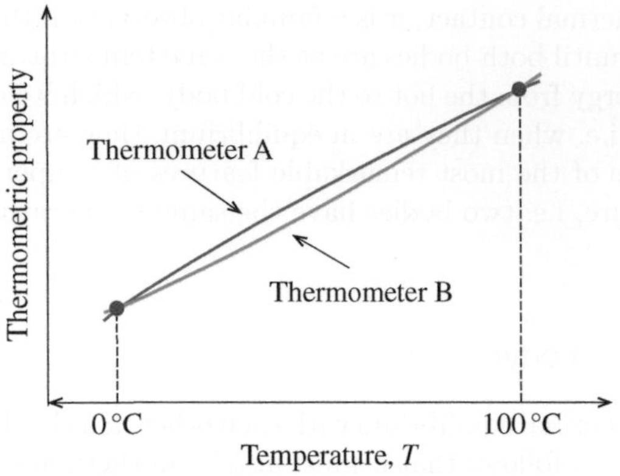

Figure 6.3 In general, when two thermometers are calibrated so that they agree at 0 and 100 °C, they will not agree exactly at other temperatures. This is because their thermometric properties are usually not exactly linearly proportional to temperature.

When we calibrate two thermometers, say a mercury thermometer and a platinum resistance thermometer, so that they agree at 0 °C and 100 °C, they will not in general agree exactly at other temperatures. This is because their thermometric properties are not exactly linearly proportional to temperature, as we saw for the platinum resistance thermometer. This effect is illustrated schematically in Figure 6.3. There is, however, one group of thermometers whose measured temperatures do agree with each other even far away from their calibration points. These are the *constant-volume gas thermometers.* These thermometers follow the ideal gas law $PV = nRT$ if the pressure of the gas is sufficiently low, where n specifies the amount of gas in moles. We thus define the *ideal gas temperature scale* as

$$T = \frac{1}{nR}PV, \quad \text{in the limit } P \to 0. \tag{6.5}$$

We write $P \to 0$ because the lower the pressure of the gas, the more exactly it follows the ideal gas law.

A schematic diagram of a constant-volume gas thermometer is shown in Figure 6.4. It consists of a glass bulb filled with dilute gas. It does not matter what the gas is, so long as it is dilute. This bulb is connected to a mercury barometer by flexible tubing. By raising or lowering the reservoir, the mercury in the left-hand arm can be brought to the zero of the scale to keep the volume of the gas constant. The pressure P of the gas is given by the difference h in height between the mercury levels in the two arms of the barometer. Figure 6.5 shows a plot of P against T for a constant-volume gas thermometer. The plot is a straight line, and when the line is extrapolated to zero pressure, it meets the temperature scale at absolute zero. This limit is the same, no matter what gas is used.

An important aspect of the constant-volume gas thermometer that follows from Equation (6.5) is that its temperature scale can be calibrated at just one point, not two as in the Celsius scale. The calibration point that is used in practice is the triple point of water T_{triple}. This is the unique combination of temperature and pressure at which water vapour, water, and ice coexist in equilibrium. It is chosen because it is much more precise and reproducible than any other calibration point. If P_{triple} is the pressure at the triple point of water, then it follows from Equation (6.5) that the temperature at any other point is given by

$$T = T_{\text{triple}}\left(\frac{P}{P_{\text{triple}}}\right). \tag{6.6}$$

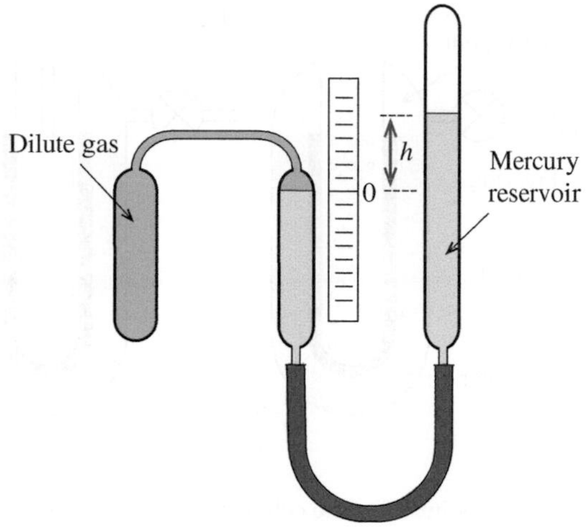

Figure 6.4 A schematic diagram of a constant-volume gas thermometer. It consists of a glass bulb filled with dilute gas. The bulb is connected to a mercury barometer by flexible tubing. By raising or lowering the reservoir, the mercury in the left-hand arm can be brought to the zero of the scale to keep the volume of the gas constant. The pressure of the gas is given by the difference in heights h between the mercury levels in the two arms of the barometer.

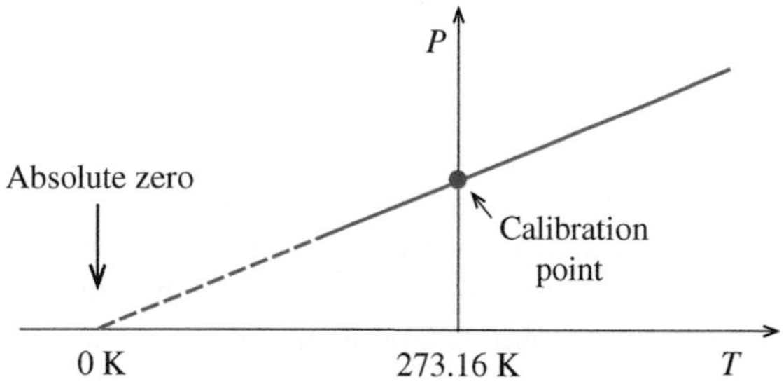

Figure 6.5 A plot of pressure P against temperature T for a constant-volume gas thermometer. The plot is a straight line. The temperature scale is calibrated by the triple point of water at 273.16 K. When the straight-line plot is extrapolated to zero pressure, it meets the temperature scale at absolute zero, 0 K. This limit is the same, no matter what gas is used.

The triple point temperature T_{triple} was chosen so that the size of the degree on the ideal gas scale equals as nearly as possible the degree Celsius, i.e. that there is a temperature difference of 100° between the ice and steam points on the ideal gas scale. This criterion led to $T_{\text{triple}} = 273.16$ K. Here T_{triple} is quoted in kelvins, which is the usual way to quote temperatures on the ideal gas scale. Hence, we have

$$T = \frac{273.16}{P_{\text{triple}}} P, \tag{6.7}$$

where T is measured in kelvins (K), and $T = 0$ K at absolute zero. The temperature scale, defined by Equation (6.7), has the advantage that any measured temperature does not depend on the properties of the

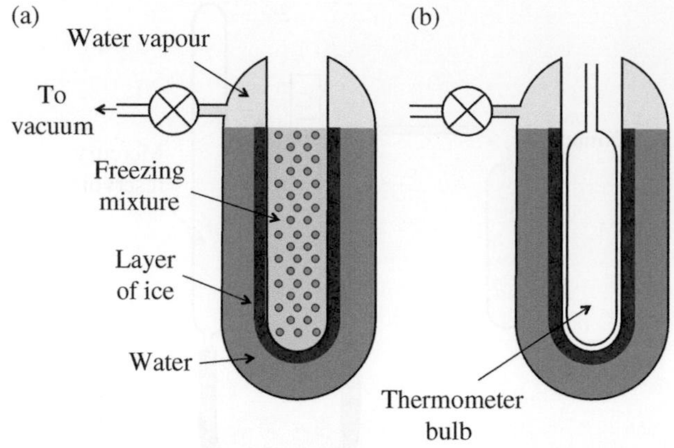

Figure 6.6 A schematic diagram of a triple-point cell. The cell is first evacuated so that all the air is removed, leaving just water vapour above the liquid water. (a) A freezing mixture is then placed into the cell so that a layer of ice is produced on the inner surface of the cell. (b) The freezing mixture is then removed and replaced by the thermometer to be calibrated. Triple-point cells can produce the triple point temperature of water with an uncertainty of less than 0.0001 K.

particular gas that is used but depends only on the general properties of gases. On the Celsius temperature scale, the triple point of water is 0.01 °C. It then follows that

$$T(\text{in kelvin}) = T(\text{in degrees Celsius}) + 273.15.$$

The triple point of water is realized by a *triple-point cell*, which is shown schematically in Figure 6.6. The cell is first evacuated so that all the air is removed and this leaves just water vapour above the liquid water. A freezing mixture is then placed in the middle of the cell so that a layer of ice is produced on the inner surface of the cell. The freezing mixture is removed and replaced by the thermometer to be calibrated. Triple-point cells can produce the triple point temperature of water with an uncertainty of less than 0.0001 K.

The procedure to measure accurate temperatures with a constant gas thermometer is extremely involved and requires many experimental checks. Therefore, its main use is in calibrating other kinds of thermometers.

In Chapter 7, we will see that the second law of thermodynamics can be used to define the *absolute temperature scale*. It is absolute in the sense that it does not depend on the properties of any particular substance, such as a dilute gas. However, we will see that the absolute temperature scale is identical to the ideal gas scale we have described.

6.2.3 Definition of the kelvin

The kelvin is one of the seven base units of the *International System* (*SI*); the other SI base units are: metre, kilogram, second, ampere, mole, and candela. In 1954, the kelvin was defined as equal to the fraction 1/273.16 of the temperature of the triple point of water. However, in November 2018, there was a dramatic revision of the SI base units. This followed a long-standing desire to define all seven base units in terms of fundamental physical constants, rather than physical objects such as the metal bars that were used to define the metre. This was to eliminate any artifact or degradation of the physical objects and ensure the long-term stability of the units. This objective was finally realized in 2018. The seven physical constants upon which the seven base units are now defined are: the speed of light in vacuum c, Planck's constant h, the elementary charge e, the Boltzmann constant k, Avogadro's number N_A, and the luminous efficacy of a specified monochromatic source. For example, the metre is now defined as the length of the path travelled by light in a vacuum in 1/299,792,458 of a second.

We have seen that the Boltzmann constant k is the link between temperature and energy. For example, the average kinetic energy of a molecule in a gas in thermal equilibrium is equal to $3/2\,kT$. It follows that if we have a precise value of k, we can define the kelvin, the unit of temperature, in terms of the joule, the unit of energy. Thus, the kelvin is now defined in terms of the joule and a *fixed* value of the Boltzmann constant k, where the value of k is that obtained from the most precise measurements of it. (The joule itself is in turn defined in terms of the kilogram, metre, and second.) More formally, *the kelvin is defined by taking the fixed numerical value of the Boltzmann constant k to be* 1.380649×10^{-23} *when expressed in the unit* J/K.

As the new definition of the kelvin has its origin in the statistical treatment of matter, where thermodynamic temperature is a measure of the average thermal energy of the constituent particles, we can interpret the new definition of the kelvin as follows: *the kelvin is the thermodynamic temperature at which particles have a mean energy of exactly* $(1/2) \times 1.380649 \times 10^{-23}$ J *per degree of freedom.*

6.3 Heat

We noted in the previous section that when two bodies at different temperatures are placed in contact, thermal energy, which we know as heat, flows between them. Hence, we can define heat as a spontaneous flow of energy from one body to another caused by a difference in temperature between them. Eventually, the two bodies attain the same temperature and at that point the flow of energy stops. We see that *heat is energy in transit*, and it is important to emphasize that a body does not contain a certain amount of heat. We will see in Section 6.4 that when we heat or cool down a body, we change its *internal energy.*

Perhaps surprisingly, the realization that heat is a form of energy was discovered relatively recently. Benjamin Thompson, also known as Count Rumford, came to the conclusion that heat is energy after noticing that the boring of cannon produced a great amount of heat. Later, James Joule put the idea on a firm basis in a classic series of experiments, as we will see in Section 6.5.

Processes by which heat may be transferred are classified into three categories according to the mechanisms involved: conduction, convection, and radiation. Conduction is the main form of heat transfer that we will consider. We explored the conduction of heat in a gas in Section 4.5. We considered a gas across which there is a temperature gradient and an imaginary plane normal to this gradient. Molecules traversing the plane from opposite sides possess different kinetic energies on average. Those molecules coming from the higher temperature side have higher average kinetic energy. When these molecules collide with the lower-energy molecules on the other side of the plane, they give up some of their energy in the process. The lower-energy molecules gain in kinetic energy and there is a net transport of energy across the plane, i.e. thermal conduction.

6.3.1 The measurement of heat

The heat evolved or absorbed in a particular process is measured by a *calorimeter*. Calorimeters find use in many applications including the measurement of the heat of chemical reactions and the heat given off in physical processes such as a phase change in a substance. Joule used calorimeters when studying the connection between heat and work.

The principle of operation of a calorimeter is as follows. The calorimeter is typically a thin-walled metal container that has a small heat capacity to minimize the amount of heat it absorbs. The calorimeter is filled with a known amount of water and is thermally insulated to minimize heat transfer with the surroundings. When an object at, say, a higher temperature than the water is immersed in the water, the water temperature increases. From the measured increase and the known specific heat of water, the amount of heat transferred to the water can be calculated.

Before the introduction of SI units, the unit of heat was the calorie (cal). It was defined as the amount of heat required to raise the temperature of 1 g of water from 14.5 to 15.5 °C. Now heat is defined in terms of the SI unit of energy, the joule, where

$$1\ \text{cal} = 4.186\ \text{J}.$$

Hence, for each 4.186 J of heat absorbed by the water in the calorimeter, there is a 1° rise in temperature. Note that the food calorie that we see written on the packages of food products is actually a kilocalorie, or 4186 J, which is the amount of heat required to raise the temperature of 1 kg of water by 1 °C.

Worked example

A calorimeter contains 100 g of water at an initial temperature of 20 °C. A lump of aluminium of mass 75 g that was heated to a temperature of 75 °C is immersed in the water. What will be the final temperature of the water? Ignore the small amount of heat that would be absorbed by the container. The specific heat of aluminium is 910 J/kg · K.

Solution

The amount of heat gained by the water must be equal to the amount of heat lost by the lump of aluminium. Therefore,

$$0.10 \times 4186 \times (T_\text{f} - 20) = 0.075 \times 910 \times (75 - T_\text{f}),$$

where T_f is the final temperature. This gives

$$T_\text{f} = 27.7°\text{C}.$$

6.4 Internal energy

Internal energy is one of the most important concepts in thermodynamics. In simple terms, internal energy U is the total energy stored in a system. The system could, for example, be a given quantity of gas or water. More fully, internal energy is the sum of the kinetic energy of all the constituent particles of the system, plus the potential energy of interaction between those particles. Clearly, we cannot know or indeed measure this sum of energy because of the huge number of particles involved; although the ideal gas provides a special case that we describe below. However, this is not a restriction because in thermodynamics, only changes ΔU

in internal energy have physical significance; we do not need to know the absolute value of the internal energy. So, for example, we may wish to determine the change in internal energy when the temperature of a given amount of gas is raised by say 50 °C.

We will see that we can produce a change ΔU in the internal energy of a system by the transfer of heat Q to it or from it, or by doing work W on it or by it. This is embodied in the first law of thermodynamics

$$\Delta U = Q + W, \tag{6.8}$$

where all the terms in this expression are energies. There is, however, an important difference between internal energy U and heat Q and work W. A system has a certain amount of internal energy U, but it does not contain a certain amount of heat or work; heat and work are different forms of energy in transit.

6.4.1 Internal energy; a function of state

In Section 6.1, we saw that the variables that depend on the state of the system are called functions of state. For a fluid, which includes gases and liquids, the variables are pressure P, volume V, and temperature T. And we saw that an important property of a state function is that it depends only on the state of the system and not on how that state was reached. The internal energy of a fluid is a unique function of only two of these variables, either $U(P, V)$, $U(P, T)$, or $U(T, V)$; the remaining variable is determined by the equation of state. The choice of the two variables depends on how we wish to specify the state of the fluid. Here we suppress the dependence on the mass of the fluid as we shall usually be considering a constant mass. As the variables P, V, and T are all functions of state, the internal energy U, which is expressed in terms of those variables, must also be a function of state. This means *the internal energy U of a particular state of a system is independent of how that state was reached.* And, hence, the change in internal energy of a system during any thermodynamic process depends only on the initial and final states of the system and not on the way the change was affected.

Given the particular pair of variables for U, we can readily obtain the differential $\mathrm{d}U$. For example, if U is a function of P and V, we have

$$\mathrm{d}U(V,P) = \left(\frac{\partial U}{\partial V}\right)_{\mathrm{P}} \mathrm{d}V + \left(\frac{\partial U}{\partial P}\right)_{\mathrm{V}} \mathrm{d}P. \tag{6.9}$$

This equation gives the change in internal energy U with respect to infinitesimal changes in V and P. Differentials that can be obtained by differentiation of some state function such as U, P, V, or T are called *exact differentials*. The importance of exact differentials will become apparent when we discuss applications of the first law of thermodynamics.

6.4.2 Internal energy of an ideal gas

For the special case of an ideal gas, we can determine its internal energy. We recall an ideal gas as one in which the molecules move randomly through the gas without any interaction with each other. The potential energy between the molecules is zero and therefore the internal energy is entirely due to the thermal motion of the molecules; that is translational motion and possibly rotational and vibrational motion. The mean translation energy of a molecule at temperature T is equal to $3/2 kT$, where k is the Boltzmann constant, one mole of gas contains Avogadro's number N_A of molecules. Hence, the contribution of the translational energy to the internal energy is $3/2 N_A kT$

$$= \frac{3}{2} \times 6.022 \times 10^{23} \times 1.381 \times 10^{-23} \times 300 = 3.74\,\text{kJ},$$

for $T = 300$ K. More generally, if a molecule has f degrees of freedom then according to the equipartition of energy theorem (Section 3.7), the contribution to the internal energy of one mole of gas is $1/2\, fN_A kT$, or $1/2\, fRT$. We emphasise that internal energy relates to the *random* motions of the molecules: translational and possibly rotational and vibrational motions. This has important consequences when we try to extract work from a gas, as we will see in Chapter 7.

We see that the internal energy of an ideal gas depends only on temperature T and we can readily understand this. In an ideal gas, the volume occupied by the molecules is negligible and so too is the mutual interaction between the molecules. Hence, the mean spacing between the molecules and therefore the volume V and pressure P of the gas do not matter. Moreover, the kinetic energies of the molecules do not depend on the distance between them. The fact that the internal energy U of a gas depends only on its temperature, so long as the gas is sufficiently dilute, was verified in an experiment by Joule. This experiment is described in Section 6.10.

As a simple example, suppose we heat one mole of a monatomic gas from 0 to 100 °C. What will be the change in its internal energy, assuming the gas behaves as an ideal gas?

We have $U(T) = N_A 3/2 kT$ as a monatomic gas has only translational motion. Then

$$\frac{dU}{dT} = N_A \frac{3}{2} k.$$

Therefore,

$$\int_{U_1}^{U_2} dU = N_A \frac{3}{2} k \int_{273}^{373} dT = N_A \frac{3}{2} k [T]_{273}^{373},$$

giving

$$\Delta U = U_2 - U_1 = 6.02 \times 10^{23} \times \frac{3}{2} \times 1.38 \times 10^{-23} \times 100 = 1250\,\text{J}.$$

Finally, we make a distinction between the internal energy of a system and its *external* energy. External energy consists of the kinetic energy of the centre of mass of the system plus any potential energy that the system might possess, such as gravitational potential energy. These are *not* included in the internal energy. For example, we do not increase the internal energy of a cup of coffee when we raise the cup to our lips. We shall only consider systems that are at rest and do not possess any potential energy due to external fields. Hence, we can ignore external energy.

6.5 Work and Joule's paddle wheel experiment

We have seen that we can change the internal energy of a system by heating or cooling it. The other way to change its internal energy is by doing work on it or by having the system perform work. Heat and work are thus equivalent ways of transferring energy into or out of a system. And we emphasize that, just as a system does *not* contain a certain amount of heat, so it does *not* contain a certain amount of work; a system only contains internal energy. The experimental evidence for the equivalence of heat and work came from the experiments of Joule. He showed that for a given quantity of water, the same rise in temperature could be produced by transferring a given quantity of energy either as heat or work.

Joule was born in Salford near Manchester and as a child, he had the good fortune to be tutored by John Dalton, a founder of the atomic theory of matter. Joule was a member of a successful brewing family, but he also performed experiments using his cellar as a laboratory. These were pioneering experiments that were instrumental in establishing the first law of thermodynamics as a universal principle of physics. In 1850, he published his detailed measurements of the mechanical equivalent of heat using his famous paddle-wheel apparatus.

A schematic diagram of the paddle-wheel apparatus is shown in Figure 6.7. It consists of a copper cylinder that is thermally insulated from its surroundings. The cylinder contains a known mass of water. There is a paddle wheel that rotates in the cylinder, which has fins fixed to its sides. The paddle wheel, which rotates between the fins, is driven by an arrangement of weights and pulleys. In the experiment, the weights are allowed to drop causing the paddle wheel to churn the water inside the copper cylinder. This action converts the potential energy of the falling weights into work performed on the water. By knowing the mass of the weights and the distance through which they fall, the amount of mechanical work delivered to the water can be determined.

The arrangement of paddle wheel and fixed fins plays a crucially important role. It prevents the water from swirling around in bulk motion en masse, i.e. from acquiring any mass motion. This means that all the work done on the water is converted into random motion of the water molecules. We recall that it is the random motion of molecules in a substance that is directly related to its internal energy. In this case, it

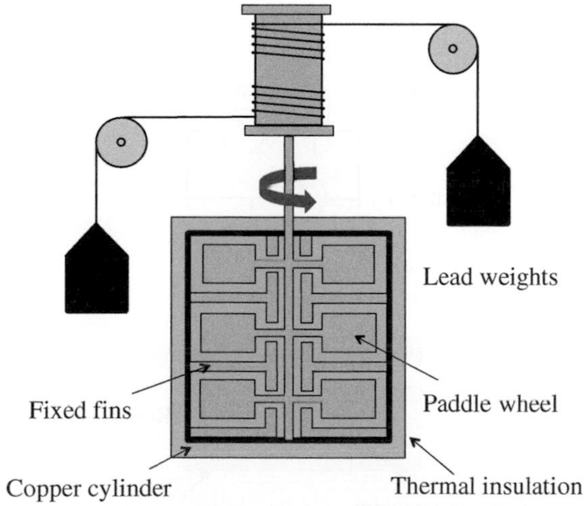

Figure 6.7 A schematic diagram of Joule's famous paddle-wheel experiment. It consists of a copper cylinder, which is thermally insulated from its surroundings and contains a known mass of water. There is a paddle wheel that rotates in the cylinder and fins that are fixed to its sides. The paddle wheel is driven by an arrangement of weights and pulleys as illustrated. In the experiment, the weights are allowed to drop causing the paddle wheel to churn the water inside the calorimeter. This action converts the potential energy of the falling weights into work performed on the water. By knowing the mass of the weights and the distance through which they fall, the amount of mechanical work delivered to the water can be determined.

is the random motion of the water molecules that is directly related to the temperature of the water, which therefore increases.

Joule measured the temperature rise of the water against the amount of work delivered to it, taking careful account of the heat absorbed by the copper cylinder and associated metalwork and of possible heat losses from the apparatus. He found that *the temperature rise was directly proportional to the amount of work performed.* Moreover, he found that the observed rise in temperature could be also achieved by adding an amount of heat Q that was proportional to the mechanical work W done:

$$W = JQ, \tag{6.10}$$

where J is Joule's constant, or the *mechanical equivalence of heat.* Using modern units, Joule found that it takes 4.2 J of mechanical energy to raise the temperature of 1 g of water by 1 °C. Joule was an exceptional experimenter. Despite the limited accuracy of the scientific equipment available to him, his result for the mechanical equivalence of heat is within 1% of the accepted value today, which is $J = 4.186$ J/cal.

Joule investigated other ways of performing work on the water. In one experiment, he immersed a resistor in water; see Figure 6.8. If a voltage V is applied across resistance R for time t, the electrical work done is VRt. The physical picture we have is of electrons moving through the resistor and scattering off the ions fixed in the crystal lattice of the conductor. This causes these ions to become vibrationally excited. There is a transfer of energy from the excited ions to the water molecules, thereby increasing their thermal energy and raising the temperature of the water. Joule found that for a given amount of electrical work performed by the resistor, the same amount of mechanical work from his paddle-wheel experiment produced the same temperature rise in the water. He thus demonstrated that for a given change in the temperature of the water, i.e. the internal energy of a system, the same amount of work is required, irrespective of the mechanism used to perform the work or indeed any combination of mechanisms.

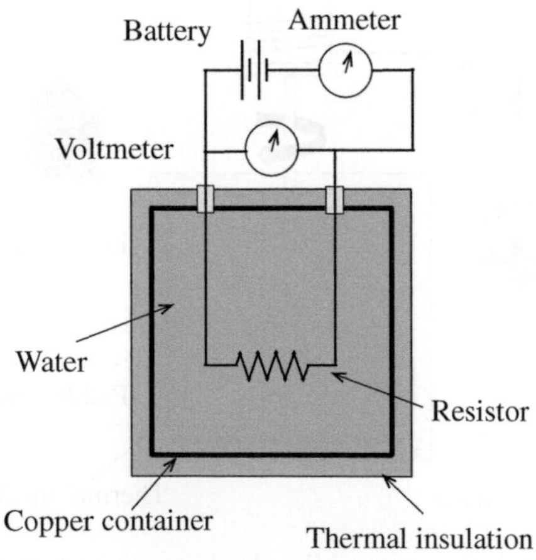

Figure 6.8 A schematic diagram of an arrangement for performing electrical work on a quantity of water. The water is contained in a thermally insulated cylinder and a resistor is immersed in the water. When a voltage is applied across the resistor, the electrical work done is VRt, where V is the voltage, R is the resistance, and t is the time for which the current passes. Joule found that for a given amount of electrical work performed, the same amount of mechanical work from his paddle-wheel experiment produced the same temperature rise in the water.

Worked example

Joule famously packed thermometers for his honeymoon in the Swiss Alps, in order to measure the difference in temperature between the top and bottom of a waterfall near Chamonix. Given that the waterfall is 450 m high, what would be the expected difference in temperature? Take the specific heat of water c to be 4.19 kJ/kg · K.

Solution

The kinetic energy of the water just before it hits the ground is converted into thermal energy, which causes a rise in the temperature of the water. The kinetic energy of a mass m of water is equal to its original potential energy mgh, where h is the height of the waterfall. Therefore,

$$mgh = mc\Delta T,$$

where c is the specific heat of water and ΔT is the change in temperature. Therefore,

$$mgh = mc\Delta T = \frac{gh}{c} = \frac{9.81 \times 450}{4.19 \times 10^3} = 1.05\ ^\circ\mathrm{C}.$$

Notice that the mass m of the water cancels out. This is a relatively small change in temperature and illustrates the challenge Joule had in the experiment.

6.6 First law of thermodynamics

We can increase the internal energy of a system by adding heat to it, and we can increase its internal energy by performing work on it. If we add a quantity of heat Q *to* a system and also perform work W *on* it, then we increase its internal energy U by an amount ΔU, where

$$\Delta U = Q + W. \tag{6.8}$$

This is the first law of thermodynamics and is just the conservation of energy when heat is taken into account. When heat enters a body, Q is positive and when heat leaves a body, Q is negative. And, when we perform work on a body, W is positive and when a system does work, W is negative. These signs simply indicate the direction of energy flow.

6.6.1 Paths between thermodynamic states

An important feature of Equation (6.8) is that it says the sum $Q + W$ must be equal to ΔU, but it says nothing about the individual values of Q and W. These can be chosen arbitrarily as we now illustrate.

The experiments of Joule showed that for a given amount of water, he could obtain the same rise in temperature by performing mechanical or electrical work on it; provided the same amount of work W was done. Alternatively, he could produce the same temperature change by applying an amount of heat $Q = W/J$. These are different *paths* between the initial and final states of the system, the first two paths involving only work and the last path involving only heat.

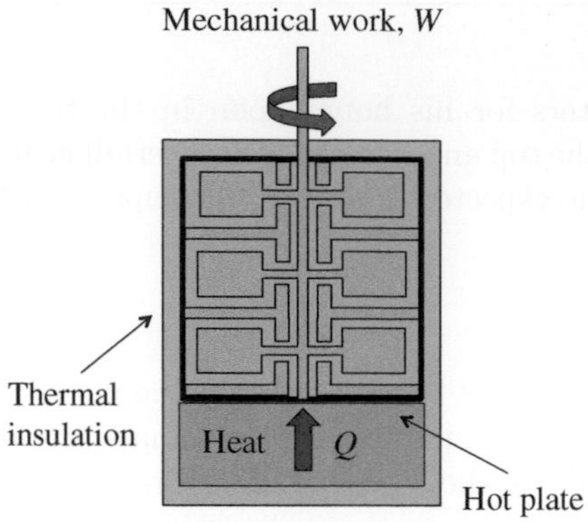

Figure 6.9 An experimental arrangement that allows mechanical work W to be performed on a system and heat Q to be transferred to it.

Consider now the experimental arrangement shown in Figure 6.9. This consists of a thermally insulated copper cylinder containing a certain amount of water. There is a paddle wheel that can be rotated so that mechanical work can be performed on the water. There is also a hot plate upon which the copper cylinder is placed. The temperature of the water can be increased by heating it or by driving the paddle wheels. Suppose that the water is initially at temperature $T_1 = 25\,°\text{C}$, with internal energy U_1. We could add heat Q_1 to the water increasing its temperature to 45 °C and then stir the water, performing work W_1 until it reaches the temperature $T_2 = 100\,°\text{C}$, with internal energy U_2. Alternatively, we could stir the water, performing work W_2, until it reaches the temperature of 50 °C and then add heat Q_2 to the water increasing its temperature to 100 °C. Although these two paths have different individual values of work W and heat Q they result in the same change $\Delta T = (T_2 - T_1)$ in temperature. We may say that

$$W_1 + Q_1 \text{ over first path} = W_2 + Q_2 \text{ over second path} = \text{constant}. \tag{6.11}$$

These are just two paths, i.e. combinations of work and heat that change the water temperature from 25 to 100 °C, but clearly there is an infinite number of paths. Although these various paths have different individual values of work W and heat Q to achieve the same rise in temperature, it is found that their sum $W + Q$ is fixed. This result shows us that a given change in temperature ΔT is determined only by the initial and final temperatures of the water and not by the path by which the change is achieved.

Here, we have characterised the internal energy of the initial and final states of the water by their temperatures. We recall from Section 6.4 that the internal energy of a fluid is a unique function of only two of the variables P, V, and T; the remaining variable is determined by the equation of state. So, we can describe the internal energy of the water in terms of temperature and volume; $U(T, V)$. If we make the assumption that the volume of the water remains constant during the temperature change, the internal energy of the water is a function of T alone: $U(T)$. In that case it follows that if the temperature change ΔT is independent of the path taken, then the internal energy change ΔU must also be independent of that path. This is a general result; *in a thermodynamic process, the change in internal energy in going from one state of the system to another is independent of the path taken between the two states.*

We can make an analogy between internal energy and the altitude of a geographical location. Yr Wyddfa (Snowdon) is the highest mountain in England and Wales at an elevation of 1085 m. There are various ways for getting to the summit of the mountain, including walking, mountain biking or by train; or by a combination of these, say, getting the train to the intermediate station Clogwyn and walking the rest of the way. But it does not matter how we get to the summit; when we arrive there, we are always at the same altitude. That is, the difference in vertical height from our starting position to the summit is always the same, no matter what path we chose. In an analogous way, it does not matter how we change a system from an initial state with internal energy U_1 to a final state with internal energy U_2, i.e. by whatever combination of heat and work, the change ΔU in internal energy is always the same:

$$\Delta U = (U_2 - U_1) = \text{constant}. \tag{6.12}$$

The fact that the change ΔU in internal energy in going from an initial to a final state is independent of the path taken between the two states is a clear demonstration that internal energy U is a function of the state of a system alone; it does not matter how that state was reached. On the other hand, the heat Q added to the system in the process does depend on the path taken. This demonstrates that a system in a particular state does not 'contain' a certain amount of heat. Similarly, the fact that the work done on the system depends on the path again demonstrates that a system does not 'contain' a certain amount of work.

6.6.2 Thermodynamic definitions of internal energy and heat

From a microscopic point of view, we cannot in general calculate the internal energy of a system because of the huge number of particles involved. However, the first law of thermodynamics enables us to define internal energy, or at least the change in internal energy in terms of macroscopic quantities that we can measure. In particular, we can define ΔU in terms of the work W performed on a system, which is usually straightforward to measure.

Consider the paddle-wheel apparatus shown in Figure 6.10. In particular, the copper cylinder is thermally insulated from its surroundings, which means that there can be no transfer of heat; $Q = 0$. Such a process is called an *adiabatic* process. We change the state of the system by doing work on it, i.e. by stirring the water. Suppose we change the system from an initial state with internal energy U_1 to a final state of internal energy U_2 by performing work W on it. As $Q = 0$, we have from the first law, Equation (6.8):

$$\Delta U = (U_2 - U_1) = W. \tag{6.13}$$

We can thus define the change in the internal energy of a system as the work performed on the system when the process is adiabatic.

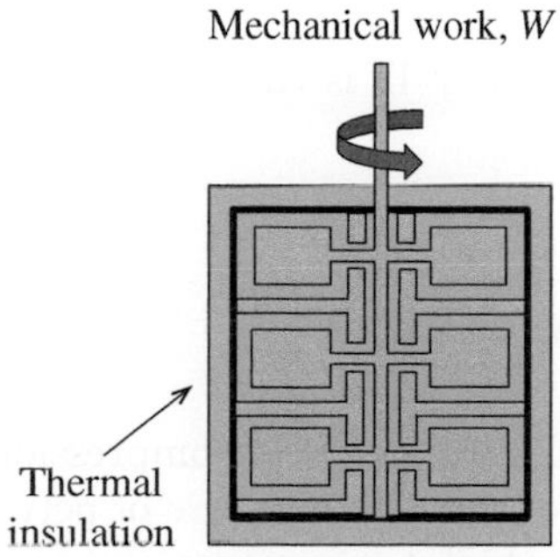

Figure 6.10 The figure illustrates a paddle-wheel apparatus. The copper cylinder is thermally insulated from its surroundings, which ensures that there is no transfer of heat to the system when mechanical work is performed on it. Such a process, where $Q = 0$, is called an *adiabatic* process. This was the condition in Joule's paddle-wheel experiment.

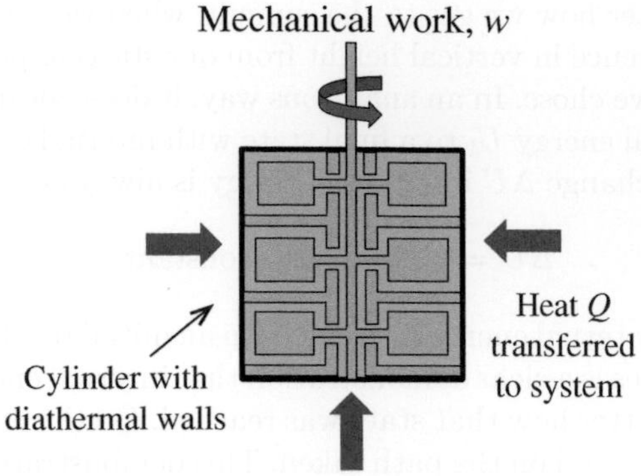

Figure 6.11 In this arrangement of a paddle-wheel experiment, the diathermal walls of the cylinder allow heat Q to be transferred between the water and the surroundings as work w is performed on the water.

We noted previously that calorimetry provides one way to measure the amount of heat transferred in a thermodynamic process. However, the first law provides an alternative way to define and measure heat transfer that may be more convenient in practice. Suppose that we repeat the above paddle-wheel experiment as in Figure 6.10, but remove the thermal insulation from the copper cylinder so that the walls allow heat transfer, i.e. are *diathermal*; see Figure 6.11. We can still effect the same change from the initial state of internal energy U_1 to the final state with internal energy U_2. In this case, however, we find that the work w needed to produce the change in internal energy $\Delta U = (U_2 - U_1)$ depends on how much heat is added or subtracted from the system by conduction through the diathermal walls. But according to the first law of thermodynamics, the sum of the work w done on the system and the net heat Q added to the system must be equal to that value of ΔU:

$$\Delta U = w + Q. \tag{6.14}$$

ΔU is the same as when the change in internal energy was made in the adiabatic process above. Substituting for $\Delta U = W$ from Equation (6.13) gives

$$Q = W - w. \tag{6.15}$$

Equation (6.15) provides us with a definition of heat and a means to measure the amount of heat transfer; w is the work done in the given process and W is the work done when the same process is conducted adiabatically.

6.7 Work done during volume changes

Many practical applications depend on the expansion or compression of a gas. Examples include the petrol engine and the steam turbine. In a petrol engine, a mixture of petrol vapour and air is ignited by a spark plug, causing it to explode. The resulting high temperature and pressure cause the gas to expand rapidly, pushing a cylinder and doing work.

To illustrate the work done during a volume change, we consider a gas confined in a cylinder by a frictionless piston, as illustrated by Figure 6.12. The piston has cross sectional area A and the pressure of the

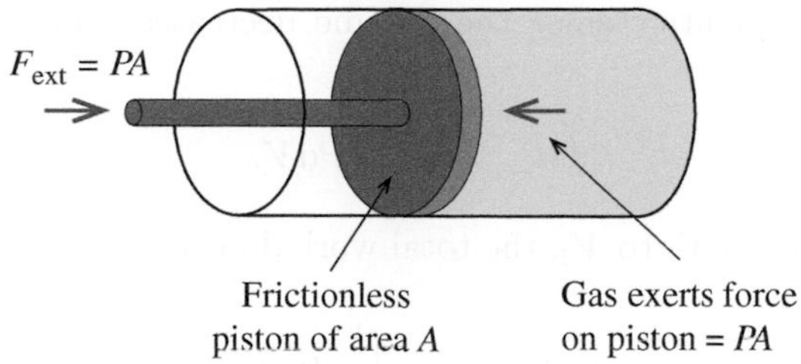

Figure 6.12 The figure shows a gas at pressure P, confined in a cylinder by a frictionless piston of cross sectional area A. The force exerted by the gas on the piston is $F = PA$. When the piston moves *out* an infinitesimal amount $\mathrm{d}x$, the work done *by* the gas as it expands is $\mathrm{d}W_{\text{by gas}} = P\mathrm{d}V$, where $\mathrm{d}V$ is the infinitesimal change in volume. When the piston moves *in* an infinitesimal amount $\mathrm{d}x$, the work done *on* the gas as it is compressed is $\mathrm{d}W_{\text{on gas}} = -P\mathrm{d}V$.

gas is P. The force exerted by the gas on the piston is $F_{\text{by gas}} = PA$. At equilibrium, the external force F_{ext} exerted by the piston on the gas must also be equal to PA, but acts in the opposite direction. If the piston moves *out* an infinitesimal amount $\mathrm{d}x$, the work done *by* the gas as it expands is

$$\mathrm{d}W_{\text{by gas}} = F_{\text{by gas}}\mathrm{d}x = PA\mathrm{d}x. \tag{6.16}$$

But

$$A\mathrm{d}x = \mathrm{d}V,$$

where $\mathrm{d}V$ is the change in volume. During the infinitesimal change in the volume of the gas, we can assume that the pressure P is constant. Hence, the work done by the gas in this volume change is

$$\mathrm{d}W_{\text{by gas}} = P\mathrm{d}V. \tag{6.17}$$

For a finite change of volume from V_1 to V_2, the total work done by the gas is

$$W_{\text{by gas}} = \int_{V_1}^{V_2} P\mathrm{d}V. \tag{6.18}$$

Of course, the pressure P of the gas may vary during a finite change in volume and to evaluate the integral in this equation, we need to know if and how the pressure varies. This depends on the constraints imposed on the gas during the expansion process.

Suppose now that the piston moves *into* the cylinder by an infinitesimal amount $\mathrm{d}x$ so that the gas is compressed. The work $\mathrm{d}W$ done *by* the external force F_{ext} acting *on* the gas is

$$\mathrm{d}W_{\text{on gas}} = F_{\text{ext}}\mathrm{d}x. \tag{6.19}$$

As $F_{\text{ext}} = PA$,

$$\mathrm{d}W_{\text{on gas}} = PA\mathrm{d}x. \tag{6.20}$$

The elemental change in the volume V of the gas is

$$\mathrm{d}V = (V - A\mathrm{d}x) - V = -A\mathrm{d}x.$$

The change dV is a negative quantity since the volume decreases. Substituting for $Adx = -dV$ in Equation (6.20), we obtain

$$dW_{\text{on gas}} = -PdV. \tag{6.21}$$

For a finite change of volume from V_1 to V_2, the total work done on the gas is

$$W_{\text{on gas}} = -\int_{V_1}^{V_2} PdV. \tag{6.22}$$

Again, to evaluate the integral in this equation, we need to know how the pressure P of the gas varies during the compression process and again this depends on the constraints imposed on the gas.

We will explore various constraints that may be imposed on a gas when it expands or contracts. In particular, we will see how the work done during a volume change is well defined by the properties of the gas when the process involved is *reversible*. And we will describe the necessary conditions for a process to be reversible.

Worked example

Calculate the work done on a gas when one mole of gas is reduced in volume by a factor of 2, at a constant temperature of 23 °C. Assume that the compression is reversible.

Solution

From Equation (6.22), we have

$$W_{\text{on gas}} = -\int_{V_1}^{V_2} PdV.$$

We have $P = RT/V$ at constant T. Substituting for P in the integral, we obtain

$$W_{\text{on gas}} = -RT\int_{V_1}^{V_2} \frac{dV}{V} = -RT\ln\left(\frac{V_2}{V_1}\right) = +RT\ln\left(\frac{V_1}{V_2}\right).$$

Note that since $V_2 < V_1$, the work done on the gas is positive and is equal to

$$8.31 \times 296 \times \ln 2 = +1700\ \text{J}.$$

6.8 Reversible processes

In everyday language, we might describe a process as reversible if, when the process is reversed, the system and its surroundings are unchanged in every respect. In thermodynamics, for a process to be reversible, it must be possible to reverse its direction by an infinitesimal change in the applied conditions. For this to be the case, two conditions must be satisfied: the process must be *quasistatic* and there must be no frictional or viscous forces involved.

6.8.1 Quasistatic processes

Suppose we have a gas contained in a cylinder by a frictionless piston and, for the sake of simplicity, the cylinder is held in contact with a heat reservoir. This ensures that the temperature of the gas is maintained at a constant temperature T during any change in the volume of the gas.

The volume V of the gas is reduced by pushing the piston further into the cylinder. This, of course, increases the pressure P. Suppose the piston is pushed in very slowly; so slowly that the gas molecules have plenty of time to redistribute themselves as the gas is compressed. Then, although the gas pressure changes continuously, the number density of molecules, i.e. the gas pressure, is uniform throughout the cylinder at all times during the compression. This means that we can consider the gas to be always in a state of equilibrium. The state of the gas is therefore defined by its equation of state $PV = RT$ throughout the compression process in terms of P and V even though these are continuously changing. Such a process in which the system remains in equilibrium throughout the process is called a quasistatic process.

Contrast this situation with the piston being pushed in so rapidly that the pressure close to the piston is higher than elsewhere in the cylinder. In that case, the pressure would not be uniform throughout the cylinder since P would vary from point to point. The gas would not be in a state of equilibrium and could not be defined by an equation of state.

6.8.2 Idealised reversible process

Consider Figure 6.13, which shows a cylinder, closed at both ends, with a frictionless piston that can move freely within the cylinder. The cylinder is connected to a heat reservoir to maintain a constant temperature. One mole of gas at pressure P is confined below the piston. Above the piston, there is a gas whose pressure P' can be varied as described later. The confined gas exerts an upward force on the piston equal to PA, while the gas above the piston exerts a downward force on the piston equal to $P'A$. Initially, the two gas pressures are equal with $P' = P$, and the piston is stationary.

The top part of the cylinder is connected to a source of compressed gas at high pressure by an entry valve. There is also an exit valve connected to vacuum. Suppose that the entry valve is opened *very slightly*

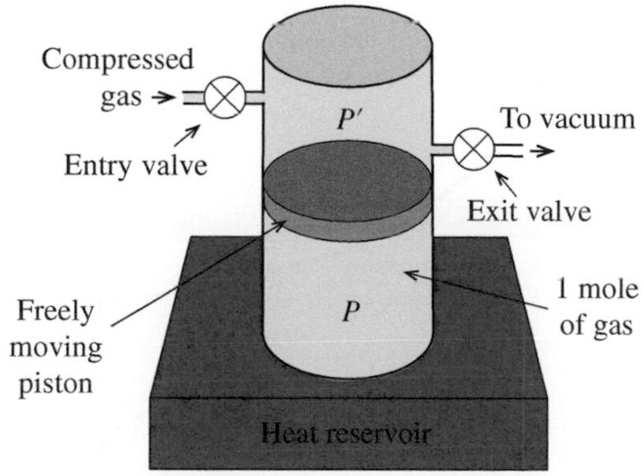

Figure 6.13 This idealised system consists of a cylinder closed at both ends, with a frictionless piston that can move freely within the cylinder. The cylinder is connected to a heat reservoir to maintain a constant temperature. Below the piston is a fixed amount of gas at pressure P. Above the piston is gas at pressure P'. The top part of the cylinder is connected to a source of compressed gas at high pressure by an entry valve, and there is also an exit valve connected to vacuum. By suitable adjustment of the entry and exit valves, P' can be varied over a wide range of pressure.

so that pressure P' increases *very slowly*. This causes the piston to move slowly downwards to equalise the pressures on either side of it. This maintains the two pressures P and P' essentially equal, and the system remains in equilibrium, i.e. it is a quasistatic process. As the pressure P of the confined gas below the piston steadily increases, its volume decreases according to $V \propto 1/P$.

We can plot the pressure P of the confined gas against its volume V on a P–V diagram. As the temperature T is constant, this is an *isothermal* compression, and we have from the equation of state $P = RT/V$. The resultant *isotherm* is shown in Figure 6.14, from initial volume V_1 to final volume V_2. Note that the point (V, P), which defines the state of the gas in terms of V and P at any stage of the compression, always lies on this isotherm.

According to Equation (6.21), the infinitesimal work done on the gas when the piston is pushed down and the volume reduces by the amount $\mathrm{d}V$ is

$$\mathrm{d}W_{\text{on gas}} = -P\mathrm{d}V.$$

and the total work done in the compression from initial volume V_1 to final volume V_2 is, from Equation (6.22),

$$W_{\text{on gas}} = -\int_{V_1}^{V_2} P\mathrm{d}V$$

Substituting for $P = RT/V$ in the integral, we obtain

$$W_{\text{on gas}} = -RT\int_{V_1}^{V_2} \frac{\mathrm{d}V}{V} = -RT\ln\left(\frac{V_2}{V_1}\right) = RT\ln\left(\frac{V_1}{V_2}\right).$$

The total work done in the compression of the confined gas is indicated by the shaded area under the curve in Figure 6.14. It is important to emphasise that the work done is defined in terms of the properties of the confined gas, its pressure-volume, and temperature.

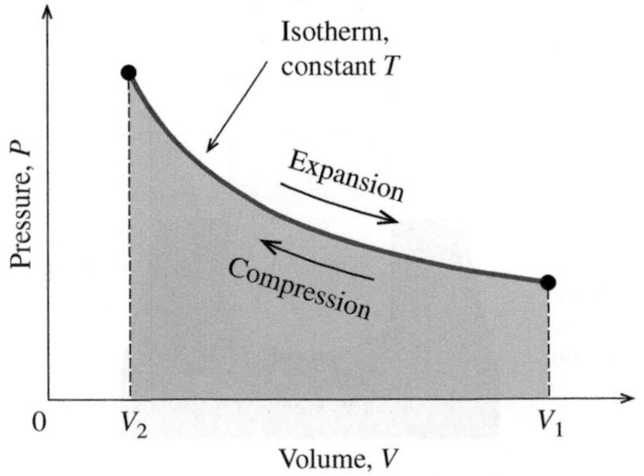

Figure 6.14 A plot of pressure P against volume V for the confined gas below the piston in Figure 6.13. As the temperature of the gas is constant, the curve is an isotherm connecting initial volume V_1 to final volume V_2. The point (V, P) defines the state of the gas at any stage of the compression, and always lies on the isotherm. The shaded area represents the work expended by the gas in the isothermal expansion.

We can reverse the above process by closing the entry valve and *slightly opening* the exhaust valve; see Figure 6.13. Now the pressure P of the confined gas pushes the piston slowly upwards, with the two pressures P and P' remaining essentially equal during the expansion process. The state of the gas is described by the same isothermal curve $P = RT/V$ but now the system moves along the curve in the opposite direction. So, this isotherm represents the reversible path for this compression/expansion cycle.

In the expansion, the confined gas does the work. The infinitesimal work done is $\mathrm{d}W_{\text{by gas}} = P\mathrm{d}V$, and the total work done is

$$W_{\text{by gas}} = \int_{V_2}^{V_1} P\mathrm{d}V = RT\ln\left(\frac{V_1}{V_2}\right). \tag{6.23}$$

The area under the curve is the same as for the compression process. Hence, the work done on the confined gas in the compression stage is equal to the work done by this gas in the expansion stage. It follows that the net work done in the compression/expansion cycle is zero.

When the confined gas is expanding, it is pushing against the gas above the piston, which is at pressure P'. If P' were equal to zero, the confined gas would do no work. This case is called *free expansion*, which we deal with in Section 6.10. When P' is greater than P by an infinitesimal amount, the piston moves downwards, but if P' is less than P by an infinitesimal amount, the piston moves upwards. Thus, an infinitesimal change in P' reverses the process, in accord with a necessary condition for a reversible process.

6.8.3 The effect of frictional forces

Consider now what happens when there is friction between the piston and the cylinder sides. To compress the confined gas, the applied pressure P' must now be greater than the pressure P by a *finite* amount to overcome the frictional forces. And in the second part of the cycle, when the confined gas expands, P' must be less than P by a *finite* amount for the same reason. As P' has to be changed by a finite amount to reverse the process, this is not a reversible process. In a cycle of compression and expansion, the relationship between P' and V is illustrated in Figure 6.15. We see that the system does not follow the original path in the reverse direction but will follow a new path. The dashed curve shows the relationship between P' and V for the idealised case when friction is not present.

With friction, the downward force acting on the piston is again $P'A$ but now $P' > P$. Then the total work done in the compression in changing from initial volume V_1 to final volume V_2, is

$$W_{\text{on gas}} = -\int_{V_1}^{V_2} P'\mathrm{d}V, \quad \text{with } P' > P. \tag{6.24}$$

Hence, the total work done in this irreversible process now depends on the magnitude of the frictional forces. This means we can no longer define the work done in terms of the properties of the confined gas. Similarly, in the expansion stage, we have

$$W_{\text{by gas}} = \int_{V_2}^{V_1} P'\mathrm{d}V, \quad \text{with } P' < P. \tag{6.25}$$

If we carry out the cycle ABCDA by first compressing and then expanding the confined gas, as depicted in Figure 6.15, the external force due to the applied pressure P' does a net amount of work, which is given by the area of the cycle ABCDA. This work in overcoming the friction is lost as thermal energy.

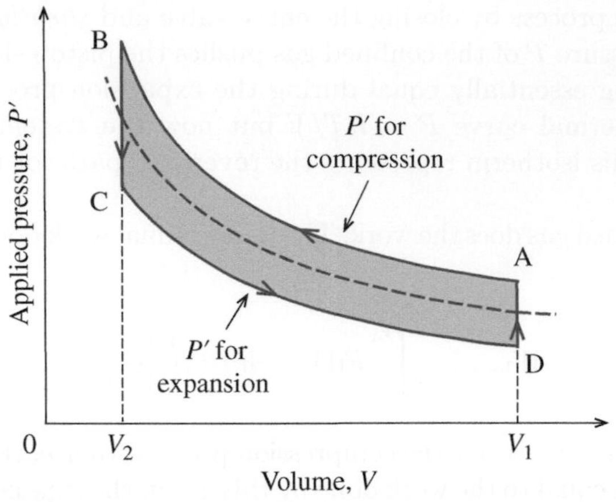

Figure 6.15 When there is friction between the piston and the cylinder sides, the applied pressure P' must be greater than the pressure P of the confined gas by a finite amount when the gas is being compressed. And when the confined gas is expanded, P' must be less than the pressure P by a finite amount. This figure shows the relationship between P' and V in a cycle of compression and expansion. We see that the system does not follow the original path in the reverse direction but will follow a new path, i.e. the process is not reversible. In carrying out the cycle ABCDA by first compressing and then expanding the gas, the applied pressure P' does a net amount of work, which is given by the shaded area. This work in overcoming the friction is lost as thermal energy. The dashed curve shows the relationship between P' and V for the idealised case when friction is not present.

We can summarise our conclusions about work in reversible and irreversible processes by the relation

$$\mathrm{d}W \geq -P\mathrm{d}V, \tag{6.26}$$

where $\mathrm{d}W$ is the work done on the system by an external agency, and P and V define the state of the system. The equality sign holds for reversible processes and the 'greater than' sign holds for irreversible processes.

6.8.4 Practical realization of a reversible process

A practical process that is close to being reversible is illustrated in Figure 6.16. The figure shows a gas confined in a cylinder by a piston. The piston is well lubricated so that it slides very easily in the cylinder and friction can be neglected. On top of the piston is a dish containing a pile of lead shot. At equilibrium, the upward force F exerted by the gas on the piston is balanced by the downward force due to the weight of the piston, the dish and lead shot and atmospheric pressure. The temperature of the gas is held constant as the volume of the gas is changed by keeping the cylinder in thermal contact with a heat reservoir.

Suppose that a single lead shot is removed from the dish. This causes the piston to move up a small distance in order to balance the opposing forces acting on the piston, increasing the volume by a small amount ΔV. Because this increase is very small, the gas within the cylinder will not be disturbed appreciably. Hence, the gas remains essentially in equilibrium with uniform pressure throughout.

If we want to make a finite change of volume from V_1 to V_2, we do this by slowly removing one lead shot at a time until the final volume V_2 is reached. During each of the resulting small changes in volume, we can assume that the gas remains in equilibrium and that each of the intermediate states of the gas can be described by the equation of state. The expansion process is quasistatic. At each step we can plot the

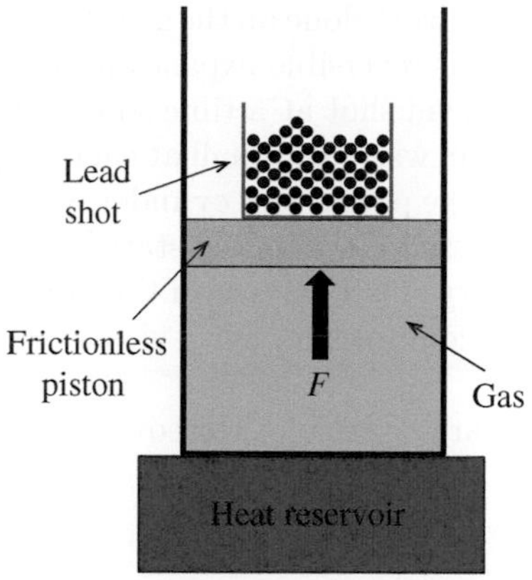

Figure 6.16 The figure shows a practical process that is close to being reversible. The gas is confined in a cylinder by a piston, which is well lubricated so that it slides very easily in the cylinder and friction can be neglected. The cylinder is in thermal contact with a heat reservoir to ensure that the temperature of the gas is held constant during any volume change. On top of the piston is a dish containing a pile of lead shot. At equilibrium, the upward force F exerted by the pressure of the gas on the piston is balanced by the downward force due to the weight of the piston, the dish and lead shot and atmospheric pressure. By adding or removing a single lead shot the downward force acting on the gas can be changed by a tiny amount, leading to a small change ΔV in the volume of the gas.

pressure P against volume V. This plot is shown in Figure 6.17. The (V, P) values on this P–V diagram always lie close to the curve $P = RT/V$. The work done at each step is given by the area of the corresponding vertical strips in the figure, $P\Delta V$. The total work done by the gas in the expansion from V_1 to V_2 is the total area of these individual strips and this is a positive quantity.

This process can be reversed by slowly adding one shot at a time to the dish. Now the (V, P) values at each step closely retrace the $P = RT/V$ curve in going from volume V_2 to V_1. The sum of the areas of the

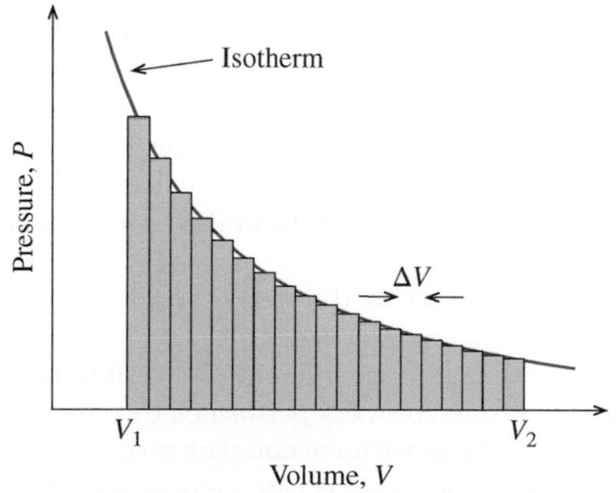

Figure 6.17 This figure shows a plot of pressure P against volume V for the practical realisation of the quasistatic expansion of gas shown in Figure 6.16. The (V, P) values on this P–V diagram always lie close to the curve $P = RT/V$. The work done $P\Delta V$ at each step is given by the area of the corresponding vertical strip in the figure. The total work done by the gas in the expansion from V_1 to V_2 is the total area of these individual strips and is a positive quantity.

resulting vertical strips then represents the work done on the gas. The total area is the same as for the expansion stage and so the net work done in this reversible expansion/compression cycle is zero.

In this process, we removed a single lead shot at a time so that the resultant difference between the forces acting on the two faces of the piston was very small at each step. And the gas stayed close to equilibrium. Moreover, the friction between the piston and cylinder was negligible. Hence, if we replaced the single lead shot, we could return the system to its original state and there would be no change to the system or to its surroundings. That is, a very small change in the applied conditions returns the system to its original state. Thus, the expansion/compression cycle closely fulfills the conditions for a reversible process.

All the processes that we will discuss are reversible. Moreover, in practice, many processes approximate to being reversible.

6.9 Expansion of gases and the first law of thermodynamics

In general, when we add heat Q to a gas, that heat appears partly as an increase in the internal energy of the gas and partly as the work W required to expand the gas. From the first law, we have

$$\Delta U = Q - W, \tag{6.27}$$

where ΔU is the *change* in the internal energy when we add an *amount* of heat Q to a gas and the gas does an *amount* of work W. The negative sign for W arises because the gas does work and so its internal energy decreases.

As we saw in Section 6.6, ΔU has a well-defined value because it depends only on the initial and final states of the system, but Q and W can have arbitrary values. This is also the case for extremely small changes, which may be represented by the equation

$$\mathrm{d}U = \delta Q - \delta W. \tag{6.28}$$

Again, the quantity $\delta Q - \delta W$ must be equal to $\mathrm{d}U$ but the first law places no constraints on the individual values of δQ and δW. However, there are some processes where δQ and δW do have well-defined values. This means that they can be treated as well-defined *differences* in Q and W, respectively, i.e. be treated as differentials $\mathrm{d}Q$ and $\mathrm{d}W$. We can do this if δW can be expressed in terms of the functions of state, say the pressure P and volume V of the gas. In that case, δW is no longer arbitrary, but has a well-defined value. As both $\mathrm{d}U$ and δW have well-defined values, it follows from Equation (6.28) that δQ must itself have a well-defined value. Treating δQ and δW as differentials $\mathrm{d}Q$ and $\mathrm{d}W$, we can write Equation (6.28) as

$$\mathrm{d}U = \mathrm{d}Q - \mathrm{d}W. \tag{6.29}$$

The necessary conditions for treating δQ and δW as differentials are obtained in a reversible process. Indeed, the importance of reversible processes is that the work performed by or on a system in a reversible process is well defined by the properties of the system, as we have noted already. As an illustration of this, we consider the work done in the reversible expansion of a gas. When a gas at pressure P expands by an infinitesimal amount $\mathrm{d}V$, the work it performs is $P\mathrm{d}V$, Equation (6.17). As we saw in Section 6.8, in a reversible process, the pressure P is uniquely defined at each stage of the expansion process by the equation of state $PV = RT$. Thus, for a given value of $\mathrm{d}V$, the term $P\mathrm{d}V$ has a well-defined value. We can thus substitute $P\mathrm{d}V = \delta W$ in Equation (6.28). $\mathrm{d}U$ also has a well-defined value and therefore it follows that δQ must also be well defined over the incremental change of volume $\mathrm{d}V$. We can therefore write it as the differential $\mathrm{d}Q$. Thus, for any infinitesimal *reversible* change in a gas, we must have

$$\mathrm{d}U = \mathrm{d}Q - P\mathrm{d}V. \tag{6.30}$$

As each of the terms in this equation is now in the form of a differential, they can be integrated exactly if the relevant path of the process is known.

Finally, we note that if either δW or δQ is zero in a given process, then from Equation (6.28), the remaining quantity must have a well-defined value because $\mathrm{d}U$ does. The remaining quantity can then be taken to be a differential. For example, in an adiabatic process $\delta Q = 0$, and we can therefore write $\mathrm{d}U = -\mathrm{d}W$. Similarly, if a volume change is made so that $\delta W = 0$, we have $\mathrm{d}U = \mathrm{d}Q$.

In Section 6.1, we noted that as the temperature T of an ideal gas can be related to its pressure P and volume V by a mathematical equation, $PV = RT$. Hence, we can readily obtain an expression for the differential $\mathrm{d}T$:

$$R\mathrm{d}T = V\mathrm{d}P + P\mathrm{d}V.$$

Differentials such as $\mathrm{d}T$ that are obtained by differentiating functions of state are called *exact differentials*. Quantities that behave like differentials, such as $\mathrm{d}Q$ and $\mathrm{d}W$ above, but cannot be obtained by differentiating the functions of state are called *inexact differentials*. In some texts, inexact differentials are denoted by a bar through the differential symbol 'd'.

6.10 The Joule effect; the free expansion of an ideal gas

We have emphasized that the internal energy U of an ideal gas depends only on its temperature T:

$$U = U(T). \tag{6.31}$$

Evidence to support this comes from an experiment performed by Joule in which he investigated the expansion of a gas into a vacuum. Such an expansion is called a *free expansion*. It is an example of an irreversible process as inequalities in temperature and pressure occur during the expansion process. However, as the initial and final states of the system are equilibrium states, we can use thermodynamics to describe the process. Essentially, the purpose of Joule's experiment was to determine whether the free expansion caused a change in the temperature of the gas.

Consider the apparatus shown schematically in Figure 6.18. It consists of two compartments connected by a valve. The two compartments and connecting valve are thermally insulated so that heat cannot enter or leave the system. Initially, the left-hand compartment is filled with gas, while the right-hand compartment is evacuated. The valve is then opened and gas rushes into the evacuated compartment so that the gas eventually fills both compartments. Any change in the temperature of the gas is measured. Joule found that within the accuracy of his measurements, the temperature of the gas did not change.

The gas does no work as it expands into the right-hand compartment because there is nothing to push against. Hence, $W = 0$. Moreover, as the system is thermally insulated, no heat can enter or leave and $Q = 0$. From the first law, we have

$$\Delta U = Q - W,$$

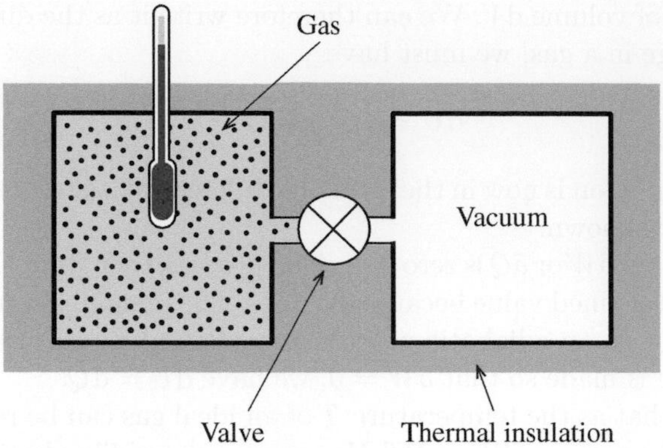

Figure 6.18 A schematic diagram of an apparatus to observe the free expansion of a gas. It consists of two compartments connected by a valve. The two compartments and the connecting valve are thermally insulated so that heat cannot enter or leave the system. Initially, the left-hand compartment is filled with gas, while the right-hand one is evacuated. The valve is then opened and gas gushes into the evacuated container so that the gas eventually fills both compartments. Any change in the temperature of the gas is measured.

giving

$$\Delta U = 0. \tag{6.32}$$

Thus, according to the first law, we may expect that the internal energy U of a gas does not change in a free expansion.

Suppose that the internal energy of the gas does depend on both temperature T and volume V, i.e. $U(T, V)$. Then considering U to be a function of T and V, we have

$$\mathrm{d}U = \left(\frac{\mathrm{d}U}{\mathrm{d}T}\right)_{\mathrm{V}} \mathrm{d}T + \left(\frac{\mathrm{d}U}{\mathrm{d}V}\right)_{\mathrm{T}} \mathrm{d}V.$$

If there is no change in the gas temperature, $\mathrm{d}T = 0$ and from Equation (6.32), $\mathrm{d}U = 0$. Hence,

$$\left(\frac{\mathrm{d}U}{\mathrm{d}V}\right)_{\mathrm{T}} \mathrm{d}V = 0, \tag{6.33}$$

i.e. U does not depend on V. Considering U to be a function of T and P, we have

$$\mathrm{d}U = \left(\frac{\mathrm{d}U}{\mathrm{d}T}\right)_{\mathrm{P}} \mathrm{d}T + \left(\frac{\mathrm{d}U}{\mathrm{d}P}\right)_{\mathrm{T}} \mathrm{d}P.$$

Again, for $\mathrm{d}T = 0$ and $\mathrm{d}U = 0$, we have

$$\left(\frac{\mathrm{d}U}{\mathrm{d}P}\right)_{\mathrm{T}} \mathrm{d}P = 0. \tag{6.34}$$

and U does not depend on P. The conclusion of the experimental observation that there is no temperature change in the gas is therefore that U is a function of only T and neither P nor V is involved.

When Joule made more accurate measurements, however, he found that there *was* a change in the temperature of the gas. The gas cooled down by a small amount. (For the free expansion of one mole of gas at standard temperature and pressure, in which the volume of the gas doubles, the change in temperature is approximately $-0.5\,°\mathrm{C}$.) The reason is that real gases are not ideal, even though dilute gases are a good approximation to an ideal gas. For a real gas, its internal energy U has contributions arising from the potential energy between the molecules as well as from their kinetic energy. The molecules in a real gas attract one another as discussed in Section 5.1. As the gas expands, the potential energy of the molecules increases as the molecules pull away from each other's attractive potentials. This is analogous to the way the gravitational potential energy of a body is increased if it is pulled further away from the centre of the earth. The increase in potential energy is balanced by a decrease of the kinetic energy of the molecules and so the gas cools down.

Joule's experiments lead to a collaboration between Joule and Lord Kelvin. They developed further experiments to investigate the cooling effects of gases when they expand, as described in Section 6.13.

6.11 Molar specific heats of an ideal gas

We discussed the specific heats of gases from a microscopic point of view in Section 3.8. There we saw that the specific heat of a gas gives information about the molecules of the gas, and, in particular, their degrees of freedom. Here we consider the heat capacities of a gas from the viewpoint of classical thermodynamics, where we are dealing with macroscopic quantities that can be measured experimentally.

We will consider one mole of gas and so obtain molar specific heats. The specific heat of a gas depends upon the mode of heating. We can keep the volume of the gas constant or we can keep the pressure constant. The molar specific heat at constant volume is denoted by C_{V} and the molar specific heat at constant pressure is denoted by C_{P}.

6.11.1 Molar specific heat at constant volume, C_{V}

In this case, we raise the temperature of one mole of gas in a rigid container of fixed volume, as illustrated by Figure 6.19. The molar specific heat at constant volume is defined as

$$C_{\mathrm{V}} = \left(\frac{\partial Q}{\partial T}\right)_{\mathrm{V}}. \tag{6.35}$$

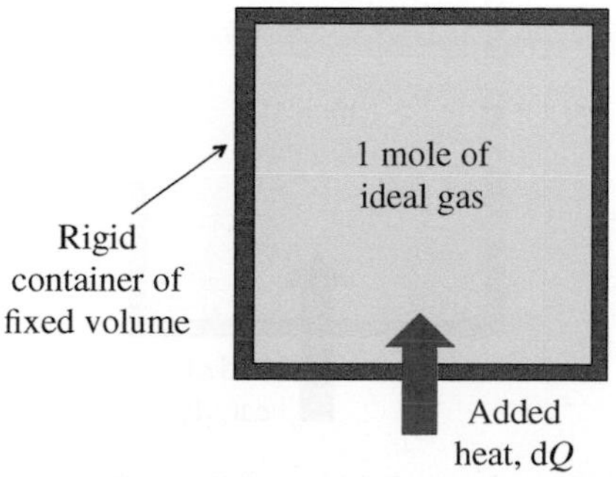

Figure 6.19 Heat dQ is added to one mole of gas that is held in a rigid container. As there is no expansion of the gas, the gas does no work and all the added heat goes into increasing the internal energy of the gas. Hence, $\mathrm{d}U = \mathrm{d}Q$.

From the first law of thermodynamics for an infinitesimal reversible process, we have

$$dQ = dU + P\,dV. \tag{6.36}$$

As the volume of the gas is constant, no work is done by the gas as it is heated; $P\,dV = 0$, and, hence,

$$dQ = dU. \tag{6.37}$$

In this case, all the added heat goes into increasing the internal energy of the gas. Differentiating Equation (6.37) with respect to T, we obtain

$$\left(\frac{\partial U}{\partial T}\right)_V = \left(\frac{\partial Q}{\partial T}\right)_V = C_V. \tag{6.38}$$

We see that the specific heat at constant volume is the rate of change of internal energy with temperature.

6.11.2 The difference in molar specific heats, $C_P - C_V$

We now consider the case of adding heat to one mole of gas at constant pressure. Figure 6.20 is a schematic diagram showing the gas contained in a cylinder with a piston. As the gas is heated, the piston moves so that the pressure of the gas remains constant. This means that the gas does work as it expands. Considering the expansion to be reversible, the state of the gas during the expansion can always be represented by the ideal gas equation $PV = RT$. For an infinitesimal change, this gives, Equation (6.2):

$$P\,dV + V\,dP = R\,dT.$$

As $dP = 0$, we have $P\,dV = R\,dT$. Substituting for $P\,dV$ in Equation (6.36), we obtain

$$dQ = dU + R\,dT.$$

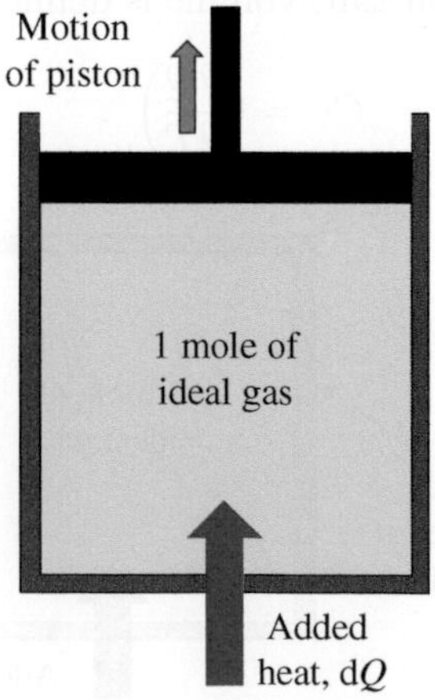

Figure 6.20 Heat dQ is added to 1 mol of gas that is held in a cylinder fitted with a frictionless piston. As the heat is added, the piston moves so that the pressure of the gas remains constant, which means that the gas does work as it pushes against the piston. In this case, $dQ = dU + P\,dV$.

Differentiating this equation with respect to T at constant pressure gives

$$\left(\frac{\partial Q}{\partial T}\right)_{\mathrm{P}} = \left(\frac{\partial U}{\partial T}\right)_{\mathrm{P}} + R. \tag{6.39}$$

The internal energy U of an ideal gas depends only on temperature T and is independent of the pressure P or volume V of the gas. It follows that the change in U depends only on the change in T and it does not matter whether the change in T takes place at constant pressure or constant volume:

$$\left(\frac{\partial U}{\partial T}\right)_{\mathrm{P}} = \left(\frac{\partial U}{\partial T}\right)_{\mathrm{V}} = \left(\frac{\partial Q}{\partial T}\right)_{\mathrm{V}} = C_{\mathrm{V}}. \tag{6.40}$$

Putting $(\partial U/\partial T)_{\mathrm{P}} = C_{\mathrm{V}}$ in Equation (6.39), we obtain

$$\left(\frac{\partial Q}{\partial T}\right)_{\mathrm{P}} = C_{\mathrm{V}} + R. \tag{6.41}$$

But

$$\left(\frac{\partial Q}{\partial T}\right)_{\mathrm{P}} = C_{\mathrm{P}}. \tag{6.42}$$

Hence,

$$C_{\mathrm{P}} = C_{\mathrm{V}} + R, \tag{6.43}$$

which is the result we obtained in Section 3.8. It follows that the molar specific heat at a constant pressure of an ideal gas is always larger than that at constant volume, the difference being equal to $R = 8.31\ \mathrm{J/mol \cdot K}$.

Although we have derived Equation (6.43) for an ideal gas, it turns out that it is closely obeyed by many real gases under normal conditions of pressure, especially for monatomic gases that have only translational degrees of freedom. A table of values of $C_{\mathrm{P}} - C_{\mathrm{V}}$ for various gases is shown in Table 6.1. Agreement is less good for diatomic and polyatomic molecules, which also have rotational and vibration degrees of freedom.

6.11.3 Reversible adiabatic expansion of an ideal gas

Having obtained expressions for the molar heat capacities, we can now consider the reversible adiabatic expansion of an ideal gas. We start with the first law, Equation (6.36):

$$\mathrm{d}Q = \mathrm{d}U + P\mathrm{d}V.$$

Table 6.1 Differences of molar heat capacities for various gases.

	Monatomic		Diatomic		Polyatomic
Gas	Ar	Ne	H_2	O_2	CO_2
$C_{\mathrm{P}} - C_{\mathrm{V}}(\mathrm{J/mol \cdot K})$	8.31	8.31	8.32	8.38	8.48

For an adiabatic process $\mathrm{d}Q = 0$, and using $\mathrm{d}U = C_V \mathrm{d}T$, we have

$$0 = C_V \mathrm{d}T + P\mathrm{d}V. \tag{6.44}$$

For an ideal gas,

$$P\mathrm{d}V + V\mathrm{d}P = R\mathrm{d}T.$$

Substituting for $P\mathrm{d}V$ in Equation (6.44) gives

$$\begin{aligned} -C_V \mathrm{d}T &= R\mathrm{d}T - V\mathrm{d}P \\ &= (C_P - C_V)\mathrm{d}T - V\mathrm{d}P. \end{aligned}$$

Hence,

$$C_P \mathrm{d}T = V\mathrm{d}P.$$

From Equation (6.44)

$$C_V \mathrm{d}T = -P\mathrm{d}V.$$

Then,

$$\frac{C_P}{C_V} = \gamma = \frac{-V\mathrm{d}P}{P\mathrm{d}V},$$

where γ is the ratio of the molar heat capacities. This gives

$$\gamma \frac{\mathrm{d}V}{V} = \frac{-\mathrm{d}P}{P}.$$

Integrating from initial to final conditions, we obtain

$$\gamma \ln \frac{V_2}{V_1} = -\ln \frac{P_2}{P_1},$$

leading to the following result for an adiabatic expansion:

$$P_1 V_1^\gamma = P_2 V_2^\gamma. \tag{6.45}$$

Figure 6.21 shows an adiabatic curve and two isotherms for one mole of an ideal gas on a P–V diagram. The adiabatic curve is shown in red and the two isotherms are shown in blue, for temperatures $T = 273$ K and $T = 473$ K, respectively. Suppose that the gas initially has volume V_1 and pressure P_1 at the temperature of 473 K. If the gas is expanded adiabatically, it follows the adiabatic curve. As it expands, it steadily gets cooler. Eventually, it crosses the 273 K isotherm, attaining that temperature at pressure P_2 and volume V_2.

The slope of an isotherm at any point on the curve is given by differentiating $PV =$ constant. This gives

$$\frac{\mathrm{d}P}{\mathrm{d}V} = -\frac{P}{V}. \tag{6.46}$$

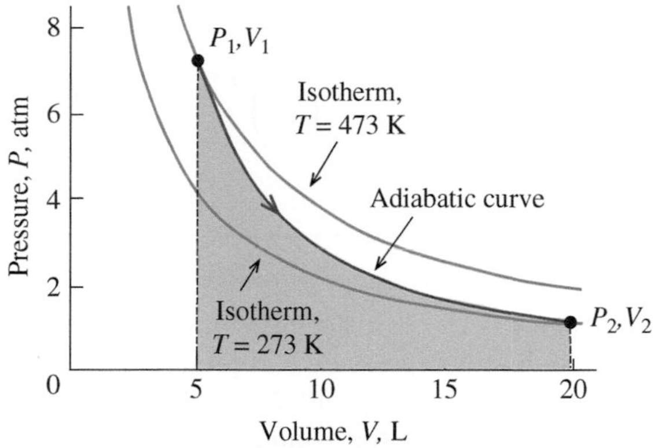

Figure 6.21 This figure shows an adiabatic curve and two isotherms for 1 mol of an ideal gas on a P–V diagram. The two isotherms, shown as blue curves, are for temperatures $T = 273$ K and $T = 473$ K, respectively. Suppose that the gas initially has volume V_1 and pressure P_1 at the temperature of 473 K. As the gas is expanded adiabatically, it follows the adiabatic curve, shown in red. As it expands, it steadily gets cooler. Eventually, it crosses the 273 K isotherm, attaining that temperature at pressure P_2 and volume V_2. We see from the figure that the adiabatic curve is steeper than the isotherms. The shaded area represents the work done by the gas in the adiabatic expansion.

The slope of the adiabatic curve at any point is given by differentiating PV^γ = constant. This gives

$$\frac{\mathrm{d}P}{\mathrm{d}V} = -\gamma \frac{P}{V}. \tag{6.47}$$

The slope of the adiabatic curve is γ times that of the isotherm at their point of intersection. Since γ always has a value greater than 1, it follows that the slope of the adiabatic curve is always greater than that of the isotherm, as seen in Figure 6.21.

The shaded area in Figure 6.21 represents the work done by the gas in the adiabatic expansion. The infinitesimal work done by the gas in expanding by $\mathrm{d}V$ is

$$\mathrm{d}W = P\mathrm{d}V.$$

At any stage in the expansion, we have

$$PV^\gamma = P_1 V_1^\gamma,$$

so that P in Equation (6.17) can be replaced by $P_1 V_1^\gamma V^{-\gamma}$. Then the work done by the gas in expanding adiabatically from V_1 to V_2 is

$$W = P_1 V_1^\gamma \int_{V_1}^{V_2} V^{-\gamma} \mathrm{d}V.$$

$$= P_1 V_1^\gamma \left[\frac{V^{1-\gamma}}{1-\gamma} \right]_{V_1}^{V_2},$$

which gives

$$W = \frac{P_1 V_1}{\gamma - 1} \left[1 - \left(\frac{V_1}{V_2} \right)^{\gamma - 1} \right]. \tag{6.48}$$

Then using $P_1V_1^{\gamma} = P_2V_2^{\gamma}$, we obtain

$$W = \frac{P_1V_1}{\gamma-1}\left[1-\left(\frac{P_2}{P_1}\right)^{(\gamma-1)/\gamma}\right]. \tag{6.49}$$

Worked example

The compression ratio of a diesel engine is 16 : 1. If the initial pressure and temperature are 1.01×10^5 Pa and 23 °C, respectively, calculate the final temperature and pressure after compression. It is at this point diesel oil is injected. Take the value of γ for air to be 1.40. Assume that the compression happens so quickly that there is no transport of heat to the surroundings and hence that the process can be considered to be adiabatic. What would be the final temperature and pressure if the compression process was isothermal?

Solution

Since the compression can be considered to be adiabatic, we have

$$P_2 = P_1\left(\frac{V_1}{V_2}\right)^{\gamma} = (1.01 \times 10^5)(16)^{1.4}$$
$$= 49.0 \times 10^5 \text{ pa} = 48.5 \text{ atm}.$$

From the equation of state, we obtain

$$\frac{T_2}{T_1} = \frac{P_2V_2}{P_1V_1} = \left(\frac{V_1}{V_2}\right)^{\gamma}\left(\frac{V_2}{V_1}\right) = \left(\frac{V_1}{V_2}\right)^{\gamma-1}.$$

Therefore,

$$T_2 = T_1\left(\frac{V_1}{V_2}\right)^{\gamma-1} = (296)(16)^{0.4}$$
$$= 897 \text{ K} = 624°\text{C}.$$

This high temperature causes the fuel to ignite spontaneously, without the need for spark plugs. If the compression had been isothermal, the final pressure would be

$$P_2 = P_1\left(\frac{V_1}{V_2}\right) = 16 \text{ atm}.$$

Because the temperature increases in adiabatic compression, the final pressure is much greater.

6.12 Enthalpy

We have discussed various processes in terms of the thermodynamic function U, which defines the internal energy of a system. However, there are many processes in which it is more convenient to use a different thermodynamic function. The simplest such special function is *enthalpy*, which is a particularly convenient choice for processes that occur at constant pressure. The enthalpy H of a system is defined as

$$H = U + PV, \tag{6.50}$$

where U is the internal energy, V is the volume the system occupies and P is the pressure of its surroundings.

Suppose we add heat Q to a system at constant pressure P, and the only work that is done is $P\mathrm{d}V$ work. From the first law,

$$\Delta U = Q - P\mathrm{d}V.$$

We therefore have,

$$\Delta U = U_2 - U_1 = Q - P\mathrm{d}V = Q - P(V_2 - V_1),$$

where U_1 and U_2 are the internal energies of the initial and final states with corresponding volumes V_1 and V_2, respectively. Gathering terms, we obtain

$$(U_2 + PV_2) - (U_1 + PV_1) = Q.$$

Then with the definition of enthalpy as in Equation (6.50), we have

$$\Delta H = H_2 - H_1 = Q. \tag{6.51}$$

The change ΔH in enthalpy is equal to the heat Q absorbed at constant pressure. We can compare this result with the case where we add heat Q to a system at constant *volume*. For that case, the first law gives

$$\Delta U = Q, \tag{6.52}$$

i.e. at constant pressure, the change in internal energy U is equal to the heat Q absorbed.

Since enthalpy H is defined in terms of the state functions U, P, and V, it follows that enthalpy is also a state function. Therefore a change in enthalpy when the system changes from one state to another is independent of the path connecting them. The result $\Delta H = Q$ is therefore very powerful. It relates a quantity we can measure, the heat transfer at constant pressure to a state function, the change in enthalpy. We do not need to consider how we get from one state to another; all that matters are the initial and final states.

If we have a process that occurs at constant pressure in a thermally isolated system, $Q = 0$. In that case, it follows from Equation (6.51) that

$$\Delta H = 0, \quad H = \text{constant}. \tag{6.53}$$

6.12.1 Specific heat at constant pressure, C_P

We can use Equation (6.51) to obtain an expression for molar specific heat at constant pressure C_P. Differentiating the equation with respect to T, at constant pressure, we obtain

$$\left(\frac{\partial H}{\partial T}\right)_\mathrm{P} = \left(\frac{\partial Q}{\partial T}\right)_\mathrm{P}.$$

$(\partial Q/\partial T)_{\mathrm{P}}$ is the specific heat at constant pressure C_{P} and hence we have

$$C_{\mathrm{P}} = \left(\frac{\partial H}{\partial T}\right)_{\mathrm{P}}. \tag{6.54}$$

We can compare this result with that for the specific heat at constant volume

$$C_{\mathrm{V}} = \left(\frac{\partial U}{\partial T}\right)_{\mathrm{V}}.$$

Many processes in the physical sciences and engineering take place at constant pressure. In such constant-pressure processes, enthalpy plays a key role in their description and analysis.

6.13 The Joule–Kelvin effect

The Joule–Kelvin effect is the change of temperature that occurs in the expansion of a gas through a *throttle* or valve from high pressure to low pressure. This is a continuous steady-state flow process as illustrated in Figure 6.22, with gas entering at constant rate from the left-hand side. The throttle acts to constrict the flow of gas and the reduction in pressure from P_i to P_f leads to a change in temperature of the gas.

An experimental arrangement to investigate the Joule–Kelvin effect is shown schematically in Figure 6.23. It consists of a thermally insulated pipe through which there is a steady flow of gas. The throttle that is fitted to the pipe is a porous plug. It constricts the flow of gas so that a pressure differential is maintained across it. There are two pistons that move in the pipe. During the expansion of the gas, the two pistons move at appropriate rates to ensure that the pressures on either side of the porous plug are maintained at the constant values P_i and P_f, respectively.

As in the case of the free expansion of a gas, the Joule–Kelvin process is irreversible. Again, however, the initial and final states are equilibrium states, so we can apply the laws of thermodynamics. Initially, the gas occupies volume V_i at pressure P_i in the left-hand side of the porous plug, as shown in Figure 6.24a, while the volume of gas to the right-hand side is zero. At the end of the expansion, all the gas has been forced through the porous plug so that the volume to the right of the constriction is V_f, at pressure P_f while the volume of gas to the left is zero, as in Figure 6.24b.

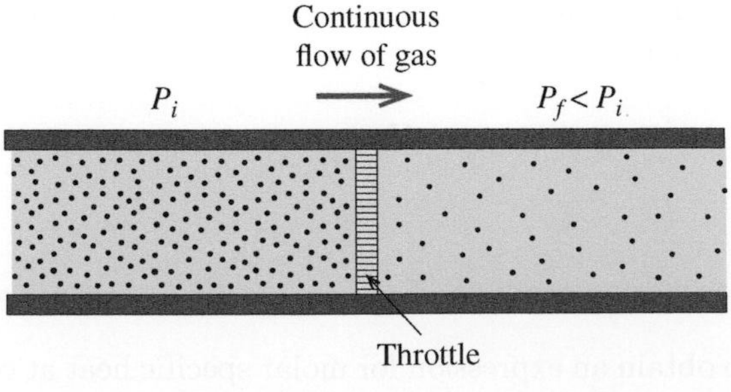

Figure 6.22 The figure illustrates the Joule–Kelvin effect, which is the change in temperature that occurs in the expansion of a gas through a throttle from high-pressure P_i to low-pressure P_f. Gas passes through the throttle at a constant rate from the left-hand side. The change in pressure from P_i to P_f leads to a change in the temperature of the gas.

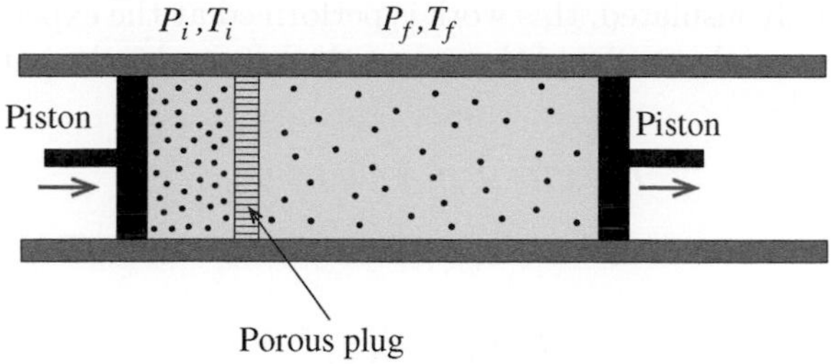

Figure 6.23 The figure shows an experimental arrangement to investigate the Joule–Kelvin effect. It consists of a pipe through which there is a steady flow of gas. The pipe is fitted with a porous plug, which constricts the flow of gas so that a pressure differential is maintained across it. There are two pistons that move in the pipe. During the expansion of the gas, the two pistons move at appropriate rates to ensure that the pressures on the two sides of the constriction stay at constant values P_i and P_f, respectively. The entire system is thermally insulated so that heat does not enter or leave it.

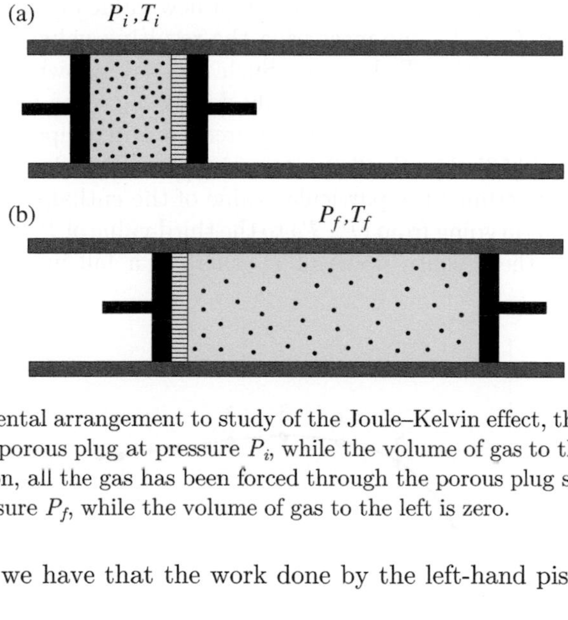

Figure 6.24 (a) In an experimental arrangement to study of the Joule–Kelvin effect, the gas initially occupies volume V_i in the left-hand side of the porous plug at pressure P_i, while the volume of gas to the right-hand side is zero. (b) At the end of the expansion, all the gas has been forced through the porous plug so that the volume to the right of the plug is V_f, at pressure P_f, while the volume of gas to the left is zero.

From Equation (6.21), we have that the work done by the left-hand piston *on* the gas in forcing it through the porous plug is

$$W_{\text{on gas}} = -\int_{V_i}^{0} P_i \mathrm{d}V = P_i V_i$$

as P_i is constant. Similarly, the work done *by* the gas on the left-hand piston is

$$W_{\text{by gas}} = \int_{0}^{V_f} P_f \mathrm{d}V = P_f V_f.$$

Hence, the net work done by the gas is

$$P_i V_i - P_f V_f. \tag{6.55}$$

As the whole system is thermally insulated, this work is performed at the expense of the internal energy of the gas. If U_i and U_f are the internal energies of the gas on the left- and right-hand sides of the porous plug, the change in internal energy is

$$U_f - U_i = P_i\, V_i - P_f\, V_f,$$

giving

$$U_f + P_f V_f = U_i + P_i V_i,$$

or in terms of enthalpy, $H = U + PV$

$$H_f = H_i. \tag{6.56}$$

Hence, in a Joule–Kelvin expansion, the enthalpy remains constant. This is typical of steady-state continuous flow processes.

Suppose that we keep the pressure P_i and temperature T_i to the left of the porous plug at the same values as before, but now set a new value of P_f This will result in a new value of T_f. And suppose we do this for a number of different values of P_f each time measuring the resulting value of T_f. We can then plot the resulting set of (P_f, T_f) points on a $P_f - T_f$ diagram. Such a plot is shown in Figure 6.25, obtained for seven successive values of P_f; from 1 to 7 with P steadily decreasing. Also included on the diagram, as a red dot, is the (P_i, T_i) point corresponding to the constant pressure and temperature to the left of the porous plug. It follows from Equation (6.56) that the line connecting the points is a curve of constant enthalpy, where the values of P_i and T_i determine the particular value of the enthalpy.

From Figure 6.25, we see that in going from (P_i, T_i) to the third value of P_f, there is a rise in temperature, but in going from (P_i, T_i) to the seventh value of P_f, there is a fall in temperature. In general, the

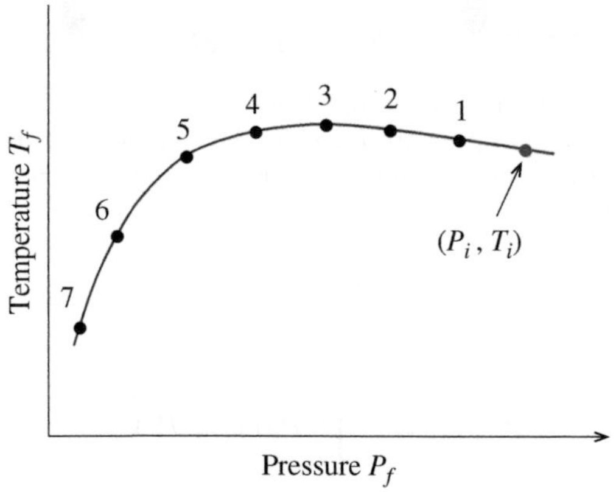

Figure 6.25 A plot of (P_f, T_f) points on a $P_f - T_f$ diagram for seven successive values of P_f from 1 to 7 with P steadily decreasing. P_f and T_f are, respectively, the final pressure and temperature of the gas after a Joule–Kelvin expansion. Also included on the diagram, as a red dot, is the (P_i, T_i) point corresponding to the pressure and temperature of the gas before the expansion. The solid line is a curve of constant enthalpy, where the values of P_i and T_i determine the particular value of the enthalpy. In going from (P_i, T_i) to the third value of P_f there is a rise in temperature, but in going from (P_i, T_i) to the seventh value of P_f there is a fall in temperature. In general, the temperature change depends on the values of P_i, T_i, and P_f and may be a decrease, increase, or there may be no change at all.

temperature change depends on the values of P_i, T_i, and P_f, and may be a decrease, increase, or there may be no change al all.

Curves of constant enthalpy can be drawn for different values of P_i and T_i. A set of such curves is shown in Figure 6.26. The slope at any point on these curves is called the *Joule–Kelvin coefficient* μ, and is defined as

$$\mu = \left(\frac{\partial T}{\partial P}\right)_{\mathrm{H}}, \tag{6.57}$$

where μ gives the change in temperature produced by an infinitesimal pressure difference. The locus of the points on the curves at which μ is zero, corresponding to the maxima of the curves, is shown as the red dotted curve. This curve is known as the *inversion curve.* Inside the inversion curve, T increases with P along any of the curves, that is $\mu = (\partial T/\partial P)_{\mathrm{H}}$ is positive. This means that if we expand the gas inside the inversion curve, i.e. reducing its pressure, the gas will be cooled. Outside the inversion curve, on the other hand, $(\partial T/\partial P)_{\mathrm{H}}$ is negative, which means that the gas warms up on expansion.

The temperature drop produced by a Joule–Kelvin expansion can be read off directly from an enthalpy diagram such as that in Figure 6.26. Given the initial condition (T_1, P_1) of the gas, (point a on the figure), and the final pressure P_2, which is usually atmospheric pressure, the final temperature T_2 lies on the same line of constant enthalpy as the point (T_1, P_1), i.e. the final state corresponds to point b on the figure. Typical values of μ are of the order of a few tenths of a degree Celsius per atmosphere. Hence, if the initial pressure is say 200 atm and the final pressure is atmospheric pressure the temperature decrease is of order 20 °C.

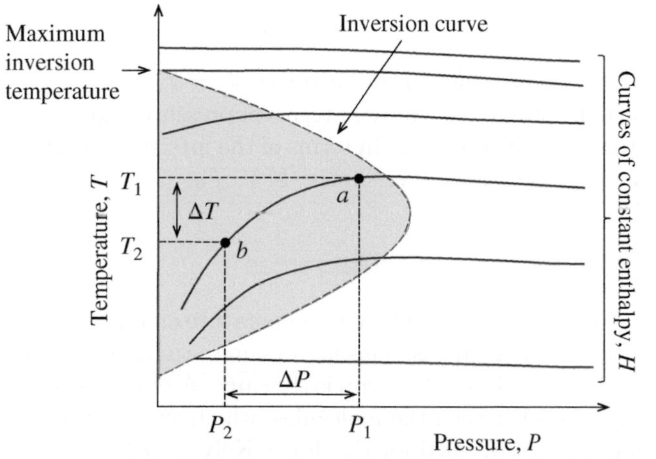

Figure 6.26 A set of curves of constant enthalpy, shown in blue. The slope at any point on these curves gives the change of temperature produced in a Joule–Kelvin expansion. The locus of the points corresponding to the maxima of the curves is known as the inversion curve. Inside the inversion curve, T decreases with decreasing P along any of the curves. This means that if we expand the gas inside the inversion curve, the gas will be cooled. Outside the inversion curve, on the other hand, T increases with decreasing P, which means that the gas warms up on expansion. The temperature change produced by a Joule–Kelvin expansion can be read off directly from the enthalpy diagram. Given the initial condition (T_1, P_1) of the gas, (point a on the figure), and the final pressure P_2, which is usually atmospheric pressure, the final temperature T_2 lies on the same line of constant enthalpy as the point (T_1, P_1), i.e. the final state corresponds to point b on the figure.

Table 6.2 The maximum inversion temperature for several gases.

Gas	He	H_2	Ar	N_2	O_2
Max. inversion temperature (K)	43	202	723	623	761

The Joule–Kelvin effect provides a practical way to liquefy gases. In order to achieve a lower temperature, it is necessary to work in the region where $\mu > 0$. In particular, the initial temperature T_i must be less than the *maximum inversion* temperature, which is indicated in Figure 6.26. Table 6.2 gives the maximum inversion temperature of several gases. The maximum inversion temperature for most gases is above room temperature. However, the maximum inversion temperature for helium is 43 K. Hence, to liquefy helium, the gas must first be cooled to below this temperature in order for the Joule–Kelvin effect to liquefy the gas.

The Joule–Kelvin effect does not result in a temperature change for an ideal gas. We can see this as follows. For an ideal gas, we have $PV = RT$, and hence

$$H = U + PV, \tag{6.50}$$

becomes

$$H = U + RT. \tag{6.58}$$

As the internal energy U of an ideal gas is a function of temperature T alone, it follows that H is also a function of temperature alone. Hence, the condition (6.56) implies

$$H_f(T_f) = H_i(T_i). \tag{6.59}$$

As the enthalpy is constant, it follows that for an ideal gas, $T_f = T_i$.

We note here that we usually specify enthalpy in terms of pressure and temperature: $H(P, T)$. By contrast, when we discuss thermodynamic processes in terms of the internal energy of a system, we usually use temperature T and volume V as the relevant variables: $U(V, T)$.

6.13.1 Joule–Kelvin effect and intermolecular forces

We have seen that the Joule–Kelvin effect results in a temperature change in a real gas, whereas there is no temperature change in the case of an ideal gas. This is the same situation as for the free expansion of a gas that we discussed in Section 6.10 and the explanation is the same. A temperature change occurs in a real gas because of the mutual interactions between the molecules, which are not present in an ideal gas.

The following expression can be obtained for the Joule–Kelvin coefficient:

$$\mu = \frac{1}{C_P}\left[T\left(\frac{\partial V}{\partial T}\right)_P - V\right], \tag{6.60}$$

where C_P is the molar specific heat at constant pressure and $(\partial V/\partial T)_P$ is the coefficient of expansion. In Section 5.3, we discussed the virial equation

$$\frac{PV}{RT} = 1 + \frac{B}{V} + \frac{C}{V^2} + \frac{D}{V^3} + \cdots, \tag{6.61}$$

which is a good approximation for the equation of state of a real gas. The constants B, C, and D are functions of temperature. If the gas pressure P is not too large, we can ignore the C and D terms and write

$$PV = RT + BP. \tag{6.62}$$

Differentiating this equation with respect to T at constant pressure and multiplying by T gives

$$T\left(\frac{\partial V}{\partial T}\right)_{\mathrm{P}} = \frac{RT}{P} + T\frac{\mathrm{d}B}{\mathrm{d}T}. \tag{6.63}$$

From Equation (6.62), we have

$$\frac{RT}{P} = V - B,$$

and substituting for RT/P in Equation (6.63) gives

$$T\left(\frac{\partial V}{\partial T}\right)_{\mathrm{P}} = V - B + T\frac{\mathrm{d}B}{\mathrm{d}T}. \tag{6.64}$$

Then, substituting for $T(\partial V/\partial T)_{\mathrm{P}}$ in Equation (6.60), we obtain

$$\mu = \frac{1}{C_{\mathrm{P}}}\left(T\frac{\mathrm{d}B}{\mathrm{d}T} - B\right). \tag{6.65}$$

In Section 5.3, we discussed the behaviour of B with respect to temperature. We described how the attractive force between molecules dominates at low temperatures when the kinetic energy of the molecules $\sim kT$ is small compared to ε, the potential energy between two molecules. At high temperature, however, it is the repulsive force between molecules that dominates. The result of this is that B has a negative value at low temperature, but steadily increases with increasing temperature and becomes positive at high temperature. As B is a steadily increasing function of T, $(\mathrm{d}B/\mathrm{d}T)$ is always positive. A plot of B against T for argon is shown in Figure 5.11.

Hence, at low temperatures where B is negative, μ is positive. And at high temperatures, where B becomes positive and sufficiently large, μ becomes negative. So, the existence of the inversion curve where μ is zero is a reflection of the competition between molecular attraction and repulsion. For an ideal gas, $B = 0$. It follows from Equation (6.65) that μ is then zero as we expect.

6.14 Thermochemistry

Thermochemistry is the study of the heat effects that accompany chemical reactions. These include the formation of solutions and changes in the state of a substance such as vapourization and melting. Usually, these processes take place at constant pressure, usually atmospheric pressure. And when the process is open to the atmosphere, a change in volume usually occurs. In that case, the system has to do work in 'pushing back' the atmosphere. Hence, the change in the internal energy of the system is not equal to the energy supplied to the system as heat, because some of the energy is used to do the expansion work. However, if we deal with the enthalpy of the system, we automatically take into account the energy lost or gained by the expansion work. Then, according to Equation (6.51), the change in the enthalpy of the system is equal to the heat supplied:

$$\Delta H = Q.$$

This result has great importance because it relates something we can measure, i.e. the heat transfer at constant pressure, to a state function. As we have emphasized, state functions are defined by the state of the system and not how that state was reached, just like the internal energy of the system. Because of this property of enthalpy, it finds wide application in thermochemistry.

6.14.1 The enthalpy of vaporization

In order to convert a liquid into its vapour, we must provide enough energy for the molecules in the liquid to overcome their mutual attraction. In addition, however, energy must also be provided for the expansion work that the resulting vapour has to do in pushing back the atmosphere. The enthalpy of vapourization is the energy that must be supplied as heat per mole at constant pressure to vapourise the liquid. As we are dealing with enthalpy, this takes into account the expansion work performed by the vapour.

Taking the enthalpy of the liquid to be $H_i = U_i + PV_i$, and the enthalpy of the vapour to be $H_f = U_f + PV_f$, the change in enthalpy in going from the liquid to vapour phase is

$$H_f - H_i = (U_f + PV_f) - (U_i + PV_i)$$

or

$$\Delta H = \Delta U + P\Delta V, \tag{6.66}$$

where ΔH is the enthalpy of vapourization and ΔV is the volume change.

6.14.2 Heats of reaction

A chemical reaction may release or absorb heat. Reactions that absorb heat are classified as endothermic, while reactions that release heat are classified as exothermic. The burning of hydrogen in oxygen is an example of an exothermic reaction:

$$\mathrm{H_2} + \frac{1}{2}\mathrm{O_2} \rightarrow \mathrm{H_2O}\ (\mathrm{gas}), \quad \Delta H = -286\,\mathrm{kJ}.$$

For reactions that occur at constant pressure, the amount of heat Q involved in a chemical reaction, called the *heat of reaction*, is directly related to the enthalpy change ΔH. For the above reaction, ΔH has a negative sign because heat is released in the reaction.

The heat of reaction Q can be measured using a calorimeter that contains a known amount of the reactants. From the known heat capacity of the calorimeter and its change in temperature, the value of Q can be calculated. The measured heat of reaction depends on the conditions under which the reaction occurs. There are two particular conditions that are used in practice because they lead to heats of reaction that are equal to changes in thermodynamic functions, either ΔH or ΔU.

The first such condition is the measurement of the heat of reaction at constant pressure, usually with the calorimeter maintained at atmospheric pressure. Then from Equation (6.51), the heat of reaction Q_P is equal to the change in enthalpy:

$$\Delta H = Q_P. \tag{6.67}$$

The second condition is the measurement of the heat of reaction at constant volume, which is conducted in a *bomb calorimeter*. In this case, no work is done by or on the system. Then, from the first law of thermodynamics, the measured heat of reaction Q_V at constant volume is exactly equal to the change in internal energy of the system:

$$\Delta U = Q_V. \tag{6.68}$$

Sometimes it is useful to deduce ΔH for a particular reaction from measurements taken with a bomb calorimeter for which $\Delta U = Q_V$. To do this, account must be taken of any volume changes that may occur in the reaction, in accordance with Equation (6.66). If the reactants and products are all solids or liquids, the value of $P\Delta V$ is usually negligible compared with ΔH or ΔU and so may be ignored. Then we can set $Q_P = Q_V$, giving $\Delta H = Q_V$. For reactions in which gases do occur, the value of $P\Delta V$ depends upon the change Δn in the number of moles of gas involved in the reaction. Using the ideal gas equation, we have $P\Delta V = \Delta nRT$, where Δn is the difference between the number of moles of gases after and before the reaction. Therefore, from Equation (6.66), we obtain

$$\Delta H = \Delta U + \Delta nRT. \tag{6.69}$$

For example, in the reaction

$$SO_2 + \frac{1}{2}O_2 \rightarrow SO_3,$$

ΔU is measured to be -97.0 kJ by a bomb calorimeter at 298 K. The value of Δn, i.e. the difference between the number of moles of gases after and before the reaction is

$$\Delta n = 1 - \left(1 + \frac{1}{2}\right) = -\frac{1}{2}\ \text{mol}.$$

Therefore,

$$\begin{aligned}\Delta H &= \Delta U - \frac{1}{2}RT \\ &= -97.0 \times 10^3 - \frac{1}{2}(8.31 \times 298) = -98.2\ \text{kJ}.\end{aligned}$$

Problems 6

6.1 Determine the number of molecules in 14 g of nitrogen gas and hence determine the total internal energy of this quantity of nitrogen gas at (i) 20 °C and (ii) 80 °C.

6.2 A gas, initially at pressure P_0, undergoes a free expansion until its volume is four times its original volume. What will be the final pressure of the gas?

6.3 A platinum resistance thermometer is calibrated by measuring its resistance at 0 and 100 °C. Other temperatures are found by linearly interpolating between these two fixed points. The resistance of a platinum wire at temperature T °C measured on the ideal gas scale is given by $R(T) = R_0(1+\alpha T+\beta T^2)$, where $\alpha = 3.8 \times 10^{-3}\,\mathrm{K}^{-1}$ and $\beta = -5.6 \times 10^{-7}\,\mathrm{K}^{-2}$. What temperature will the platinum resistance thermometer indicate when the ideal gas scale is 200 °C?

6.4 A quantity of heat Q is added to helium gas so that the gas expands at constant pressure. Show that the amount of heat that goes into the expansion work is $2Q/5$.

6.5 The figure shows a schematic diagram of Ruchardt's method of determining γ, the ratio of specific heats. The two spherical glass vessels are joined by a glass tube. The volume of each vessel is V_0 and the pressure in both is P_0. The spherical mass m is a close fit in the tube of cross sectional area A. At equilibrium, the mass sits in the middle of the tube. Show that when the mass is displaced slightly, it performs simple harmonic motion with a period given by $\tau = 2\pi\sqrt{(mV_0/[2\gamma P_0 A^2])}$. Assume that the changes involved are adiabatic. It was found in one experiment with molecular nitrogen in the vessel that $\tau = 0.73$ seconds for $m = 26 \times 10^{-3}$ kg, $V_0 = 9.2 \times 10^{-3}\,\mathrm{m}^3$ and $A = 2.5 \times 10^{-4}\,\mathrm{m}^2$. Use these data to obtain the value for nitrogen. Take P_0 to be 1.01×10^5 Pa.

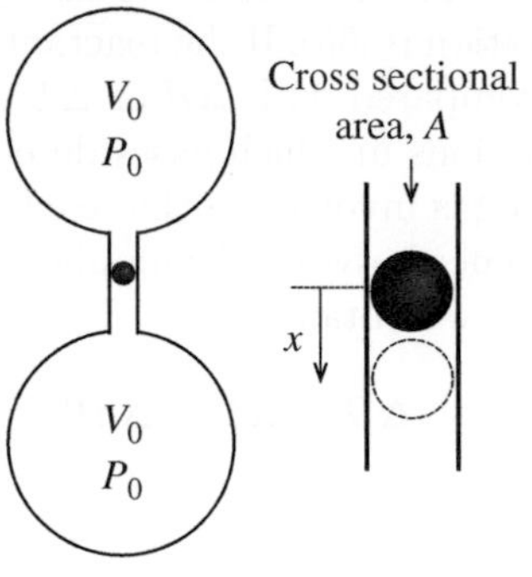

6.6 Oxygen gas of mass 0.16 kg at atmospheric pressure and temperature 20 °C is heated to 100 °C. (a) How much heat must be supplied if the volume of the gas is kept constant during the heating? (b) How much heat must be supplied if the pressure of the gas is kept constant during the heating? (c) What is the change in the internal energy of the gas for these two processes? (d) How much work does the gas do in the constant pressure case (b).

6.7 (a) One mole of a diatomic gas at temperature of 22 °C is expanded *isothermally* to twice its volume. Calculate (i) the work done by the gas and (ii) the amount of heat supplied to the gas. (b) One mole of the same gas at an initial temperature of 22 °C is expanded *adiabatically* to twice its volume. Calculate (i) the final temperature, (ii) any change in the internal energy of the gas, and (iii) the work done by the gas.

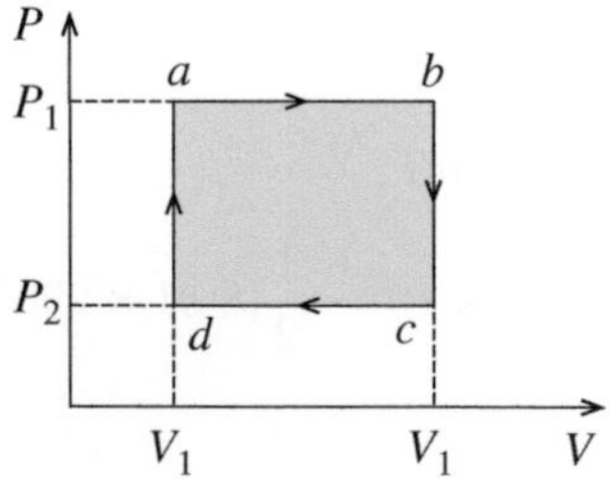

6.8 A gas undergoes the cycle shown in the figure: $a \rightarrow b$, $b \rightarrow c$, $c - d$ and $d \rightarrow a$. Write down an expression for the work done by the gas over the cycle in terms of the pressures and volumes indicated in the figure. What is the change in internal energy of the gas after completing the cycle? How much energy is absorbed by the gas during the complete cycle?

6.9 Suppose that in the figure of Problem 6.8, $P_1 = 3$ atm, $P_2 = 1$ atm, $V_1 = 1$ L, and $V_2 = 2.5$ L. For each of the four steps in the cycle, calculate the work done by the gas and the amount of heat that is absorbed over a complete cycle. If the amount of gas is 0.1 mol, determine the temperatures at points a, b, c, and d.

6.10 Calculate (i) the work done when 1 mol of water at atmospheric pressure and 100 °C is converted into water vapour at the same pressure and temperature, and (ii) the increase in internal energy of the water. The molar volume of the liquid and the vapour are $1.88 \times 10^{-4}\,\mathrm{m^3/mol}$ and $3.02 \times 10^{-2}\,\mathrm{m^3/mol}$, respectively and the enthalpy of vaporization is 4.06×10^4 J/mol.

6.11 As we will see in Chapter 9, the velocity υ of propagation of a longitudinal wave, like a sound wave, is given by

$$\upsilon = \sqrt{\frac{\text{elastic modulus}}{\rho}},$$

where ρ is the density of the medium and the elastic modulus is appropriately chosen. For sound waves in air, the appropriate modulus is the bulk modulus B, which may be defined as $B = (-\mathrm{d}P)/(\mathrm{d}V/V)$, where a change $\mathrm{d}P$ in pressure produces the fractional change in volume $\mathrm{d}V/V$. (a) Show that if the propagation of sound is an isothermal process, $\upsilon = \sqrt{P/\rho} = \sqrt{\overline{u_x^2}}$, where u_x is the x-component of velocity of an air molecule. (b) Show that if the propagation of sound is an adiabatic process, $\upsilon = \sqrt{\gamma P/\rho} = \sqrt{\gamma \overline{u_x^2}}$. (c) Obtain a value of $\overline{u_x^2}$ for air and use the expressions for υ obtained in (a) and (b) to obtain a value for the velocity of sound in air, assuming a molecular weight for air of 30 u. Compare your results with the accepted value of 343 m/s for the velocity of sound in air.

6.12 Along a certain line of constant enthalpy, the Joule–Kelvin coefficient μ for air varies as $\mu = (a - bP)$ K/atm, where $a = 0.266$ and $b = 8.95 \times 10^{-4}$. What is the amount of cooling from an initial pressure of 20 atm to a final pressure of 1 atm?

7

The second law of thermodynamics

7.1 Introduction

Chapter 6 dealt with the first law of thermodynamics. That law deals with the energy balance in thermodynamic processes. It is essentially the conservation of energy when heat is taken into account. There are, however, processes that never happen even though they conserve energy. It is a familiar observation that heat flows from a hotter body to a colder body. The heat never flows from a colder to a hotter body, although this is not forbidden by the first law. The fact that it does not happen is a consequence of the second law of thermodynamics. So, while the first law deals with the energy balance in a process, the second law determines in what *direction* a process will occur. Because of this, the second law is sometimes called the 'arrow of time'.

The subject of thermodynamics came to prominence in the eighteenth century for a very practical reason. The steam engine had been invented and was playing a central role in the industrial revolution. Steam engines were converting heat from the burning of coal into mechanical work to drive emerging industries. A very practical consideration was the maximum amount of work that a given steam engine could produce. Various designs of the steam engine were evolving with increasing efficiency, and it was important to know if there was a limit to the maximum possible efficiency.

It was a young engineer Sadi Carnot who came up with the answer. He did not deal with the details of a particular steam engine but considered the problem in a much wider sense. Carnot obtained an expression for the maximum efficiency that a steam engine could have, or indeed any device that converts heat into work; so called *heat engines.*

Carnot's analysis of heat engines led eventually to the work of Rudolf Clausius who was a contemporary of Joule and Kelvin. Clausius developed the basic principles underlying the second law of thermodynamics and in 1865, he introduced the new concept of *entropy.* By introducing entropy, he was able to distinguish between energy conservation, as expressed by the first law, and the direction of natural processes as expressed by the second law. All natural processes, like the flow of heat from a hotter body to a colder body, are irreversible processes and occur in only one direction. Clausius showed that the changes in energy that occur within a system do not determine the direction of a process. Instead, it is the change in entropy that determines the direction of irreversible processes.

Physics of Matter, First Edition. George C. King.

We will define and discuss entropy in the following sections but here we state a form of the *entropy principle*: *In an irreversible process, the entropy of a system plus its surroundings always increases.* Only processes that are in accord with this stringent requirement can occur. Thus, when a hotter body transfers heat to a colder body, the overall entropy increases and so the process is allowed by the second law. On the other hand, the transfer of heat from a colder body to a hotter body would involve a decrease in overall entropy and so this process is forbidden. The entropy of the system and its local surroundings is sometimes referred to as the 'universe'. This leads to the statement that the entropy of the universe always increases and never decreases. Notice here a very importance difference between energy and entropy. Energy is conserved but entropy is not.

Reversible processes are idealisations of real processes. As we saw in Section 6.8, reversible processes are only approximately attainable under special circumstances. For the special case of reversible processes: *there is no net change in entropy.* Entropy, therefore, provides a quantitative measure to distinguish between reversible and irreversible processes.

It was the intense interest in steam engines that led to the formulation and understanding of the second law of thermodynamics. However, the reach of thermodynamics expanded rapidly in the nineteenth century. This is because the generality of the laws of thermodynamics makes them applicable to all natural processes – physical, chemical, and biological. Indeed, it is striking that a very practical investigation into steam engines led to one of the fundamental laws of nature; the direction of natural processes.

Entropy, as introduced by Clausius, is a concept that is expressed in terms of the macroscopic properties of a system, such as heat and temperature. It was the remarkable achievement of Boltzmann in the 1870s to relate this macroscopic concept of entropy to the microscopic behaviour of the constituent molecules of matter. As the number of molecules in any ordinary amount of matter is huge, Boltzmann's approach is statistical in nature. This statistical version of the second law provides an explanation of the direction of natural processes in terms of probabilities. Essentially, in any natural process, the final state of the system will be more probable than the initial state. It does not say the reverse process is impossible, it simply says that it is highly unlikely; so unlikely that it would not happen in many times the age of the universe.

We will discuss the second law and the concept of entropy from the two different approaches mentioned earlier. The first approach, based on the macroscopic quantities of a system, is fundamentally connected to the efficiency of heat engines, and that is where we will start our discussion.

7.2 Heat engines

It is straightforward to transform work into heat. We do this when we rub our hands together. Moreover, the conversion of work into heat may be achieved with up to 100% efficiency. The conversion of heat into work, however, is not so straightforward, and it is very challenging to achieve this conversion with high efficiency. For example, a petrol combustion engine may typically run at an efficiency of 30–40%. A heat engine is a device that converts heat *partially* into work. The word partially is important because no practical heat engine can convert heat into work with 100% efficiency.

Suppose that we have two blocks of copper, where one block is at a higher temperature than the other. And suppose that the two blocks are connected by a copper rod, as illustrated in Figure 7.1. Heat flows from the hot to the cold block as long as there is a temperature difference between them. The heat that flows from the hot block is dissipated as a random, thermal motion of the atoms of the cold block. In a sense, the available heat from the hot block is 'wasted' because it does not do any useful work. In the arrangement of a heat engine, there is also a flow of heat between a hot and cold body. However, the heat engine is placed in the path of the heat flow and a fraction of the heat is converted into useful work. The fraction is determined by the efficiency of the heat engine.

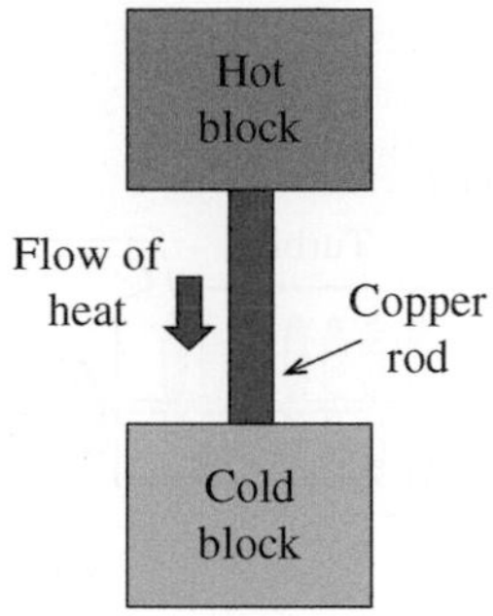

Figure 7.1 The figure illustrates two blocks of copper, one at a higher temperature than the other. They are connected by a copper rod through which heat flows from the hot to the cold block. The heat from the hot block is dissipated as random, thermal motion of the atoms of the cold block. In a sense, the heat from the hot block is 'wasted' because it does not do any useful work.

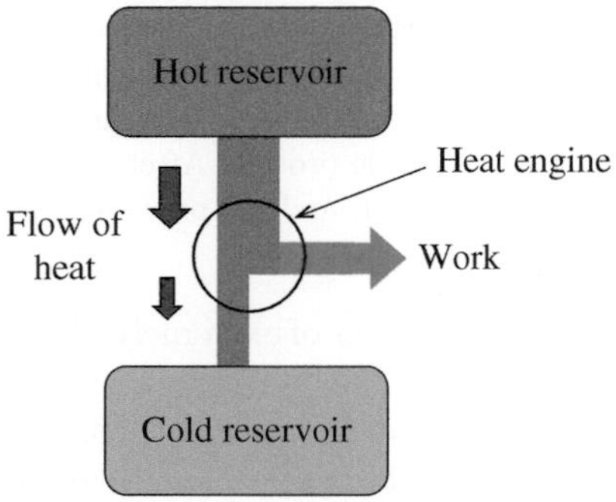

Figure 7.2 This figure illustrates the arrangement for a heat pump. There is a source of heat, called the hot reservoir and a cold reservoir at a lower temperature. There is a working substance that transports heat between the two reservoirs, usually a gas or vapour. The heat engine is positioned between the two reservoirs. The working substance passes through the heat engine, which converts a portion of the transported heat into useful work. Heat that is not converted passes into the cold reservoir.

The arrangement for a heat engine is illustrated in Figure 7.2. There is a source of heat called the *hot reservoir* and there is a *cold reservoir* at a lower temperature. The two reservoirs are connected thermally, usually by a flow of gas or vapour between them. The substance that transports the heat between the two reservoirs is called the *working substance*. The heat engine is positioned between the hot and cold reservoirs so that the working substance passes through it. The action of the heat engine is to convert a portion of the transported heat into useful work. The heat that is not extracted from the working substance is discarded to the cold reservoir. For a heat engine to be useful, it must be able to operate continuously. This means that the working substance must undergo a cyclic process in which it is returned to its original state at the end of each cycle.

7.2.1 The steam turbine

Steam turbines are used in the majority of power stations to provide the mechanical work to drive the electricity generators. They provide a good example of the workings of a heat engine. A schematic diagram of a steam turbine is shown in Figure 7.3. In the turbine, there is a series of blades that are mounted on a shaft.

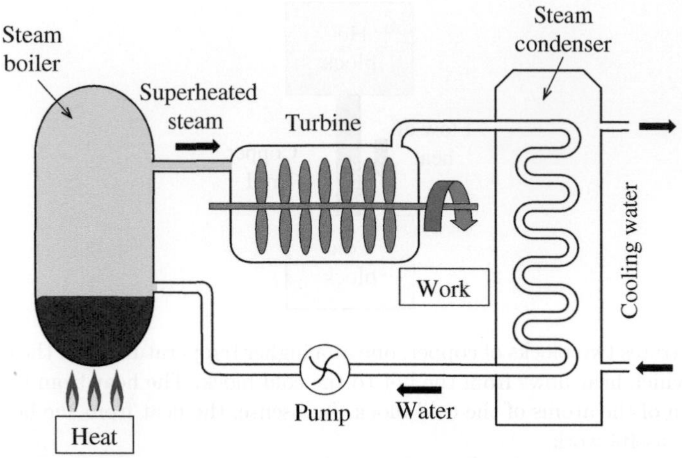

Figure 7.3 A schematic diagram of a steam turbine. In the turbine, there is a series of blades that are mounted on a shaft. This shaft is coupled to an electrical generator. Water of extremely high purity is boiled to produce steam at high pressure and temperature, typically 500 °C. The high-pressure steam is expanded into the turbine and impacts the turbine blades causing them to spin. A series of turbine blades is used as the more gradual expansion of the steam in the turbine results in a closer approach to an ideal reversible process. After passing through the turbine, the steam is condensed to water, which is then returned to the boiler, and the process repeats.

This shaft is coupled to an electrical generator. Water of extremely high purity is boiled to produce steam at high pressure and temperature, typically 500 °C. The high-pressure steam is expanded into the turbine and impacts the turbine blades causing them to spin. A series of turbine blades is used as the more gradual expansion of the steam in the turbine results in a closer approach to an ideal reversible process. After passing through the turbine, the steam is condensed to water, which is then returned to the boiler, and the process repeats.

The steam boiler is the hot reservoir and the condenser is the cold reservoir, and water in the form of steam is the working substance that flows through the turbine. The turbine is the heat engine that converts heat from the steam into mechanical energy. This is a cyclic process in which the working substance, the water, is returned to its original state.

A schematic representation of a heat engine is shown in Figure 7.4. The temperatures of the hot and cold reservoirs are T_H and T_C, respectively and these are assumed to be constant. In each cycle of operation of the engine, Q_H is the amount of heat absorbed from the hot reservoir by the working substance, Q_C is the amount of heat that is discarded to the cold reservoir and W is the amount of work produced.

After each complete cycle of operation of the heat engine, the working substance is returned to its original state and hence there is no change in its internal energy; $\Delta U = 0$. From Figure 7.4, which illustrates the flow of energy into and out of the heat engine, we see from energy balance that

$$W = Q_H - Q_C. \tag{7.1}$$

The *thermal efficiency* ε of a heat engine is defined as the ratio of the work delivered by the heat engine and the heat extracted from the hot reservoir. Therefore,

$$\varepsilon = \frac{W}{Q_H} = \frac{Q_H - Q_C}{Q_H}, \tag{7.2}$$

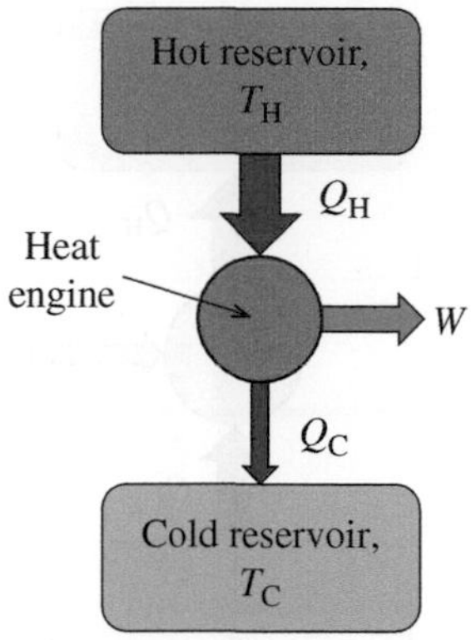

Figure 7.4 A schematic representation of a heat engine. It shows the flow of energy into and out of the heat engine. T_H and T_C are the temperatures of the hot and cold reservoirs respectively, which are assumed to be constant. Q_H is the amount of heat absorbed from the hot reservoir in a single cycle of operation of the engine, and Q_C is the amount of heat that is discarded to the cold reservoir. In each cycle of operation, the engine produces an amount of work $W = Q_H - Q_C$.

or

$$\varepsilon = 1 - \frac{Q_C}{Q_H}. \tag{7.3}$$

We see that the efficiency of a heat engine can never be greater than 1, and could only be 1 if $Q_C = 0$. No heat engine is able to convert the heat absorbed from a hot reservoir into work with 100% efficiency, i.e. Q_C can never be equal to zero. Some of the absorbed heat *must* be discarded to a reservoir at a lower temperature. This is a consequence of the second law of thermodynamics, as we will see in Section 7.7. In practice, the best heat engines operate with close to 40% efficiency.

One version of the second law of thermodynamics, called the Kelvin statement, follows from the empirical observation that Q_C can never be equal to zero:

> *It is impossible for an engine that operating in a cycle will produce no other effect than the extraction of heat from a reservoir and the performance of an equivalent amount of work.*

If this were not true, it would be possible, for example, to extract heat from the ocean to power a ship. There is a huge amount of thermal energy in the oceans of the world, but without the availability of a heat reservoir at a lower temperature, it cannot be converted into useful work.

7.2.2 Refrigerators and heat pumps

We can think of a refrigerator as a heat engine working in reverse. In the operation of a heat engine, heat is taken from a hot reservoir and heat is discarded to a cold reservoir. In the case of a refrigerator, heat is removed from a cold reservoir and heat is discarded to a hot reservoir. To achieve this, there must be an *input* of work to drive the engine.

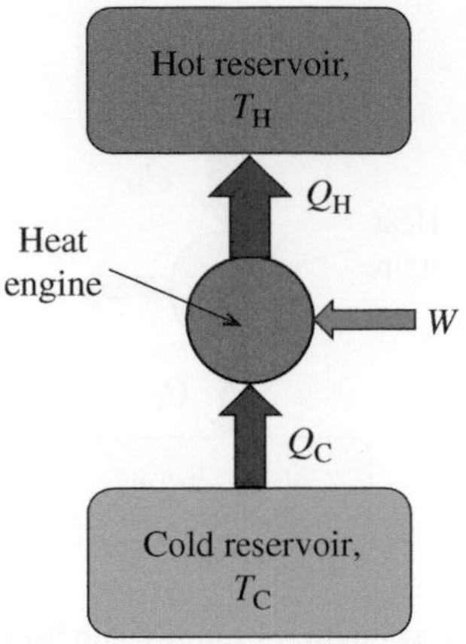

Figure 7.5 A schematic diagram of a refrigerator, where a heat engine acts in reverse, extracting heat from a cold reservoir and delivering heat to a hot reservoir. In this case, energy balance gives $W + Q_C = Q_H$, where W is the work that is input to the refrigerator, Q_C is the amount of heat extracted from the cold reservoir and heat Q_H is the amount of heat that is delivered to the hot reservoir.

A schematic representation of a refrigerator is shown in Figure 7.5. This figure illustrates the flow of energy through the refrigerator. In this case, energy balance gives

$$W + Q_C = Q_H, \tag{7.4}$$

where W is the work that is input to the refrigerator, Q_C is the amount of heat removed from the cold reservoir and heat Q_H is the amount of heat discarded to the hot reservoir, in each cycle of operation. As Figure 7.5 indicates, the heat discarded to the hot reservoir is *always* greater than the heat removed from the cold reservoir.

The efficiency of a refrigerator is called the *coefficient of performance*, $\mathrm{CoP}_{\mathrm{fridge}}$. It is the ratio of the heat removed from the cold reservoir and the work performed to achieve this:

$$\mathrm{CoP}_{\mathrm{fridge}} = \frac{Q_C}{W} = \frac{Q_C}{Q_H - Q_C}. \tag{7.5}$$

As $Q_H > Q_C$, $\mathrm{CoP}_{\mathrm{fridge}}$ is always greater than 1. In a domestic refrigerator, the work W is usually supplied by an electric motor. The refrigerator cabinet acts as the cold reservoir and is cooled as heat Q_C is removed from it. The heat Q_H is discarded to the surrounding room, which acts as the hot reservoir. A domestic refrigerator might have a value of $\mathrm{CoP}_{\mathrm{fridge}}$ of about 3.

Heat flows from a hotter body to a colder body. The reverse process never happens spontaneously. If we want to transfer heat from a colder to a hotter body, as is done in a refrigerator, then work must be provided to do this. This empirical observation leads to Clausius' version of the second law:

It is impossible to construct a device that, operating in a cycle, will produce no effect other than the transfer of heat from a colder to a hotter body.

At first sight, the Clausius and Kelvin–Planck statements of the second law may seem to be unconnected. However, they can be shown to be equivalent in all respects. Perhaps the Clausius statement is the more obvious as it is simply a statement of the observation that heat flows spontaneously only from hot to cold.

A *heat pump* works in an analogous way to a refrigerator. Again heat is taken from a cold reservoir and delivered to a hot reservoir. But now the objective is different. In the case of a heat pump, the objective is to deliver heat to a hot body rather than cooling down a cold body. In that case, what is important is the ratio of the amount of heat Q_H delivered to the hot body and the work W required to do this. This is called the coefficient of performance $\text{CoP}_{\text{ht pump}}$ of the heat pump

$$\text{CoP}_{\text{ht pump}} = \frac{Q_H}{W} = \frac{Q_H}{Q_H - Q_C}. \tag{7.6}$$

Clearly, this is a number greater than 1. So, the heat delivered to the hot body is greater than the equivalent amount of work that is delivered to the heat pump.

Heat pumps are used to transfer heat from the cold air outside a building to the warmer air inside. In a practical system for heating a house, the value of $\text{CoP}_{\text{ht pump}}$ might typically be about 5. This means that 1 kW of electrical power delivered to the heat pump delivers 5 kW of heat. This clearly has economic advantages and it is expected that heat pumps will play a significant role in space heating in the future.

7.3 The Carnot cycle

Following the intense interest in the development of steam engines, Carnot investigated the general problem of how to produce mechanical work from a source of heat. And in 1824, he published a famous monograph describing his pioneering investigations. The principles he laid down applied not just to the steam engine, but to any engine that converts heat into work. Carnot's work is an outstanding example of the discipline of engineering giving rise to a fundamental law of physics; the second law of thermodynamics.

Carnot devised a hypothetical engine to represent the operation of an *idealised* heat engine. It consists of an ideal gas contained in a cylinder by a frictionless piston. The gas is taken through several stages of expansion and compression in a *Carnot cycle*. In this cycle, the gas absorbs heat from a hot reservoir, converts part of this heat into work and discards the remaining heat to a cold reservoir. Carnot showed (i) that the efficiency ε of a Carnot engine depends only on the temperatures T_H and T_C of the hot and cold reservoirs respectively, i.e. $\varepsilon = f(T_H, T_C)$, and (ii) that the efficiency of a Carnot engine is the maximum possible efficiency that any engine can have for given values of T_H and T_C.

Before describing the details of the Carnot cycle, we recall some features of reversible and irreversible processes. In particular, if we place a hot body in contact with a second body that is colder than the first by a *finite* amount, the heat transfer between the two is irreversible. The finite temperature difference leads to dissipation, i.e. a loss of useful heat. However, if the hot body is hotter by an *infinitesimal* amount, the direction of heat transfer can be reversed by decreasing the temperature of the hot body by an infinitesimal amount. This process is reversible and there is no dissipation of heat. Therefore, to avoid unnecessary heat losses during any heat transfer process in the Carnot cycle, it is crucially important that there are no finite temperature differences. Thus, when absorbing heat from the hot reservoir, the ideal gas (the working substance) and the reservoir remain at the same temperature, i.e. the process is isothermal. Similarly, when the gas discards heat to the cold reservoir, the process is isothermal at the temperature of the cold reservoir.

The Carnot cycle also involves adiabatic expansion and compression of the gas, during which the temperature of the gas changes. However, by thermally isolating the gas cylinder during these processes, there is no heat exchange between the gas and either of the two reservoirs; heat exchanges that would make the processes irreversible. Instead, the gas is always brought to the temperature of the hot reservoir before being put in contact with it and likewise with the cold reservoir. This again prevents any wasteful flow of heat between bodies at different temperatures. Reversible processes are also required to be quasistatic. This means that all the expansion and compression stages in the Carnot cycle are conducted sufficiently slowly so that the gas always remains close to a state of equilibrium. In this case, the ideal gas law is applicable in all the stages of the cycle.

By having only reversible, quasistatic stages in the Carnot cycle, all the possible processes that could lead to unnecessary losses of heat are eliminated. Hence, the heat absorbed from the hot reservoir is used in the most advantageous way to deliver useful work. It follows that a Carnot engine is the most efficient engine that is possible for given values of the temperatures of the hot and cold reservoirs. This is expressed by Carnot's theorem, which says:

No heat engine operating between two given reservoirs can be more efficient than a Carnot engine operating between the same two temperatures.

Carnot devised some ingenious combinations of separate Carnot engines to prove this theorem formally.

7.3.1 Stages of the Carnot cycle

The cycle considered by Carnot consists of two isothermal and two adiabatic reversible processes that operate on n moles of an ideal gas. These four stages are illustrated in Figure 7.6. They are also shown on a P–V diagram in Figure 7.7. The four stages of the Carnot cycle are as follows:

1. The gas cylinder is held in contact with the hot reservoir at temperature T_{H} and the gas is expanded isothermally from V_a to V_b, absorbing heat Q_{H} from the hot reservoir.
2. The cylinder is removed from the hot reservoir and the gas is expanded adiabatically from V_b to V_c, until its temperature drops to T_{C}.
3. The cylinder is held in contact with the cold reservoir at temperature T_{C} and the gas is compressed isothermally from V_c to V_d, discarding heat Q_{C} to the cold reservoir.

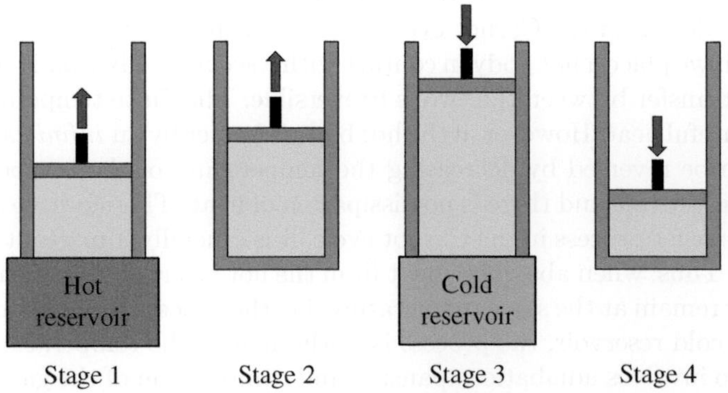

Figure 7.6 The four stages of a Carnot cycle using an ideal gas as the working substance.

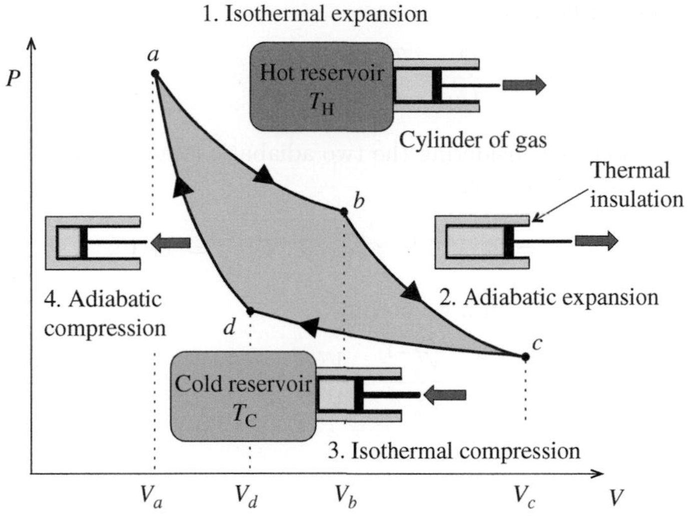

Figure 7.7 The four stages of a Carnot cycle represented on a P–V diagram.

4. The cylinder is removed from the cold reservoir and the gas is compressed adiabatically from V_d to V_a, until its temperature rises back to T_{H}.

After these four stages, the gas returns to its original state, and the entire process constitutes a complete cycle. However, some of the heat absorbed from the hot reservoir has been converted into work.

For the isothermal expansion of the gas in stage 1, we have from the first law,

$$\Delta U = Q_{\mathrm{H}} - W_{\mathrm{bygas}},$$

where ΔU is the change in internal energy of the gas, Q_{H} is the heat absorbed from the reservoir and W_{bygas} is the work done *by* the gas in the expansion. As the expansion is isothermal, the internal energy of the gas remains constant; $\Delta U = 0$, and $Q_{\mathrm{H}} = W_{\mathrm{bygas}}$. We see that the work the gas performs in the expansion is provided by an equivalent amount of heat that is absorbed from the hot reservoir. Then, for an expansion from volume V_a to V_b, the amount of heat taken from the reservoir is

$$Q_{\mathrm{H}} = W_{\mathrm{bygas}} = \int_{V_a}^{V_b} P \mathrm{d}V. \tag{7.7}$$

Substituting for $P = nRT_{\mathrm{H}}/V$, we obtain

$$Q_{\mathrm{H}} = nRT_{\mathrm{H}} \int_{V_a}^{V_b} \frac{\mathrm{d}V}{V} = nRT_{\mathrm{H}} \ln \frac{V_b}{V_a}. \tag{7.8}$$

Similarly, in the isothermal compression of the gas during stage 3, the amount of heat delivered to the cold reservoir is

$$Q_{\mathrm{C}} = nRT_{\mathrm{C}} \ln \frac{V_c}{V_d}. \tag{7.9}$$

The ratio of the two amounts of heat is

$$\frac{Q_C}{Q_H} = \frac{T_C \ln(V_c/V_d)}{T_H \ln(V_b/V_a)}. \tag{7.10}$$

We can simplify this expression by considering the two adiabatic stages in the cycle. For each adiabatic stage, we have $TV^{\gamma-1} = \text{const}$, and hence,

$$T_H V_b^{\gamma-1} = T_C V_c^{\gamma-1} \quad \text{and} \quad T_H V_a^{\gamma-1} = T_C V_d^{\gamma-1}.$$

Dividing the first expression by the second, we obtain

$$\frac{V_b^{\gamma-1}}{V_a^{\gamma-1}} = \frac{V_c^{\gamma-1}}{V_d^{\gamma-1}},$$

which gives

$$\frac{V_b}{V_a} = \frac{V_c}{V_d}. \tag{7.11}$$

Hence, the two logarithms in Equation (7.10) are equal and the equation reduces to

$$\frac{Q_C}{Q_H} = \frac{T_C}{T_H}. \tag{7.12}$$

This is the result we have been seeking. The amount of heat discarded to the cold reservoir at T_C compared to the amount of heat absorbed at the hot reservoir at T_H is just equal to the ratio T_C/T_H.

The total work done *by* the gas in a complete cycle is given by the area under the curves corresponding to the two expansion stages, i.e. under the curves *ab* and *bc*. The total work done *on* the gas in a complete cycle is given by the area under the curves corresponding to the two compression stages, i.e. under the curves *cd* and *da*. It follows that the total work done by the gas in each cycle is given by the orange-shaded area enclosed by the above curves.

7.3.2 Thermal efficiency of a Carnot engine

In Section 7.2, we found that the thermal efficiency of a heat engine is

$$\varepsilon = 1 - \frac{Q_C}{Q_H}. \tag{7.3}$$

Substituting for Q_C/Q_H from Equation (7.12), we obtain

$$\varepsilon = 1 - \frac{T_C}{T_H}. \tag{7.13}$$

This expression for thermal efficiency does not involve the working substance and we conclude that the thermal efficiency of a Carnot engine is independent of the particular working substance, and depends only on the ratio of the temperatures of the two reservoirs. This expression for thermal efficiency is much more useful than one involving Q_H and Q_C as temperatures T_H and T_H are more easily known.

In practice, the temperature of the cold reservoir is limited to that of the surroundings of the heat engine, which means that T_C will not be far from room temperature. Most effort, therefore, goes into having T_H as high as possible. For example, in a steam turbine, a high value of T_H is obtained by working with pressurised steam at high temperature. This is also why special, novel materials are being developed for jet engines so that the engines can operate at even higher temperatures. And, for the same reason, water at say 80 °C, that is a side product of an industrial process, is of little use for producing mechanical work.

Because the Carnot cycle is reversible in all its stages, it is possible to run the cycle backwards. In this case, work W is put into the heat engine to extract heat Q_C from a cold reservoir and deposit heat Q_H in a hot reservoir. This, of course, is the operation of a refrigerator and a heat pump. In Section 7.2, we saw that the coefficient of performance of a heat pump,

$$\text{CoP}_{\text{ht pump}} = \frac{Q_H}{W} = \frac{Q_H}{Q_H - Q_C}.$$

Substituting for $Q_C/Q_H = T_C/T_H$, we obtain

$$\text{CoP}_{\text{ht pump}} = \frac{T_H}{T_H - T_C}. \tag{7.14}$$

Strikingly, and in contrast to the case of a heat engine, this equation shows that heat pumps are more efficient for smaller temperature differences. Hence, they are more useful in moderate climates than cold climates. The equation also explains why radiators fed by a heat pump tend to run at lower temperatures than ones fed by a conventional gas boiler.

Worked example

Earlier types of nuclear reactors used uranium metal as fuel and operated at a core temperature of 350 °C. Later types used uranium oxide as fuel. Because of the higher melting point of the oxide, these reactors can run at a higher temperature of 650 °C. (a) Compare the maximum thermal efficiencies of the two types of reactor. Assume that the excess heat is discarded to a river at a temperature of 15 °C. (b) If the uranium-oxide reactor generates 900 MW of electricity, take the value of its maximum thermal efficiency to estimate the rise in temperature of the water if it flows at the rate of 50 m^3/s. The specific heat c of water is 4.2 kJ/kg.

Solution

(a) The thermal efficiency of the uranium-fired reactor is

$$1 - \frac{288}{623} = 54\%,$$

and that of the uranium oxide-fired reactor is

$$1 - \frac{288}{923} = 69\%.$$

Hence, by changing to the uranium oxide fuel, the thermal efficiency increases by 15%. Note, however, that even at a higher efficiency, at least 31% of the total heat produced by the reactor must be discarded to the river as waste heat. Moreover, no improvements in reactor design can increase the thermal efficiency above 69%.

(b) Taking the value of 69% for the thermal efficiency of the reactor, the amount of power discarded to the river is $900 \times 0.31 = 279\,\text{MW}$. The mass of water passing the power plant per second is $50 \times 1000 = 5 \times 10^4\,\text{kg/s}$.
(c) Therefore, the temperature rise of the water is

$$\Delta T = \frac{\Delta Q/\text{s}}{c \times m/\text{s}} = \frac{279 \times 10^3}{4.2 \times 5 \times 10^4} = 1.3°\text{C}.$$

The Carnot engine we have considered is an ideal heat engine. All unnecessary losses of heat are eliminated and the engine is taken to be free from any frictional forces. Therefore, Equation (7.13) gives an upper limit to the possible thermal efficiency. In practice, of course, a real engine does dissipate heat through friction and loses appreciable amounts of heat to the surroundings by conduction, convection, and radiation. This is very evident in a motorcar engine. Nevertheless, it is still very useful to know the theoretical maximum value of thermal efficiency.

7.3.3 The Kelvin or absolute temperature scale

The ideal gas temperature scale was defined in Section 6.2 in terms of the properties of ideal gases, i.e. how the pressure of a fixed amount of dilute gas varies with temperature. Ideally, we would have a temperature scale that was independent of the properties of a particular substance, i.e. an absolute temperature scale. In our discussion of the Carnot engine, we found the following relationship:

$$\frac{\text{The heat discarded to cold reservoir}}{\text{The heat absorbed from hot reservoir}} = \frac{\text{Temperature of cold reservoir}}{\text{Temperature of hot reservoir}}.$$

Although we obtained this relationship for the particular example of a Carnot engine based on an ideal gas, it applies to *any* reversible Carnot engine and is independent of the working substance. Kelvin thus proposed that this relationship be used to establish an absolute temperature scale, i.e. to *define* the ratio of the temperatures of the cold and hot reservoirs as the ratio of the heat discarded at the cold reservoir and the heat absorbed at the hot reservoir. As the relationship applies to any reversible Carnot engine, this provides a temperature scale that is truly independent of the properties of any substance. It is called the *Kelvin temperature scale* or the *absolute temperature scale.*

Using the above relationship, and designating temperatures on the Kelvin scale by the symbol Θ, we can write

$$\frac{\Theta_\text{C}}{\Theta_\text{H}} = \frac{Q_\text{C}}{Q_\text{H}}, \tag{7.15}$$

where Θ_H and Θ_C are the temperatures of the hot and cold reservoirs, respectively, on the Kelvin scale, Q_H is the heat absorbed from the hot reservoir, and Q_C is the heat discarded to the cold reservoir. Equation (7.15) applies to any reversible Carnot engine. Our analysis of the *particular* case of a Carnot engine that was based on an *ideal gas* gave the result

$$\frac{Q_\text{C}}{Q_\text{H}} = \frac{T_\text{C}}{T_\text{H}}. \tag{7.12}$$

Here, the temperatures T_C and T_H are measured on the ideal gas scale. This is clearly the case because Equation (7.12) was obtained using the ideal gas law, $PV = nRT$, and the relationship for adiabatic expansion $PV^\gamma = \text{const}$. It follows from Equations (7.12) and (7.15) that

$$\frac{\Theta_C}{\Theta_H} = \frac{T_C}{T_H}. \tag{7.16}$$

This equation gives the relationship between the Kelvin and the ideal gas temperature scales. To complete the definition of the Kelvin temperature scale, we assign Θ_{triple} to the triple point of water. Then, if T and Θ refer to any temperature and T_{triple} and Θ_{triple} refer to the triple point of water, Equation (7.16) gives

$$\frac{\Theta}{\Theta_{\text{triple}}} = \frac{T}{T_{\text{triple}}}. \tag{7.17}$$

As $\Theta_{\text{triple}} = T_{\text{triple}} = 273.16\,\text{K}$, it follows that

$$\Theta = T, \tag{7.18}$$

i.e. *that the Kelvin and ideal gas temperature scales are identical.*

7.4 Entropy

From the analysis of Carnot's heat engine, we have the following relationship:

$$\frac{Q_C}{T_C} = \frac{Q_H}{T_H}. \tag{7.19}$$

The term on the right-hand side corresponds to the heat Q_H absorbed from the hot reservoir at temperature T_H. Similarly, the term on the left-hand side corresponds to the heat Q_C discarded to the cold reservoir at temperature T_C. Following on from the work of Carnot, Clausius introduced the concept of entropy. He described entropy as that quantity that increases by Q/T whenever heat Q enters a system at temperature T. It is heat, of course, that flows from the hot reservoir. However, it may be helpful to think of entropy 'flowing' from the heat reservoir, and entropy 'flowing' through the heat engine into the cold reservoir.

An entropy change $\mathrm{d}S$ is defined through the relationship

$$\mathrm{d}S = \frac{\mathrm{d}Q}{T}. \tag{7.20}$$

Put into words, this definition states that if we supply an infinitesimal amount of heat $\mathrm{d}Q$ to a system at temperature T, the entropy of the system increases by $\mathrm{d}Q/T$. Note that Equation (7.20) *only applies if* $\mathrm{d}Q$ *is transferred reversibly*, i.e. that the system is at the same temperature T as the source of heat. There must be no temperature gradients as that would result in irreversible heat conduction. We see from Equation (7.20) that the units of entropy are joules/degree (J/K). As T is always positive, the sign of $\mathrm{d}S$ is the same as that of $\mathrm{d}Q$. When heat enters a system, its entropy increases and when heat leaves, its entropy decreases.

It follows from Equation (7.20) that for a *reversible* process that takes a system from state 1 to state 2, the change in entropy between the two states can be written as

$$S_2 - S_1 = \int_1^2 \mathrm{d}S = \int_1^2 \frac{\mathrm{d}Q}{T}. \tag{7.21}$$

We emphasise that the process connecting the two states must be reversible. This means that the process must be brought about by having a sequence of near-equilibrium states between the initial and final states. In that case, the temperature in the integral is well defined at all stages of the process, and the heat $\mathrm{d}Q$ absorbed between adjacent equilibrium states is a measurable quantity. As the left-hand side of Equation (7.21) refers only to the initial and final states, it follows that the integral on the right-hand side must be independent of the reversible process used to evaluate the integral, i.e. the equation applies for *any* reversible process connecting the two states.

7.4.1 The measurement of changes in entropy

The difference in entropies of two states 1 and 2 of a system can be found by considering any reversible process that connects the two states and evaluating the integral in the equation

$$\Delta S = S_2 - S_1 = \int_1^2 \frac{\mathrm{d}Q}{T}. \tag{7.22}$$

Suppose that the state of the system is specified only by its temperature, i.e. that all its other parameters such as volume and pressure are kept constant. Then we can relate $\mathrm{d}Q$ to T through the specific heat C of the system, which is defined by

$$C = \frac{\mathrm{d}Q}{\mathrm{d}T}.$$

Substituting for $\mathrm{d}Q = C\mathrm{d}T$ in Equation (7.22) gives

$$\Delta S = \int_{T_1}^{T_2} \frac{C}{T}\,\mathrm{d}T. \tag{7.23}$$

Hence, if we know the specific heat of a system over the temperature range T_1 to T_2, we can find the corresponding entropy change. Examples of determining entropy changes in this way are presented in Section 7.5.

Substituting $\mathrm{d}Q = T\mathrm{d}S$ in $C = \mathrm{d}Q/\mathrm{d}T$ gives

$$C = T\frac{\mathrm{d}S}{\mathrm{d}T}, \tag{7.24}$$

which provides some further useful relationships. For two particular examples, we have at constant volume:

$$C_\mathrm{V} = T\left(\frac{\partial S}{\partial T}\right)_\mathrm{V}, \tag{7.25}$$

and at constant pressure:

$$C_\mathrm{P} = T\left(\frac{\partial S}{\partial T}\right)_\mathrm{P}. \tag{7.26}$$

7.4.2 Entropy as a state function

Equation (7.21) implies that entropy S is a state function, which it is. We recall from Section 6.4 that state functions are determined entirely by the state of the system, and are not determined by the path by which that state was reached. Hence, the difference in the entropies of two states 1 and 2 depends only on these states. If the system is changed from state 1 to state 2 by different processes, then the entropy of the system will always change by the *same* amount. However, the entropy of the system's surroundings will in general change by *different* amounts. A reversible process is distinguished by the fact that the changes in the system are accompanied by compensating changes in the system's surroundings, such that the entropy of the system plus the surroundings remains the same.

We can illustrate that entropy is a state function using the example of a reversible process in an ideal gas. As we recall from Section 6.8, we can achieve this by conducting the process in a series of small steps during which we can consider the system to remain close to equilibrium. The first law for reversible processes is

$$\mathrm{d}U = \mathrm{d}Q - P\mathrm{d}V. \tag{6.30}$$

Then, using $\mathrm{d}U = C_V\mathrm{d}T$ from Equation (6.38), we obtain

$$\mathrm{d}Q = P\mathrm{d}V + C_V\mathrm{d}T. \tag{7.27}$$

As it stands, we cannot integrate this equation because Q is not a state function and so the value of $\mathrm{d}Q$ depends on the path taken between the initial and final states. However, by substituting $P = RT/V$ from the ideal gas law and rearranging, we obtain

$$\frac{\mathrm{d}Q}{T} = R\frac{\mathrm{d}V}{V} + C_V\frac{\mathrm{d}T}{T}. \tag{7.28}$$

Both terms on the right-hand side of this expression can now be integrated. Hence, for a change from state 1 to state 2, we have

$$\int_1^2 \frac{\mathrm{d}Q}{T} = R\int_{V_1}^{V_2} \frac{\mathrm{d}V}{V} + C_V\int_{T_1}^{T_2} \frac{\mathrm{d}T}{T}. \tag{7.29}$$

From Equation (7.22), the term on the left-hand side of the equation is just the change in entropy ΔS. Then evaluating the integrals on the right-hand side, we obtain

$$\Delta S = R\ln\frac{V_1}{V_2} + C_V\ln\frac{T_1}{T_2}. \tag{7.30}$$

Hence, the change in entropy depends only on the state functions V and T of the initial and final states of the ideal gas, and hence does not depend on the way the change was made. We conclude that entropy is also a state function. We did not specify a particular process in obtaining Equation (7.30), except that it was reversible. Hence, it must apply to any reversible process that takes the gas from an initial state 1 to a final state 2. Notice that by dividing Q, which is not a state function by T, we obtain the function Q/T, which is a state function. The factor $1/T$ is called an *integrating factor*.

7.5 Entropy changes in reversible processes

In this section, we consider examples of entropy changes in reversible processes. We recall that these are special processes that can only be approximately obtained under carefully controlled conditions.

7.5.1 Reversible processes in an ideal gas

A particularly straightforward process for which to determine entropy change is the reversible, isothermal expansion of an ideal gas. Here, the gas is contained in a cylinder by a frictionless piston, and this system is held in contact with a heat reservoir at constant temperature T as the gas is expanded. We encountered this process in the first stage of the Carnot cycle.

Suppose that we have one mole of an ideal gas that undergoes a reversible isothermal expansion from V_1 to V_2. The work done by the gas in the expansion is

$$W = \int_{V_1}^{V_2} P\mathrm{d}V = RT\int_{V_1}^{V_2} \frac{\mathrm{d}V}{V} = RT\ln\frac{V_2}{V_1}.$$

To keep the temperature of the gas constant during the expansion, heat must be transferred from the reservoir to the gas. The amount of heat Q is exactly equal to the work W done by the gas. From Equation (7.22), we have

$$\Delta S = \frac{1}{T}\int \mathrm{d}Q = \frac{Q}{T}, \tag{7.31}$$

where we have taken T outside the integral because it is constant. Hence, with $Q = RT\ln(V_2/V_1)$, we find

$$\Delta S = R\ln\frac{V_2}{V_1}. \tag{7.32}$$

As heat is added to the gas during the expansion, Q is positive and so the entropy of the gas increases.

The entropy change of the gas is $+Q/T$. The same amount of heat Q leaves the heat reservoir at temperature T. Hence the change in entropy of the reservoir is $-Q/T$. The net change in entropy of the gas plus the reservoir is therefore zero. This is a general result:

In a reversible process, the entropy change of the system plus its surroundings is zero.

In the case of a reversible, *adiabatic* process, no heat enters or leaves the system. Hence, $\Delta Q = 0$, and there is no change in entropy: $\Delta S = 0$. Every reversible adiabatic process is a constant entropy process. For this reason, reversible, adiabatic processes are called *isentropic* processes.

7.5.2 Water and ice mixture

Suppose that we have a mixture of water and ice in a thermally insulated container. And suppose that the temperature of the water is infinitesimally *above* the freezing point of water. In that case, the ice will steadily melt. We take the amount of water to be much larger than the amount of ice so that the water temperature remains constant. Hence, we can assume that the melting process takes place at a constant temperature.

This process is reversible because if the water temperature were infinitesimally *below* 0 °C, the water would freeze to ice. This is in sharp contrast to say putting an ice cube in a glass of water at room temperature. In that case, there is a finite temperature difference and the process is irreversible.

During the melting process, heat must be supplied to the ice and this is given by $Q = mL$, where m is the mass of the ice and L is the latent heat of fusion. The change in entropy of the ice is therefore

$$\Delta S_{\text{ice}} = \frac{mL}{T} \tag{7.33}$$

If, for example, we have 15 g of ice, the change in entropy is

$$\Delta S_{\text{ice}} = \frac{0.015 \times 334 \times 10^3}{273.15} = +18.3\,\text{J/K},$$

where we have taken L to be 334 kJ/kg and the melting point of ice to be 273.15 K. The change in entropy of the ice cube is positive. The water delivers heat to the ice cube and so its entropy decreases by

$$\Delta S_{\text{water}} = \frac{-\left(0.015 \times 334 \times 10^3\right)}{273.15} = -18.3\,\text{J/K}.$$

Hence, the net entropy change of the ice-water mixture is zero. This is an isolated system, and illustrates the general result:

For a reversible process in an isolated system, the entropy change is zero.

7.5.3 The Carnot cycle

The Carnot cycle consists of four reversible stages. There is no entropy change during the adiabatic compression or expansion. During the isothermal expansion at temperature T_H, an amount of heat Q_H passes to the working substance, whose entropy, therefore, increases by Q_H/T_H. During the isothermal compression at temperature T_C, an amount of heat Q_C is removed from the working substance, whose entropy, therefore, decreases by Q_C/T_C. Hence, the net change in entropy of the working substance is

$$\Delta S_{\text{ws}} = \frac{Q_H}{T_H} - \frac{Q_C}{T_C}.$$

Using Equation (7.12) to relate Q_H and Q_C:

$$\frac{Q_H}{T_H} = \frac{Q_C}{T_C},$$

gives

$$\Delta S_{\text{ws}} = \frac{Q_H}{T_H} - \frac{Q_C}{T_C} = 0. \tag{7.34}$$

There is no change in the entropy of the working substance and therefore of the heat engine during its cycle of operation. We also have that the entropy loss of the hot reservoir $-Q_H/T_H$ is exactly equal to the entropy gain $+Q_C/T_C$ of the cold reservoir during the operation. Hence, the entropy change for the heat engine plus the two heat reservoirs over the complete cycle is also zero. This is a general result:

The total entropy change during any reversible cycle is zero.

Worked example

A Carnot engine takes 2500 J of heat from a reservoir at 500 K, and discards some heat to a reservoir at 300 K during one cycle. Find the net entropy change of the working substance.

Solution

The amount of heat taken in by the working substance during the isothermal expansion, $Q_H = 2500$ J. This results in an increase in the entropy of the working substance equal to $+2500/500 = +5$ J/K. The amount of heat removed from the working substance during the isothermal compression is

$$Q_C = \frac{T_C}{T_H} Q_H = \frac{300}{500} \times 2500 = 1500 \text{ J}.$$

This results in a decrease in the entropy of the working substance equal to $-1500/300 = -5$ J/K. Hence, the net change in entropy of the working substance is zero.

Temperature-entropy (T–S) *diagrams*

The four separate stages of the Carnot cycle were represented on a P–V diagram in Figure 7.7. Now that we have introduced entropy, we can also represent the four stages on a T–S diagram. This shows the relationship between the entropy S and temperature T of the working substance during the Carnot cycle. Such a T–S diagram is shown in Figure 7.8, where the four stages of the cycle are represented by the paths *ab*, *bc*, *cd*, and *da*, respectively.

(i) Path *ab* corresponds to the isothermal expansion of the gas at the temperature T_H of the hot reservoir. During this expansion, the entropy of the gas increases from S_1 to S_2 as heat is absorbed from the reservoir. (ii) Path *bc* corresponds to the adiabatic expansion of the gas during which its temperature changes from T_H to T_C, but its entropy stays constant as there is no exchange of heat and $\Delta S = 0$. (iii) Path *cd* corresponds to the isothermal compression at the temperature T_C of the cold reservoir, during which the entropy of the gas returns to its initial value S_1 (iv) Path *da* corresponds to the adiabatic compression during which the temperature of the gas returns to T_H but the entropy stays constant.

It follows from Equation (7.20),

$$dS = \frac{dQ}{T},$$

and inspection of Figure 7.8, that the amount of heat absorbed by the gas during a cycle of operation is

$$Q_H = T_H(S_2 - S_1), \tag{7.35}$$

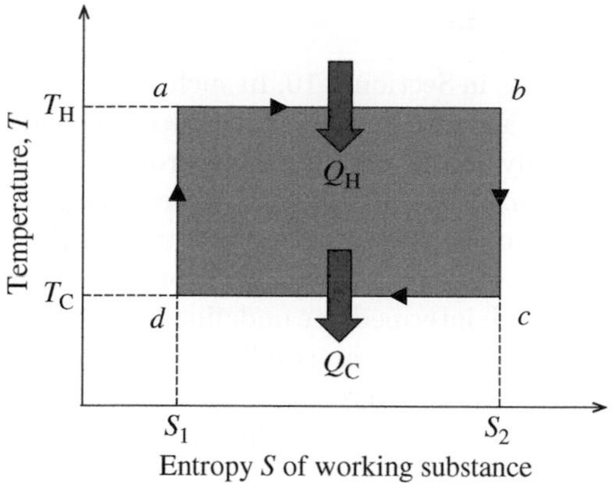

Figure 7.8 The four stages of a Carnot cycle represented on a T–S diagram. This shows the relationship between the entropy S and temperature T of the working substance during the Carnot cycle.

and that the amount of heat discarded by the gas to the cold reservoir is

$$Q_C = T_C(S_2 - S_1). \tag{7.36}$$

Hence, the shaded area within the rectangle *abcd* is equal to the net amount of heat $Q_H - Q_C$, taken in by the working substance during a complete cycle. This is also equal to the amount of work W delivered by the heat engine:

$$W = Q_H - Q_C.$$

It follows from Equations (7.35) and (7.36) that $Q_H/Q_C = T_H/T_C$ in agreement with Equation (7.12).

7.6 Entropy changes in irreversible processes

In our discussion of entropy changes in Section 7.4, we saw the following result for *reversible* processes:

$$\Delta S = S_2 - S_1 = \int_1^2 \frac{dQ}{T}.$$

We emphasised that the integral can only be evaluated for reversible processes. This is because for such processes, we can consider the system to pass through a sequence of equilibrium states for which the temperature T is always well defined as also is the heat absorbed dQ between adjacent equilibrium states. These conditions are not obtained for irreversible processes. During an irreversible process, the system will invariably pass through intermediate, unknown, and undefined non-equilibrium states, so that the integral in Equation (7.21) cannot be evaluated.

However, we also saw that entropy S of a system is a state function and does not depend on the way in which the state was arrived at. Thus, when dealing with an irreversible process connecting particularly initial and final states, we imagine a hypothetical *reversible* process that takes the system from the same initial state to the same final state. We then determine the entropy change for that reversible process, which will be the same as that for the irreversible process. In this section, we illustrate this method for several examples of irreversible processes.

7.6.1 Free expansion of an ideal gas

We discussed the free expansion of a gas in Section 6.10. In such an expansion, a gas is initially confined in one compartment of a container that is connected by a valve to a second, evacuated compartment. The whole system is rigid and is thermally insulated from its surroundings so that no heat can flow in or out, and no work is done by the system. When the valve is opened, the gas rushes into the evacuated compartment. If the gas is ideal, there is no change in the temperature of the gas.

This is clearly an irreversible process; a gas never spontaneously confines itself to a smaller volume. In the free expansion, the gas passes through intermediate undefined states that we do not know. However, the initial state, of volume V_1, before the expansion is an equilibrium state with a well-defined entropy, say S_1, and the state of the gas when equilibrium has been re-established after an expansion to volume V_2 also has a well-defined entropy, say S_2. Furthermore, the entropy change $\Delta S = S_2 - S_1$ is independent of the process that connects the initial and final states.

The obvious hypothetical, reversible process to use in this example is the isothermal expansion of the gas from the same initial volume V_1 to the same final volume V_2. We solved this situation in the previous section. We saw that for one mole of gas, the change in entropy in the expansion of the gas from V_1 to V_2 is

$$\Delta S = R \ln \frac{V_2}{V_1}.$$

This then is also the change in entropy for the free expansion of one mole of the gas from volume V_1 to V_2.

In this example, the system is isolated from its surroundings. So, in this irreversible process, the entropy of the isolated system increases. This is a general result:

For an irreversible process in an isolated system, the entropy of the system increases

There is no change to the surroundings in this free-expansion process and hence the entropy change of the gas plus the surroundings also increases. So, referring to the system plus the surroundings as the universe, we may say that the entropy change of the universe increases in an irreversible process. In the free expansion of a gas, the entropy of the universe increases because $V_2 > V_1$. If $V_2 < V_1$, the entropy of the universe would decrease. But a gas never spontaneously confines itself to a smaller volume. This illustrates the general principle:

The entropy of the universe never decreases.

Note that in the free expansion of a gas, the entropy increases despite the fact that there is no transfer of heat. Sometimes entropy is described in terms of the degree of disorder of a system. A somewhat different interpretation of entropy is that when entropy increases, our knowledge of the system decreases. Thus, in the free expansion of a gas, entropy increases not because there is a transfer of heat but because we have less knowledge of the spatial positions of the molecules after they have filled the evacuated space.

7.6.2 Temperature equalisation

Suppose that we have two identical blocks of copper of mass 1.5 kg. One is initially at a temperature of 20 °C and the other at 60 °C. The two blocks are brought into contact and reach equilibrium at 40 °C. We want to determine the net entropy change that occurs in this process. We suppose that the two blocks are isolated from the surroundings.

This, of course, is an irreversible process, but we can find the entropy change by considering a reversible process that produces the same changes in the temperatures of the two blocks. We imagine a process in which the temperature of each of the blocks is changed in infinitesimal steps by placing it in contact with a succession of heat reservoirs of steadily changing temperature until the final temperature is reached. As the temperature difference is infinitesimal, we can consider each of these steps to be reversible. We can then use

$$\Delta S = \int_1^2 \frac{\mathrm{d}Q}{T},$$

and integrate all the steps to calculate the change in entropy.

The heat required to carry out each infinitesimal step $\mathrm{d}T$ is $\mathrm{d}Q = mc\mathrm{d}T$, where m is the mass of the block and c is the specific heat of copper. Hence,

$$\Delta S_1 = mc\int_{T_1}^{T_2} \frac{\mathrm{d}T}{T} = mc \ln \frac{T_2}{T_1}. \tag{7.37}$$

We can readily say that the final temperature of both blocks will be 40 °C. Therefore, for the first block, at an initial temperature of 20 °C, the entropy change is

$$\Delta S_1 = 1.5 \times 385 \times \left(\ln \frac{313}{293}\right) = +38.1 \text{ J/K},$$

where we have taken c to be 385 J/kg · K. Similarly, the change in entropy of the second block is

$$\Delta S_2 = mc\ln \frac{T_2}{T_1} = 1.5 \times 385 \times \left(\ln \frac{313}{333}\right) = -35.8 \text{ J/K}.$$

ΔS_2 has a negative sign as heat leaves that block. The net change in entropy is

$$\Delta S_1 + \Delta S_2 = +38.1 - 35.8 = +2.3 \text{ J/K}.$$

Notice that the entropy of the second block decreases in this irreversible process. However, the increase in entropy of the first block more than compensates for this decrease. This is in accord with the general principle:

The entropy in an irreversible process always increases when we take all parts of the system into account

7.6.3 Heating water

Suppose that we heat a mass m of water from temperature T_1 to T_2. There would inevitably be a large temperature difference between the water and the source of heat, say a hot plate. This makes the process irreversible. However, because entropy is a state function, the entropy change for the water depends only on the initial and final states of the water and not on how the change was made. Hence, as in the previous example, we imagine a process in which the temperature of the water is increased in infinitesimal steps by placing the water in contact with a succession of heat reservoirs of steadily increasing temperature. As the temperature difference is infinitesimal, we can consider each of these steps to be reversible. Hence, as in the previous example, we have

$$\Delta S = mc\int_{T_1}^{T_2} \frac{\mathrm{d}T}{T} = mc \ln \frac{T_2}{T_1},$$

where c is the specific heat of water that we assume to be constant.

Worked example

(a) 1 kg of water at 20 °C is placed in contact with a heat bath at 80 °C. Determine the total change in entropy of the water plus heat bath when equilibrium has been reached. The specific heat of water is 4.2 kJ/kg.

(b) What would be the change in total entropy if the 1 kg of water at 20 °C is heated in two stages by successively placing the water in contact with water baths at temperatures of 50 °C and 80 °C respectively, allowing equilibrium to be reached at each stage?

Solution

(a) The corresponding reversible process is to heat the water using a succession of heat baths of steadily increasing temperature. Using the usual symbols, we have

$$\Delta S_{\text{water}} = mc\int_{T_1}^{T_2} \frac{\mathrm{d}T}{T} = mc \ln \frac{T_2}{T_1}.$$

$$= 1.0 \times 4.2 \times 10^3 \times \left(\ln \frac{353}{293}\right) = +\ 782.4\ \mathrm{J/K}.$$

The total amount of heat absorbed by the water is

$$Q = mc\Delta T = 1.0 \times 4.2 \times 10^3 \times (80 - 20) = 2.52 \times 10^5\ \mathrm{J}.$$

This is delivered by the heat bath at the constant temperature of 353 K. Hence, the entropy change of the heat bath is

$$\Delta S_{\text{bath}} = \frac{Q}{T} = \frac{-\ 2.52 \times 10^5}{353} = -\ 713.9\ \mathrm{J/K},$$

with a minus sign because heat flows out of the heat bath. Hence, the total change in entropy of the water plus heat bath is

$$\Delta S_{\text{water}} + \Delta S_{\text{bath}} = +\ 68.5\ \mathrm{J/K}.$$

(b) For the change from 20 to 50 °C, the total change in entropy is

$$+\ 409.4 - 390.1 = +\ 19.3\ \mathrm{J/K},$$

and for the change from 50 to 80 °C, the total change in entropy is

$$+\ 373.3 - 356.9 = +\ 16.1\ \mathrm{J/K}.$$

Adding these two results together, the total entropy change when the water is heated in two stages is +35.4 J/K, which is less than the entropy change when the water is heated in a single step. If the water is heated in six separate stages from 20 to 80 °C at intervals of 10 °C, the total entropy change of the water plus the six heat baths is +12 J/K.

We see that taking smaller intermediate steps in temperature between the initial and final temperatures results in a smaller overall increase in entropy. By raising the temperature of the water using an infinite number of heat baths, the overall change in entropy would be zero, i.e. we would obtain the reversible heating of water.

7.7 Entropy and the second law

Our examples of entropy changes for various processes illustrate the following general principle:

The total entropy of the system plus its surroundings, i.e. the universe, increases for irreversible processes and remains constant for reversible processes:

$$\Delta S_{\text{sys}} + \Delta S_{\text{surr}} \geq 0. \tag{7.38}$$

This is the *entropy principle*. It is an alternative statement of the second law of thermodynamics. It is equivalent to both the Clausius and Boltzmann statements of the second law that we encountered in Section 7.1.

The Clausius statement of the second law says that *heat cannot by itself pass from a colder to a hotter body*. Consider Figure 7.9, which represents a cold body at temperature T_C and a hot body at temperature T_H, where $T_H > T_C$. Suppose that it was possible for the spontaneous transfer of heat Q from the cold to the hot body. Entropy would 'flow' out of the cold body and its change in entropy would be $-Q/T_C$. On the other hand, the change in entropy of the hot body would be $+Q/T_H$. The net entropy change, therefore, would be

$$\Delta S = Q\left(\frac{1}{T_H} - \frac{1}{T_C}\right).$$

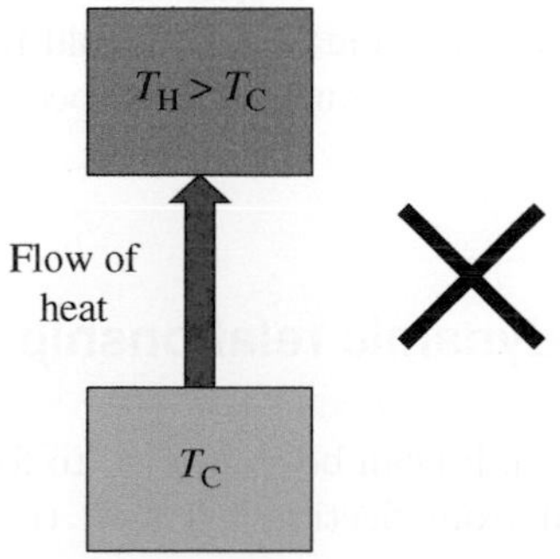

Figure 7.9 The spontaneous flow of heat from a colder body to a hotter body cannot occur. This would lead to a decrease in the total entropy of the system, which would violate the entropy principle.

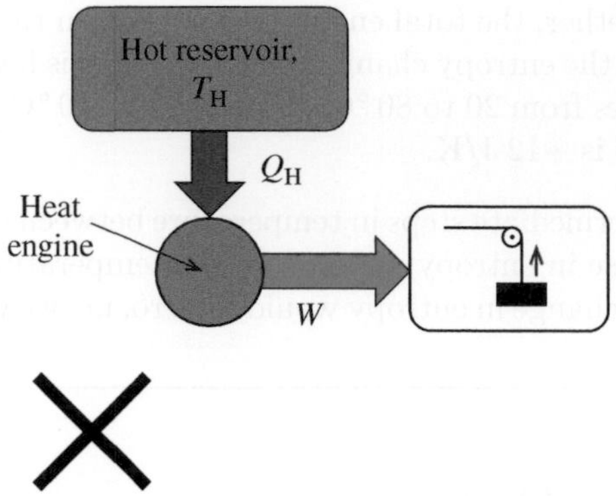

Figure 7.10 A perfect engine can never exist because it would lead to a decrease in the total entropy of the system, which would violate the entropy principle. A certain amount of entropy 'flows' from the hot reservoir and this entropy must be absorbed somewhere. This function must be provided by a cold reservoir.

However, as $T_H > T_C$, this would be a decrease in entropy. This would violate the entropy principle that says entropy never decreases. Thus, the observation that heat never flows spontaneously from a colder body to a hotter body is equivalent to requiring that the spontaneous flow of heat must be accompanied by an increase in entropy.

The Kelvin statement of the second law says that *a cyclic process whose only effect is the complete conversion of heat into work cannot occur*. Imagine a 'perfect' heat engine that, operating in a cycle, extracts heat Q_H from a hot reservoir at temperature T_H and converts it completely into work W, as illustrated by Figure 7.10. This work is used to perform some action on an external device, such as lifting a weight. As the heat engine must return to its original state at the end of each cycle, its entropy must be unchanged. Moreover, although the heat engine delivers work to the external device, this does not involve a change in the entropy of the device. There is, however, a change in the entropy of the hot reservoir, which is $-Q_H/T_H$. Hence, the total change in entropy of the complete system (heat engine plus reservoirs plus external device) that results from the heat extracted from the reservoir is

$$\Delta S = -\frac{Q_H}{T_H} < 0.$$

This result again violates the entropy principle. So, in accord with the Kelvin statement, the entropy principle says that it is not possible to build a 'perfect' engine that completely converts heat into work. This explains why every heat engine must discard some heat to a cold reservoir. A certain amount of entropy 'flows' from the hot reservoir and this entropy must be absorbed somewhere. This function is provided by the cold reservoir.

7.8 The fundamental thermodynamic relationship

The first and second laws of thermodynamics can be combined to form what is known as the *fundamental thermodynamic relationship*. We recall from Section 6.9 that the first law for infinitesimal reversible changes is

$$dU = dQ - PdV. \tag{6.30}$$

We also have for infinitesimal reversible changes,

$$\mathrm{d}S = \frac{\mathrm{d}Q}{T}.$$

Substituting $\mathrm{d}Q = T\mathrm{d}S$ in Equation (7.30) gives

$$\mathrm{d}U = T\mathrm{d}S - P\mathrm{d}V. \tag{7.39}$$

This is the fundamental thermodynamic relationship. We have obtained it for the case of reversible changes. However, the fundamental thermodynamic relationship is always valid for infinitesimal changes, whether reversible or not. We can see this from the fact that it only involves state functions or differentials of state functions.

The fundamental relationship is of great importance because we now have a relationship for a change in internal energy $\mathrm{d}U$ in terms of state functions that do not depend on the path of a process, instead of the path-dependent quantities $\mathrm{d}Q$ and $\mathrm{d}W$. This means that the fundamental thermodynamic relationship is the starting point for the application of thermodynamics to many practical problems.

Suppose that we know the internal energy of a system as a function of S and V: $U(S, V)$. Then

$$\mathrm{d}U = \left(\frac{\partial U}{\partial S}\right)_{\mathrm{V}} \mathrm{d}S + \left(\frac{\partial U}{\partial V}\right)_{\mathrm{S}} \mathrm{d}V. \tag{7.40}$$

Comparing this equation with the fundamental thermodynamic relationship, and requiring them to be identical, we obtain

$$\left(\frac{\partial U}{\partial S}\right)_{\mathrm{V}} = T, \tag{7.41}$$

and

$$\left(\frac{\partial U}{\partial V}\right)_{\mathrm{S}} = -P. \tag{7.42}$$

Hence, if we know the internal energy of a system as a function of entropy and volume, $U(S, V)$, we can at once find T and P. This means that we would have complete thermodynamic knowledge of the system.

We will see applications of the fundamental thermodynamic relationship in the coming sections. And in Section 7.11, we will see how it leads to other useful thermodynamic relationships.

7.9 Phase changes and the Clausius–Clapeyron equation

We discussed phase changes in Section 5.5 and their representation on a phase diagram. Such a phase diagram shows the pressure P of a substance as a function of temperature T, as illustrated in Figure 7.11. Along the solid lines, called *phase boundaries*, two phases of the substance are in equilibrium. The Clausius–Clapeyron equation gives the relationship between P and T at a phase boundary. We will consider the liquid–vapour phase boundary, which is also called the *vapourisation curve*, but the results we obtain apply to any phase boundary.

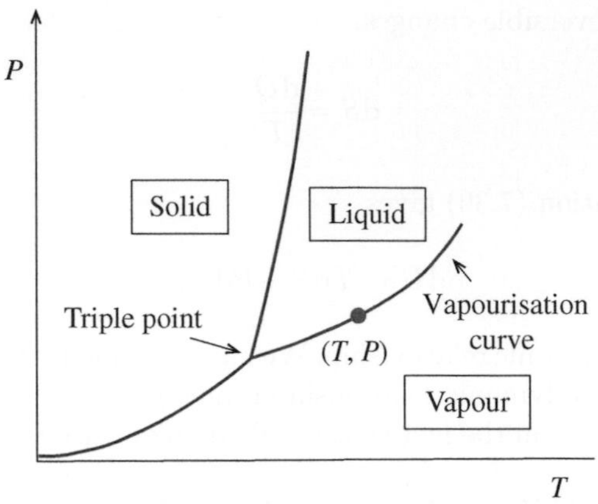

Figure 7.11 A schematic diagram of the phase diagram of a substance. Along the solid lines, called phase boundaries, two phases of the substance are in equilibrium. The Clausius–Clapeyron equation gives the relationship between P and T at any point on a phase boundary.

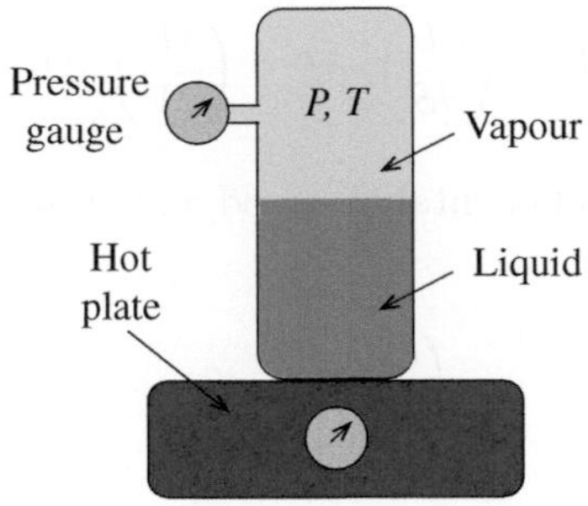

Figure 7.12 An experimental arrangement for determining the vapourisation curve of a liquid.

Figure 7.12 illustrates how a vapourisation curve can be determined. The figure shows a sealed flask containing a quantity of liquid that is in equilibrium with its vapour at well-defined values of temperature T and pressure P. The flask is in contact with a hot plate that controls the temperature. If we heat the container, more liquid will pass into the vapour phase and the pressure will increase. By measuring pressure P as a function of temperature T, we obtain a plot of the vapourisation curve.

We consider a point (T, P) on the vapourisation curve as indicated by the red dot in Figure 7.11. At this combination of P and T, the liquid and vapour phases are in equilibrium and the substance can convert freely between the two phases. As these phase transitions occur at constant temperature and pressure, we can assume a transition between the two phases to be reversible.

We consider a transition from the liquid phase to the vapour phase for a fixed amount of substance. We let U_l, S_l, and V_l be the internal energy, entropy, and volume of the liquid phase and U_v, S_v, and V_v be the respective values for the vapour phase. And to this phase transition, we apply the fundamental thermodynamic relationship:

$$dU = TdS - PdV.$$

As T and P are constant, we can integrate this equation for the present case in which the substance starts completely in the liquid phase and ends up completely in the vapour phase:

$$\int_{U_l}^{U_v} \mathrm{d}U = T\int_{S_l}^{S_v} \mathrm{d}S - P\int_{V_l}^{V_v} \mathrm{d}V. \tag{7.43}$$

This gives

$$U_v - U_l = T(S_v - S_l) - P(V_v - V_l),$$

or

$$U_v - TS_v + PV_v = U_l - TS_l + PV_l. \tag{7.44}$$

The quantity $U - TS + PV$ is called the *Gibbs free energy* G and is defined accordingly:

$$G = U - TS + PV. \tag{7.45}$$

For an infinitesimal, reversible process,

$$\mathrm{d}G = \mathrm{d}U - T\mathrm{d}S - S\mathrm{d}T + P\mathrm{d}V + V\mathrm{d}P.$$

Substituting for $\mathrm{d}U$ from the fundamental relationship, Equation (7.39), we obtain

$$\mathrm{d}G = -S\mathrm{d}T + V\mathrm{d}P. \tag{7.46}$$

As a phase change takes place at constant T and P, we have $\mathrm{d}G = 0$, and

$$G = \text{constant}. \tag{7.47}$$

Hence, if two phases of a given amount of a substance transform into each other at constant temperature and pressure, the Gibbs free energy of one phase must equal the Gibbs free energy of the other. For the present case,

$$G_l(T, P) = G_v(T, P) \tag{7.48}$$

where $G_l(T, P)$ and $G_v(T, P)$ are the Gibbs free energies of the liquid and vapour phases respectively at temperature T and pressure P.

As described earlier and as is evident from Figure 7.11, the liquid-to-vapour phase transition can occur at other combinations of T and P. And at any point on the phase boundary, $G_l(T, P) = G_v(T, P)$. Indeed, this condition determines the relationship between P and T at the phase boundary and hence the shape of the vapourisation curve.

We consider two points (T, P) and $(T + \mathrm{d}T, P + \mathrm{d}P)$ on the vapourisation curve an infinitesimal distance apart, as shown on Figure 7.13. As we move up the curve, the Gibbs free energy will change. However, at each of the two points, there must be equality between the Gibbs free energies. At the point (T, P), we have

$$G_l(T, P) = G_v(T, P) \tag{7.49}$$

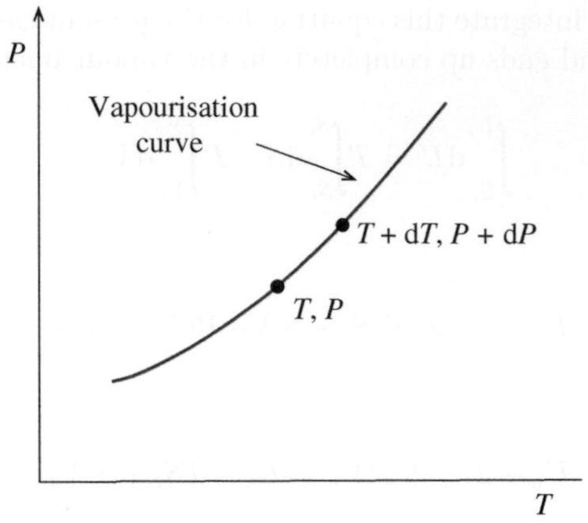

Figure 7.13 The liquid–vapour phase equilibrium curve along which $G_l(T, P) = G_v(T, P)$.

and at the point $(T+\mathrm{d}T,\ P+\mathrm{d}P)$,

$$G_l(T+\mathrm{d}T, P+\mathrm{d}P) = G_v(T+\mathrm{d}T, P+\mathrm{d}P). \tag{7.50}$$

Subtracting Equation (7.49) from Equation (7.50) gives

$$G_l(T+\mathrm{d}T, P+\mathrm{d}P) - G_l(T,P) = G_v(T+\mathrm{d}T, P+\mathrm{d}P) - G_v(T,P). \tag{7.51}$$

Letting

$$G_l(T+\mathrm{d}T, P+\mathrm{d}P) - G_l(T,P) = \mathrm{d}G_l,$$

where $\mathrm{d}G_l$ is the infinitesimal change in Gibbs free energy for the liquid phase, and similarly,

$$G_v(T+\mathrm{d}T, P+\mathrm{d}P) - G_v(T,P) = \mathrm{d}G_v,$$

It follows from Equation (7.51) that

$$\mathrm{d}G_l = \mathrm{d}G_v. \tag{7.52}$$

Using Equation (7.46), we can write for the liquid phase

$$\mathrm{d}G_l = -S_l\mathrm{d}T + V_l\mathrm{d}P, \tag{7.53}$$

and for the vapour phase

$$\mathrm{d}G_v = -S_v\mathrm{d}T + V_v\mathrm{d}P. \tag{7.54}$$

Hence,

$$-S_l\mathrm{d}T + V_l\mathrm{d}P = -S_v\mathrm{d}T + V_v\mathrm{d}P, \tag{7.55}$$

and upon rearrangement,

$$\frac{\mathrm{d}P}{\mathrm{d}T} = \frac{S_v - S_l}{V_v - V_l} \tag{7.56}$$

or

$$\frac{\mathrm{d}P}{\mathrm{d}T} = \frac{\Delta S}{\Delta V}. \tag{7.57}$$

This is the Clausius–Clapeyron equation. In the present case, it says that the slope of the vapourisation curve at the point on the curve specified by temperature T and pressure P is determined by the entropy change ΔS and the volume change ΔV that occurs during the transition from liquid to vapour phase. We have obtained the Clausius–Clapeyron equation for the case of the vapourisation curve, but the equation applies to any phase boundary.

A phase change that is accompanied by a change ΔS in entropy involves the absorption or emission of heat ΔQ. If L_{12} is the latent heat of transformation when a given quantity of substance is transformed from phase 1 to phase 2, at the temperature T, then the entropy change is

$$\Delta S = \frac{\Delta Q}{T} = \frac{L_{12}}{T}. \tag{7.58}$$

Using this result in Equation (7.57) gives

$$\frac{\mathrm{d}P}{\mathrm{d}T} = \frac{L_{12}}{T\Delta V}, \tag{7.59}$$

which is an alternate form of the Clausius–Clapeyron equation.

Most liquids absorb heat when they are vapourised and the volume of the vapour is greater than that of the liquid. Hence, ΔS and ΔV are both positive, which means that $\mathrm{d}P/\mathrm{d}T$ for the vapourisation curve is usually positive. Similarly, most solids absorb heat on melting and turn into liquids of larger volume and so ΔS and ΔV are again both positive. This means that $\mathrm{d}P/\mathrm{d}T$ for the solid–liquid phase boundary is usually positive and the melting temperature goes up as the pressure goes up. An important exception to this is water. When ice is melted, its volume decreases. i.e. ΔV is negative. Hence, the solid–liquid phase boundary has a negative slope as seen in the phase diagram of water shown in Figure 5.16. It follows that the melting temperature of ice decreases when the applied pressure increases. The lowering of the melting temperature of ice under pressure plays a role in the motion of glaciers. Ice at the bottom of the glacier melts under the weight of the ice above it, facilitating the slide of the glacier.

Worked example

Use the Clausius–Clapeyron equation to find the boiling point of water at the summit of Mount Everest. Take the atmospheric pressure at the summit to be 3.50×10^4 Pa, which is about 35% of its value at sea level. The latent heat of vapourisation of water is 2.26×10^3 kJ/kg, and the volumes per kilogram in the liquid and vapour phase at $T = 100\,°\mathrm{C}$ and $P = 1.01 \times 10^5$ Pa are, respectively, $1.04 \times 10^{-3}\ \mathrm{m^3/kg}$ and $1.673\ \mathrm{m^3/kg}$.

Solution

From Equation (7.59),

$$\frac{\mathrm{d}P}{\mathrm{d}T} = \frac{L_{12}}{T\Delta V} = \frac{2.26 \times 10^6}{373 \times \left(1.673 - 1.04 \times 10^{-3}\right)}$$
$$= 3.62 \times 10^3\ \mathrm{J/m^3K} \equiv 3.62 \times 10^3\ \mathrm{Pa/K}.$$

$$\Delta P = \left(3.50 \times 10^4 - 1.01 \times 10^5\right) = -6.6 \times 10^4 \text{ Pa}.$$

Hence,

$$\Delta T = -\frac{6.6 \times 10^4}{3.62 \times 10^3} = -18\ ^\circ\text{C}.$$

Water boils at about 80 °C at the height of Mount Everest.

7.10 Gibbs free energy

Phase changes take place at constant temperature and pressure. And because of this, as we saw in Section 7.9, Gibbs free energy is the appropriate choice of thermodynamic function to describe phase changes. In fact, many processes take place at constant T and P, including the majority of chemical and biological reactions.

And here again, Gibbs free energy is the natural choice of thermodynamic function. Clearly, when dealing with processes like chemical reactions, we are dealing with systems that are not isolated but which interact with their local surroundings. And so more generally, Gibbs free energy is appropriate to specify systems that are in contact with, and can exchange energy with, their surroundings.

Gibbs free energy is defined as

$$G = U - TS + PV,$$

in terms of the variables U, T, S, P, and V. These are all functions of state and it follows that G is also a function of state. We also have the relationship for enthalpy:

$$H = U + PV, \tag{6.50}$$

and so we can also write

$$G = H - TS. \tag{7.60}$$

7.10.1 Physical interpretation of Gibbs free energy

One way to interpret Gibbs free energy is as follows. Imagine that we created from nothing a system in a state that has volume V, entropy S, and is in equilibrium with its surroundings at temperature T and pressure P. The system could be a quantity of gas. How much energy would we need to create the system? We would need to supply the internal energy U of the system itself. We would also need to supply energy to 'push back' the atmosphere to accommodate the system. This is the expansion work PV.

But, suppose that the system is not isolated but is in contact with its surroundings so that heat can flow between the two. In that case, the surroundings can supply, in the form of heat, some of the energy required to create the system. The amount of heat provided *by* the surroundings is just equal to TS; the entropy of the system times the temperature of the surroundings. Hence, the amount of energy in the form of work that we would need to supply is $U + PV - TS$, which we recognise as the Gibbs free energy G.

Conversely, if we completely destroyed the system, the amount of work that we would obtain, i.e. the 'free' energy, would be its internal energy U plus the PV work from the compression of the atmosphere, *minus* the heat equal to TS that would have to be returned to the surroundings to dispose of the system's entropy. Hence, the amount of work obtained is $U + PV - TS = G$, i.e. the Gibbs free energy.

Usually, of course, we are dealing not with the creation or destruction of an entire system but rather with processes that the system undergoes. In that case, we are interested not in G but in the change ΔG that occurs in the Gibbs free energy. Because ΔG is such a useful quantity, values of ΔG have been tabulated for a huge range of processes and chemical reactions.

7.10.2 The example of a lead acid battery

As an example of the use of Gibbs free energy, we consider a lead-acid battery. The chemical reaction occurring in the battery is

$$Pb + PbO_2 + 4H^+ + 2SO_4^{2-} \rightarrow 2PbSO_4 + 2H_2O.$$

We are interested in the change ΔG in Gibbs free energy that occurs in the reaction, as ΔG is equal to the work delivered by the battery. From Equation (7.60), we obtain

$$\Delta G = \Delta H - T\Delta S - S\Delta T. \tag{7.61}$$

As the temperature T of the surroundings is constant,

$$\Delta G = \Delta H - T\Delta S. \tag{7.62}$$

From Equation (7.50), we obtain,

$$\Delta H = \Delta U + P\Delta V + V\Delta P. \tag{7.63}$$

As there is no compression $P\Delta V$ work involved in the chemical reaction, and P is constant,

$$\Delta H = \Delta U. \tag{7.64}$$

The change ΔH in enthalpy is equal to the change ΔU in the internal energy of the system, which is the chemical energy from the reaction.

The values of ΔG, ΔH and also ΔS for the above reaction have been tabulated. For each mole of Pb, the values are: $\Delta G = -394$ kJ, $\Delta H = -316$ kJ and $\Delta S = +263$ J/K. So, even before the battery is constructed, a prediction can be made from the tabulated data of the amount of electrical work that the battery would deliver, as this is equal to the value of ΔG. This value is negative because work is delivered *by* the battery. Note that the magnitude of ΔH is less that the magnitude of ΔG. The difference is 78 kJ. So the energy delivered by the chemical reaction is actually less than the work delivered by the battery. This extra energy comes from heat absorbed from the surroundings. The entropy of the products is greater than that of the reactants and so the absorbed heat is equal to (263 J/K × 298 K) = 78 kJ,where T has been taken to be 298 K. We see that part of the work delivered by the battery is actually supplied by the surroundings. The energy flow in the battery is illustrated by Figure 7.14.

When the battery is charged, this reaction occurs in reverse. 394 kJ of energy must be supplied by electrical work. Of this, 316 kJ is stored as chemical energy and 78 kJ goes into the surroundings as heat to dispose of the excess entropy of the reaction products. In practice, it may be noticed that a battery does become warm when it is put on charge.

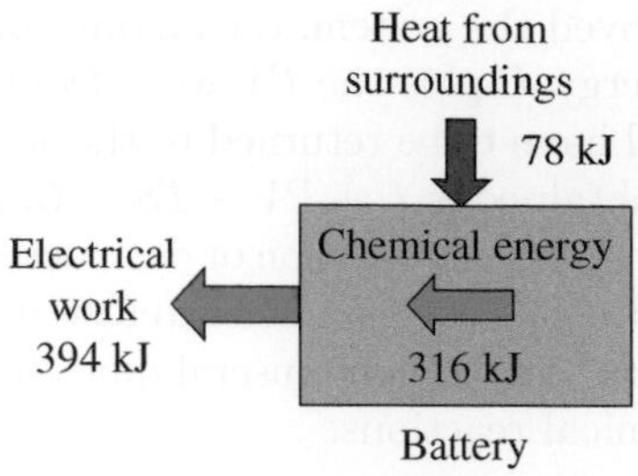

Figure 7.14 Energy flow in a lead-acid battery. Of the 394 kJ of electrical work delivered by the battery, 316 kJ comes from the chemical reaction and 78 kJcomes from heat supplied by the surroundings.

Worked example

A hydrogen fuel cell uses the following reaction to produce electrical power:

$$H_2 + O \rightarrow H_2O.$$

The reaction takes place in the cell in a controlled way at room temperature. For each mole of H_2, the enthalpy change for the reaction is $\Delta H = -286$ kJ and the entropy change is $\Delta S = -163$ J/K. Determine the efficiency of an ideal hydrogen fuel cell.

Solution

We have from Equation (7.62), $\Delta G = \Delta H - T\Delta S$, as the temperature of the cell is constant. Therefore, the electrical work delivered by the cell for each mole of H_2 is

$$\Delta G = [-286 - 298 \times (-0.163)]\,\text{kJ} = -237\,\text{kJ},$$

with $T = 298$ K. We see that not all the 286 kJ of energy produced by the chemical reaction is converted into electrical work. The total entropy of the reactants $H_2 + O$ is greater than the entropy of the water and the excess entropy must be disposed of as waste heat to the surroundings. This waste heat is equal to $298 \times (-0.163) = -49$ kJ/mol. This gives the efficiency of an ideal hydrogen fuel cell to be $237/286 = 83\%$. In practice, the waste heat will be more and the efficiency therefore less. Nevertheless, the efficiency of a typical fuel cell, which may be up to 60%, is greater than that of other engines such as the internal combustion engine.

7.10.3 Gibbs free energy and spontaneous processes

We know from the entropy principle that during any process or indeed chemical reaction, the sum of the entropies of the system plus the surroundings must increase, or in the case of reversibility, remain the same:

$$\Delta S_{sys} + \Delta S_{surr} \geq 0.$$

If heat ΔQ is taken from the surroundings at temperature T during the process, the entropy change of the surroundings is

$$\Delta S_{surr} = -\frac{\Delta Q}{T}. \tag{7.65}$$

This is the case whether the process is reversible or not; the surroundings, at a constant temperature, do not care. Hence, from Equation (7.38), we have

$$\Delta S_{\text{sys}} - \frac{\Delta Q}{T} \geq 0. \tag{7.66}$$

Substituting for ΔQ from the first law:

$$\Delta U = \Delta Q - P\Delta V, \tag{7.67}$$

we obtain

$$\Delta U + P\Delta V - T\Delta S \leq 0. \tag{7.68}$$

where the equals sign is for the special case of reversible processes. From the definition of Gibbs free energy, $G = U - TS + PV$, we obtain for constant T and P:

$$\Delta U + P\Delta V - T\Delta S = \Delta G \tag{7.69}$$

Comparing Equations (7.68) and (7.69), we see that

$$\Delta G \leq 0. \tag{7.70}$$

Thus, the condition $\Delta S_{\text{sys}} + \Delta S_{\text{surr}} \geq 0$, leads to the condition $\Delta G \leq 0$.

We can interpret Equation (7.70) with respect to chemical reactions as follows. For spontaneous, i.e. irreversible, processes $\Delta S_{\text{sys}} + \Delta S_{\text{surr}} > 0$,which means that $\Delta G < 0$ for these processes. Hence, for a chemical reaction to be spontaneous, the change ΔG in the reaction must be negative. This is the case for the chemical reaction in a lead-acid battery. If ΔG is positive for a given chemical reaction, that reaction will not be spontaneous and external energy must be provided for it to occur. The case $\Delta G = 0$ means that the chemical reaction is reversible, i.e. that the reactants and resulting products are in equilibrium with each other. As we saw in Section 7.9, this is similarly the case for the state of equilibrium at a phase boundary.

7.11 Thermodynamic identities

The fundamental thermodynamic relationship

$$dU = TdS - PdV$$

gives rise to a host of other useful thermodynamic relationships. As we saw in Section 7.8, if we know the internal energy of a system as a function of entropy and volume, $U(S, V)$, we can find T and P simply by differentiation:

$$\left(\frac{\partial U}{\partial S}\right)_V = T.$$

$$\left(\frac{\partial U}{\partial V}\right)_S = -P.$$

And from these two equations, we can find other relationships between thermodynamic variables. We do this by making use of a mathematical identity for partial double differentiation.

A note on partial differentiation If we have a function $f(x, y)$ of two variables x and y, the order of differentiation in double differentiation is immaterial:

$$\frac{\partial}{\partial x}\left(\left(\frac{\partial f}{\partial y}\right)_x\right)_y = \frac{\partial}{\partial y}\left(\left(\frac{\partial f}{\partial x}\right)_y\right)_x,$$

or more concisely

$$\frac{\partial^2 f}{\partial x \partial y} = \frac{\partial^2 f}{\partial y \partial x}. \tag{7.71}$$

As a simple example, suppose $f(x, y) = x^2 y^3$. Then,

$$\left(\frac{\partial f}{\partial x}\right)_y = 2xy^3; \quad \left(\frac{\partial f}{\partial y}\right)_x = 3x^2 y^2.$$

$$\frac{\partial}{\partial y}\left(2xy^3\right)_x = \frac{\partial^2 f}{\partial y \partial x} = 6xy^2; \quad \frac{\partial}{\partial x}\left(3x^2 y^2\right)_y = \frac{\partial^2 f}{\partial x \partial y} = 6xy^2.$$

Using the above identity for double differentiation, and taking U as a function of S and V, we have

$$\frac{\partial^2 U}{\partial V \partial S} = \frac{\partial^2 U}{\partial S \partial V}.$$

Differentiating Equation (7.41) with respect to V, we obtain

$$\frac{\partial}{\partial V}\left(\left(\frac{\partial U}{\partial S}\right)_{\mathrm{V}}\right)_{\mathrm{S}} = \left(\frac{\partial T}{\partial V}\right)_{\mathrm{S}}. \tag{7.72}$$

Similarly, differentiating Equation (7.42) with respect to S, we obtain

$$\frac{\partial}{\partial S}\left(\left(\frac{\partial U}{\partial V}\right)_{\mathrm{S}}\right)_{\mathrm{V}} = -\left(\frac{\partial P}{\partial S}\right)_{\mathrm{V}}. \tag{7.73}$$

Hence,

$$\left(\frac{\partial T}{\partial V}\right)_{\mathrm{S}} = -\left(\frac{\partial P}{\partial S}\right)_{\mathrm{V}}. \tag{7.74}$$

7.11.1 Maxwell's relations

Equation (7.74) is one of the so-called *Maxwell's relations.* These do not refer to a particular process but express relationships that hold for *any* equilibrium state of a system. And this makes them very useful. To derive Equation (7.74), we started with the fundamental thermodynamic relationship for U in terms of the variables S and V, which are the so-called *natural variables* for internal energy. We have found, however, that it is sometimes convenient to use an alternative thermodynamic function to specify a system. For example, in our discussion of the Joule–Kelvin effect (Section 6.13), we found that it was more convenient to use enthalpy H:

$$H = U + PV,$$

and in discussing phase transitions, it was convenient to use Gibbs free energy G:

$$G = U - TS + PV.$$

The other relevant thermodynamic function that is appropriate to mention here is the *Helmholtz free energy* F. We have not encountered it previously, but it is defined by

$$F = U - TS. \tag{7.75}$$

Helmholtz free energy is the most appropriate thermodynamic function for processes that occur at constant pressure P and constant volume V.

As internal energy U gives rise to one of Maxwell's relations, each of the other three functions gives rise to one of the remaining Maxwell's relations. For example, starting with the Gibbs free energy

$$G = U - TS + PV.$$

We have

$$\mathrm{d}G = \mathrm{d}U - T\mathrm{d}S - S\mathrm{d}T + P\mathrm{d}V + V\mathrm{d}P.$$

Substituting for $\mathrm{d}U$ from the fundamental thermodynamic relationship, Equation (7.39), we obtain

$$\mathrm{d}G = -S\mathrm{d}T + V\mathrm{d}P. \tag{7.76}$$

This equation shows that the natural variables for G are T and P, from which

$$\mathrm{d}G = \left(\frac{\partial G}{\partial T}\right)_{\mathrm{P}} \mathrm{d}T + \left(\frac{\partial G}{\partial P}\right)_{\mathrm{T}} \mathrm{d}P. \tag{7.77}$$

A comparison of Equations (7.76) and (7.77) gives

$$\left(\frac{\partial G}{\partial T}\right)_{\mathrm{P}} = -S, \quad \left(\frac{\partial G}{\partial P}\right)_{\mathrm{T}} = V. \tag{7.78}$$

Using the identity

$$\frac{\partial^2 G}{\partial P \partial T} = \frac{\partial^2 G}{\partial T \partial P},$$

we obtain

$$\left(\frac{\partial S}{\partial P}\right)_{\mathrm{T}} = -\left(\frac{\partial V}{\partial T}\right)_{\mathrm{P}}, \tag{7.79}$$

which is another of the Maxwell relations.

We can use exactly the same procedure for H and F when it will become apparent that the natural variables for H are S and P and those for F are V and T. Enthalpy H gives rise to the Maxwell relation

$$\left(\frac{\partial T}{\partial P}\right)_{\mathrm{S}} = \left(\frac{\partial V}{\partial S}\right)_{\mathrm{P}}. \tag{7.80}$$

and Helmholtz free energy F gives rise to the Maxwell relation

$$\left(\frac{\partial S}{\partial V}\right)_{\mathrm{T}} = \left(\frac{\partial P}{\partial T}\right)_{\mathrm{V}}. \tag{7.81}$$

Equations (7.74), (7.79), (7.80), and (7.81) are the four Maxwell relations.

The importance of the Maxwell relations is that they provide relationships between quantities that are difficult to measure to quantities that are easy to measure or determine. This is particularly the case for relations (7.79) and (7.81). The quantities on the left-hand side of these relations are hard to measure, while the quantities on the right-hand side can be determined from the equation of state. As an example of the use of the Maxwell relations, we derive an equation for the Joule–Kelvin coefficient μ that we could only state in Section 6.13.

Worked example

Obtain the following expression for the *Joule–Kelvin coefficient*

$$\mu = \frac{1}{C_{\mathrm{P}}}\left[T\left(\frac{\partial V}{\partial T}\right)_{\mathrm{P}} - V\right],$$

where

$$\mu = \left(\frac{\partial T}{\partial P}\right)_{\mathrm{H}}.$$

Solution

The appropriate thermodynamic function for a Joule–Kelvin expansion is enthalpy: $H = U + VP$. For infinitesimal changes, we have

$$\mathrm{d}H = \mathrm{d}U + V\mathrm{d}P + P\mathrm{d}V.$$

Substituting for $\mathrm{d}U$ from the fundamental thermodynamic equation, $\mathrm{d}U = T\mathrm{d}S - P\mathrm{d}V$, we obtain

$$\mathrm{d}H = T\mathrm{d}S + V\mathrm{d}P.$$

As the enthalpy remains constant in a Joule–Kelvin expansion, $\mathrm{d}H = 0$ and hence

$$T\mathrm{d}S = -V\mathrm{d}P.$$

The natural variables to describe the Joule–Kelvin effect are T and P: and hence we express the change in entropy $\mathrm{d}S$ in terms of infinitesimal changes in T and P. For $S = S(T, P)$, we have

$$\mathrm{d}S = \left(\frac{\partial S}{\partial T}\right)_{\mathrm{P}} \mathrm{d}T + \left(\frac{\partial S}{\partial P}\right)_{\mathrm{T}} \mathrm{d}P.$$

Substituting for $\mathrm{d}S = -V\mathrm{d}P/T$ gives

$$T\left(\frac{\partial S}{\partial T}\right)_{\mathrm{P}} \mathrm{d}T + T\left(\frac{\partial S}{\partial P}\right)_{\mathrm{T}} \mathrm{d}P = -V\mathrm{d}P.$$

Now use is made of the Maxwell relation (7.79)

$$\left(\frac{\partial S}{\partial P}\right)_{\mathrm{T}} = -\left(\frac{\partial V}{\partial T}\right)_{\mathrm{P}},$$

giving

$$T\left(\frac{\partial S}{\partial T}\right)_{\mathrm{P}} \mathrm{d}T - T\left(\frac{\partial V}{\partial T}\right)_{\mathrm{P}} \mathrm{d}P = -V\mathrm{d}P.$$

Substituting for $T(\partial S/\partial T)_{\mathrm{P}} = C_{\mathrm{P}}$ from Equation (7.26) gives

$$C_{\mathrm{P}}\mathrm{d}T = \left[T\left(\frac{\partial V}{\partial T}\right)_{\mathrm{P}} - V\right]\mathrm{d}P.$$

Hence, for the conditions of constant enthalpy, we can write

$$\left(\frac{\partial T}{\partial P}\right)_{\mathrm{H}} = \mu = \frac{1}{C_{\mathrm{P}}}\left[T\left(\frac{\partial V}{\partial T}\right)_{\mathrm{P}} - V\right].$$

This gives an expression for the Joule–Kelvin coefficient in terms of the volume of the gas, its molar specific heat C_{P}, and its expansion coefficient, all of which are easy to measure experimentally.

7.12 A statistical approach to the second law of thermodynamics

We have discussed the second law of thermodynamics in terms of macroscopic observables such as temperature, pressure, and volume. This is the approach of classical thermodynamics. We now discuss an alternative approach in which we look at the microscopic behaviour of a system. As we have noted previously, any reasonable amount of matter will contain a huge number of molecules, $\sim 10^{23}$, and, consequently, we must use statistical methods to deal with their behaviour. This is the approach followed by Boltzmann, Gibbs, Planck, and other workers and takes us into the realm of *statistical mechanics.*

We will see that the statistical version of the second law shows that the equilibrium state of a system is that state that occurs with maximum probability. It does not forbid a state spontaneously moving to a state of lower entropy; rather it maintains that such a change would be highly unlikely; so unlikely that typically it would never happen in many ages of the universe. This approach to the second law thus provides an explanation of why natural, irreversible processes proceed in a particular direction. We begin our discussion with some elementary results for permutations and combinations.

7.12.1 Permutations and combinations

Suppose we have six books that we want to put on a bookshelf, as illustrated schematically in Figure 7.15. How many different arrangements or *permutations* of these books are there? For the first book, say the one on the left side of the bookshelf, there is a choice of six books. For the next position, we are left with a choice of five books. Thus, for *each* of the first six books, there are five possible choices for the second position. This means that for the first two positions, there are 6×5 possible arrangements. For each of these 6×5 arrangements, there is a choice of four books for the third position. That means that for the first three positions, there are $6 \times 5 \times 4$ possible arrangements. Extrapolating this to all six books, the total number of possible arrangements is $6 \times 5 \times 4 \times 3 \times 2 \times 1 = 6! = 720$. (For 10 books, there are $10! = 3.6$ million possible arrangements!). Generalising this result, the number of possible arrangements or permutations of n different things is $n!$

Now suppose that we want to choose two of the six books to take on holiday. How many combinations of two books are there? We have six books to make our first choice and the remaining five books to make our second choice. That sounds like 6×5 combinations. However, this counts the combination of the same two books twice. The 30 combinations would include, for example, the pairings (book B and book C) and also (book C and book B), which, of course, is the same combination. To determine the number of distinct combinations, we have to divide the above number 30 by the number of ways of arranging two books. This is just 2! Hence, the number of distinct combinations is $(6 \times 5)/2! = 15$. Generalising this result: The number of distinct combinations of N different things taken n at a time is

$$\frac{N(N-1)(N-2)\cdots(N-n+2)(N-n+1)}{n!}.$$

We can simplify this result by writing the numerator as $N!/(N-n)!$, for which all the factors of the factorials cancel, except for the first n factors. For example, with $N = 10$ and $n = 4$,

$$\frac{10!}{(10-4)!} = \frac{10 \times 9 \times 8 \times 7 \times 6 \times 5 \times 4 \times 3 \times 2 \times 1}{6 \times 5 \times 4 \times 3 \times 2 \times 1} = 10 \times 9 \times 8 \times 7.$$

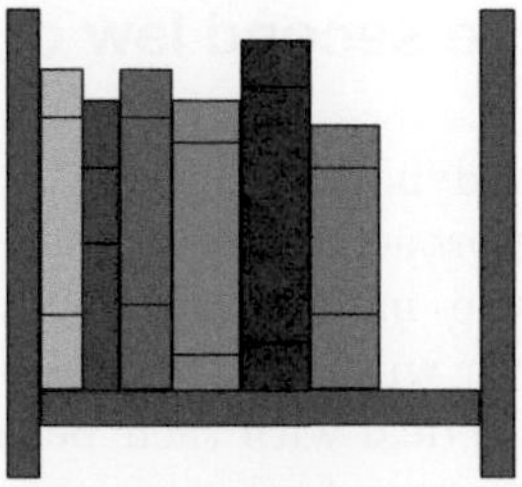

Figure 7.15 A particular arrangement of six books on a bookshelf. The total number of possible arrangements is 720.

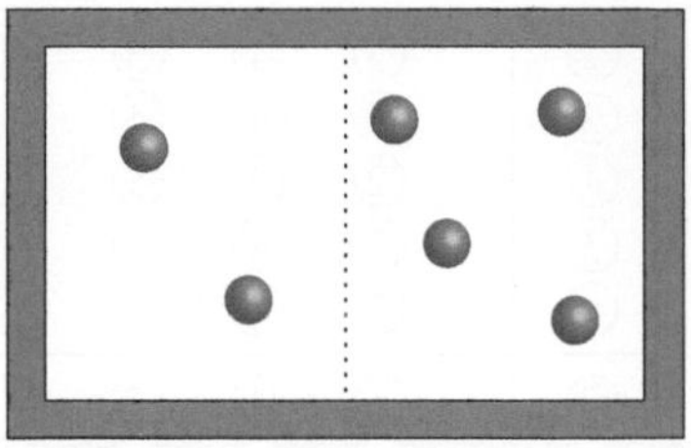

Figure 7.16 A schematic diagram showing six molecules in a box.

Hence, the number of distinct combinations of N things taken n at a time is

$$\frac{N!}{(N-n)!n!}. \tag{7.82}$$

This is the result that we need.

Spatial configuration of molecules in a box

Imagine now that we have six molecules in a box, as illustrated schematically in Figure 7.16. The six molecules move continuously within the box, and at any instance of time, any particular molecule will be in either the left or the right half of the box. As the two halves have the same volume, a molecule has the same probability of being in either half. We might intuitively expect that it is most likely that there will be three molecules in each half. More generally, we wish to determine the probability that there are n molecules in say the left half. The molecules are indistinguishable, of course, but it is useful for our purpose to denote then one to six so that we do not count the combination of the same molecules more than once. For example, the combination of (molecule 1 and molecule 2) being in the left half is the same combination as (molecule 2 and molecule 1) being in the left half; the ordering of the molecules does not matter.

There is only one possible combination for the case in which there are no molecules in the left half and all six molecules in the right half. There are six possible combinations where there is just one molecule in the left half. To determine the number of combinations for the case where there are two molecules in the left half we use Equation (7.82). For $N = 6$, $n = 2$, the number of combinations is

$$\frac{6!}{2! \times 4!},$$

which gives 15 possible combinations. These 15 combinations are shown in Figure 7.17. The number of combinations for three molecules occupying the left half is similarly,

$$\frac{6!}{3! \times 3!} = 20.$$

From the symmetry of the system, the number of combinations for 4, 5, and 6 molecules in the left half are respectively 15, 6, and 1. Hence, the total number of combinations that arise from the different ways to distribute 6 molecules between the two halves of the box is 64.

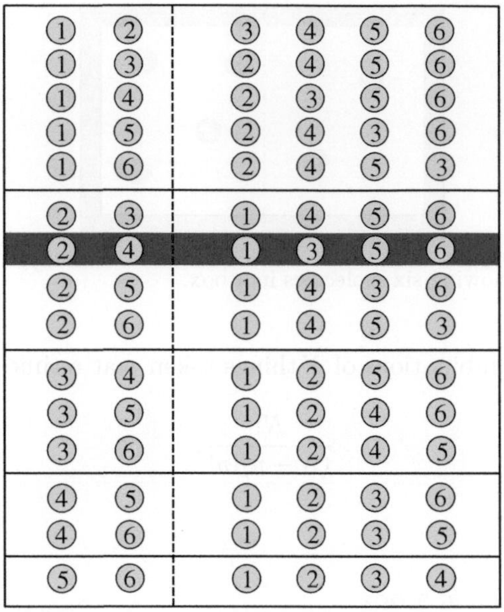

Figure 7.17 The 15 possible combinations for having two molecules in the left half of the box shown in Figure 7.16. The highlighted combination corresponds to molecules 2 and 4 being in the left half.

We now make the assumption that all the possible 64 combinations are equally likely to occur. Thus, suppose it were possible to take a series of snapshots of the six molecules as they move around the box, and then count the number of times each of the possible combinations occurred. We would find that all 64 combinations occur equally often, i.e. the system would spend, on average, the same amount of time in each of them. Hence, the probability of finding three molecules in the left half of the box is $20/64 = 0.313$. When we calculate and plot the probabilities for all seven values of n, from 0 to 6, we get the plot of probability shown in Figure 7.18. There we see that indeed the most probable combination is that in which there are three molecules in each half. We also see that the probability of there being no molecules in the left half is relatively small, but it is not zero.

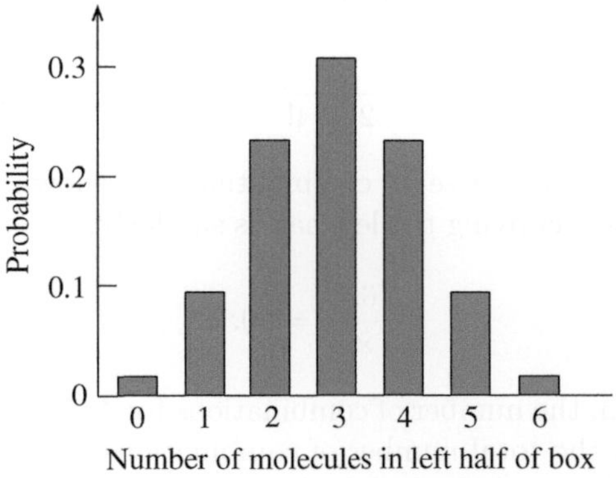

Figure 7.18 A plot of the probability of finding n molecules in the left half of the box shown in Figure 7.16, where $n = 0$ to 6.

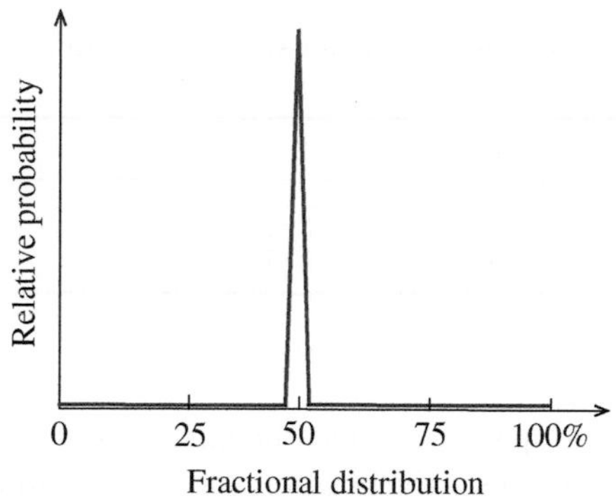

Figure 7.19 A schematic diagram of a probability plot for an ordinary amount of gas in a container with $\sim 10^{23}$ molecules. It plots probability against the fractional distribution of molecules in the two halves of the container. It shows that the overwhelming probability is that the molecules are equally shared between the two halves of the container. The figure is schematic in that the width of the peak of probability is far too narrow to be properly indicated on the diagram.

We can determine the probability of *all* the molecules being in one half of the box very readily in an alternative way. The chance of a molecule being in the left half is $1/2$. Hence, the probability of all six molecules being in the left half is $(1/2)^6 = 1/64$, as we saw previously. Imagine then that we have one mole of gas in a container. The probability of all the molecules being in the left half of the container is $(1/2)^{6.02 \times 10^{23}}$, which is a vanishingly small number. We could confine the gas to one half of the container by a partition. If we remove that partition, the gas would rapidly fill the whole container. However, the reverse process would never be observed. The probability of all the molecules returning to fill just half the container is utterly negligible. This event would never occur in many times the age of the universe.

Figure 7.19 is a schematic diagram of a probability plot for an ordinary amount of gas in a container with $\sim 10^{23}$ molecules. It plots probability against the percentage of molecules being in say the left half of the container. It shows that the overwhelming probability is that molecules are equally shared between the two halves. The figure is schematic in that the width of the peak in probability is far too narrow to be properly indicated on the diagram.

Microstates and macrostates

We showed in Figure 7.17 the 15 possible combinations for having two molecules in the left half of the box. Each combination is a detailed specification of the locations of the individual molecules. Such a detailed specification is called a *microstate* of the system. One of the possible microstates is highlighted in red in the figure. It says that molecules 2 and 4 are in the left half, and molecules 1, 3, 5, and 6 are in the right half. We can also give a more general, less detailed specification of the six-molecule system. Thus we can specify the system by saying there are n molecules in the left half. This more general specification of a system is called a *macrostate*. For this system, there are seven possible macrostates, for $n = 0$ to 6. And each macrostate has a number of microstates that give rise to that microstate. For the $n = 2$ macrostate, there are 15 microstates. The number of distinct microstates giving the same macrostate is called the *multiplicity* w of the macrostate. The multiplicities for all seven macrostates are shown in Table 7.1. We note that the total number of microstates is much larger than the total number of macrostates.

Table 7.1 The seven possible macrostates, ($n = 0 - 7$) for six molecules in a box together with the multiplicity w and entropy S of these macrostates.

Macrostate, n	0	1	2	3	4	5	6
Multiplicity, w	1	6	15	20	15	6	1
Entropy, S, 10^{-23} J/K.	0	2.47	3.74	4.13	3.74	2.47	0

More generally, a microstate is the specification at the microscopic level of the individual particles of a system in a particular state. Classically, this is the position and momentum of the all the particles. Quantum mechanically, it is the quantum states of all the particles. And a macrostate is the more general specification of the system in terms of a few macroscopic variables of the system; for example, in terms of pressure, temperature, and volume. Clearly, the number of microstates of a system is invariably much larger than the number of macrostates. For example, any ordinary amount of gas contains a huge number of molecules, $\sim 10^{23}$, with a correspondingly high number of microstates. However, we can specify its macrostate entirely by just its pressure, volume, and the amount of gas.

Equilibrium of an isolated system and the postulate of a priori probabilities

In our discussion of the location of the six molecules in the box, we made the assumption that all possible combinations of the locations were equally probable. This is in accord with a postulate introduced by Boltzmann. This is the postulate of equal *a priori* probabilities. This says that *for an isolated system, all possible microstates are equally likely to occur, i.e. to have equal a priori probabilities.* As a result of this postulate, the probability that a system is in a given macrostate is

$$\frac{\text{The number of microstates for that macrostate}}{\text{Total number of microstates for all the macrostates of the system}}.$$

There is a second postulate that says *equilibrium corresponds to the state of maximum probability*, i.e. the macrostate with the largest multiplicity. We can thus define equilibrium microscopically in terms of equally probable microstates.

If a system is in an equilibrium state, it is unlikely to move spontaneously to a state of lower multiplicity and hence lower probability. Even very small fluctuations away from equilibrium will occur extremely rarely, and substantial deviations from equilibrium will, for practical purposes, never occur unless the system is subjected to an external disturbance. If the system is especially prepared to be in a state different from equilibrium, it will return to equilibrium in a process in which the multiplicity increases to its maximum value. We have a statistical version of the second law: the state of an isolated thermodynamic system in equilibrium is the one with the greatest multiplicity.

Worked example

Imagine that there are 100 molecules in a box. (a) Compare the number of microstates for the cases where (i) all 100 molecules are in the left half of the box and (ii) there are 50 molecules in each half of the box. (b) Determine the probability that there will be 50 molecules in each half of the box.

Solution

(a) For the macrostate corresponding to the combination $n_1 = 100$, $n_2 = 0$, where $n_2 = (N - n_1)$, the number of microstates is

$$w_{100,0} = \frac{N}{n_1!n_2!} = \frac{100!}{100!0!} = 1,$$

as $0! = 1$. For the macrostate corresponding to the combination $n_1 = 50$, $n_2 = 50$, the number of microstates is

$$w_{50,50} = \frac{100!}{50!50!} = 1.01 \times 10^{29}.$$

By comparing the respective number of microstates, we see that the probability of all the molecules being in one half of the box is utterly negligible.

(b) The probability of all the molecules being in the left half is $(1/2)^{100}$. The number of microstates for this macrostate is 1. We have that the probability of a particular macrostate is

$$\frac{\text{Number of microstates corresponding to that macrostate}}{\text{Total number of microstates for all possible macrostates}}.$$

Hence, the total number of microstates for all possible macrostates is $1/(1/2)^{100} = 2^{100}$. Then the probability of there being 50 molecules in each half of the box is

$$\frac{1.01 \times 10^{29}}{2^{100}} = 0.078 = 7.8\%.$$

It is a striking result that the number of possible microstates increases by 2^N when the volume occupied by the N molecules doubles. It arises from the following standard formula

$$\sum_{n=0}^{N} \frac{N!}{(N-n)!n!} = 2^N.$$

7.12.2 Probability and entropy; Boltzmann's equation

We can state the second law of thermodynamics due to Clausius in the following way. During natural (as distinct from idealised, reversible) processes, the entropy of an isolated system always increases. Clausius defined entropy S in macroscopic terms through the equation

$$\mathrm{d}S = \frac{\mathrm{d}Q}{T}.$$

Then in 1877, Boltzmann made a remarkable advance. He established the relationship between the macroscopically observable entropy S of a system and the microscopic properties of that system. Boltzmann's equation is

$$S = k \ln w, \tag{7.83}$$

where S is the entropy of a system in a particular macrostate, w is the multiplicity of that macrostate, and k is the Boltzmann constant. The importance of the equation is that it relates a macroscopic quantity that we can measure to the microscopic properties of a system. This famous equation is engraved on Boltzmann's gravestone.

It is not surprising that entropy S and multiplicity w are related by a logarithmic function. The total entropy of two systems is the *sum* of their separate entropies:

$$S = S_1 + S_2.$$

And, as we have seen, we can think of w as representing the probability for a system being in particular macrostate of the system. The probability of the occurrence of two independent systems is the *product* of their separate probabilities:

$$w = w_1 \times w_2.$$

Accordingly, the entropy is proportional to the logarithm of the probability, as in Boltzmann's equation.

Boltzmann's equation shows that entropy can never be negative. It also shows that the smallest possible value of S is zero. We can use Equation (7.83) to compute the entropies of the various possible macrostates for our example of six molecules in a box. Their values are presented in Table 7.1, where we see that the macrostate with $n = 3$, corresponding to three molecules in each half of the box, has the greatest entropy, and also the greatest multiplicity.

We can also use Equation (7.83) to determine entropy differences between one state and another. If we have a system that changes from a state with multiplicity w_1 to one with multiplicity w_2, the change in entropy is

$$\Delta S = S_2 - S_1 = k \ln w_2 - k \ln w_1 = k \ln \frac{w_2}{w_1}. \tag{7.84}$$

Probability and the non-reversibility of natural processes

The statistical approach to the second law provides an explanation for the non-reversibility of natural processes. It shows that this apparent non-reversibility is not due to any natural law, but simply that the probability of the reverse process occurring is negligibly small. Thus, when an isolated system undergoes a natural, spontaneous process, it moves to a macrostate that has a greater number of microstates and hence a greater probability of occurring. As the number of microstates increases, so too does the entropy of the system according to Boltzmann's equation (7.83), in agreement with Clausius' principle of the increase in entropy. On the other hand, the system can essentially never spontaneously undergo a process to a macrostate in which the number of possible microstates is reduced because the probability reduces with the decrease in the number of microstates. We may say that Nature prefers the more probable macrostates to the less probable and so natural processes take place in the direction of greater probability.

Change of entropy in the free expansion of a gas

In Section 7.6, we used classical thermodynamics to determine the entropy change in the free expansion of a gas. We now consider this change using the approach of statistical mechanics. To do this, we consider two configurations of a gas in a container. In the first configuration, we consider all the gas to be confined to one half of the container. In the second, we consider the gas to fill the whole container. We determine the multiplicity of these two configurations and their respective entropies to obtain the difference between the two entropies.

Initially, all N molecules are in one half of the container and none in the other, with $n_1 = N$ and $n_2 = 0$. Hence, the initial multiplicity is

$$w_i = \frac{N!}{n_1!\,n_2!} = \frac{N!}{N!\,0!} = 1.$$

From Equation (7.83), the initial entropy is therefore

$$S_i = k \ln 1 = 0.$$

Finally, the N molecules occupy the full volume of the container, and $n_1 = n_1 = N/2$. This gives the final multiplicity

$$w_f = \frac{N!}{(N/2)!\,(N/2)!},$$

and hence, the final entropy is

$$S_f = k \ln w_{\mathrm{f}} = k \ln(N!) - 2k \ln[(N/2)!]. \tag{7.85}$$

N, the number of molecules will be a very large number and $N!$ will be an unmanageably large number. However, we can deal with $N!$ using *Stirling's approximation*. For large N, Stirling's approximation is

$$\ln N! = N \ln N - N,$$

Then using Stirling's approximation in Equation (7.85), we find

$$\begin{aligned} S_f &= k[N(\ln N) - N] - 2k[(N/2)\ln(N/2) - (N/2)] \\ &= k[N(\ln N) - N - N\ln(N/2) + N] \\ &= k[N(\ln N) - N(\ln N - \ln 2)] = Nk \ln 2. \end{aligned}$$

For one mole of gas $N = N_{\mathrm{A}}$, and substituting for $N_{\mathrm{A}}k = R$, where R is the gas constant, we obtain

$$S_f = R \ln 2.$$

The change in entropy between the initial and final configurations of the m moles gas is therefore

$$\Delta S = S_f - S_i = R \ln 2 - 0 = R \ln 2.$$

This result is in accord with Equation (7.32) that was obtained using classical thermodynamics.

Problems 7

7.1 An ideal heat engine takes 2000 J of heat from a hot reservoir at 207 °C, does some work, and deposits some heat to a cold reservoir at 15 °C. How much work does it do, how much heat is discarded and what is the efficiency of the heat engine?

7.2 (a) The temperature of 1 mol of a monatomic gas is raised reversibly from 300 to 400 K with its volume kept constant. What is the change in entropy of the gas? What is the change in entropy of the gas if the pressure is kept constant?

(b) A copper block of mass 1 kg is heated from 300 to 400 K. What is its change of entropy? The specific heat of copper is 385 J/kg · K.

7.3 One mole of nitrogen gas is shaken vigorously in a thermally insulated flask so that the temperature of the gas increases from 22 to 50 °C. Does the entropy of the gas increase? If so, by how much? How much work is done?

7.4 An ice cube of mass 35 g and at temperature 0 °C is put into a large bucket of water whose temperature is infinitesimally above 0 °C. Calculate the change in entropy of the ice cube and of the water in the bucket and the overall change in entropy. The latent heat of fusion of water is 333×10^3 J/kg · K.

7.5 An ice cube of mass 35 g and at temperature −10 °C is put into a large bucket of water whose temperature is 15 ° C. Calculate the change in entropy of the ice cube and the water in the bucket and the overall change in entropy. The latent heat of fusion of water is 333×10^3 J/kg, the specific heat of ice is 2.2×10^3 J/kg · K and the specific heat of water is 4.2 kJ/kg · K

7.6 Two ice cubes of mass 25 g at 0 °C are rubbed together until completely melted. The water is then heated up to 20 °C by being placed in contact with a reservoir at 20 °C. Calculate the change in entropy of the ice cubes plus melted water, of the reservoir, and of the universe. The latent heat of fusion of ice is 333×10^3 J/kg, and the specific heat of water is 4.2 kJ/kg · K.

7.7 1 kg of water at 20 °C is heated to 80 °C in six stages by successively placing the water in contact with heat baths at temperatures of 30 °C, 40 °C, 50 °C, 60 °C, 70 °C, and 80 °C, respectively, allowing equilibrium to be attained at each stage. Determine the total entropy change of the water plus the six heat baths. The specific heat of water is 4.2 kJ/kg · K.

7.8 Ocean thermal energy conversion uses ocean thermal gradients between cooler deeper water and warmer shallow or surface water to run a heat engine. It offers a continuously available, renewable energy source. In parts of the world, temperature differences of 20 °C or more are available. Estimate the maximum thermodynamic efficiency of such a system for $T_H = 300$ K, $T_C = 280$ K. If the heat engine is to provide 5 kW of work, calculate the rate at which the waste heat must be removed for the engine to operate at maximum efficiency.

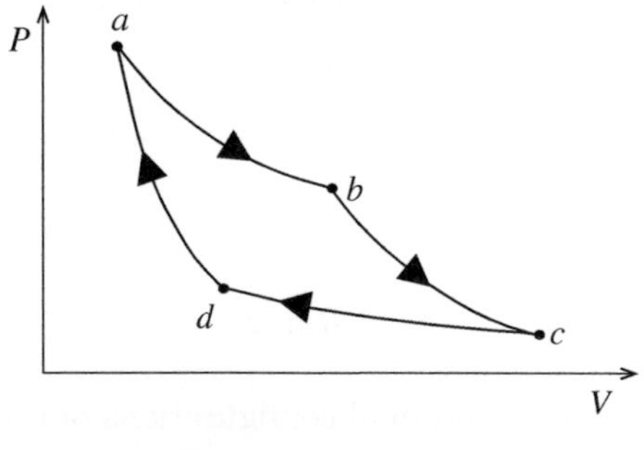

7.9 0.5 mol of an ideal diatomic gas is taken through a Carnot cycle as in the figure, which is not to scale. The temperatures of the hot and cold reservoirs are 227 °C and 27 °C, respectively. The pressure P_a at point a in the P–V diagram is 10 atm. The volume is doubled during the isothermal expansion $a \rightarrow b$. (a) Calculate the pressure and volume of the gas at points b, c, and d. (b) Find Q, W, and ΔU for each of the four steps and for the complete cycle. (c) Find the efficiency of the heat engine from your results for Q and W. Compare this value with the value obtained from the equation for the efficiency in terms of the temperatures of the hot and cold reservoirs.

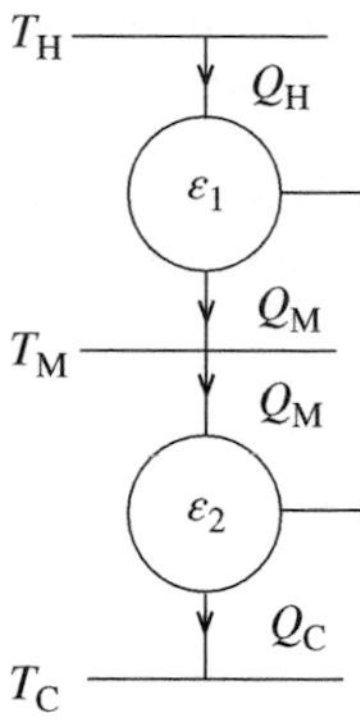

7.10 The figure illustrates two ideal heat engines connected in series, where the heat discarded by the first is used as the heat input for the second. The efficiencies of the two engines are ε_1 and ε_2, respectively. (a) Show that the overall efficiency ε of the combined engines is given by

$$\varepsilon = \varepsilon_1 + (1 - \varepsilon_1)\varepsilon_2.$$

(b) The first engine operates between temperatures T_H and T_M and the second operates between temperatures T_M and T_C. Show that the combination of the two heat engines is equivalent to a single engine operating between temperatures T_H and T_C.

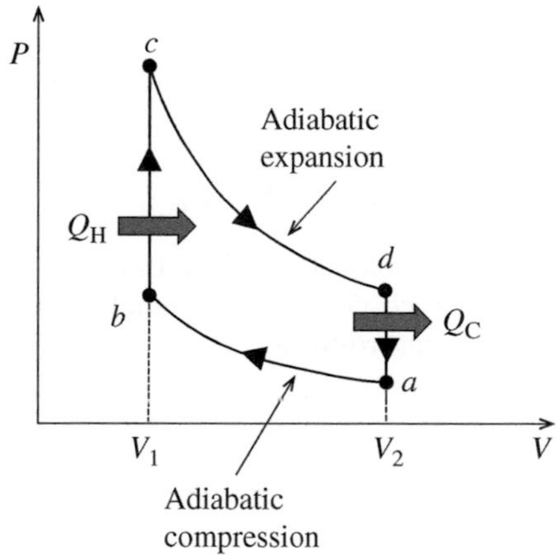

7.11 The figure shows a P–V diagram for the *Otto cycle*, which is an idealised model of a petrol engine. The petrol enters the cylinder at point a and is ignited at point b. The four stages of the cycle are:

1. $(a \to b)$ adiabatic compression from volume V_2 to V_1.
2. $(b \to c)$ pressure increases at constant volume V_1.
3. $(c \to d)$ adiabatic expansion from volume V_1 to V_2: the power stroke.
4. $(d \to a)$ pressure reduces at constant volume V_2.

Determine the efficiency of the cycle in terms of the compression ratio $r = V_2/V_1$. Hence, deduce the maximum efficiency of a petrol engine that has a compression ratio $r = 8$. Take $\gamma = 1.3$ for air/petrol vapour.

7.12 Use the second law, in the form of entropy increase, to show that the efficiency of a heat engine cannot be greater than $(1 - T_C/T_H)$, and that the coefficient of performance of a heat pump cannot be greater than $(1 - T_C/T_H)^{-1}$, where T_H and T_C are respectively the temperatures of the hot and cold reservoirs involved.

7.13 A system in thermal equilibrium at temperature T consists of N particles, which have two energy levels that are separated by energy ε. The energy of the lower state can be taken to be zero. (a) Obtain expressions for the number of particles in each state at temperature T. What will be the populations of the two levels when (i) $T \rightarrow 0$, and (ii) $T \rightarrow \infty$? (b) What is the total energy E of the system at temperature T? What is the total energy for (i) $T \rightarrow 0$, and $T \rightarrow \infty$? (c) Show that the specific heat $\mathrm{d}E/\mathrm{d}T$ of the system is $(N\varepsilon^2)/(2kT^2)$, at high temperature, i.e. when $T \gg \varepsilon/k$, where k is the Boltzmann constant.

7.14 One mole of a monatomic gas in equilibrium is well described by the van der Waals equations:

$$\left(P + \frac{a}{V^2}\right)(V - b) = RT, \quad U = \frac{3}{2}RT - \frac{a}{V},$$

where a and b are constants. Using the fundamental thermodynamic relationship, obtain an expression for the entropy of the gas as a function of volume V and temperature T.

7.15 Tin undergoes a phase transition from its grey form to its white, metallic form at 18 °C and 1 atm. The latent heat for this transition is 2.2×10^3 kJ/mol. What is the transition temperature at a pressure of 50 atm? The densities of grey and white tin are $5.75 \times 10^3\ \mathrm{kg/m^3}$ and $7.30 \times 10^3\ \mathrm{kg/m^3}$, respectively. The atomic weight of tin is 118.7 u.

7.16 Starting with the definition of enthalpy H, obtain the Maxwell Equation

$$\left(\frac{\partial T}{\partial P}\right)_S = \left(\frac{\partial V}{\partial S}\right)_P.$$

Starting with the definition of Helmholtz free energy F, obtain the Maxwell Equation

$$\left(\frac{\partial P}{\partial T}\right)_V = \left(\frac{\partial S}{\partial V}\right)_T.$$

7.17 Consider the situation where we have four molecules in a box. List the possible ways these four molecules can be arranged in the two halves of the box. Hence, deduce (i) the number of macrostates, (ii) the multiplicity of the macrostates, and (iii) the entropy of the macrostates.

8

Solids

In Chapter 2, we discussed why a substance occurs in gaseous, liquid, or solid form. We saw that it was due to a competition between the binding energy of the constituent molecules and their thermal, kinetic energy. In gases, the kinetic energy dominates and the molecules are essentially free to move around their container, unaffected by their neighbours except for elastic collisions. Solids lie at the other extreme. In solids, the binding energy dominates and the molecules or atoms are tightly bound and closely packed together rigid. This results in the most characteristic property of solids. They have appreciable stiffness and maintain their shape.

Solids appear in a wide variety of forms. Of these, crystalline solids provide an ideal form to understand the structure and properties of solids. This is because of their high degree of regularity; a reoccurring pattern of atomic positions that extends over many atoms. Consequently, we will focus most of our attention on the crystalline state of matter. Nearly everything we know about crystal structures has been learnt from diffraction experiments. In this chapter, we introduce the principles of X-ray crystallography and how it is used to determine crystal structure. We also relate the properties of solids to the forces acting between their constituent atoms. This follows on from our discussion of interatomic forces in Chapter 2.

8.1 Types of solids

Solids may be classified as *crystalline*, *amorphous*, or *polymeric*. We may distinguish between crystalline and amorphous solids as follows. At sufficiently low temperatures, most substances will condense to form a solid. If the substance is cooled sufficiently slowly, the atoms have time to arrange themselves into a regular array with *long-range order*. By this, we mean that there is a well-defined spatial relationship between atoms that are far from each other, i.e. much further than the mean distance between the atoms. Metals and rare gas solids are examples of crystalline substances. If, on the other hand, a substance is cooled very rapidly, the atoms may not have sufficient time to arrange themselves into an extended, regular array. The resulting solid is described as amorphous. The atoms in an amorphous solid are again closely packed together and are bound to their nearest neighbours, resulting in a rigid shape. But it is a *short-range order* extending over a relatively few atoms. In the same way that liquids have no long-range order, we may think

Physics of Matter, First Edition. George C. King.

Figure 8.1 Pure, single crystals of silicon like the one shown here are grown for the semiconductor industry and have a length greater than 1 m. They are sliced into thousand of thin wafers onto which the layers of an integrated circuit are etched. Source: Targray Technology International Inc.

of the structure of an amorphous solid to be like an instantaneous snapshot of a liquid. Indeed, amorphous structures are sometimes referred to as super-cooled liquids. Glass is an example of an amorphous solid.

The long-range order of crystals may extend over an enormous number of atoms. A vivid example of this is the high-purity silicon crystals that are manufactured for the semiconductor industry. These single crystals have a length greater than 1 m; see Figure 8.1. Along the length of such a crystal, there will be $\sim 3 \times 10^8$ silicon atoms in a regular array, in which the arrangement of the last few atoms is in perfect step with the first. Crystals may also appear naturally as large single crystals, as, for example, a calcite crystal or in a diamond ring. However, most common solids are a composite of very many small crystals in which case they are called *polycrystalline*. For example, if a rock is broken into two, we may see that the rock is formed of many small crystals, less than 1 mm in size.

The essential feature of polymeric solids or *polymers* is that they typically consist of long chains of molecules. These chains are formed by the coming together of many smaller subunits called *monomers*, which typically bind together by covalent bonds. It is this tendency of the monomers to link up and form long chains that lead to the special properties of polymers. Wood and rubber are examples of polymers that occur naturally, while proteins are an example of biological polymers or *biopolymers*, which are fundamental to biological structure and function. The early study of polymers was directed towards naturally occurring polymers. However, it was discovered that monomers could be combined in ways that did not occur naturally to produce tailor-made artificial polymers. This led to a huge industry that now produces many millions of tonnes annually of plastics, synthetic fibres, and related products. An example is polyethylene, also known as polythene. This polymer has the structure shown in Figure 8.2. The repeating monomer is CH_2 while the chain is terminated by hydrogen atoms. Most of the polymers that are encountered in everyday life contain the elements carbon and hydrogen.

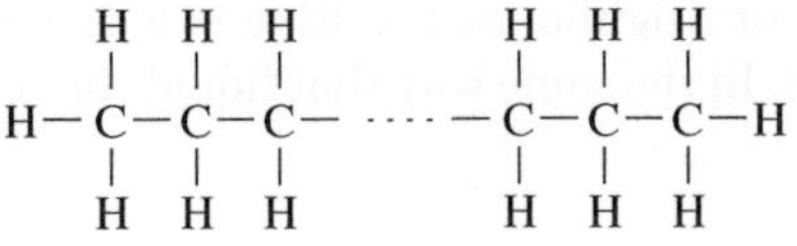

Figure 8.2 Polyethylene is an example of a polymer. Its structure consists of a chain of CH_2 monomers that is terminated by hydrogen atoms.

8.2 Crystal structure

The external forms of naturally occurring crystals such as calcite give an indication of the regularity of the internal arrangement of atoms in a crystal. Crystals are observed to have relatively flat surfaces with sharp edges called *crystal faces* and it is found that the angles between the various faces are well defined. And if a crystal is cleaved, it is found that the planes of fracture are parallel to the outer crystal faces, again indicating a regularity to the structure of the crystal. Robert Hooke was one of the scientists who studied the shapes of natural crystals. He concluded that a crystal must owe its regular shape to the systematic packing of minute spherical particles. Perhaps he was guided by the arrangement of a pile of cannon balls, like that shown in Figure 8.3. Indeed, the forces between the atoms in a solid are such that, to a good approximation, the atoms do behave like attracting rigid spheres. Such a picture of the regular arrangement of minute spherical particles is the starting point of our discussion of crystals.

8.2.1 Close packing of atoms in a crystal

The atoms in a solid attract one another until their outer electron shells start to overlap significantly and so indeed they behave like rigid spheres. Consequently, the separation between the centres of the atoms, i.e. the nucleus–nucleus separation, is about the same as the diameter of the atoms themselves, $\sim$0.1 nm. In order to minimise the total potential energy of the crystal, the atoms are packed together as closely as possible. A close-packed layer of atoms can be formed as shown in Figure 8.4. This arrangement of atoms gives the maximum utilisation of space. We will refer to this layer as layer A. A second close-packed layer (layer B) can be formed by placing atoms in the triangular depressions of layer A as shown in Figure 8.5. In this way, each atom in layer B will touch three of the atoms in layer A and again the packing will be as close as possible. The atoms in layer B have been given a slightly lighter shade for the sake of clarity.

A third layer of atoms (layer C) can be added in two distinct ways, both of which result in the same degree of close packing. To illustrate the two ways, Figure 8.5 shows two different sets of triangular depressions in layer B, indicated by red and black dots, respectively. The red dots lie directly above the atoms in layer A. If the atoms in layer C are placed over the red dots, they reproduce the arrangement of atoms in layer A. Hence, the resulting pattern is described as ABA. This pattern is shown in Figure 8.6. The black dots lie directly above those triangular depressions in layer A that are not occupied

Figure 8.3 A pile of cannon balls provides a clue to the arrangement of atoms in a crystal. Source: Markobe/ Adobe Stock.

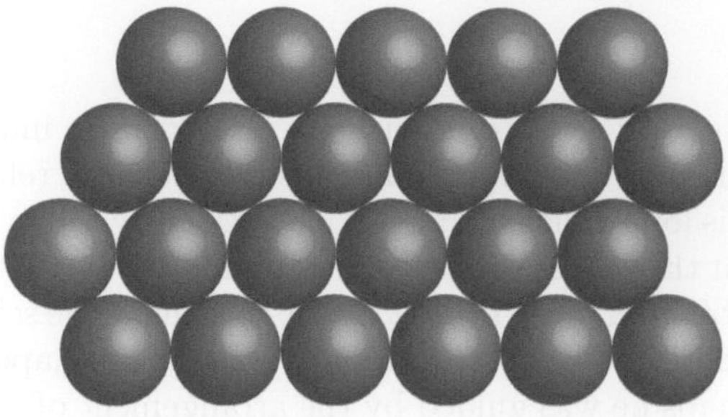

Figure 8.4 This figure shows how a close-packed layer of atoms (layer A in the text) can be formed, giving the maximum utilisation of space.

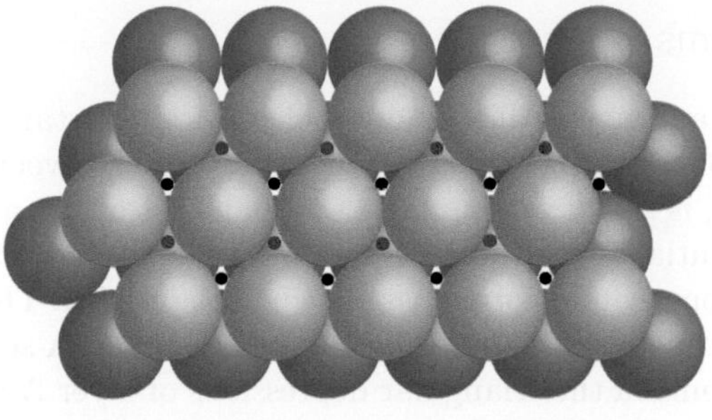

Figure 8.5 This figure shows how a close-packed layer of atoms (layer B in the text) can be added to layer A, shown in Figure 8.4. It also shows two different sets of triangular depressions in layer B. The red dots lie directly above the atoms in layer A. The black dots lie directly above those triangular depressions in layer A that are not occupied by atoms in layer B.

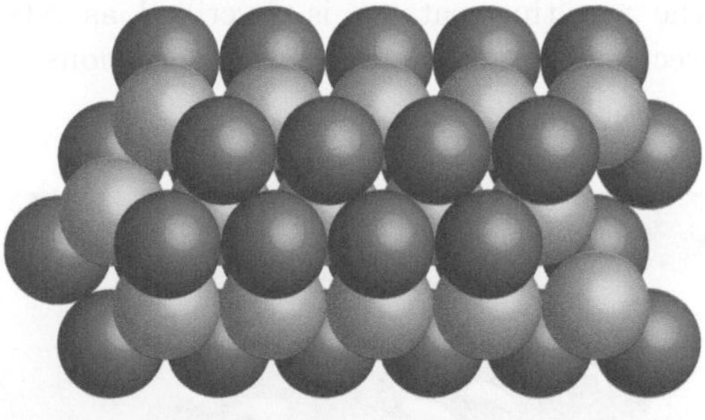

Figure 8.6 This figure shows one way a layer of atoms (layer C in the text) can be added to layer B, shown in Figure 8.5, where the atoms in layer C are placed over the red dots shown in Figure 8.5. This reproduces the arrangement of atoms in layer A. The resulting pattern is described as ABA.

by atoms in layer B. Placing the atoms in layer C above the black dots gives the pattern that is shown in Figure 8.7 and is called the ABC pattern.

The number of neighbouring atoms immediately surrounding an atom is called the *coordination number* n. For both close-packed patterns discussed earlier, each atom has 12 nearest neighbours. This is the

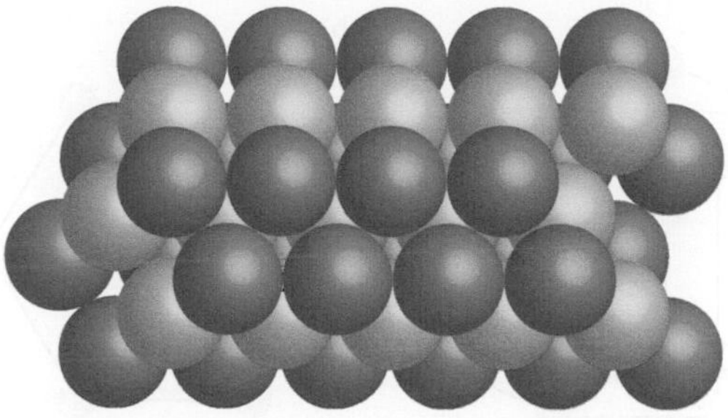

Figure 8.7 This figure shows a second-way layer C can be added to layer B. In this case, the atoms in layer C are placed over the black dots shown in Figure 8.5. The resulting pattern is described as ABC.

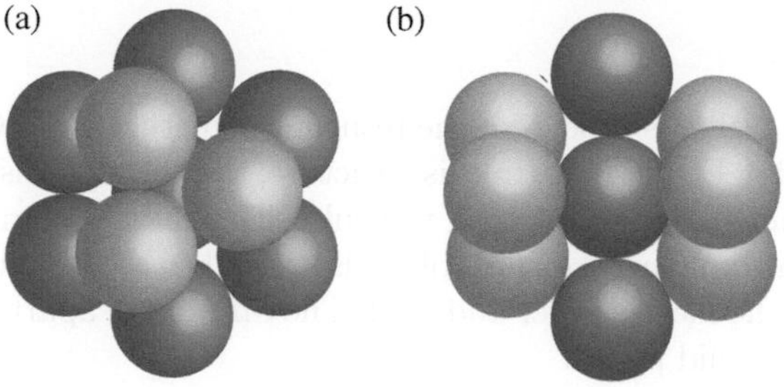

Figure 8.8 (a) The view looking down on a close-packed layer of atoms. (b) A side view of the close-packed layer. Each atom has 12 nearest neighbours, which is the maximum number that is possible.

maximum number that is possible and the atoms are packed together with the least waste of space. Figure 8.8a is the view looking down on a close-packed layer of atoms while Figure 8.8b is a side view. The *packing fraction* is the fraction of space occupied by the atoms. For both structures, this is 0.74, which means that 74% of the available space is filled. The fact that so many metals are close-packed accounts for their common characteristic of high density.

8.2.2 Some common crystal structures

A number of crystal structures can be described using the concept of close packing. In particular, the two kinds of stacking patterns described above lead to two different types of crystal structure: *hexagonal close-packed* (hcp) and face-centred cubic (fcc), also called *cubic close packed* (ccp). About 45% of the elements in the periodic table adopt either one or the other of these two kinds of structure.

Hexagonal close-packed structure

If the ABA pattern described above is repeated, the resulting structure is ABABABA.... This produces the hcp structure illustrated in Figure 8.9. The atoms with the lighter shading match those in Figure 8.6. To illustrate the hcp structure more clearly, and also the other structures that will be presented, the atoms are not drawn to scale; as noted earlier, in practice the atoms are essentially touching each other. From the six-fold symmetry of the hcp structure, we can readily see why it is called hexagonal. Crystals with the hcp structure include the metals beryllium and cadmium.

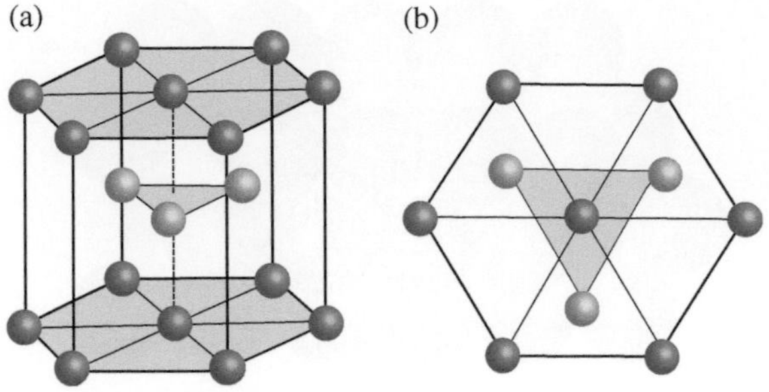

Figure 8.9 (a) The hexagonally close-packed (hcp) structure resulting from the ABAB ... pattern of atomic layers. (b) The six-fold symmetry gives the structure its name.

Face-centred cubic structure

If the ABC pattern described earlier is repeated, the resulting structure is ABCABC.... This produces the fcc structure that is illustrated in Figure 8.10. In this structure, there are atoms at each corner of a cube and also atoms at the centre of each of the six faces of the cube. The shaded plane shows the orientation of a close-packed layer of atoms in the structure. This plane is normal to a body diagonal of the cube. Crystals with the fcc structure include the metals silver and gold. The rare gases, apart from helium, also adopt this structure when they are in the solid phase.

Body-centred cubic structure

The ABABAB ... and ABCABC ... structures exhaust the ways we can combine close-packed layers of atoms. A slightly less packed structure than the hcp and fcc structures is the *body-centred cubic structure* (bcc). In this structure, there are atoms at each corner of a cube and an atom at the centre of the cube, as

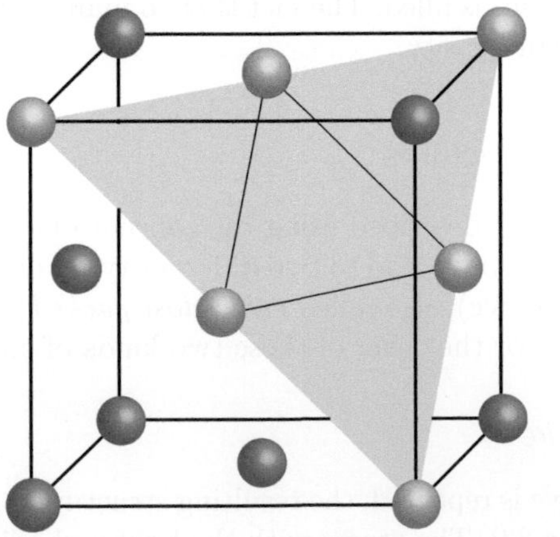

Figure 8.10 The face-centred cubic (fcc) structure resulting from the ABCABC ... pattern. In this structure, there are atoms at each corner of a cube and also atoms at the centre of each of the six faces of the cube. The shaded plane shows the orientation of a close-packed layer of atoms in the structure. This plane is normal to a body diagonal of the cube.

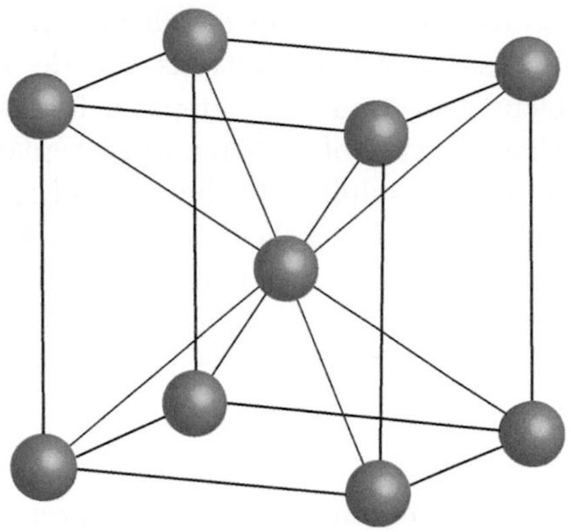

Figure 8.11 The body-centred cubic structure (bcc). In this structure there are atoms at each corner of a cube and an atom at the centre of the cube. An atom thus has eight nearest neighbours, with coordination number $n = 8$. Notable examples that have the bcc structure are the alkali metals, Li, Na, K, Rb and Cs.

shown in Figure 8.11. An atom thus has eight nearest neighbours, with coordination number $n = 8$. Notable examples that have the bcc structure are the alkali metals, Li, Na, K, Rb, and Cs.

Diamond and zinc blende structures

Carbon has four electrons in its outermost shell, and a carbon atom can form a covalent bond with each of the four adjacent carbon atoms. These covalent bonds are strongly *directional* and this leads to the crystal structure of diamond; a carbon atom with adjacent carbon atoms at the corners of a regular tetrahedron. The diamond structure is illustrated in Figure 8.12. We distinguish two different types of carbon atoms, as indicated by the colours blue and green in the figure. They differ only in the orientation of the bonds to the

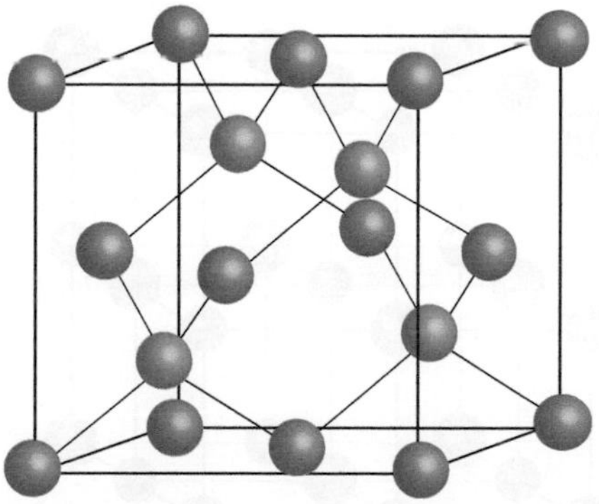

Figure 8.12 This figure shows the diamond/zinc blende structure. In a diamond, a carbon atom has four adjacent carbon atoms that are situated at the corners of a regular tetrahedron. There are two different types of carbon atoms, as indicated by the colours blue and green in the figure. They differ only in the orientation of the bonds to the nearest neighbours. The zinc blende structure differs from the diamond structure only in that one type of carbon atom is replaced by one type of atom and the other by a different type of atom. In the case of zinc sulphide, the zinc atoms are represented in green and the sulphur atoms in blue.

nearest neighbours. The coordination number is 4, which indicates that this structure is far from being close-packed. Silicon and germanium also have four electrons outside a closed shell and they also adopt the diamond structure. The zinc blende structure, as adopted, for example, by zinc sulphide, differs from the diamond structure only in that one type of carbon atom is replaced by one type of atom, e.g. zinc, and the other by a different type of atom, e.g. sulphur. In the case of Figure 8.12, the zinc atoms are represented by the green spheres and the sulphur atoms by the blue spheres. The zinc blende structure is adopted by some important semiconducting compounds such as gallium arsenide GaAs, which explains its importance in Solid State Physics.

8.2.3 Ionic crystals

Ionic crystals consist of positively and negatively charged ions of different elements. A sodium chloride crystal, for example, contains an equal number of positively charged Na^+ ions (cations) and negatively charged Cl^- ions (anions). These ions again approximate to attracting rigid spheres and so ionic crystals are usually dominated by close-packing considerations. The structure of a NaCl crystal is shown in Figure 8.13. We see that each ion has six nearest neighbours. We can consider this structure to be two interlocking fcc structures of Na^+ and Cl^- ions, in which these ions are mutually displaced by one half the edge width of the cube. We can see this more clearly by hiding the Na^+ ions as in Figure 8.14. An alternative structure for ionic crystals is exemplified by the structure of caesium chloride, which is shown in Figure 8.15. This structure contains equal numbers of Cl^- and Cs^+ ions, which form a bcc structure. A Cs^+ ion is at the centre of the cube, while there is a Cl^- ion at each of the corners. Both the Cl^- and Cs^+ ions are surrounded by eight ions of the opposite charge.

Effect of the relative sizes of the ions in a crystal

We see that different ionic substances adopt different crystal structures, as exemplified by caesium chloride and sodium chloride, which have coordination numbers of 8 and 6, respectively. The crystal structure a substance adopts is a subtle balance between the energy of interaction of the ions and the relative sizes

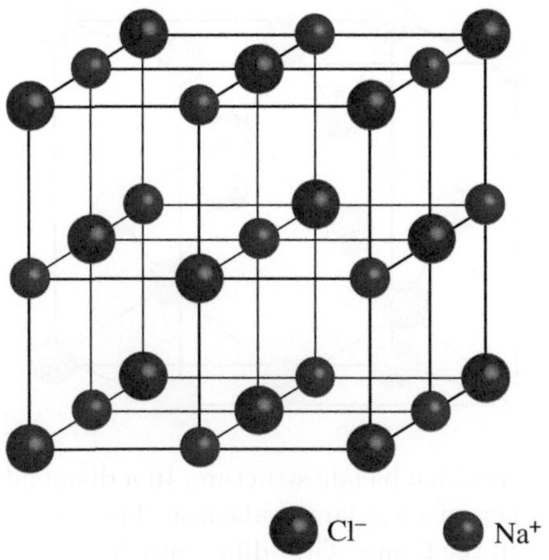

Figure 8.13 The structure of the sodium chloride crystal. Each ion has six nearest neighbours. This structure can be considered to be two interlocking fcc structures of Na^+ and Cl^- ions, in which these ions are mutually displaced by one half the edge width of the cube.

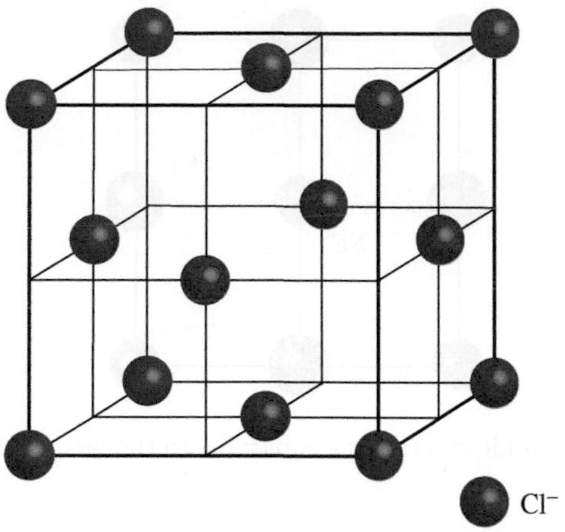

Figure 8.14 In this figure, the Na^+ ions of the sodium chloride crystal are hidden so that the fcc structure of the Cl^- ions can be seen more clearly.

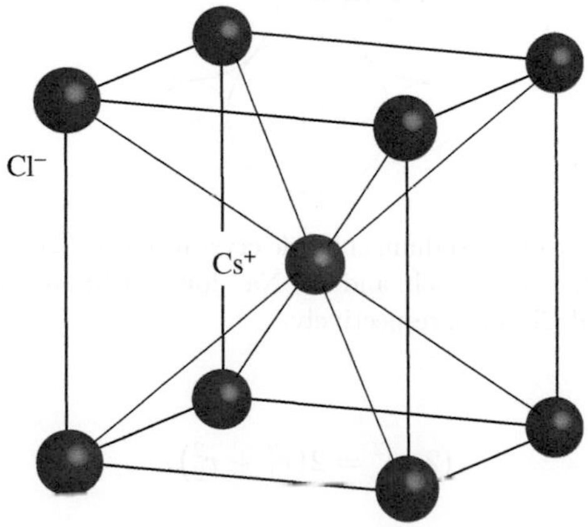

Figure 8.15 The structure of caesium chloride. This structure contains equal numbers of Cl^- and Cs^+ ions that form a bcc structure. A Cs^+ ion is at the centre of the cube, while there is a Cl^- ion at each of the corners. Both the Cl^- and Cs^+ ions are surrounded by eight ions of the opposite charge.

of the cation and anion. The electrons in the negatively charged anions, such as the Cl^- ions in sodium chloride, are, in general, less tightly bound than those in the positively charged cation, the Na^+ in this case. Consequently, the anion is usually larger. The structure of an ionic crystal is then likely to be determined by the number of larger anions that will pack around the smaller cation.

To illustrate the role played by the relative sizes of the cation and anion, we consider the case of the sodium chloride crystal. Figure 8.16 shows one of the planes of the crystal that is parallel to the face of the cubic structure. Suppose we have a situation where adjacent Cl^- ions along an edge of the cube touch, and the Na^+ ion just fits in the space between the Cl^- ions. This arrangement is illustrated by Figure 8.17, where we take the radii of the Na^+ and Cl^- ions to be r_1 and r_2, respectively. Then, by the Pythagoras theorem, we have

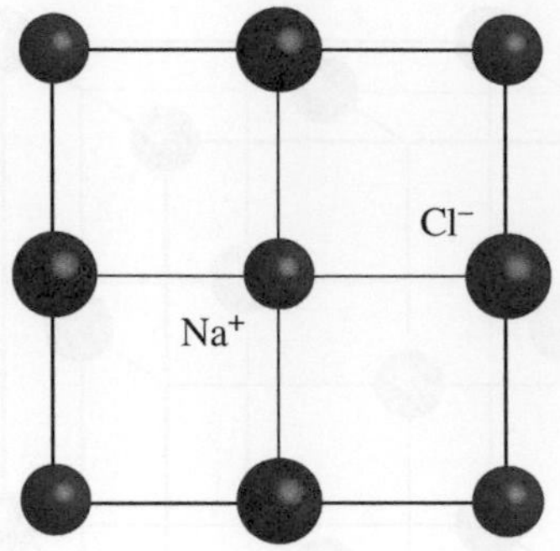

Figure 8.16 A plane of the sodium chloride crystal that is parallel to the face of the cubic structure of the crystal.

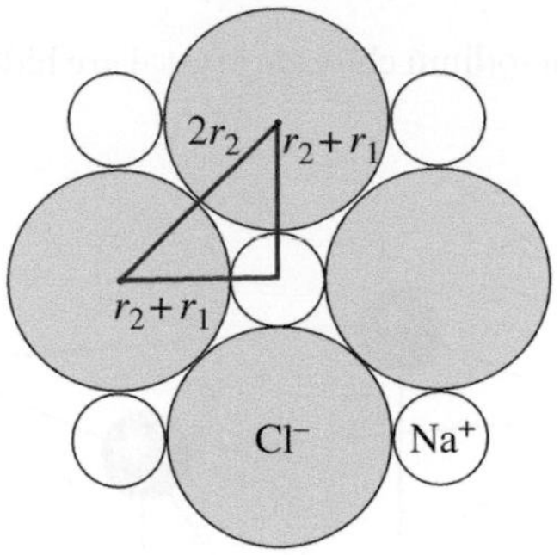

Figure 8.17 A representation of a plane of the sodium chloride crystal that is parallel to the face of the cubic structure. Adjacent Cl^- ions along an edge of the cube touch, and the Na^+ ion just fits in the space between the Cl^- ions. r_1 and r_2 are the radii of the Na^+ and Cl^- ions, respectively.

$$(2r_2)^2 = 2\left(r_1^2 + r_2^2\right).$$

This gives $2r_2 = \sqrt{2}(r_1 + r_2)$, from which

$$\frac{r_1}{r_2} = \sqrt{2} - 1 = 0.414.$$

This simple geometric consideration gives a certain limit to the ratio of radii of the two ions. If the ratio were any smaller than 0.414, the Na^+ ion would 'rattle' in the cage formed by the Cl^- ions and would not be touching them. This suggests that ionic crystals that have the sodium chloride structure have values of $r_1/r_2 > 0.414$. In fact, this is generally borne out in practice. For example, NaCl, NaI and KBr have the sodium chloride structure and have the values for r_1/r_2 of 0.56, 0.46, and 0.68, respectively.

As the size of the cation increases relative to that of the anion, it becomes possible for the cation to accommodate more nearest neighbours and then these ionic crystals adopt the caesium chloride crystal structure. Similar geometric considerations to the above for the caesium chloride crystal give the limit $r_1/r_2 > 0.732$. And indeed, it is found that ionic crystals with larger values than 0.732 adopt the caesium chloride structure. For example, the CsCl, CsBr, and CsI all adopt the caesium chloride crystal structure and have values of the ratio of 0.93, 0.87, and 0.78.

Worked example

What are the relative sizes of the radii of the Cs^+ and Cl^- ions for which the Cs^+ ion just fits between the Cl^- ions in the CsCl crystal?

Solution

Figure 8.18 shows the relevant plane of the CsCl crystal that lies diagonally across the cubic structure. This plane contains Cl^- ions at the corners of the cube at positions A, B, C, and D. The Cs^+ ion lies in the middle of this plane. If the side of the crystal cube has length a, the distance between B and C is a, the distance between C and D is $\sqrt{2}\,a$, and the distance between D and B is $\sqrt{3}\,a$. The plane of interest is also shown in Figure 8.19. There we see that the ions at A and D will touch when $2r_2 = a$, where r_2 is the radius of the

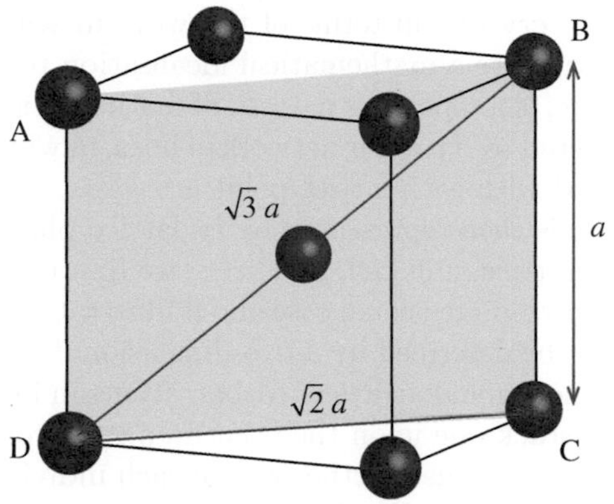

Figure 8.18 A plane of the caesium chloride crystal that lies diagonally across the cubic structure. This plane contains Cl^- ions at the corners of the cube at positions A, B, C and D, while the Cs^+ ion lies in the middle of this plane. The side of the crystal cube has length a.

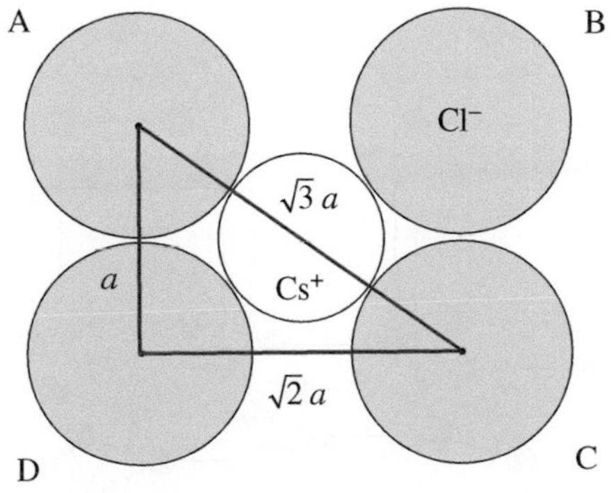

Figure 8.19 A representation of a plane of the caesium chloride crystal that lies diagonally across the cubic structure. a is the length of the cube side, r_1 is the radius of the Cs^+ ion and r_2 is the radius of the Cl^- ion. The Cl^- ions at A and D touch when $2r_2 = a$.

Cl^- ion. Then, the separation of ions C and D will be $\sqrt{2}\,a = \sqrt{2} \times 2r_2$, and the separation of ions at A and C will be

$$\sqrt{3}\,a = \sqrt{3} \times 2r_2 = 2(r_1 + r_2),$$

where r_1 is the radius of the Cs^+ ion. Therefore,

$$r_1 = r_2\left(\sqrt{3} - 1\right), \text{and } \frac{r_1}{r_2} = 0.732.$$

8.3 The crystal lattice, unit cell, and basis

We have described the structure of crystals in terms of the ways in which the constituent atoms pack together in the crystal. Now we introduce a mathematical idealisation to describe crystal structure. This idealisation, called the crystal *lattice*, is a repeating pattern of mathematical points that extends throughout space. This lattice can be connected by a regular network of lines, in which the lattice is broken up into a number of *unit cells*. And associated with each lattice point is a *basis*, which may represent a single or a group of atoms. A crystal structure is then represented by its lattice plus the associated basis.

To describe the concepts of the lattice, unit cell, and basis, we first consider the two-dimensional case. This is easier to visualise than the three-dimensional case and it illustrates all three concepts. Graphite provides us with a real crystal that can be described by a two-dimensional lattice. Graphite is a form or *allotrope* of carbon in which there are hexagonal arrays of carbon atoms in individual layers, as illustrated in Figure 8.20. The individual layers stack one upon the other in a series of equally spaced parallel layers. Strong covalent bonds bind the carbon atoms together within each individual layer, while adjacent layers are bonded together by relatively weak van der Waals forces. This means that the layers can easily slide past each other, which is the reason that graphite can be used as a lubricant.

The arrangement of carbon atoms in one layer of a graphite crystal is illustrated in Figure 8.21. The atoms are situated at the corners of regular hexagons. To describe a crystal structure, we need to establish a set of coordinate axes for the crystal. It is usual to cite the origin of the coordinate system on one of the

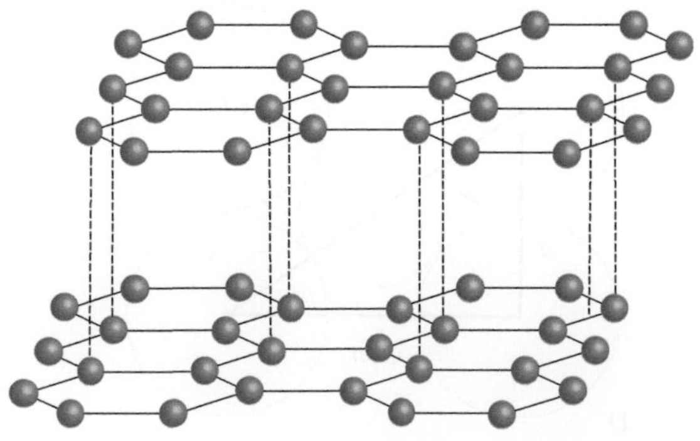

Figure 8.20 Graphite is a form of carbon in which there are hexagonal arrays of carbon atoms in individual layers. The individual layers stack one upon the other in a series of equally spaced parallel layers. Strong covalent bonds bind the carbon atoms together within each individual layer, while adjacent layers are bonded together by relatively weak van der Waals forces.

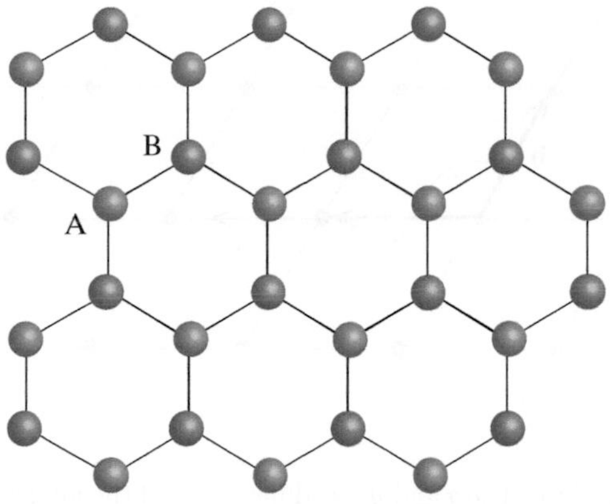

Figure 8.21 The arrangement of atoms in one layer of a graphite crystal. The atoms are situated at the corners of regular hexagons. Atom A is at the origin of the coordinate system that is used to describe the crystal structure. The positions occupied by all the carbon atoms shown in green are identical to that at the origin. On the other hand, the positions occupied by those carbon atoms in blue are not. The set of identical positions identified in this way, i.e. the positions of the carbon atoms shown in green are the lattice points and define the crystal lattice.

atoms, say atom A in Figure 8.21. We then identify all the positions in the crystal that are identical in all respects to the origin. This means that an observer situated at any of those positions has exactly the same view as an observer at the origin. The positions occupied by all the carbon atoms shown in green are identical to that at the origin. On the other hand, the positions occupied by those atoms in blue, for example, atom B, are not; the neighbouring atoms around atom B do not lie in the same directions as those around atom A. The set of identical positions identified in this way, i.e. the positions of the carbon atoms shown in green are the *lattice points* and define the crystal lattice.

The crystal lattice of graphite is shown in Figure 8.22. The x and y coordinate axes of the lattice are obtained by joining the lattice point at the origin, atom A, to *two* of the neighbouring lattice points. Note that the coordinate axes of a crystal do not need to be orthogonal. The distances and directions of the nearest lattice points along the x and y axes are identified by the *lattice vectors* **a** and **b**, respectively. The lattice is completely determined by the lengths of **a** and **b** and the angle γ between them. For graphite, $a = b = 0.246$ nm, and $\gamma = 120°$. The positions of all the lattice points of the graphite crystal are reached by drawing all possible vectors of the form

$$\mathbf{r} = u\mathbf{a} + v\mathbf{b}$$

from the origin, where u and v take on all integer values, positive, negative, and zero. That the crystal appears identical when viewed from all the positions given by this equation is an indication that it possesses the important property of *translational invariance*. Comparison of Figures 8.21 and 8.22 illustrates that the crystal lattice is not in general the same as the crystal structure.

8.3.1 Types of crystal lattice and the unit cell

It is found that there are five possible ways of arranging a two-dimensional array of lattice points so as to fill all the available space. This gives five possible types of two-dimensional lattices. These are illustrated in Figure 8.23, and are (a) the square lattice, (b) the rectangular lattice, (c) the centred rectangular lattice,

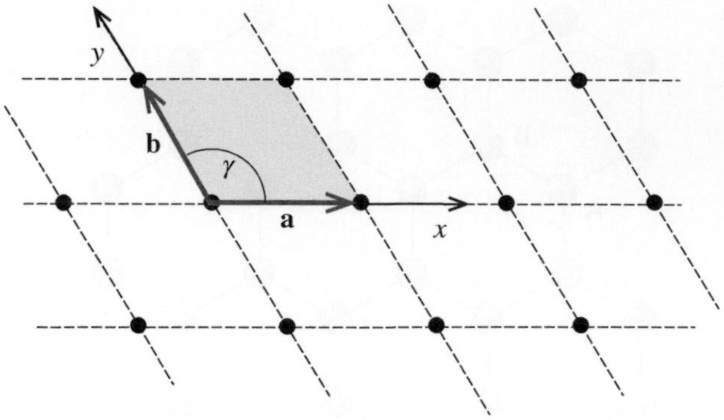

Figure 8.22 The crystal lattice of graphite. The x and y coordinate axes of the lattice are obtained by joining the lattice point at the origin to two of the neighbouring lattice points. The lattice is completely determined by the lengths of the lattice vectors **a** and **b** and the angle γ between them. The unit cell of the lattice is defined by these lattice vectors and is the parallelogram shaded in blue. The entire crystal lattice may be constructed by stacking such cells together, as indicated by the broken lines in the figure.

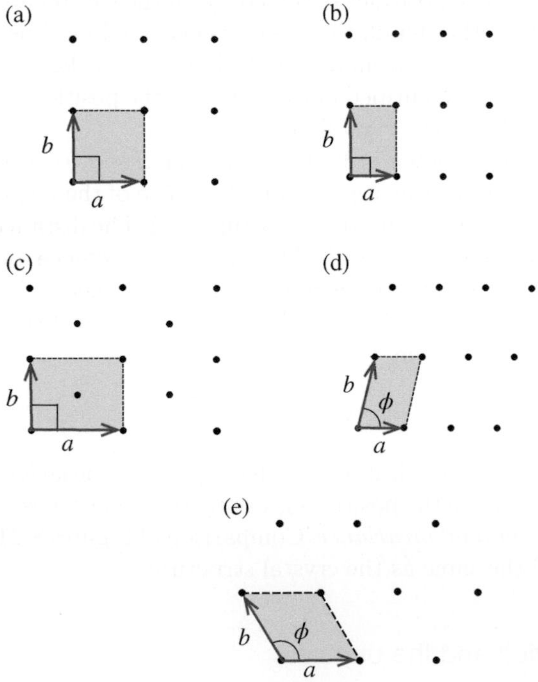

Figure 8.23 Unit cells for a two-dimensional lattice. (a) Square lattice, $a = b$, $\phi = 90°$, (b) rectangular lattice, $a \neq b$, $\phi = 90°$, (c) centred rectangular lattice, $a \neq b$, $\phi = 90°$, (d) oblique lattice, $a \neq b$, $\phi \neq 90°$ and (e) hexagonal lattice, $a = b$, $\phi = 120°$.

(d) the oblique lattice, and (e) the hexagonal lattice. These five types of lattice are distinguished by the kinds of symmetry they possess. This includes their *rotation symmetry*. For example, the square lattice (a) is unaltered by rotating it through 90° about an axis that passes through a lattice point and is perpendicular to the lattice plane. The square lattice thus possesses four-fold rotational symmetry. On the other hand, the hexagonal lattice (e) is unaltered on rotating it about a lattice point through 90°, and hence it has six-fold rotational symmetry. We emphasise that lattices with different symmetry properties are fundamentally different from each other. Not only are the symmetry properties of a crystal lattice reflected in the physical appearance of the crystal, sharp crystal faces at fixed angles to each other, but the symmetry properties can determine its physical properties. These include the optical and electrical characteristics of the crystal.

It is the case that two-dimensional lattices must have either one, two, four or six-fold rotation axes. A two-dimensional lattice, for example, cannot have five-fold rotational symmetry. The reason for this is that such a lattice would not fill all the available space. For the same reason, it is possible to cover a floor with tiles that have the shapes of parallelograms, squares, or regular hexagons, but is not possible with tiles that have the shape of a regular pentagon; see Figure 8.24.

In practice, a lattice is usually specified in terms of a *unit cell*. We may describe a unit cell as a small unit of space that when many identical units are stacked together, they reconstruct the full lattice. For the example of the two-dimensional graphite lattice, the unit cell is the parallelogram defined by the lattice vectors **a** and **b** in Figure 8.22. The unit cell is shaded in blue in the figure. The entire lattice may be constructed by stacking such cells together, as indicated by the broken lines in the figure. The unit cells for the five possible types of two-dimensional lattices are shown in Figure 8.23, where again they are shaded in blue.

We make a distinction here between *primitive* unit cells and *non-primitive* unit cells. And we make the point that more than one unit cell can be constructed for a given lattice. To do this, we consider the centred-rectangular unit cell shown in Figure 8.23c. For this lattice, we can construct the two alternative unit cells that are shown in Figure 8.25. They are defined by lattice vectors **a** and **b** and lattice vectors **a**′ and **b**′, respectively. For the unit cell defined by lattice vectors **a**′ and **b**′, there is a lattice point at each corner of the cell. But as the figure illustrates, each of these lattice points is shared between four adjacent unit cells. We may say that a given lattice point contributes a quarter of a lattice point to each of the four cells. It follows that the unit cell defined by lattice vectors **a**′ and **b**′ has just one lattice point. We also see that this unit cell has the smallest possible area that a unit cell could have for this lattice. Hence, this unit cell is called the *primitive cell*, which we can define as one that contains exactly one lattice point and which has the smallest possible area. For the second unit cell, defined by lattice vectors **a** and **b**, there is again a lattice point at each of its corners, and again these together contribute one lattice point to the unit cell. But there is also a

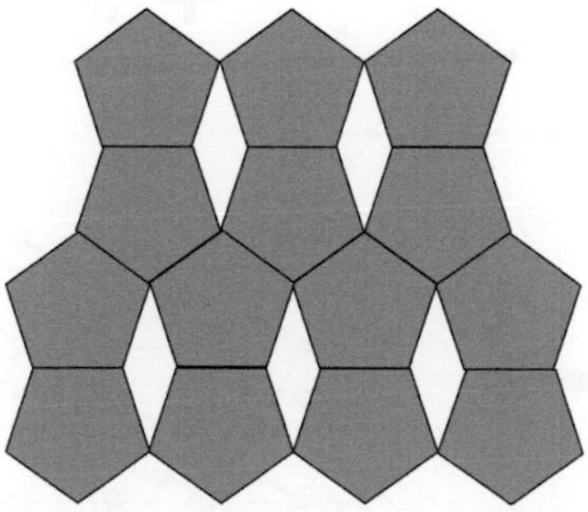

Figure 8.24 This figure illustrates that it is not possible to cover a floor with tiles that have the shape of a regular pentagon.

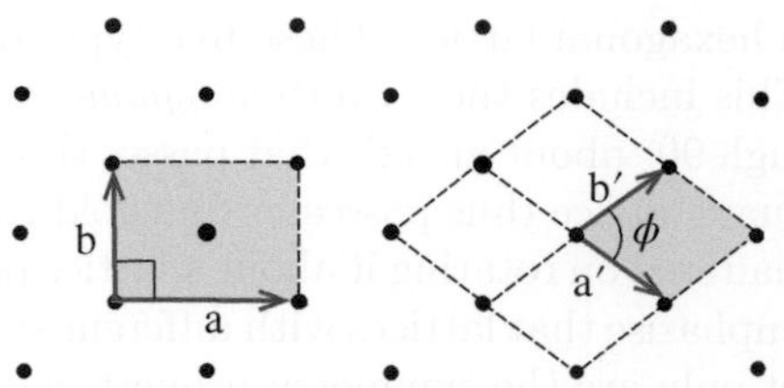

Figure 8.25 The unit cell defined by lattice vectors $\mathbf{a}'$ and $\mathbf{b}'$ contains a single lattice point and it has the smallest possible area that a unit cell could have for this lattice. Hence, this unit cell is called the primitive unit cell of the lattice. The unit cell defined by lattice vectors **a** and **b** contains two lattice points and its area is twice that of the primitive unit cell. Consequently, it is called a non-primitive unit cell.

lattice point at the centre of the cell. Hence, the total number of lattice points for this unit cell is two. Moreover, its area is twice that of the primitive unit cell. Consequently, the unit cell defined by lattice vectors **a** and **b** is called a non-primitive cell. Often, however, it is more convenient to describe a lattice in terms of a non-primitive cell.

8.3.2 The basis

Once the crystal lattice has been determined in the way described above and used to identify suitable coordinate axes and the unit cell, the description of the crystal structure is completed by specifying the atom or group of atoms that are associated with each lattice point. This may be a single atom as in the case of a crystal of copper, or it may consist of thousands of atoms as, for example, for a crystallised protein. This arrangement of atoms is known as the *basis* of the crystal lattice.

We again use the graphite crystal as an example. In Figure 8.21, two particular types of carbon atoms are distinguished. As noted previously, atom A is positioned at a lattice point, whereas atom B is not. Thus, a suitable choice of basis for the graphite crystal lattice is obtained by associating atom B with atom A, including their spatial arrangement. Consider now Figure 8.26. Figure 8.26a shows the lattice of graphite

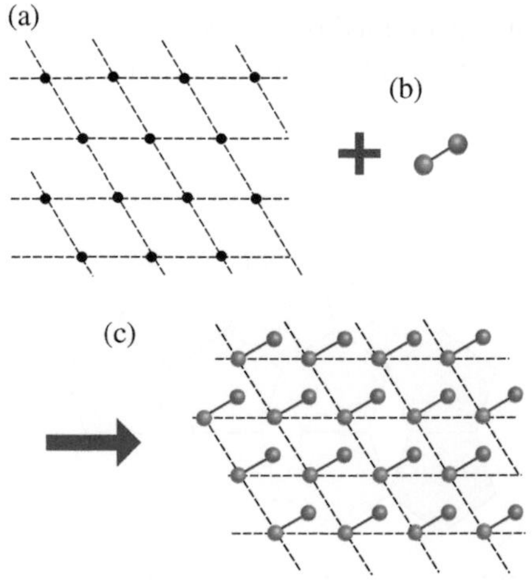

Figure 8.26 (a) The crystal lattice of graphite. (b) The basis of the unit cell consisting of two carbon atoms. (c) If a 'green' carbon atom of the basis is placed at each lattice point, the crystal structure of graphite is obtained.

and Figure 8.26b shows the basis consisting of atoms A and B. If we place the 'green' carbon atom of the basis at each lattice point, we generate the crystal structure of graphite as shown in Figure 8.26c. Hence, the complete description of a crystal structure includes both the crystal lattice and the basis:

$$\text{Crystal structure} = \text{Crystal lattice} + \text{Basis}$$

We can use such a combination of lattice plus basis to give an alternative, more correct description of the structure of, for example, the sodium chloride crystal, see Figure 8.13. We can describe this structure as an fcc lattice in which the basis is one chlorine ion at each lattice point and one sodium ion half a cube length above it.

8.3.3 Graphene

We have used a single layer of carbon atoms in graphite as an example of a crystal with a two-dimensional lattice. Such a single layer of carbon atoms had been postulated by scientists for a number of decades. However, it was Andre Geim and Kostya Novoselvo who first isolated and investigated such a single layer of carbon, and they did this at the University of Manchester. They called this allotrope of carbon *graphene* and they were both awarded the Nobel Prize in Physics in 2010 for their work. Surprisingly, their method for preparing graphene was remarkably simple. They pulled layers of graphene off a sample of graphite with common adhesive tape. The graphene flakes obtained from this adhesive-tape method were then transferred onto a silicon dioxide wafer for further study. Graphene has remarkable physical properties. It is the thinnest material that is known and is incredibly strong; being 200 times stronger than steel. It is also an excellent conductor of electricity and heat.

To get an understanding of the special properties of graphene, we must look at how the carbon atoms bond together in the graphene crystal. The ground state of the carbon atom is $1s^22s^22p^2$. The two 1s electrons are tightly bound to the carbon nucleus and take no part in the bonding. The bonding is done by the 2s and 2p electrons. Atomic orbitals from these four electrons *mix* in a process called *hybridisation.* This mixing of the orbitals results in new orbitals. Three of these orbitals, called sp^2, lie in the plane of the crystal lattice at an angle of 120° with respect to one another, and each contains one of the four electrons. The others are half-filled p-orbitals that lie perpendicular to the plane. The orbitals are illustrated in Figure 8.27, where they are shown sited on a single carbon atom of graphene for the sake of clarity. Each of the three sp^2 orbitals forms a covalent bond with an adjacent carbon atom. This is a very strong bond and accounts for the extremely high strength of graphene. This bond is stronger, for example, than the tetrahedral carbon–carbon bond in a diamond. Despite the large strength of the carbon–carbon bond, a sheet of graphene can be readily stretched, by up to 20% of its original length. The stretching just changes the angles between the bonds. It is also possible to twist the carbon–carbon bonds so that a sheet of graphene can be curved to a certain extent without the bonds breaking.

The p-orbitals that lie perpendicular to the lattice plane account for the electronic behaviour of graphene. The p-orbitals from adjacent carbon atoms overlap so that electrons can readily move through the lattice. The way these electrons interact with the crystal lattice has important consequences. These follow from the quantum physics of the electron/lattice system, which is outside the scope of this book. However, the resultant effects are first, that the electrons behave as if they have zero mass and therefore move at high speed through the crystal lattice under an applied voltage. And second, the electrons travel long distances in the lattice without being scattered by ions in the lattice or by lattice *defects*, i.e. irregularities in the crystal lattice. It is these scattering processes that limit the electrical conductivity of a crystal. Because of the two effects, zero mass and minimal scattering of the electrons, graphene has a much higher conductivity than say silver or copper. Thermal conductance also depends on the flow of electrons through a material. So it is

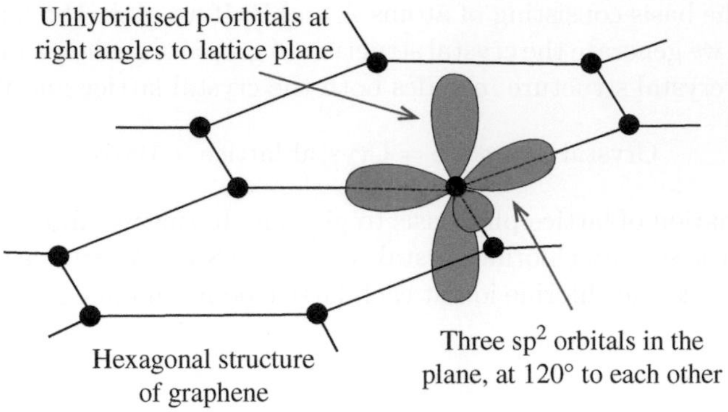

Figure 8.27 Atomic orbitals from the 2s and 2p electrons of carbon mix in a process called hybridisation. This mixing of the orbitals results in new orbitals, which are sited at each carbon atom in the crystal lattice. Three of these orbitals, called sp^2, lie in the plane of the lattice at an angle of 120° with respect to one another. The others are p-orbitals that lie perpendicular to the plane. Each of the three sp^2 orbitals forms a covalent bond with an adjacent carbon atom, which accounts for the extremely high strength of graphene. The p-orbitals from adjacent carbon atoms overlap so that electrons can readily move through the lattice.

not surprising that graphene also has extremely high thermal conductance. Because graphene is just one atom thick, it has very high transparency for visible light, ∼98%. It is thus a flexible, highly transparent material with high electrical conductivity. With its many special properties, graphene has many practical applications.

Perhaps it is worth remarking that whenever we draw a line with a pencil, there will be some flakes of graphene in that pencil line. This remarkable allotrope of carbon with so many extraordinary properties has been under our noses for a long time.

8.3.4 The three-dimensional lattice

We can readily extend our discussion to the three-dimensional crystal lattice. Again, an origin, based on a suitable atomic site, is chosen and all the positions within the crystal that are identical to it are identified. The set of identical positions identified in this way constitutes the three-dimensional lattice. The directions of the x, y, and z coordinate axes are then defined by joining the lattice point at the origin to *three* of the neighbouring lattice points; see Figure 8.28. The distances and directions of the nearest lattice points along the axes x, y, and z are specified by three lattice vectors $\mathbf{a}$, $\mathbf{b}$, and $\mathbf{c}$. The lattice is completely specified by the lengths of the vectors $\mathbf{a}$, $\mathbf{b}$, and $\mathbf{c}$ and the angles α, β, and γ between them as specified in Figure 8.28. The positions of all lattice points are reached by drawing all possible vectors of the form

$$\mathbf{r} = u\mathbf{a} + v\mathbf{b} + w\mathbf{c}$$

from the origin, where u, v, and w are integers. The ability to express the positions of lattice points in this way with a suitable choice of $\mathbf{a}$, $\mathbf{b}$, and $\mathbf{c}$ may be taken as the definition of a lattice in crystallography. The three-dimensional shape in Figure 8.28 defined by the vectors $\mathbf{a}$, $\mathbf{b}$, and $\mathbf{c}$ is called a parallelepiped and is the three-dimensional unit cell for a three-dimensional lattice.

We saw previously that a two-dimensional lattice can be classified as one of five possible types. The French physicist Auguste Bravais deduced that any three-dimensional lattice could be classified as

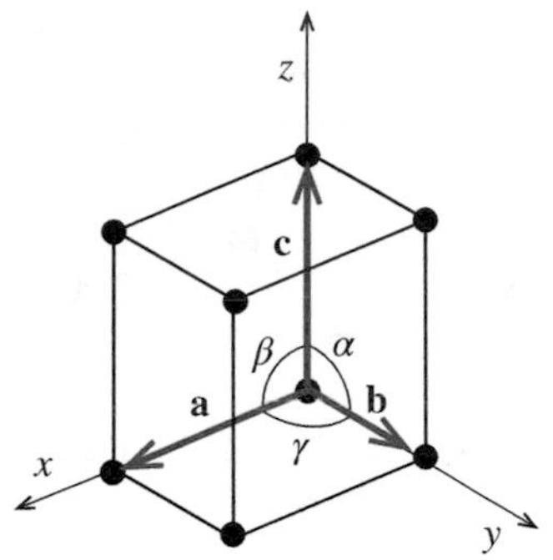

Figure 8.28 The three-dimensional shape defined by the vectors **a**, **b**, and **c** is called a parallelepiped and is the three-dimensional unit cell for a three-dimensional lattice. The lattice is completely specified by the lengths of **a**, **b**, and **c** and the angles α, β, and γ as specified in the figure.

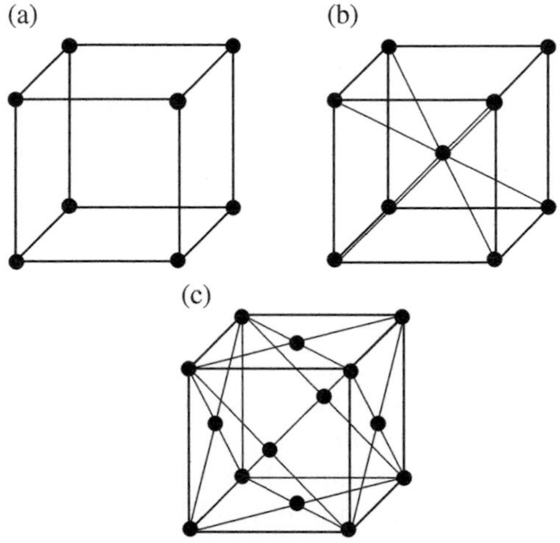

Figure 8.29 The figure shows three examples of the 14 possible Bravais lattices. These are (a) the simple cubic, (b) the body-centred cubic, and (c) the face-centred cubic, respectively. The simple cubic unit cell is a primitive unit cell. There is a lattice point at each corner of the cube, but each lattice point is shared between four unit cells, so that the number of lattice points per unit cell is one. The other two unit cells shown are non-primitive cells containing more than one lattice point per cell.

one of 14 so-called *Bravais lattices.* The significance of 14 Bravais lattices is that they are the 14 possible ways of arranging points in a regular array so as to fill all the available space. Again, the Bravais lattices are distinguished by their respective symmetry properties. Three examples of the 14 possible Bravais lattices are shown in Figure 8.29. These are (a) the simple cubic, (b) the body-centred cubic, and (c) the face-centred cubic, respectively. We note that the simple cubic unit cell is a primitive unit cell. There is a lattice point at each corner of the cube, but each lattice point is shared between four unit cells, and so the number of lattice points per unit cell is exactly one. The other two unit cells shown are non-primitive cells containing more than one lattice point per cell but again, despite being non-primitive, it is often convenient to use them to describe a crystal structure. For the sake of completeness, the primitive cell for the face-centred cubic structure is shown as the unit cell marked in red in Figure 8.30.

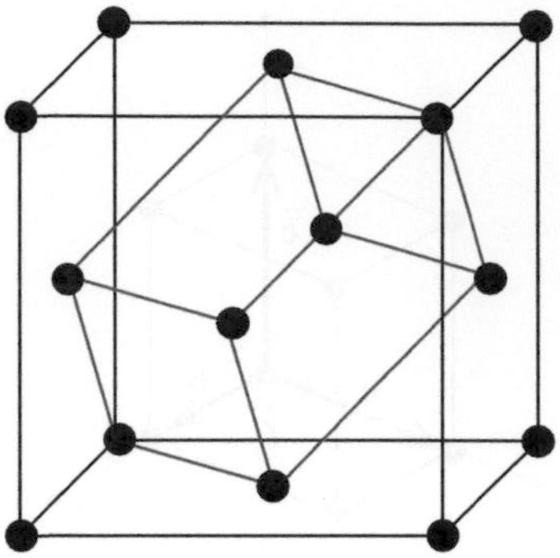

Figure 8.30 The primitive cell for the face-centred cubic structure is shown as the unit cell marked in red.

8.4 X-ray crystallography

In 1913, Max von Laue showed that X-rays are diffracted by crystals and form characteristic patterns on photographic film. He demonstrated in a single experiment that X-rays have a wave-like nature and that crystals have a regular structure. For his discovery of the diffraction of X-rays, von Laue was awarded the Nobel Prize in Physics in 1914. Shortly afterwards, father and son, W.H. Bragg and W.L. Bragg demonstrated the practical application of X-ray crystallography by determining the structure of crystals such as sodium chloride. For their work, they received the Nobel Prize in Physics in 1915. W.L. Bragg was just 25 years of age when he received the prize.

Since the discovery of the diffraction of X-rays by crystals, the technique has developed into an indispensable tool for material scientists and structural biologists. It has become the most common technique for the determination of three-dimensional crystalline structures at the atomic scale. Indeed, more than 20 Nobel prizes have been awarded for scientific achievements directly related to, or involving the use of crystallographic techniques. In particular, *protein crystallography* is now widely used to obtain the three-dimensional structure of proteins. For example, Roger Kornberg won the Nobel Prize in Chemistry using protein crystallography to solve the structure of the RNA polymerase, which contains over 30,000 individual atoms; see Figure 8.31.

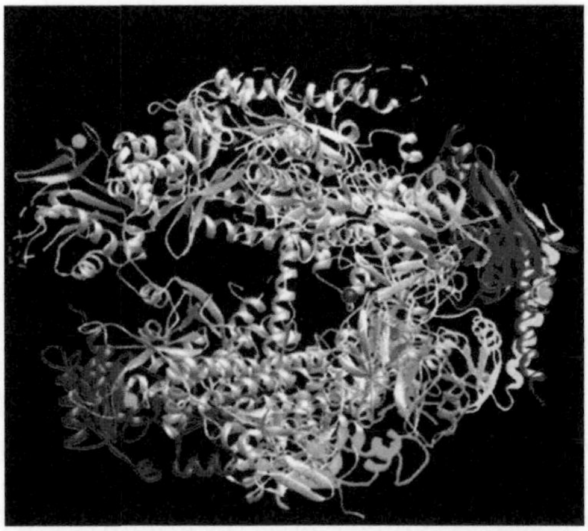

Figure 8.31 The structure of the RNA polymerase, which contains over 30,000 individual atoms. Roger Kornberg won the Nobel Prize in Chemistry for solving this structure by crystallography. 10.1126/Science.1059493.

8.4.1 The Bragg law

X-rays that are incident on a crystal may be scattered by the atoms of the crystal. More correctly, it is the electrons of the atoms that scatter the X-rays. The incident X-rays cause the electrons to oscillate at the same frequency. These oscillating electrons act as sources of secondary waves that are in phase with the incoming X-rays. The X-rays are thus scattered, i.e. redirected in all directions.

An obvious feature of the crystal structures we have described is that the atoms are arranged in well-defined planes. At certain angles of incidence, the X-rays scattered from a stack of planes in the crystal undergo constructive interference resulting in intensity maxima. In an optical experiment, the spacing of the ruled lines on a diffraction grating can be deduced from the separations of the diffraction maxima observed on a screen. And by measuring the relative intensities of the different orders of diffraction, information about the profile of the individual lines on the grating can be deduced. In an exactly similar way, the separations of the X-ray diffraction maxima from a crystal allow the distance between lattice planes and hence the size and shape of the unit cell to be deduced. Moreover, the relative intensities of the maxima provide information about the basis of the unit cell, i.e. the arrangement of the atoms in the unit cell.

W.L. Bragg proposed a simple but powerful equation that gives the connection between the wavelength of the X-rays, the distance between the lattice planes and the angles at which diffraction maxima are observed. This is the *Bragg law.* Strictly speaking, the X-ray scattering is associated with lattice points rather than atoms because it is the basis associated with each lattice point that is the true repeat unit of the crystal. However, for the case of just one atom in the basis, as we will discuss, the lattice points coincide with the positions of these atoms.

We first consider the scattering of X-rays from a single plane of atoms, as illustrated by Figure 8.32, which is an edge-wise view of the plane. X-rays are incident at a *glancing angle* θ to the plane. Contrary to the convention used in light optics, θ is defined relative to the plane rather than the normal to the plane. The scattered waves from the atoms at points A and B will be in phase and therefore interfere constructively if the distances AC and DB are equal. This occurs when the glancing angle of reflection θ is equal to the glancing angle of incidence. Hence, when a beam of X-rays is incident on a plane of atoms, there is a maximum in the direction of specular reflection and we can speak of the X-rays being *reflected* from the plane.

Consider now reflection from two adjacent planes of atoms, as illustrated by Figure 8.33. There will be constructive interference if the path difference of X-rays reflected from the two planes is an integral number of wavelengths. From Figure 8.33, we see that constructive interference will occur if

$$2d\sin\theta = n\lambda, \tag{8.1}$$

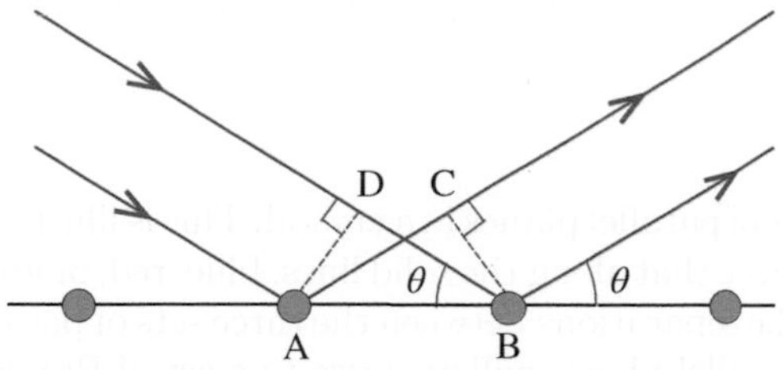

Figure 8.32 When a beam of X-rays is incident on a single plane of atoms, there is a maximum in the direction of specular reflection and we can speak of the X-rays being reflected from the plane.

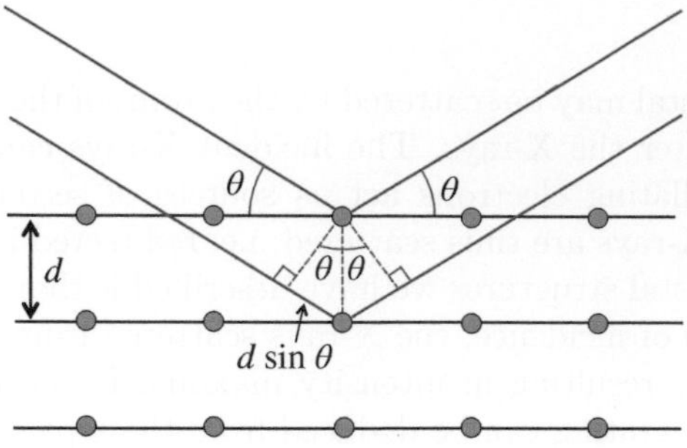

Figure 8.33 There will be constructive interference if the path difference of X-rays reflected from two parallel planes of atoms is an integral number of wavelengths. This occurs if $2d\sin\theta = n\lambda$, where d is the spacing of the planes, θ is the glancing angle, λ is the wavelength of the X-rays and n is an integer. X-rays of other wavelengths interfere destructively.

where d is the spacing of the planes, θ is the glancing angle, λ is the wavelength of the X-rays, and the integer n is the *order of reflection*. This is the Bragg law. Whereas a single plane of atoms reflects X-rays of *any* wavelength specularly at a given angle of incidence θ, a *stack* of planes only reflects a strong intensity at this angle if, in addition, the Bragg condition, Equation (8.1) is satisfied.

We emphasise that X-ray diffraction by a crystal must involve reflection from multiple planes. For a single plane, the reflected intensity is relatively weak because individual atoms scatter X-rays only weakly. However, when reflections occur from many successive planes, there are reflected waves from all the planes. At the Bragg angle, they are all in phase and their amplitudes add constructively. This results in an intense Bragg reflection. The scattering of X-rays by an individual atom is small and X-rays pass through a crystal with little attenuation. This means that all the atomic planes in the crystal become involved in the diffraction process and this requirement of multiple planes is obtained.

If the incident X-rays are not at the Bragg angle, the phase difference between the waves reflected from the first and third planes will be twice the phase difference between waves reflected from the first and second planes. Similarly, the waves reflected from the first and fourth planes will be three times the phase difference between waves reflected from the first and second planes. And so on. The point is that the reflected waves from all the planes will not be in phase with each other and hence will not interfere constructively. The result is that the intensity of reflected X-rays from a crystal falls off sharply as the glancing angle moves away from the Bragg angle.

By itself, a single Bragg reflection does not allow n to be identified. Of course, n must be an integer and n will be small at low glancing angles. However, by correlating the reflections from different sets of planes in a crystal, see the following section, the value of n for a particular reflection can be deduced.

8.4.2 Crystal planes

In general, there are different sets of parallel planes in a crystal. This is illustrated by Figure 8.34, which is an edge-wise view of the crystal. We see that along the solid lines, blue, red, or green, there are atoms positioned at regular intervals. Moreover, the separations between the three sets of parallel planes are different. We see that $d_1 > d_2 > d_3$. Each set of parallel planes will give rise to a set of Bragg reflections.

Consider now the simple cubic structure shown in Figure 8.35. Three important sets of planes of the crystal are shown. One set of planes, shown in Figure 8.35a, lies parallel to the sides of the cube. In this

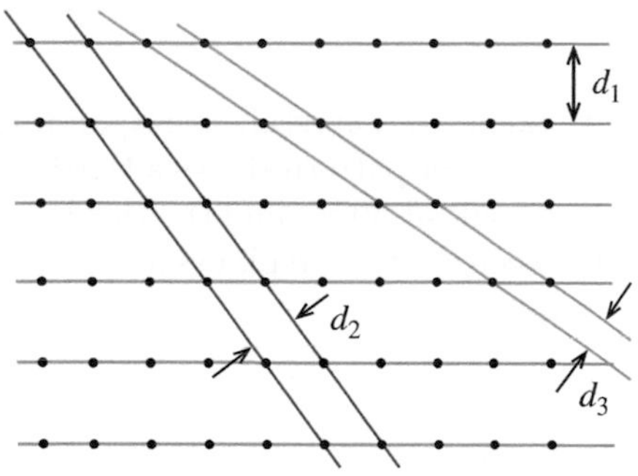

Figure 8.34 In general, there are different sets of parallel planes in a crystal. This figure represents an edge-wise view of a crystal. We see that along the solid lines, blue, red or green, there are atoms positioned at regular intervals. Moreover, the separations between the three sets of parallel planes are different; $d_1 > d_2 > d_3$. Each set of parallel planes gives rise to a set of diffraction maxima according to the Bragg law.

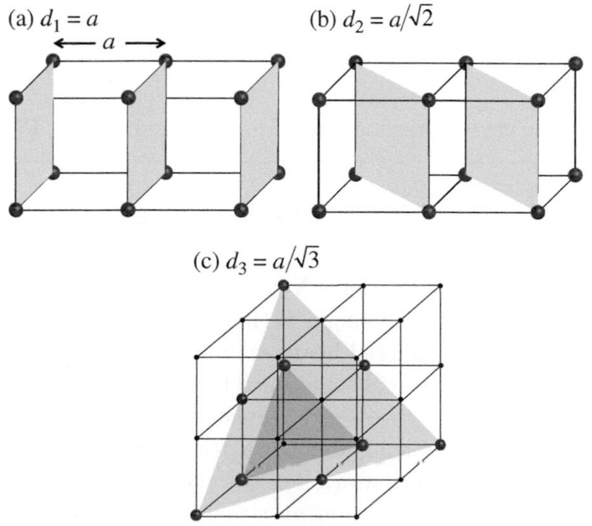

Figure 8.35 The figure shows three important sets of planes in a simple cubic structure. (a) One set of planes lies parallel to the sides of the cube. In this case, the separation of adjacent planes d_1 is equal to the distance a between the atoms. (b) In this case, the planes pass through the opposite edges of the cube, where the separation of adjacent planes d_2 is $a/\sqrt{2}$. (c) In this case, the separation of adjacent planes d_3 is $a/\sqrt{3}$, where for the sake of clarity only those atoms that are contained in an individual plane are shown, coloured either blue or red.

case, the separation of adjacent planes d_1 is equal to the distance a between the atoms. The other two sets of planes are not parallel to the sides of the cube. In Figure 8.35b, the planes pass through the opposite edges of the cube. In this case, Pythagoras' theorem gives the separation of adjacent planes d_2 to be $a/\sqrt{2}$. Figure 8.35c shows the third set of parallel planes, where the separation d_3 of adjacent planes is $a/\sqrt{3}$. For this third set of planes, atoms in the two adjacent planes are indicated either blue or red, for the sake of clarity. All three sets of planes give rise to Bragg reflections. And by observing the respective Bragg angles, the three-dimensional structure of the crystal can be deduced.

Worked example

Three different diffraction patterns are taken for a crystal, corresponding to three different sets of planes, using X-rays of wavelength 0.0587 nm. The first pattern shows a third-order Bragg angle at 32.5°, the second pattern shows a second-order Bragg angle at 30.5°, and the third pattern shows a second-order Bragg angle at 38.3°. Calculate the spacing between the crystal planes for each Bragg angle. What can be deduced about the crystal structure from these data?

Solution

For the first set of planes, we have

$$d = \frac{n\lambda}{2\sin\theta} = \frac{3 \times 0.0587}{2\sin 32.5°} = 0.164\,\text{nm}.$$

For the second and third sets of planes, we have $d = 0.116$ nm, and $d = 0.095$ nm, respectively. The spacings are in the ratios

$$0.164 : 0.116 : 0.095 = 1.73 : 1.22 : 1 \approx 1 : \frac{1}{\sqrt{2}} : \frac{1}{\sqrt{3}}.$$

Hence, the data indicate that the crystal has a cubic structure as shown in Figure 8.35.

8.5 Experimental techniques of X-ray crystallography

In an X-ray diffraction experiment, a narrow beam of X-rays of well-defined wavelength is incident upon the crystal, as in Figure 8.36. At some glancing angle θ that satisfies the Bragg condition, a diffraction maximum is observed. Notice that the diffracted beam makes an angle of 2θ with respect to the incident X-ray beam. It is unlikely that the Bragg law will be satisfied for arbitrary orientation of a *single* crystal with respect to an incident X-ray beam of fixed wavelength. However, if the crystal is rotated about a fixed axis perpendicular to the incident beam, the glancing angle θ varies for those sets of planes that are not perpendicular to the rotation axis. And, for some orientation of the crystal, a set of such planes is likely to satisfy the Bragg condition. This is the basis of the *rotating crystal method*, which is shown schematically in Figure 8.37. The diffraction maxima are observed as spots on the flat-plate detector. The single crystal is rotated in angular steps and at each step, the position and intensity of every diffraction spot are recorded.

We emphasise that Bragg's law only allows the lattice to be determined; it relates to the *positions* of the diffraction spots. Bragg's law cannot be used to deduce the nature of the basis that is associated with each lattice point. However, the observed *intensities* of the spots do reveal information from which the basis can be deduced.

An alternative method of ensuring that there are sets of lattice planes in the crystal specimen that do satisfy the Bragg law is to have the specimen in the form of many finely divided crystals, called *crystallites*. This is the basis of the *powder diffraction method*, which is illustrated in Figure 8.38a. The crystallites are held in a suitable container in the path of the incident X-ray beam. If the orientation of the crystallites is random, then for a given set of lattice planes, some of these crystallites will be orientated at the Bragg angle θ to the incident beam, as shown in Figure 8.38b. Again, the diffracted beam makes an angle of 2θ with respect to the incident beam direction. There is rotational symmetry about the incident beam direction with respect to the various orientations of the individual crystallites that lie at the Bragg angle to the

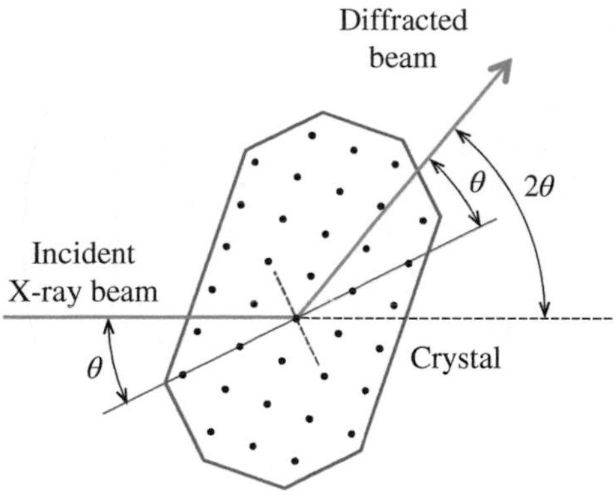

Figure 8.36 In an X-ray diffraction experiment, a narrow beam of X-rays of a well-defined wavelength is incident upon the crystal. At a certain angle θ, a diffraction maximum is observed. Notice that the diffracted beam is scattered through a total angle of 2θ with respect to the direction of the incident X-ray beam.

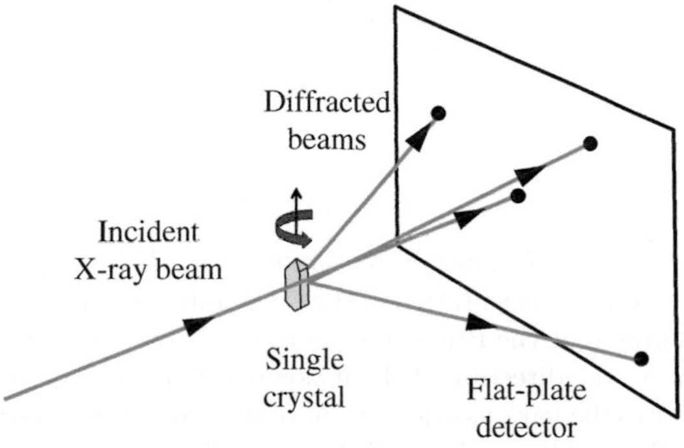

Figure 8.37 A schematic diagram of the rotating crystal method. A collimated beam of monochromatic X-rays is incident on a single crystal, which is rotated about a fixed axis perpendicular to the incident beam direction. The glancing angle varies for those sets of planes that are not perpendicular to the rotation axis. At a certain orientation of the crystal, a set of such planes satisfies the Bragg law and a set of diffraction spots is observed on the flat-plate detector.

incident beam. Hence, the diffracted beams from all the crystallites aligned at the Bragg angle will give rise to a cone of scattered radiation. If this radiation is detected on a flat-plate detector, the diffraction pattern appears as circle, as in Figure 8.38a. For several sets of planes, there would be a corresponding number of concentric circles observed on the detector. One advantage of the powder method is that the X-ray beam samples many crystallites and so the intensities of the diffraction maxima are relatively large, compared to say that obtainable from a single crystal in the rotating crystal method.

8.5.1 X-ray sources

The conventional source of X-rays is an X-ray tube. In such a device, electrons of high-energy ~30 keV, strike a metal anode, and this results in the emission of X-rays from the anode. The emitted radiation is a combination of sharp lines called *characteristic lines* and a continuous background of radiation that varies

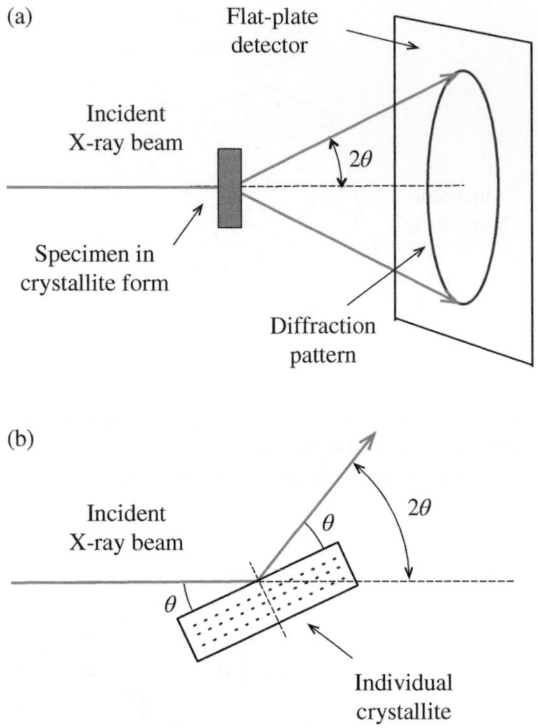

Figure 8.38 (a) A schematic diagram of the powder diffraction method. The sample is in the form of many finely divided crystals, called crystallites. (b) If the orientation of the crystallites is random, then for any set of lattice planes, some of the crystallites will be orientated at the Bragg angle θ to the X-ray beam, like the one shown. The diffracted rays make an angle 2θ with respect to the direction of the incident beam. There is rotational symmetry about the incident beam direction with respect to the various orientations of the individual crystallites that lie at the Bragg angle to the incident beam. Hence, the diffracted beams from all the crystallites aligned at the Bragg angle will give rise to a cone of scattered radiation. If this radiation is detected on a flat-plate detector, the diffraction pattern appears as a circle, as in Figure 8.38a.

smoothly with wavelength. The X-ray spectrum emitted from a copper anode is illustrated schematically in Figure 8.39. The characteristic lines are due to electronic transitions between discrete energy levels in the copper and hence these lines contain a narrow range of wavelengths. In the case of copper, the 0.154 nm radiation is due to electron transitions from the $n = 2$ to the $n = 1$ energy levels and the 0.139 nm radiation is due to electron transitions from the $n = 3$ to the $n = 1$ levels; see also Section 1.3.1. The continuous background, called *bremsstrahlung* radiation, arises because the incident electrons suffer large decelerations, and charged particles that decelerate emit electromagnetic radiation. By using suitable filters, one of the characteristic lines can be isolated providing a monochromatic beam of X-rays. Different metals can be used for the anode to obtain different X-ray wavelengths. For example, a tungsten anode provides radiation at a shorter wavelength of 0.0213 nm.

X-rays of much higher intensity are provided by a particle accelerator called a *synchrotron*. In a synchrotron, electrons travelling at almost the speed of light circulate around the accelerator in a series of straight and curved sections. In both types of section, the electrons are made to follow curved trajectories and since these charged particles are therefore accelerated, they emit electromagnetic radiation. The

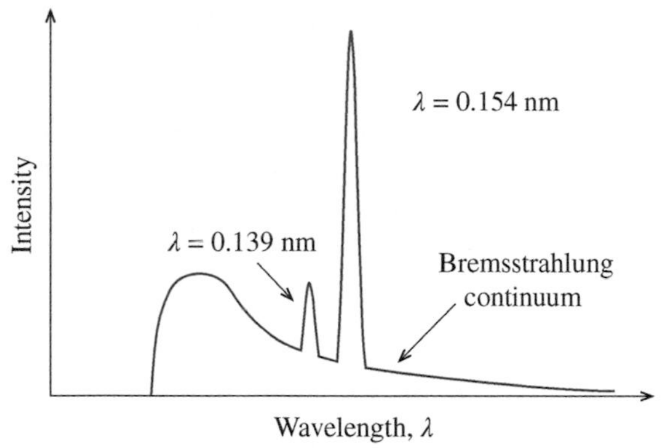

Figure 8.39 Schematic diagram of the X-ray spectrum emitted from a copper anode. The characteristic lines are due to electron transitions between discrete energy levels in the copper. The 0.154 nm radiation is due to electron transitions from the $n = 2$ to the $n = 1$ energy levels and the 0.139 nm radiation is due to electron transitions from the $n = 3$ to the $n = 1$ levels. The continuous background, called bremsstrahlung radiation, arises because the incident electrons suffer large decelerations, and charged particles that decelerate emit electromagnetic radiation. By using suitable filters, one of the characteristic lines can be isolated providing a monochromatic beam of X-rays.

radiation produced by a synchrotron covers a huge spectral range and extends into the X-ray region. Synchrotrons are large machines, perhaps 50 m in diameter, and they are expensive to build. Consequently, they are usually national facilities. The main UK synchrotron is called *Diamond* and is situated in Oxfordshire. It is widely used for X-ray diffraction studies.

To increase the intensity of the X-rays produced by a synchrotron, a device called an *undulator* is inserted into one of its straight sections. A schematic diagram of an undulator is shown in Figure 8.40. It consists of a periodic array of dipole magnets of alternating polarity with a *magnetic period* λ_0. The magnetic period may typically be 5 cm, and the total length of the magnetic array may be several metres. The resulting magnetic field between the poles of the magnets varies sinusoidally with distance and so the trajectories of the electrons that pass between the poles are also sinusoidal, i.e. the electrons are 'wiggled' by the action of the undulator. Each wiggle is an individual source of X-rays, just as the slits in a diffraction grating act as individual sources of light. And, as in the case of a diffraction grating, these individual sources of X-rays interfere constructively at specific angles. If there are N dipole magnets, the resultant intensity

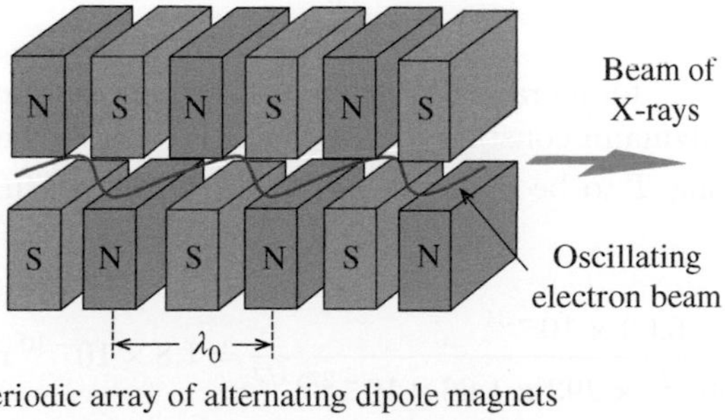

Figure 8.40 A schematic diagram of an undulator that produces a high-intensity beam of X-rays.

from the undulator is N^2 larger than that from a single wiggle. This means that the intensity of the emitted radiation from an undulator is enormously large; many orders of magnitude larger than that obtainable from a conventional X-ray tube. Moreover, the emitted radiation is concentrated into a relatively narrow band of wavelengths and is essentially quasi-monochromatic. The peak emission wavelength depends on the magnetic period λ_0 and the electron energy, but also on the strength of the magnetic field produced by the dipole magnets. Hence, by changing the gap between the magnet pole pieces, the peak emission wavelength can be tuned. Further wavelength selection of the X-ray beam, if necessary, is achieved by passing it through an X-ray monochromator.

8.5.2 Collection and analysis of diffraction patterns

A photographic plate has been traditionally used to record the diffraction patterns produced by crystals. But these are being replaced by electronic devices such as the *charge-coupled device* (CCD). These electronic devices have a flat-plate geometry and provide spatial information just like a photographic plate but the information is recorded digitally. The X-ray intensity is then measured as the number of counts per second detected.

For simple crystal structures, like that of sodium chloride, the diffraction patterns are relatively simple and easy to relate to the crystal structure using the Bragg law. However, the diffraction patterns obtained for more complicated structures, like those of proteins, are very complex. And a full data set may consist of hundreds of separate images taken at different orientations of the crystal. Consequently, sophisticated techniques using powerful computer programs are used to analyse the diffraction patterns. These techniques use the methods of *Fourier analysis*. The basic principle is that the *Fourier transform* of the detected X-ray intensity can be related to the lattice structure and the basis.

8.6 Neutron scattering

Any type of wave can be used for diffraction studies of crystals, so long as the wavelength is comparable to the separation of the lattice planes. And the physical principles governing the diffraction are exactly the same as for X-ray diffraction.

Microscopic particles, such as electrons and neutrons, exhibit wave-like properties. The wavelength λ of a particle is related to its momentum p by the *de Broglie relationship*:

$$\lambda = \frac{h}{p}, \tag{8.2}$$

where h is Planck's constant. At temperature T, neutrons have a kinetic energy due to their thermal energy $\sim kT$, where k is the Boltzmann constant. Taking $1/2mv^2 = kT$, where m is the mass of a neutron, $p = mv = \sqrt{2kTm}$. Then, taking T to be room temperature, 293 K, we find that the wavelength of the neutron is

$$\frac{6.63 \times 10^{-34}}{(2 \times 1.38 \times 10^{-23} \times 293 \times 1.66 \times 10^{-27})^{1/2}} = 1.8 \times 10^{-10}\ \text{m} = 0.18\ \text{nm}.$$

This is about the same size as the lattice spacing in crystals. Hence, the structure of a crystal can also be studied by observing the scattering of such neutrons from the crystal. This is the technique of *neutron diffraction* and the neutrons are called *thermal neutrons*.

For an electron to have a wavelength of say 0.18 nm, its energy would be ∼50 eV. Electron beams of this energy only penetrate a short distance into a crystal, ∼0.5 nm, corresponding to several atomic diameters. Consequently, the technique of *low-energy electron diffraction* finds application in the study of crystal surfaces.

Neutron scattering has some important advantages. The diffraction of X-rays results from the scattering of the X-rays from the electrons of the atoms in the crystal. Consequently, low-Z atoms like hydrogen scatter X-rays relatively weakly. This means it can be difficult to locate hydrogen atoms in a crystal as the diffraction patterns they produce can be overshadowed by the patterns of heavier atoms. The scattering of neutrons, on the other hand, is due to the scattering of the atomic nuclei. And hydrogen nuclei scatter neutrons strongly. Also, because the neutrons scatter off the nuclei, neutron scattering provides more accurate positioning of the atoms in a crystal. A further advantage of the neutron diffraction technique is that neutrons have a permanent magnetic dipole moment that will interact with permanent magnetic moments, if any, of the atoms. Hence, neutron scattering can be used to study the magnetic properties of crystals. On the other hand, neutron sources are characterised by low fluxes compared to X-ray sources. This means that relatively large amounts of the crystal specimen are usually needed, which is a difficulty for investigating specimens that are only available in small amounts, such as biological crystals. In these cases, the powder diffraction technique is employed.

The source of neutrons for crystal structure studies may be a small nuclear reactor using low-enriched uranium fuel. Alternatively, a *spallation source* may be used, in which high-energy protons strike a heavy metal target such as tantalum. This causes the emission of about 20–30 neutrons for each proton that strikes a metal nucleus. The UK has a spallation source of neutrons called ISIS. Neutrons that are produced by either method have very high energies, ∼MeV. As noted earlier, neutron energies of kT ∼25 meV are required for diffraction studies of crystals. The reduction in their energy is achieved by passing the high-energy neutrons through a *moderator*, such as a block of graphite. The neutrons make numerous collisions with the nuclei of the moderator material losing energy to those nuclei in each collision. This *thermalizes* the neutrons so that they end up at the same temperature as the moderator, which can be room temperature.

8.7 Interatomic forces in solids

In Chapter 2, we discussed the forces that operate between atoms and saw how these forces bind atoms together. There we obtained an expression for the lattice energy of an ionic crystal, i.e. the amount of energy required to separate a crystalline solid into its constituent ions. In ionic crystals, the interaction between the ions is dominated by the long-range Coulomb force. In this section, we describe other properties of solids that can be explained in terms of interatomic forces. This time, we will concentrate our discussion on solids in which the binding is due to short-range van der Waals forces, such as solid argon. However, the methods we will use can readily be adapted for the case of long-range forces.

8.7.1 Heat of sublimation

Imagine that we heat a van der Waals solid from absolute zero until all the atoms evaporate to form a dilute gas. We would need to supply sufficient energy to break all the bonds between all the atoms. This is the *heat of sublimation* L_S. As in Chapter 2, we consider any pair of atoms to be bound together by binding energy ε; the amount of energy required to separate the atoms to infinity. If each atom has n nearest neighbours, the energy required to break the bonds in one mole of the solid is

$$L_S = \frac{1}{2} N_A n \varepsilon, \tag{8.3}$$

where N_A is Avogadro's number. The familiar factor of ½ is included so that each bond is not counted twice. For close-packed crystal structures, such as in solid argon, $n = 12$. Hence,

$$L_S = 6 N_A \varepsilon. \tag{8.4}$$

The heat of sublimation of argon has been measured to be 7.7 kJ/mol. Using Equation (8.4), we obtain a value for ε of

$$\frac{7.7 \times 10^3}{6 \times 6.02 \times 10^{23}} = 2.15 \times 10^{-21}\ \text{J} \equiv 0.013\ \text{eV}.$$

Considering the simplicity of this calculation, this value for ε is in good agreement with the accepted value of 0.011 eV for solid argon.

8.7.2 Surface energy of a crystal

Imagine that we have a crystal that is cleaved in two by some means, as represented schematically by Figure 8.41. This increases the surface area of the crystal. To do this, energy must be supplied to break the bonds between the atoms on either side of the break. The *surface energy* of a crystal is the amount

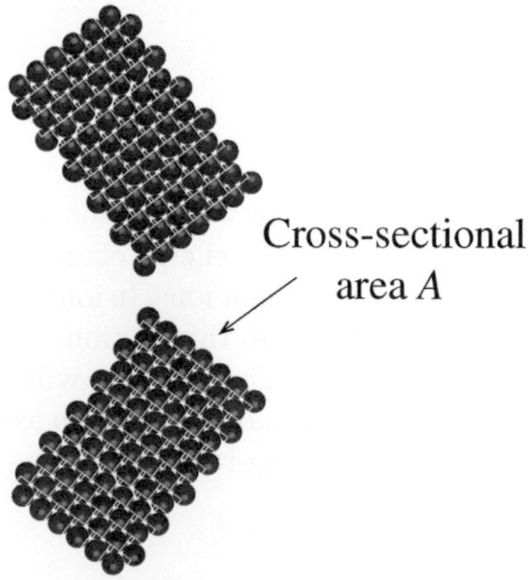

Figure 8.41 When a crystal is cleaved in two, the surface area of the crystal increases. To cleave the crystal, energy must be supplied to break the bonds between the atoms on either side of the break. The surface energy of a crystal is the amount of energy that is required to create a unit area of a new surface.

of energy that is required to create unit area of new surface. An atom in the bulk of a crystal is surrounded by atoms on all sides and experiences the attraction of all these atoms. An atom at the surface, however, only experiences the attraction of atoms that lie within the crystal, i.e. from a single hemisphere. We need to break the bonds to make the new surface whilst keeping intact those bonds connected to the atoms in the bulk of the crystal, i.e. we need to break half the bonds. If there are N_S atoms per unit surface area, the number of atoms involved is AN_S, where A is the cross-sectional area of the cleaved surface. And if an atom in the bulk of the crystal has n nearest neighbours, $1/2(nAN_S)$ bonds must be broken. If the binding energy between a pair of atoms is ε, the amount of energy required to break all the bonds is $1/2(nAN_S\varepsilon)$. This produces $2A$ of new surface area and hence, the energy required per unit area is $1/4(nN_S\varepsilon)$. If the diameter of an atom is a_0, the number of atoms per unit area $N_S \sim 1/a_0^2$. Then, substituting for N_S, we find the surface energy to be

$$\sim \frac{n\varepsilon}{4a_0^2}. \tag{8.5}$$

With $n = 12$ for solid argon, we find the surface energy to be $\sim 3(\varepsilon/a_0^2)$. Taking, $\varepsilon = 1.7 \times 10^{-21}$ J and $a_0 = 1.8 \times 10^{-10}$ m, the surface energy of solid argon is found to be $\sim(3 \times 1.7 \times 10^{-21})/(3.7 \times 10^{-10})^2 \sim 40$ mJ/m^2. This value of the surface energy, from our simple model, is in accord with more detailed calculations.

8.8 Vibrations in crystals

The atoms in a crystal are arranged in regular arrays in which the mean distance between atoms is fixed. However, the atoms are not static; they vibrate about their equilibrium positions. We can use a simple, classical picture to estimate the expected amplitude of these vibrations compared to the inter atomic mean separation and to estimate the frequency of the vibrations. Despite the simplicity of this model, it yields values for the amplitude and frequency that are correct to within an order of magnitude. Again, we consider crystals in which the atoms are held together by short-range van der Waals forces.

We first recall some relationships for simple harmonic motion (SHM). This motion arises for systems, such as a mass on a spring, that are subject to a restoring force F that is directed towards the origin and is proportional to the displacement x:

$$F = -\alpha x, \tag{8.6}$$

where α is the *spring constant.* Such a system vibrates in SHM with frequency

$$\nu = \frac{1}{2\pi}\sqrt{\frac{\alpha}{m}}\,\text{Hz}, \tag{8.7}$$

where m is the mass in motion. The potential energy of the system is given by

$$V = \frac{1}{2}\alpha x^2; \tag{8.8}$$

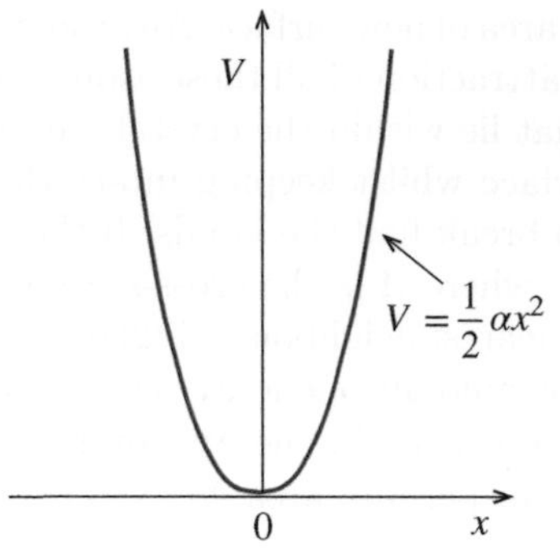

Figure 8.42 A parabolic potential well.

a function that has given rise to a parabolic potential well; see Figure 8.42. Differentiating Equation (8.8) twice with respect to x, we obtain

$$\frac{d^2 V}{dx^2} = \alpha. \tag{8.9}$$

We see that the second derivative of V can be identified with the spring constant α of the system. Physically, $d^2 V/dx^2$ gives a measure of the curvature of the potential well, where curvature at any point on a curve is defined as the reciprocal of the radius of curvature. At the minimum of a parabolic curve, the curvature is simply $d^2 V/dx^2$. Hence, α for a parabolic potential well is given by the curvature of the bottom of the well.

Invariably, the potential wells encountered in physical situations are not accurately parabolic. They may be more asymmetric as illustrated by Figure 8.43. However, if the actual potential well has a reasonable shape that goes through a minimum, it will not be very different in shape from a parabola near the minimum. Then so long as the amplitude of vibration is not too great, we can approximate the actual potential well to a parabolic potential well. Essentially, the system behaves like a simple harmonic oscillator for small amplitude vibrations. Moreover, the spring constant of the system is given by $d^2 V(r)/dr^2$ evaluated at the

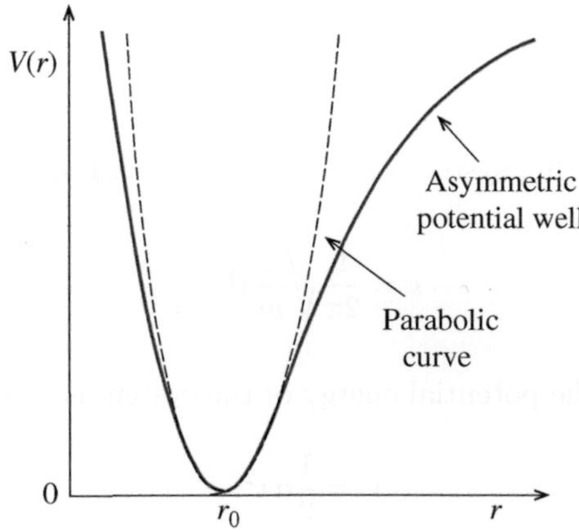

Figure 8.43 An asymmetric potential well. Close to the minimum of the well, its shape is closely parabolic.

bottom of the well. Then, if the system is displaced slightly from equilibrium, it will oscillate in SHM with frequency

$$\nu = \frac{1}{2\pi}\sqrt{\frac{(\mathrm{d}^2 V(r)/\mathrm{d}r^2)}{m}}. \tag{8.10}$$

In Section 2.2, we saw that the potential energy between two atoms bound by van der Waals forces can be represented by the Lennard–Jones potential

$$V(r) = \varepsilon\left[\left(\frac{a_0}{r}\right)^{12} - 2\left(\frac{a_0}{r}\right)^{6}\right],$$

where a_0 is the equilibrium separation. Differentiating this expression twice gives

$$\frac{\mathrm{d}^2 V(r)}{\mathrm{d}r^2} = \varepsilon\left[156\frac{a_0^{12}}{r^{14}} - 84\frac{a_0^6}{r^8}\right]. \tag{8.11}$$

Hence, at $r = a_0$,

$$\frac{\mathrm{d}^2 V(r)}{\mathrm{d}r^2} = \frac{72\varepsilon}{a_0^2}, \tag{8.12}$$

and the spring constant α for this potential is $72\varepsilon/a_0^2$.

We imagine a solid in which all the atoms are fixed at their equilibrium positions except for one, which is free to vibrate. As a first approximation, we consider this atom to be in a linear arrangement between its two nearest neighbours and that the vibration takes place along the line joining the atoms, as shown schematically in Figure 8.44a. When the atom moves away from its equilibrium position, it experiences repulsion from one of its neighbours and attraction from the other. A mechanical analogue of this situation is illustrated in Figure 8.44b. The central mass is connected by springs to fixed masses on either side of it. When the central mass moves say to the left, the left-hand spring acts to push it back to its equilibrium position, while the right-hand spring acts to pull it back. The mass thus experiences twice the restoring force due to a single spring and the resulting spring constant is twice as large.

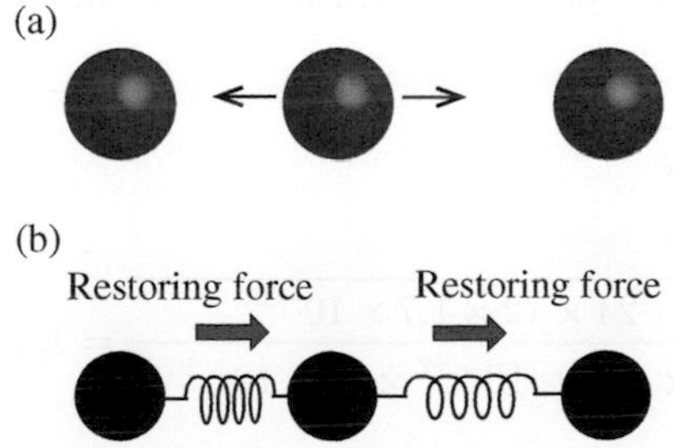

Figure 8.44 (a) Linear vibrations of an atom with two nearest neighbours. (b) A mechanical analogue in which a mass is connected to two fixed masses. When the central mass moves to the left say, the left-hand spring acts to push it back to its equilibrium position, while the right-hand spring acts to pull it back to its equilibrium position. The mass thus experiences twice the restoring force due to a single spring.

Similarly, in the case of the vibrating atom in Figure 8.44a, the resulting spring constant is

$$2 \times \frac{72\varepsilon}{a_0^2} = \frac{144\varepsilon}{a_0^2}. \tag{8.13}$$

In fact, an atom is surrounded by n nearest neighbours that cause an increase in the restoring force and spring constant. However, an increase by a factor of n in the spring constant would be an overestimate as these n atoms are distributed in three directions, while we are considering the atom to be moving in one direction. Hence an increase by a factor of $n/3$ is a reasonable estimate. We thus obtain a spring constant of

$$\frac{n}{3} \times \frac{72\varepsilon}{a_0^2} = \frac{24n\varepsilon}{a_0^2}. \tag{8.14}$$

Using this result for α in Equation (8.8), we find the potential energy of the vibrating atom to be

$$V(r) = \frac{12n\varepsilon}{a_0^2}(r - a_0)^2, \tag{8.15}$$

where $(r - a_0)$ is the displacement from its equilibrium position. We can expect that the largest amplitude of vibration, corresponding to $(r - a_0)_{\text{max}}$ to occur at the melting temperature T_{M} of the solid. The energy of an atom at temperature T is $\sim kT$. Then putting $V(r - a_0)_{\text{max}} = kT_{\text{M}}$, we obtain

$$\frac{(r - a_0)_{\text{max}}}{a_0} = \sqrt{\frac{kT_{\text{M}}}{12n\varepsilon}}. \tag{8.16}$$

For solid argon, $T_{\text{M}} = 84$ K, $n = 12$ and $\varepsilon = 1.7 \times 10^{-21}$ J. This gives

$$\frac{(r - a_0)_{\text{max}}}{a_0} = \sqrt{\frac{1.38 \times 10^{-23} \times 84}{12 \times 12 \times 1.7 \times 10^{-21}}} = 6.8 \times 10^2 \sim 7\%.$$

This result is of the correct order of magnitude. Indeed the melting temperature of a solid is sometimes defined as the temperature at which the amplitude of vibration is 10% of the inter-atomic mean separation. This result also supports our use of a parabolic potential well, as the relative amplitude of vibration is small.

Using the assumption of a parabolic potential well, we have from Equation (8.10) that the frequency of vibration of the atom is,

$$\begin{aligned} \nu &= \frac{1}{2\pi}\sqrt{\frac{24n\varepsilon}{ma_0^2}} \\ &= \frac{1}{2\pi}\sqrt{\frac{24 \times 12 \times 1.7 \times 10^{-21}}{40 \times 1.67 \times 10^{-27} \times (3 \times 10^{-10})^2}} = 1.4 \times 10^{12}\text{Hz}. \end{aligned}$$

This is indeed the correct order of magnitude for the vibrational frequencies of atoms in a solid. This frequency is called the *Einstein frequency*, and we will return to it when we discuss the thermal properties of solids in Chapter 10.

Worked example

Consider the asymmetric potential well in Figure 8.43, described by the function $V(r)$. Use a Taylor expansion to show that a system in this potential well behaves like a simple harmonic oscillator for small amplitude vibrations about the equilibrium position $r = r_0$.

Solution

Expanding $V(r)$ about $r = r_0$ using a Taylor expansion, we have

$$V(r) = V(r_0) + (r - r_0)\left(\frac{\mathrm{d}V(r)}{\mathrm{d}r}\right)_{r=r_0} + \frac{(r - r_0)^2}{2}\left(\frac{\mathrm{d}^2V(r)}{\mathrm{d}r^2}\right)_{r=r_0},$$

where we have neglected higher terms as displacements from equilibrium are small. From Figure 8.43 we see that $V(r_0) = 0$. The second term in the equation is also equal to zero as the first derivative of $V(r)$ must vanish at $r = r_0$, where $V(r)$ is a minimum. Hence, we have

$$V(r) = \frac{(r - r_0)^2}{2}\left(\frac{\mathrm{d}^2V(r)}{\mathrm{d}r^2}\right)_{r=r_0}.$$

This expression has the same form as Equation (8.8), which gives the potential energy of a simple harmonic oscillator, with $\alpha = \left(\left[\mathrm{d}^2V(r)\right]/\mathrm{d}r^2\right)_{r=r_0}$. The same result would apply if $V(r)$ was finite at $r = r_0$.

8.8.1 Thermal expansion

It is a familiar observation that solids expand when heated. If the length of a solid at 0°C is l_0, its length l at θ° C is given by the equation

$$l = l_0(1 + \alpha\theta). \tag{8.17}$$

Here, α is the coefficient of linear expansion and has a value ~10^{-5} per degree. We can obtain a physical picture of thermal expansion of a solid by considering the potential energy between two atoms as a function of their separation.

Figure 8.45 represents the asymmetric potential energy curve $V(r)$ for two neighbouring atoms. The vertical axis is the thermal energy E of the atoms. The horizontal lines indicate the range of vibrational motion of the atoms interacting with this potential for a number of different thermal energies. Where a horizontal line meets the solid curve $V(r)$, corresponding to the turning point of the motion, the energy E is entirely potential. As the thermal energy increases, the range of vibrational motion increases. However, the mean position of vibration also increases because of the asymmetry of the potential curve, i.e. the mean separation of the atoms increases as indicated by the dashed red line. The thermal energy increases with temperature. Hence, with increasing temperature, the mean separation of the atoms increases. This happens throughout the solid, which therefore expands when it is heated.

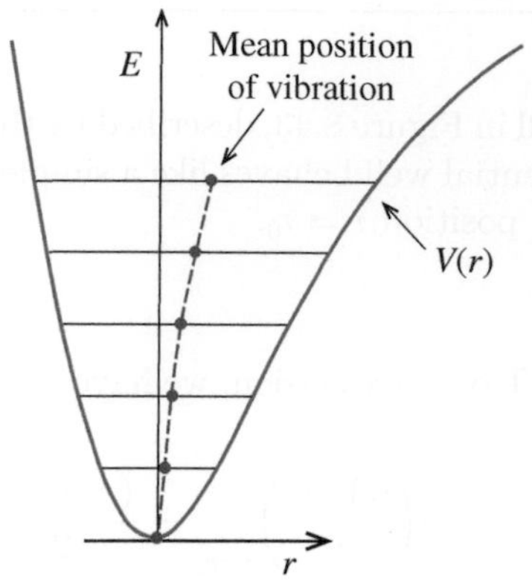

Figure 8.45 Representation of the asymmetric potential energy curve $V(r)$ for two neighbouring atoms. The vertical axis is the thermal energy E of the atoms and the horizontal lines represent different values of E. These lines also indicate the range of vibrational motion of the atoms at the respective value of E. Where a horizontal line meets the solid curve $V(r)$, corresponding to the turning point of the motion, the energy E is entirely potential. As the thermal energy increases, the range of vibrational motion increases. However, the mean position of vibration also increases because of the asymmetry of the potential curve, i.e. the mean separation of the atoms increases as indicated by the dashed red line.

Problems 8

8.1 The atomic radii of caesium and bromine are 0.167 nm and 0.196 nm, respectively. What type of crystal structure would you expect CsBr to adopt?

8.2 The heat of sublimation of a diamond is 715 kJ/mol. Calculate the binding energy of the covalent bond between two carbon atoms in the diamond.

8.3 The crystal of sodium fluoride is a cube in which alternative sites are occupied by Na^+ and F^- ions. The lattice parameter a of NaF is 0.231 nm. (a) Show that the molar volume of NaF is $2N_A a^3$. (b) The density of NaF is 2780 kg/m^3 and the atomic weights of sodium and flourine are 23 u and 19 u, respectively. Use these data to obtain a value for Avogadro's number.

8.4 What is the maximum fraction of the available volume that can be filled by solid spheres when these spheres are arranged as a simple cubic lattice?

8.5 Show (i) that the separation of the crystal planes in Figure 8.35b is $d = a/\sqrt{2}$, where a is the distance between atoms and (ii) that the separation of the crystal planes in Figure 8.35c is $d = a/\sqrt{3}$.

8.6 A hydrostatic pressure that is applied to a solid produces a volume change in the solid. If an applied pressure P produces a change ΔV in the volume of the solid, the bulk modulus B of the solid is defined as $B = -P(V/\Delta V)$. Calculate the mean change in the energy of each interatomic bond when copper is subjected to a pressure of 100 atm. Copper has a bulk modulus of 1.4×10^{11} Pa, a density of 8.9×10^3 kg/m^3, its atomic mass is 63.5 u and it has a face-centred cubic structure.

8.7 (a) A particular set of crystal planes in the NaCl crystal has a spacing of 0.282 nm. What is the maximum wavelength of an X-ray beam that will show a diffraction maximum in the $n = 3$ order of reflection? (b) When an X-ray beam of wavelength 0.154 nm is incident on a particular plane of silicon, it is found that a diffraction maximum in the $n = 1$ order is observed at an angle $\theta = 35°$ with respect to the planes. (i) What is the spacing of the planes? (ii) Would higher-order diffraction maxima occur at larger values of θ?

8.8 Diffraction maxima are observed from a sample of powdered tungsten at *scattering* angles of 40.2°, 58.3° and 73.3° for an X-ray wavelength of 0.154 nm. All the maxima correspond to the $n = 1$ order of reflection. Determine the separations of the crystal planes for these reflections, and obtain the ratios between these separations in terms of integers of low value.

8.9 (a) Through what voltage should electrons be accelerated so that they have the same de Broglie wavelength as that of a beam of neutrons at temperature of 22°C?

(b) The energies of neutrons emitted from nuclear fission reactions are typically ~2 MeV. Explain whether such energetic neutrons would be useful for determining the crystalline structure of solids.

8.10 The expression for the de Broglie wavelength of a relativistic particle is

$$\lambda = \frac{h}{p} = \frac{hc}{\sqrt{K(K + 2mc^2)}},$$

where K is the kinetic energy of the particle and m is its mass. The 'size' of a proton is about 1 fm. What kinetic energy, in eV, would an electron need to have to investigate the structure of the proton? You may assume that this kinetic energy is much greater than the rest mass energy of the electron.

8.11 Crystals of sodium chloride show strong absorption of electromagnetic radiation at a wavelength of 60 μm. Assuming this is due to simple harmonic vibrations of the sodium atoms, calculate (a) the frequency ν of the vibrations, (b) the force constant α, (c) the potential energy $V(r)$ of a sodium atom as a function of its distance r from its equilibrium position, (d) the probability function $P[r]$ and (e) the rms value of r at $T = 400$ K. The atomic weight of sodium is 23 u.

$$\text{Note,} \quad \int_{-\infty}^{+\infty} e^{-ax^2} = \sqrt{\frac{\pi}{a}}.$$

8.12 The interatomic potential between atoms in solid argon can be described by the expression

$$V(r) = \varepsilon\left[\left(\frac{a_0}{r}\right)^{12} - 2\left(\frac{a_0}{r}\right)^{6}\right],$$

where ε is the binding energy, r is the separation and a_0 is the equilibrium separation. For a displacement x from the equilibrium separation, the potential energy increases by ΔV. Use a Taylor expansion about a_0 to show that

$$\Delta V = 36\varepsilon\left(\frac{x}{a_0}\right)^2 - 252\varepsilon\left(\frac{x}{a_0}\right)^3,$$

when the Taylor expansion is taken up to and including the term in x^3.

9

The elastic properties of solids

Solids are characterised by having a rigid shape. However, we can deform the shape of a solid object by applying an appropriate force to it. A spring provides a familiar example. When we pull on a spring, it extends. The distance x by which a spring extends is described by Hooke's law:

$$F = \alpha x, \tag{9.1}$$

where F is the applied force and the constant of proportionality α is the spring constant. This is an empirical law and applies so long as the extension x is small compared to the length of the spring. Moreover, when we release the spring, it returns to its original length. A solid that returns to its original shape when the applied force is removed is called *elastic*. The property of elasticity is generally observed in solids. It is usually the case that a solid returns to its original shape when an applied force is removed, so long as the deformation is relatively small, with the amount of deformation being proportional to the force.

In this chapter, we discuss the behaviour of solids under the influence of external forces in various circumstances; how they may stretch, compress or twist. And we relate this behaviour to the interatomic forces and energies of the atoms in the solid.

9.1 Stress, strain, and elastic moduli

For solids, there are three different forms of deformation, which are illustrated in Figure 9.1. The dashed lines represent the original shape of the solid, and note that the deformations are greatly exaggerated for the purpose of illustration. We will usually consider situations where the amount of deformation is very much less than the dimensions of the solid. In Figure 9.1a, a bar is stretched; in Figure 9.1b, a rectangular block is *sheared*, i.e. twisted; and in Figure 9.1c, a spherical object is compressed uniformly on all sides. What the three forms of deformation have in common is that a deforming force called *stress* produces a deformation called *strain*. For stretching in just one direction, the stress is called *tensile stress*; for twisting, it is called *shear stress*; and for volume compression, it is called *bulk stress*. So long as the deformation is relatively

Physics of Matter, First Edition. George C. King.

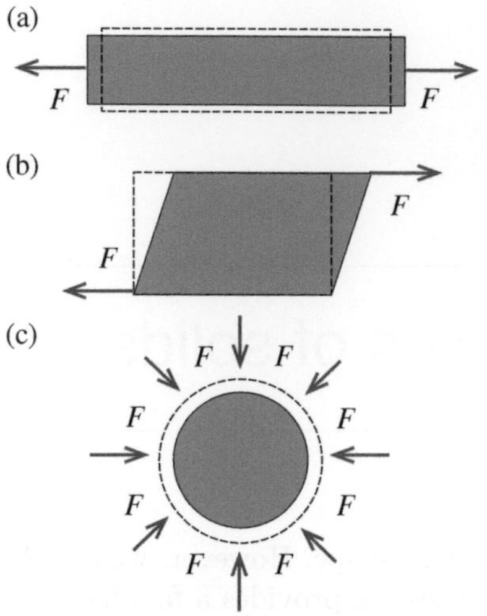

Figure 9.1 The three different forms of deformation for solids. The dashed lines represent the original, un-deformed shape of the solid and the arrows indicate the applied forces. In (a) a bar is stretched; in (b) a rectangular block is twisted; and in (c) a spherical object is compressed uniformly on all sides. The deformations are greatly exaggerated for the purpose of illustration.

small, it is found empirically that stress and strain are linearly proportional to each other. The constant of proportionality is called the *modulus of elasticity*, so that

$$\text{Stress} = \text{Elastic modulus} \times \text{Strain}. \tag{9.2}$$

We make a distinction here between a modulus of elasticity and the spring constant of Hooke's law. Hooke's law when applied to a particular object such as a spring deals with the details of the object in its entirety; its shape, size, the material it is made from and how that material was prepared. By contrast, an elastic modulus depends *only* on the material of which the object is made.

9.1.1 Tensile and compressional stress and strain

Figure 9.2 shows a bar of material that initially has length l and cross-sectional area A. It is secured to a rigid wall at one end and a longitudinal force F is applied to the other end. There is a reactive force at the wall that has the same magnitude as the applied force but acts in the opposite direction. This ensures that the bar has no tendency to move left or right, although it is in tension. When an object is stretched in only one direction, the tensile stress is defined as the ratio of the applied force F to the cross-sectional area A:

$$\text{Tensile stress} = \frac{F}{A}. \tag{9.3}$$

This is a scalar quantity because F is the magnitude of the force. The unit of stress is the pascal (Pa).

The bar in Figure 9.2 stretches to a length $l + \Delta l$, under tension. (The bar also suffers a contraction in the direction at right angles to the stretch, as described in Section 9.2.) We emphasise that the stress acts

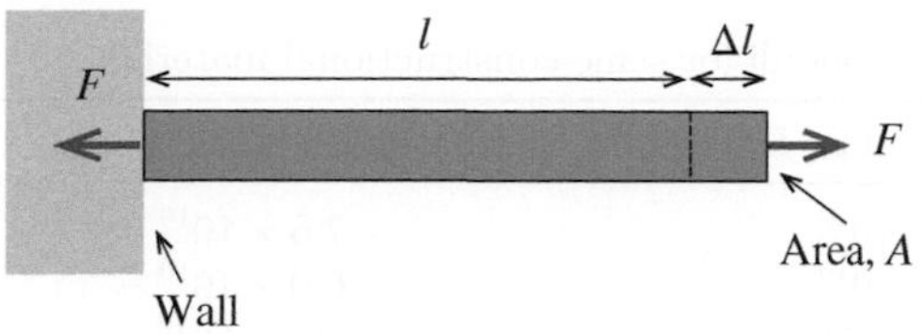

Figure 9.2 The figure shows a bar of material that initially has length l and cross-sectional area A. It is fixed to a rigid wall at one end and a longitudinal force F is applied to the other end. There is a reactive force F at the wall that has the same magnitude as the applied force but acts in the opposite direction. The applied force increases the length of the bar to $l + \Delta l$. The ratio of the applied force F to the cross-sectional area A is the tensile stress. The fractional change $\Delta l/l$ is the tensile strain. A stretched bar also suffers a contraction, not shown, in the direction at right angles to the stretch, as described in Section 9.2.

uniformly along the bar and, similarly, the strain occurs uniformly along the bar. The tensile strain is defined as the fractional change in the length $\Delta l/l$ of the bar:

$$\text{Tensile strain} = \frac{\Delta l}{l}. \tag{9.4}$$

As tensile strain is a ratio of two lengths, it is a dimensionless number. If the value of $\Delta l/l$ is small, typically less than 1% or 0.01%, the strain is linearly proportional to the stress, and the constant of proportionality is called *Young's modulus E*, where

$$\frac{F}{A} = E\frac{\Delta l}{l}, \tag{9.5}$$

or

$$E = \frac{\text{Tensile stress}}{\text{Tensile strain}} = \frac{F/A}{\Delta l/l}. \tag{9.6}$$

As $\Delta l/l$ is a dimensionless ratio, the units of Young's modulus are the same as those of F/A, i.e. pascals. Note that we can rearrange Equation (9.6) as

$$F = \left(\frac{AE}{l}\right)\Delta l, \tag{9.7}$$

when it has an analogous form to Hooke's law.

In previous discussions, we saw that a solid is composed of atoms that are held together by interatomic forces, which give a solid its rigidity. And we modelled these bonds as Hooke's law springs. Roughly speaking, when we stretch a solid, we are pulling the atoms of the solid further apart. Then when the stress is removed, the atoms return to their equilibrium separations.

Typical values of Young's modulus for some constructional materials are shown in Table 9.1, together with values of their other elastic moduli. For example, Young's modulus for steel is 20×10^{10} Pa. If a car of mass 1.5×10^{3} kg is suspended from a factory roof by a steel rod of length 2.0 m and diameter 30 mm, the rod extends by the amount

$$\Delta l = \frac{Fl}{EA} = \frac{1.5 \times 10^{3} \times 9.8 \times 2.0}{20 \times 10^{10} \times \pi(15 \times 10^{-3})^{2}} = 2.1 \times 10^{-4}\,\text{m} = 0.21\,\text{mm},$$

Table 9.1 Typical values of elastic moduli for some constructional materials.

Material	Young's modulus, E (Pa)	Bulk modulus, B (Pa)	Shear modulus, G (Pa)
Aluminium	7.3×10^{10}	7.6×10^{10}	2.5×10^{10}
Brass	9.0×10^{10}	6.0×10^{10}	4.0×10^{10}
Copper	13×10^{10}	15×10^{10}	4.5×10^{10}
Glass	6.5×10^{10}	5.5×10^{10}	2.6×10^{10}
Steel	20×10^{10}	16×10^{10}	7.7×10^{10}
Rubber	$\sim 1 \times 10^{5}$	$\sim 5 \times 10^{11}$	$\sim 5 \times 10^{5}$
Graphene	1.0×10^{12}	—	1.1×10^{9}
Diamond	1.2×10^{12}	4.5×10^{11}	5.3×10^{11}

or by $\sim$0.01% of its original length. In engineering applications, tensile strain rarely exceeds ± 0.1 %.

If the forces acting on the bar in Figure 9.2 push rather than pull the bar, the bar is said to be under *compressional stress*. Compressional strain of the bar is defined in the same way as the tensile strain, Equation (9.4), but Δl now corresponds to a reduction in length and is a negative quantity.

9.1.2 Strength of solid materials

The strength of a solid material is essentially a measure of how much stress it can withstand without fracture. More specifically, the *ultimate tensile strength* or just *tensile strength* of a material is the stress that will cause failure in a specimen of that material if it is stretched. Similarly, the *ultimate compressive strength* or just *compressive strength* of a material is the stress that will cause failure in a specimen of that material if it is compressed. Importantly, these ultimate strengths of a material are much smaller than its Young's modulus, typically by two or more orders of magnitude. This is due to tiny cracks and imperfections in the crystalline structure of the material that greatly reduce its strength.

A material may have similar values of tensile and compressive strength or these may be very different. Stone and concrete, for example, are very strong in compression but relatively weak in tension; the tensile strength of concrete is 4 MPa and the compressive strength is 40 MPa. Hence, stone structures are designed to avoid tensile stress. This is evident, for example, in the design of a Roman arch, which is illustrated in Figure 9.3. Here the stones of the arch compress the stones beneath them and none of the stones is under

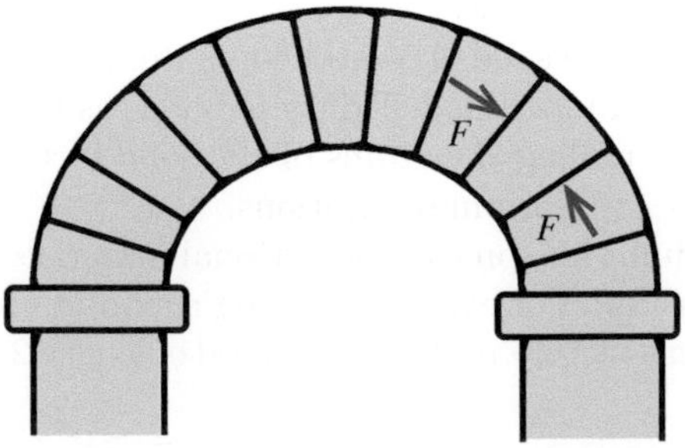

Figure 9.3 The stones of a Roman arch are arranged so that they compress the stones beneath them and none of the stones is under tension.

tension. Reinforced concrete is now a common building material. It combines the high compressive strength of concrete with the high tensile strength of steel: 400 MPa. In Section 8.3, we noted the remarkably high strength of graphene due to the strong carbon–carbon bonding. This is borne out by the tensile strength of graphene, which is 130 GPa, i.e. 1.3×10^{11} Pa, making it 300 times stronger than steel.

9.1.3 Shear stress and strain

Shear stress and strain are illustrated in Figure 9.4, where the stress is due to forces of equal magnitude but opposite directions that act tangentially on a solid. Shear stress is again a force per unit area, but now the force is applied in a direction that lies in the plane of the area, rather than perpendicular to it. Shear stress is thus defined as the force F acting tangentially to the surface divided by A the area on which it acts:

$$\text{Shear stress} = \frac{F}{A}. \tag{9.8}$$

The shear stress causes one face of the solid to be displaced by distance Δx. The shear strain is defined as the dimensionless ratio of the displacement Δx to the transverse dimension h:

$$\text{Shear strain} = \frac{\Delta x}{h}. \tag{9.9}$$

The shear strain can also be defined as the angle θ shown in Figure 9.4 where θ is measured in radians. The action of shear stress may be likened to pushing one of the covers of a book, as illustrated by Figure 9.5.

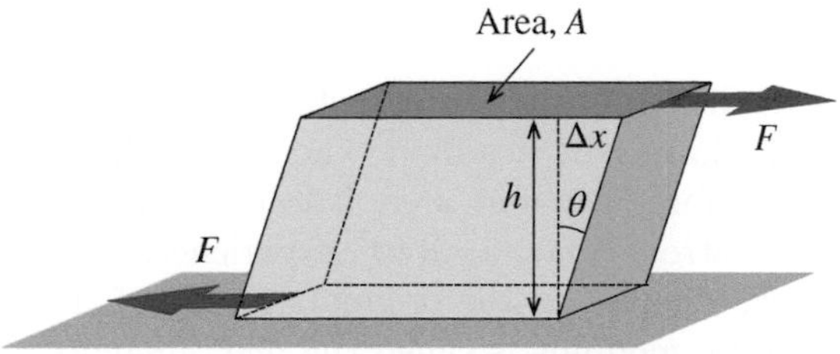

Figure 9.4 A solid under shear stress, where forces of equal magnitude F but of opposite direction act tangentially on an object. The shear stress is defined as the force F acting tangentially to the surface divided by A the area on which it acts. The shear strain is defined as the ratio of the displacement Δx to the transverse dimension h, or equivalently to the angle θ.

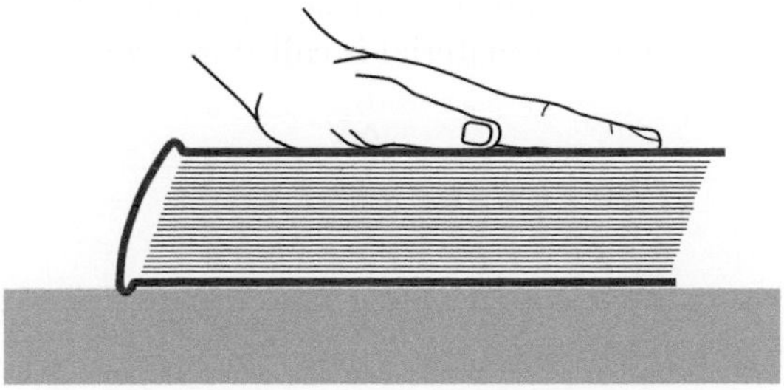

Figure 9.5 The action of shear stress acting on a solid may be likened to that of pushing the cover of a book. The pages slide past each other just as layers of atoms move over one another when an object is sheared.

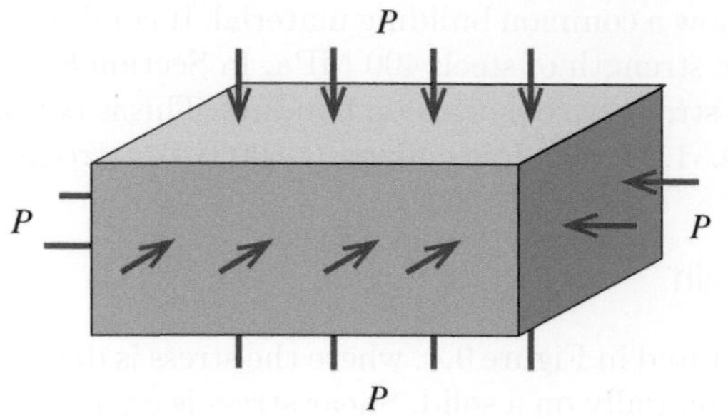

Figure 9.6 In the case of bulk stress, force is applied uniformly to all surfaces of an object, as occurs when an object is immersed in a fluid. As pressure is force per unit area, it follows that the bulk stress is the pressure P exerted on the object.

If the shear stress is sufficiently small, the shear strain is directly proportional to the shear stress and we have

$$G = \frac{\text{Shear stress}}{\text{Shear strain}} = \frac{F/A}{\Delta x/h}, \tag{9.10}$$

where G is called the *shear modulus.*

9.1.4 Bulk stress and strain

In the case of bulk stress, force is applied uniformly to all surfaces of a solid, as illustrated by Figure 9.6. This occurs, for example, where a solid is immersed in a fluid, when the solid is said to be under *hydraulic compression.* Recalling that pressure is force per unit area, it follows that the bulk stress is the pressure P exerted on the solid. It is the stress that a diver experiences underwater. The bulk strain is the dimensionless quantity $\Delta V/V$, where V is the original volume of the object and ΔV is the change in volume. The corresponding elastic modulus is called the *bulk modulus* B where, for small changes in volume, we have

$$B = -\frac{P}{\Delta V/V} = -P\frac{V}{\Delta V}. \tag{9.11}$$

In compression, ΔV is a negative quantity and so as to make B positive, this expression for B has a negative sign. The reciprocal of the bulk modulus of a material is called its *compressibility* β, where

$$\beta = -\frac{\Delta V}{PV}. \tag{9.12}$$

Worked example

A solid steel ball of diameter 150 mm falls to the bottom of the ocean to a depth of 3000 m. By what amount does the radius reduce, given that the bulk modulus for steel is 16×10^{10} Pa.

Solution

For a sphere of radius r, volume V is $(4/3)\pi r^3$. For a small change Δr in r, i.e. $\Delta r/r \ll r$, we can write

$$\Delta V = \frac{\mathrm{d}}{\mathrm{d}r}\left(\frac{4}{3}\pi r^3\right)\Delta r = 4\pi r^2 \Delta r,$$

giving

$$\frac{\Delta V}{V} = 3\frac{\Delta r}{r}.$$

Using $B = -P(V/\Delta V)$, we obtain, $\Delta r = -Pr/3B$.

Neglecting the small contribution from atmospheric pressure, and taking the density ρ of sea water to be 1036 kg/m^3, the pressure at 3000 m of sea water is

$$P = 3000\rho g = 3000 \times 1036 \times 9.8 = 3.05 \times 10^7 \text{ Pa}.$$

Hence,

$$\Delta r = -\frac{Pr}{3B} = -\frac{3.05 \times 10^7 \times 75 \times 10^{-3}}{3 \times 16 \times 10^{10}} = -4.8 \times 10^{-6} \text{ m}$$
$$= -4.8 \times 10^{-3} \text{ mm}.$$

9.2 Poisson's ratio

When a solid is *stretched* in one direction, it *contracts* in a direction at right angles to the stretch. The ratio

$$\frac{\text{Lateral contraction/original width}}{\text{Longitudinal extension/original length}} = \frac{\text{Lateral strain}}{\text{Longitudinal strain}} = \sigma \quad (9.13)$$

is called *Poisson's ratio*. Like Young's modulus, Poisson's ratio is a property of the material of the solid and does not depend on its shape. And for ordinary materials it has a positive value.

Consider Figure 9.7 which shows a block of material of length l, width w and height h. Suppose that the block is subject to a uniform compressional stress on all its faces, i.e. to hydrostatic pressure P. The compressional stress acting on the two ends of the block, Figure 9.7a, results in a compressional strain in the longitudinal direction given by

$$\frac{\Delta l_1}{l} = \frac{\text{Force/unit area}}{\text{Young's modulus}} = -\frac{P}{E}, \quad (9.14)$$

where the minus sign arises because the length l reduces. Similarly, the compressional stress acting on the two sides of the block, Figure 9.7b, results in the compressional strain

$$\frac{\Delta w}{w} = -\frac{P}{E}. \quad (9.15)$$

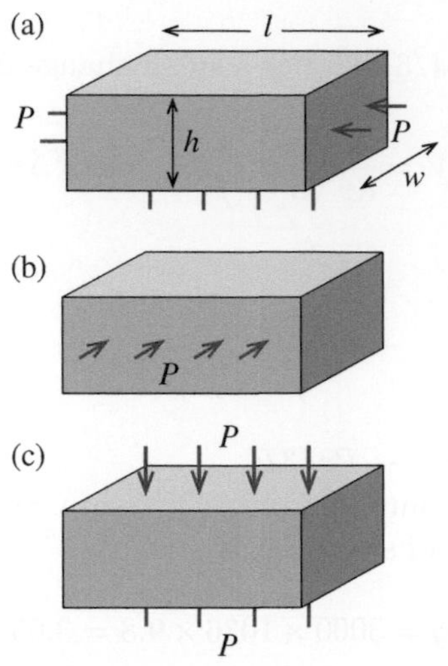

Figure 9.7 (a) Compressional stress acting on the two ends of the block results in compressional (negative) strain in the longitudinal direction. (b) Compressional stress acting on the two sides of the block results in lateral compressional (negative) strain but longitudinal (positive) strain. (c) Compressional stress acting on the top and bottom sides once again results in longitudinal (positive) strain.

The compressional stress acting on the two sides results in an extension Δl_2 in the length of the block, where the longitudinal strain $\Delta l_2/l$ and the lateral strain $\Delta w/w$ are related by Poisson's ratio, Equation (9.13):

$$\frac{\Delta l_2/l}{-\Delta w/w} = \sigma, \tag{9.16}$$

and hence,

$$\frac{\Delta l_2}{l} = +\sigma\frac{P}{E}. \tag{9.17}$$

Similarly, for the top and bottom sides of the block, Figure 9.7c, the compressional stress is once again $-P/E$, which results in a compressional strain $\Delta h/h = -P/E$, and a longitudinal strain

$$\frac{\Delta l_3}{l} = +\sigma\frac{P}{E}. \tag{9.18}$$

We can sum the individual values of strain $\Delta l_1/l$, $\Delta l_2/l$, and $\Delta l_3/l$ to obtain the total longitudinal strain $\Delta l/l$:

$$\frac{\Delta l}{l} = -\frac{P}{E} + \sigma\frac{P}{E} + \sigma\frac{P}{E} = -\frac{P}{E}(1-2\sigma). \tag{9.19}$$

In exactly the same way, we obtain

$$\frac{\Delta w}{w} = -\frac{P}{E}(1-2\sigma) \tag{9.20}$$

and

$$\frac{\Delta h}{h} = -\frac{P}{E}(1-2\sigma). \tag{9.21}$$

For values of $\Delta l/l$, $\Delta w/w$ and $\Delta h/h \ll 1$, we can write

$$\begin{aligned}\Delta V &= \left(\frac{\partial V}{\partial l}\right)\Delta l + \left(\frac{\partial V}{\partial w}\right)\Delta w + \left(\frac{\partial V}{\partial h}\right)\Delta h \\ &= wh\Delta l + lh\Delta w + lw\Delta h,\end{aligned} \tag{9.22}$$

giving

$$\frac{\Delta V}{V} = \frac{\Delta l}{l} + \frac{\Delta w}{w} + \frac{\Delta h}{h}. \tag{9.23}$$

Then from Equations (9.19), (9.20), and (9.21), we obtain

$$\frac{\Delta V}{V} = -\frac{3P}{E}(1-2\sigma). \tag{9.24}$$

We have from Equation (9.11), $B = -PV/\Delta V$, and hence

$$B = \frac{E}{3(1-2\sigma)}. \tag{9.25}$$

So, if we know the values of Young's modulus and Poisson ratio for a given material, we can determine its bulk modulus.

We see from Equation (9.25) that Poisson's ratio must be less than ½. If it were not, the bulk modulus would be negative and the solid would expand under increasing pressure. For most solids, Poisson's ratio has a value between ¼ and ⅓. Rubber is a notable exception. It has a Poisson's ratio of nearly 0.5, which means there is little change in volume when it is stretched.

Worked example

Consider a shear stress applied to a solid cube to obtain a relationship for shear modulus G in terms of Young's modulus E and Poisson's ratio σ.

Solution

We first consider the combination of tensile stress and compressional stress on a solid cube that has sides of length l, as illustrated in Figure 9.8. The cube is stretched in the horizontal direction and compressed in the vertical direction.

The horizontal strain due to the tensile stress F/l^2 is $\Delta l_1/l = F/El^2$.

The vertical strain due to the compressional stress is $-F/El^2$. This compressional stress increases the horizontal strain as given by Poisson's ratio:

$$\frac{\Delta l_2}{l} = -\sigma\left(-\frac{F}{El^2}\right) = \frac{\sigma F}{El^2}.$$

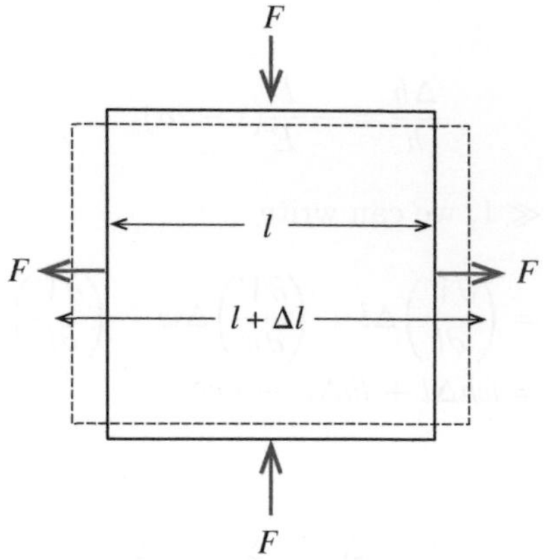

Figure 9.8 A solid cube that is stretched in one direction and compressed in the orthogonal direction.

Hence, the total strain along the horizontal direction is

$$\frac{\Delta l}{l} = \frac{\Delta l_1}{l} + \frac{\Delta l_2}{l} = \frac{F}{El^2}(1+\sigma) = \frac{F}{A}\frac{(1+\sigma)}{E},$$

where area $A = l^2$.

Consider now that the cube is subject to shear stress. We first note that the conventional way of showing shear stress as in Figure 9.4 is incomplete. If there were just the two forces shown in that figure the solid would not be in a state of equilibrium; it would rotate. To maintain a solid under equilibrium when it is subjected to shear stress, a pair of orthogonal forces must also exist. And these must produce an equal torque on the solid that acts in the opposite direction. This second set of forces is included in Figure 9.9a where for the case of a cubic solid, $F' = F$.

Imagine now that we cut the cube by a plane through the diagonal AB, shown in Figure 9.9a. Under our normal condition of small deformations, we can take the line AB to be at an angle of 45° with respect to the bottom side of the cube. Then, each of the four forces F and F' have components parallel to and normal to the plane, and these components all have magnitude $F\cos 45°$ or $F/\sqrt{2}$. It follows that the total force acting across the plane is $2 \times F/\sqrt{2} = \sqrt{2}F$, as illustrated by Figure 9.9b. This force acts over the area of the plane, which is $\sqrt{2}l \times l = \sqrt{2}l^2$. Therefore, the tensile stress normal to the plane is just F/l^2. Similarly, for the plane that passes through CD in Figure 9.9b, there is a compressional stress equal to $-F/l^2$. *We see that the shear stress on the cube is equivalent to a combination of tensile stress and compressional stress acting at right angles to each other.* This is a general result for pure shear stresses.

The tensile and compressional stresses act on the cube in the same way as we saw above in the case of the cube being stretched in one direction and compressed in the other. Then using

$$\text{Tensile strain} = \text{Tensile stress} \times \frac{(1+\sigma)}{E},$$

where the tensile stress is F/l^2 and the tensile strain is $\Delta D/D$, where ΔD is the increase in the length of the diagonal D of the cube, as shown in Figure 9.9. Therefore,

$$\frac{\Delta D}{D} = \frac{F}{l^2}\frac{(1+\sigma)}{E}.$$

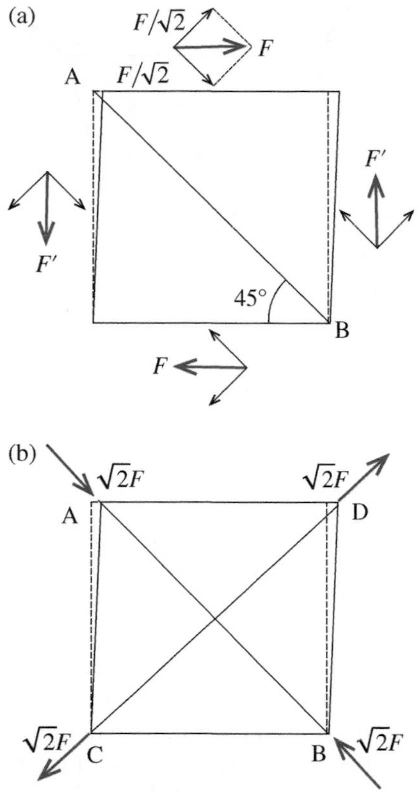

Figure 9.9 (a) A solid cube subject to a shear force F. To maintain an object under shear stress in equilibrium, a pair of orthogonal forces $F' = F'$ must also exist, as shown. This additional set of forces is not shown in the conventional representation of shear stress, as in Figure 9.4. For small deformations, the line AB can be assumed to make an angle of 45° with respect to the bottom side of the cube. In that case, each of the components of the four applied forces has magnitude $F\cos 45°$or $F/\sqrt{2}.$ (b) The shear stress on the cube is equivalent to a combination of tensile stress and compressional stress of equal strength acting at right angles to each other.

From Figure 9.10, where the deformation Δx is greatly exaggerated, we have

$$(D + \Delta D)^2 = l^2 + (l + 2\Delta x)^2.$$

Expanding this expression, ignoring the negligible terms $(\Delta D)^2$ and $(\Delta x)^2$ and using $D^2 = 2l^2$, we obtain $\Delta D/D = \Delta x/2l$. Substituting for $\Delta D/D$ here gives

$$\frac{\Delta x}{2l} = \frac{F}{l^2}\frac{(1+\sigma)}{E}.$$

With respect to Figure 9.10, we recall that shear stress

$$G = \frac{F/l^2}{\Delta x/l}.$$

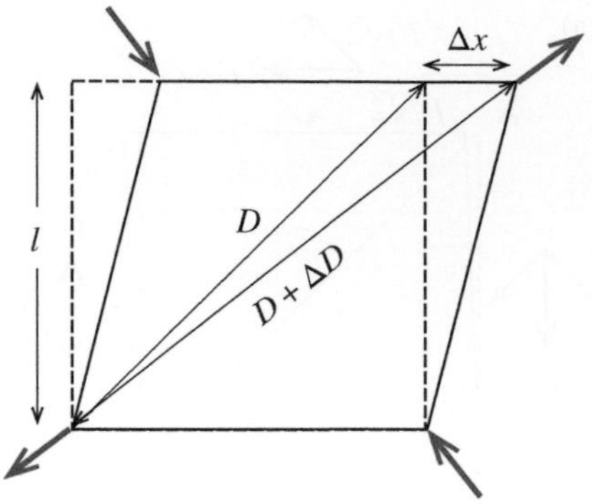

Figure 9.10 Shear stress causes one diagonal of the cube to be *slightly* elongated and the other diagonal to be *slightly* shortened.

Then substituting $\Delta x/l = F/Gl^2$ gives, on rearrangement

$$G = \frac{E}{2(1+\sigma)}. \tag{9.26}$$

Equation (9.26) is a general result. If we know the values of E and σ for a material, we can determine its shear modulus G. We have already seen that we can determine the bulk modulus B from E and σ, as $B = E/3(1 - 2\sigma)$. Hence, knowing E and σ for a material, we can know *all* its three elastic moduli. It also follows from these expressions for B and G that the three moduli for a given material are of a similar magnitude. This is generally the case as evidenced by Table 9.1. Rubber is again the exception, for which the bulk modulus is very much greater than the other two elastic moduli. It is not surprising that rubber has very different elastic properties from most other solids. Rather than being crystalline in nature, rubber consists of long-chain hydrocarbon molecules. These long chains are coiled up and tangled together, and the result of applying tensile stress is to pull these chains into partial alignment.

9.3 The velocity of sound in a thin wire

When a longitudinal wave passes down a long thin wire, each segment of the wire experiences periodic forces of compression and tension. By considering the elastic properties of the wire, we can derive an expression for the force acting on a segment of the wire and thereby deduce an expression for the velocity of the wave along the wire.

Figure 9.11a shows a small segment of the wire of length δx at position x, along an undisturbed length of the wire. When a longitudinal wave passes by, this segment is displaced in the x-direction under the influence of forces that are varying in both distance and time. We consider an instant of time where the wave causes the segment to be slightly extended; at a slightly different time, the wave would cause the segment to be slightly compressed. And, we introduce the variable ξ that measures the displacement of the segment from its undisturbed position. Figure 9.11b shows the segment when it is displaced from its equilibrium position. At time t, the left-hand end of the segment is at position $x + \xi$ and the right-hand end is at position $x + \delta x + \xi + \delta\xi$. The length of the segment when extended is therefore

$$x + \delta x + \xi + \delta\xi - (x + \xi) = \delta x + \delta\xi. \tag{9.27}$$

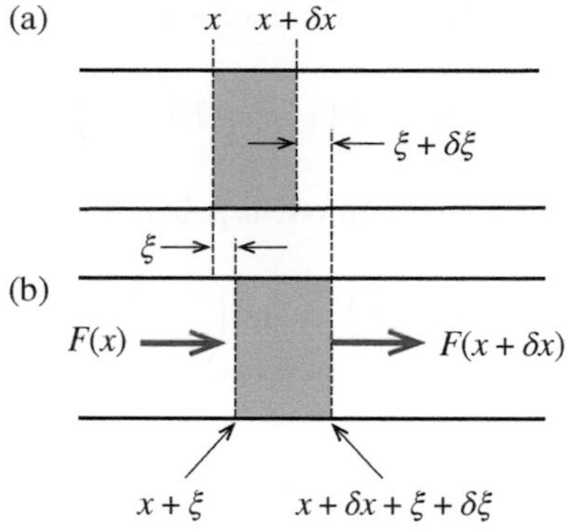

Figure 9.11 (a) A small segment of a long thin wire of length δx on an undisturbed length of the wire. (b) The segment when it is displaced from its equilibrium position by a longitudinal wave.

Hence, the increase in the segment's length is $\delta\xi$. We let the force on the left-hand edge of the segment at x be $F(x)$. We have

$$E = \frac{F(x)/A}{\delta\xi/\delta x},$$

where E is Young's modulus and A is the cross-sectional area of the wire, which gives

$$F(x) = AE\frac{\delta\xi}{\delta x}. \tag{9.28}$$

As $\delta\xi$ is made smaller and smaller, $\delta\xi/\delta x$ becomes a differential. Moreover, as the force changes with time as well as position, we write it as a partial differential, giving

$$F(x) = AE\frac{\partial\xi}{\partial x}. \tag{9.29}$$

Over the short length of the segment, the force will vary by only a small amount and we can write

$$F(x + \delta x) = F(x) + \left(\frac{\partial F}{\partial x}\right)\delta x, \tag{9.30}$$

where $F(x + \delta x)$ is the force on the right-hand edge of the segment. From Equation (9.29), we have

$$\frac{\partial F}{\partial x} = AE\left(\frac{\partial^2\xi}{\partial x^2}\right), \tag{9.31}$$

and hence,

$$F(x + \delta x) = F(x) + AE\left(\frac{\partial^2\xi}{\partial x^2}\right)\delta x, \tag{9.32}$$

where $\partial^2\xi/\partial x^2$ is evaluated at distance x. The net force on the segment is therefore

$$F(x+\delta x) - F(x) = AE\left(\frac{\partial^2\xi}{\partial x^2}\right)\delta x. \tag{9.33}$$

If the density of the wire is ρ, the mass of the segment is $\rho A\delta x$. Then from Newton's second law, we have

$$\rho A\delta x\left(\frac{\partial^2\xi}{\partial t^2}\right) = AE\left(\frac{\partial^2\xi}{\partial x^2}\right)\delta x, \tag{9.34}$$

giving

$$\left(\frac{\partial^2\xi}{\partial t^2}\right) = \frac{E}{\rho}\left(\frac{\partial^2\xi}{\partial x^2}\right). \tag{9.35}$$

This equation has the form of the one-dimensional wave equation

$$\left(\frac{\partial^2\phi}{\partial t^2}\right) = v^2\left(\frac{\partial^2\phi}{\partial x^2}\right), \tag{9.36}$$

where v is the velocity of the wave. In the present case, ϕ is the displacement ξ of the segment. Hence,

$$\left(\frac{\partial^2\xi}{\partial t^2}\right) = v^2\left(\frac{\partial^2\xi}{\partial x^2}\right). \tag{9.37}$$

By comparing Equations (9.35) and (9.37), we see that the velocity of the waves down the wire is

$$v = \sqrt{E/\rho}. \tag{9.38}$$

So, by measuring the velocity of waves down a thin wire, we can determine its Young's modulus. Indeed the most precise values of the elastic properties of materials are provided by acoustic measurements. Young's modulus for copper is 13×10^{10} Pa and its density is 8940 kg/m^3. This gives the velocity of sound in copper wire to be 3.8 km/s. By comparison, the velocity of sound in air is 330 m/s.

We have obtained the expression for the velocity of longitudinal waves on a wire. More generally, the velocity of waves in solid media of density ρ is given by

$$v = \sqrt{\frac{\text{Elastic modulus}}{\rho}}, \tag{9.39}$$

when the appropriate elastic modulus is used. For example, for transverse waves in a solid medium, the appropriate modulus is the shear modulus G.

9.4 Torsional stress and strain

If a solid rod is securely clamped at one end and a torque τ is applied to the unclamped end, every part of the rod is subjected to shear stress, also called torsional stress. In Figure 9.12, the torque comes from the two tangential forces F applied to the unclamped end. The result is that this end of the rod becomes twisted through an angle ϕ, where ϕ is related to the applied torque by

$$\tau = c \times \phi. \tag{9.40}$$

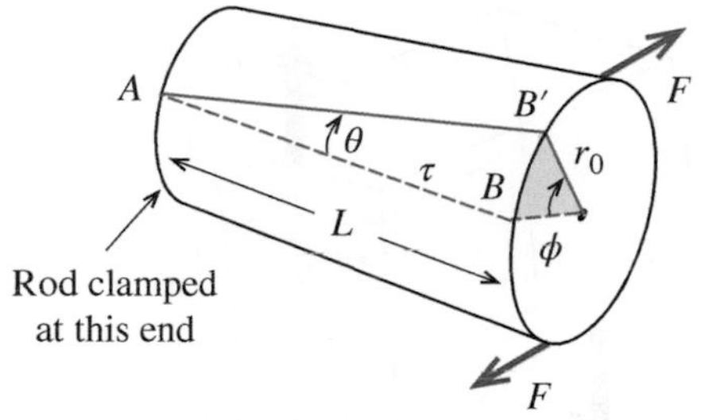

Figure 9.12 A solid rod that is securely clamped at one end is subjected to shear or torsional stress by a torque τ. The torque is due to two tangential forces F applied to the rod. The result is that the free end of the rod becomes twisted through the angle ϕ. The dashed line AB on the surface of the rod is parallel to its longitudinal axis. When the torsional stress is applied, the end B becomes displaced to B'. The torsional strain at the surface of the rod is equal to the angle θ.

The constant c is the opposing couple per unit angle of twist. It is called the *torsional constant*, and depends on the shear modulus G of the rod.

In Figure 9.12, the dashed line AB on the surface of the rod is parallel to its longitudinal axis. When the torsional stress is applied, the end B becomes displaced to B'. The torsional strain at the surface of the rod is equal to the angle θ. From Figure 9.12, we see that

$$\theta = \frac{BB'}{L} = \frac{r_0\phi}{L}, \tag{9.41}$$

where r_0 is the radius of the rod and L is its length. Similarly, the torsional strain at any other radial distance r within the rod is

$$\theta = \frac{r\phi}{L}. \tag{9.42}$$

We see that the torsional strain and hence the torsional stress are not constant within the solid rod, but increase linearly with r.

Figure 9.13 shows a thin cylindrical shell of the solid rod. It has length L, and thickness $\mathrm{d}r$ and radius r, where $0 \leq r \leq r_0$. The figure also shows an elemental segment of this cylindrical shell, with length $\mathrm{d}l$ and

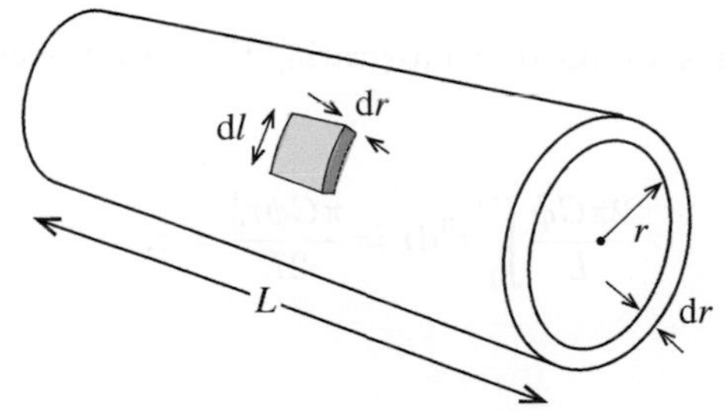

Figure 9.13 A thin cylindrical shell within the solid rod shown in Figure 9.12. The shell has length L, thickness $\mathrm{d}r$, and radius r, where $0 \leq r \leq r_0$. The figure also shows an elemental segment of the shell that has length $\mathrm{d}l$ and thickness $\mathrm{d}r$.

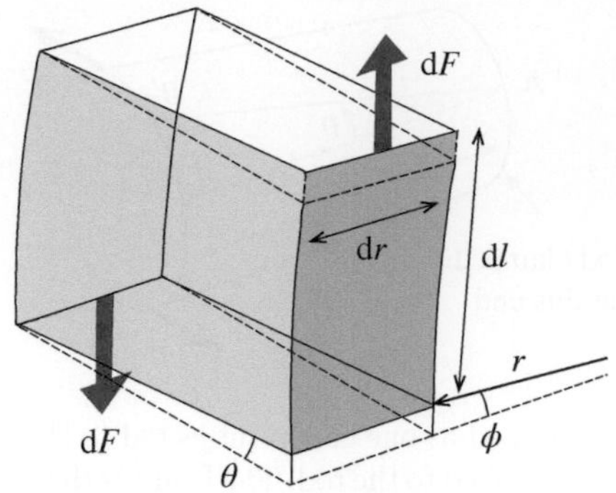

Figure 9.14 An enlarged picture of the elemental segment of the cylindrical shell shown in Figure 9.13. The dashed lines show the un-deformed shape of the segment and the solid lines shows its shape under torsional stress.

thickness dr. Figure 9.14 shows an enlarged picture of the segment and the shearing forces that act on it. The dashed lines show the original shape of the segment and the solid lines shows its shape under torsional stress.

The segment is subject to a shearing force dF, which results in a shear strain θ, where $\theta = r\phi/L$. We have shear modulus $G =$ shear stress/shear strain, and hence,

$$G = \frac{\text{Shear stress}}{\text{Shear strain}} = \frac{\mathrm{d}F/(\mathrm{d}l \times \mathrm{d}r)}{\theta} = \frac{L \times \mathrm{d}F/(\mathrm{d}l \times \mathrm{d}r)}{r\phi},$$

giving

$$\mathrm{d}F = \frac{G\phi r \mathrm{d}l \mathrm{d}r}{L}. \tag{9.43}$$

This force acting on the segment contributes a torque about the longitudinal axis of the rod equal to $r \times \mathrm{d}F$. The total torque acting on the cylindrical shell is the sum of such torques around a complete circumference $2\pi r$ of the shell:

$$\frac{G\phi r^2 \mathrm{d}r}{L}\int_0^{2\pi r} \mathrm{d}l = \frac{2\pi G\phi r^3 \mathrm{d}r}{L}. \tag{9.44}$$

The total torque acting on the rod is obtained by integrating this result over r from 0 to r_0. Hence, the total torque acting on the rod is

$$\frac{2\pi G\phi}{L}\int_0^{r_0} r^3 \mathrm{d}r = \frac{\pi G\phi r_0^4}{2L} = \tau, \tag{9.45}$$

giving

$$\tau = \frac{\pi G r_0^4}{2L}\phi. \tag{9.46}$$

We see that the torque is proportional to the fourth power of the radius of the rod. This means that a rod that is twice as thick is 16 times as stiff for torsion. From Equations (9.40) and (9.46), the torsional constant is

$$c = \frac{\pi G r_0^4}{2L}. \tag{9.47}$$

By having a large value of L and a small value of r_0 we can obtain an exceedingly small value of c; the restoring force per unit angle of rotation. And this enables extremely small forces to be measured. The classic example of this is the *torsion balance* used in Cavendish's apparatus for measuring the gravitational constant. The torsion balance has a long thin wire of quartz or phosphor bronze, and has sufficient sensitivity to measure the tiny gravitational force between two laboratory-sized objects.

9.5 Elastic moduli and interatomic forces and potential energies

We have the physical picture that when we stretch a solid we pull its atoms further apart. Similarly, when we compress a solid, we push the atoms closer together. We now develop this picture and relate elastic moduli to the forces between the atoms and their potential energies.

We have our familiar representation of the potential energy $V(r)$ between two neutral atoms as a function of their separation r, as shown by the dashed curve in Figure 9.15. The relative positions of the two atoms, which are not drawn to scale, are shown above the curve. For the sake of simplicity, the position of one of the atoms is fixed at $r = 0$. At equilibrium, the atoms lie at their equilibrium separation $r = a_0$, where the repulsive and attractive forces between the atoms balance.

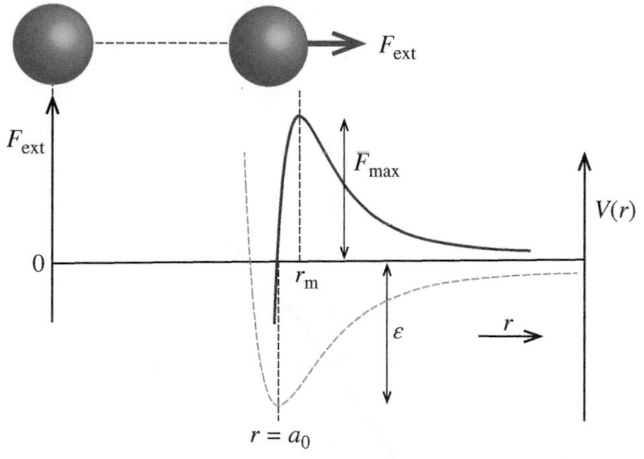

Figure 9.15 The dashed curve represents the potential energy $V(r)$ between two neutral atoms as a function of their separation r. The relative positions of the two atoms, which are not drawn to scale, are shown above the curve. For the sake of simplicity, the position of one of the atoms is fixed at $r = 0$. At equilibrium, the atoms lie at their equilibrium separation $r = a_0$. The blue curve is a plot of the external force F_{ext} required to change the interatomic separation r. The vertical axes for F_{ext} and $V(r)$ are different; the former has units of force and the latter has units of energy. However, the horizontal axes for both curves, corresponding to the interatomic separation are the same and are aligned with respect to each other.

Suppose we want to increase the separation of the two atoms by an amount δr, thereby increasing their potential energy by δV. This requires an external force F_{ext}, where the work done is $F_{ext}\delta r$, which is equal to the change in potential energy δV. Hence, $F_{ext}\delta r = \delta V$, and in the limit $\delta r \to 0$, we have

$$F_{ext} = \frac{dV(r)}{dr}. \tag{9.48}$$

F_{ext} is plotted in Figure 9.15, as the blue curve. Of course, the vertical axes for the F_{ext} and $V(r)$ curves are different; the former has units of force and the latter has units of energy. However, the horizontal axes for both curves, corresponding to interatomic separation r, are the same and are aligned with respect to each other.

We interpret the F_{ext} curve as follows. For positive values of F_{ext}, the atoms are pulled apart, while for negative values of F_{ext}, they are pushed together and $F_{ext} = 0$ at $r = a_0$. If F_{ext} is steadily increased in the positive direction, the curve eventually reaches a maximum value F_{max} at $r = r_m$. The external force F_{ext} required to increase the separation beyond r_m reduces as the force of attraction between the atoms reduces. We interpret F_{max} as the maximum external force that the pair of atoms can sustain. Hence, if the external force is greater than this value, the two atoms are pulled completely apart.

Figure 9.16 shows an enlarged view of the F_{ext} curve close to the equilibrium separation $r = a_0$. We can take this curve to be linear for small changes in the separation r of the two atoms; see also discussion in Section 8.6. From this figure, we see that if we want to increase the interatomic separation by an amount $(r - a_0) = \Delta r$, we need to apply an external force of magnitude ΔF_{ext}, where

$$\frac{\Delta F_{ext}}{\Delta r} = \text{slope of the } F_{ext} \text{ curve at } r = a_0.$$

This gives

$$\Delta F_{ext} = \Delta r \left(\frac{dF_{ext}(r)}{dr} \right)_{r=a_0}. \tag{9.49}$$

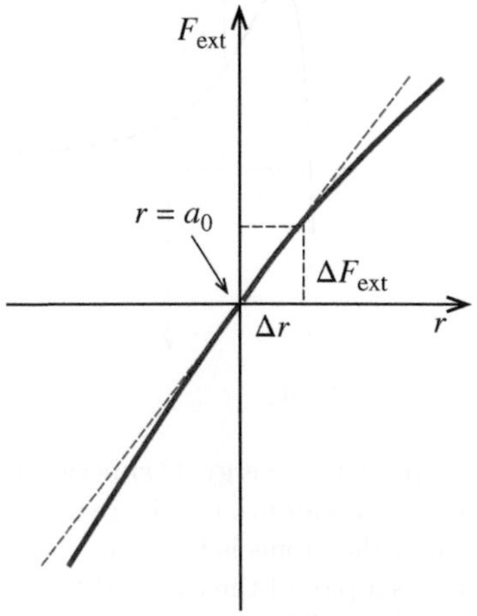

Figure 9.16 An enlarged view of the F_{ext} curve close to the equilibrium separation $r = a_0$. The curve is approximately linear for small changes in the separation of the two atoms.

Substituting for F_{ext} from Equation (9.48), we obtain

$$\Delta F_{ext} = \Delta r \left(\frac{d^2 V(r)}{dr^2} \right)_{r=a_0}. \tag{9.50}$$

This expression gives the force ΔF_{ext} required to change the inter-nuclear separation by Δr in terms of the potential energy $V(r)$ of the atoms.

9.5.1 Young's modulus

We now consider a simple model for the stretching of a solid due to tensile stress. We consider the solid to be a simple cubic array of atoms that are separated from each other by distance a_0. Figure 9.17a shows three planes of atoms. We assume that an atom only interacts with its nearest neighbour in front of it and its nearest neighbour behind it. And we model the forces between those atoms by Hooke's-law springs. This gives individual lines of interacting atoms as shown in Figure 9.17a.

Figure 9.17b shows the three planes when force F is applied across the entire solid. This increases the separation of the atomic planes to $a_0 + \Delta r$. If the cross sectional area of the solid that is perpendicular to the applied force is A, the number of atoms in the planes is A/a_0^2, giving A/a_0^2 lines of connected atoms. We may therefore say that the force f that is applied to an individual line of connected atoms is the applied force F divided by the total number of lines of connected atoms:

$$f = F \div A/a_0^2 = \frac{Fa_0^2}{A}. \tag{9.51}$$

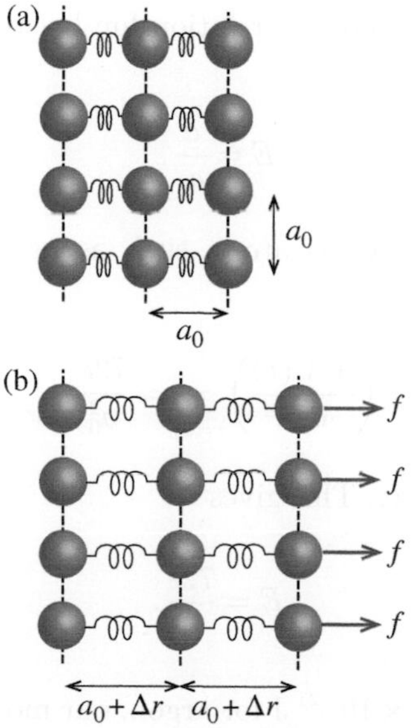

Figure 9.17 A simple model for the stretching of a solid due to tensile stress. (a) Three planes of atoms in the un-deformed state of the solid. (b) The three planes when a force is applied across the entire solid that increases the separation of the planes to $a_0 + \Delta r$. f is the external force that is applied across each line of connected atoms.

The force f increases the separation of each pair of atoms in an individual line by Δr, where, from Equation (9.50), we have

$$f = \Delta r\left(\frac{\mathrm{d}^2 V(r)}{\mathrm{d}r^2}\right)_{r=a_0}. \tag{9.52}$$

If the length of the solid is l, the number of atoms in a line is l/a_0. Therefore, the total increase Δl in the length of the solid is $\Delta r l/a_0$. Substituting for $\Delta r = a_0 \Delta l/l$ and $f = F a_0^2/A$ in Equation (9.52), we obtain

$$\frac{F}{A} = \frac{1}{a_0}\frac{\Delta l}{l}\left(\frac{\mathrm{d}^2 V(r)}{\mathrm{d}r^2}\right)_{r=a_0}, \tag{9.53}$$

or

$$\frac{F/A}{\Delta l/l} = \frac{1}{a_0}\left(\frac{\mathrm{d}^2 V(r)}{\mathrm{d}r^2}\right)_{r=a_0}. \tag{9.54}$$

$(F/A)/(\Delta l/l)$ is just Young's modulus, and hence,

$$E = \frac{1}{a_0}\left(\frac{\mathrm{d}^2 V(r)}{\mathrm{d}r^2}\right)_{r=a_0}. \tag{9.55}$$

In Section 9.3, we identified $\left(\mathrm{d}^2 V(r)/\mathrm{d}r^2\right)_{r=a_0}$ with the spring constant α, i.e. the strength of the bond between the two atoms. This gives the following relationship between Young's modulus E and the bond strength:

$$E = \frac{\alpha}{a_0}. \tag{9.56}$$

For the particular case of a van der Waals atom, such as argon, we also saw, Equation (8.12) in Section 8.8, that

$$\left(\frac{\mathrm{d}^2 V(r)}{\mathrm{d}r^2}\right)_{r=a_0} = \frac{72\varepsilon}{a_0^2},$$

where ε is the interatomic binding energy. This gives

$$E = \frac{72\varepsilon}{a_0^3}. \tag{9.57}$$

Taking $a_0 = 3.7 \times 10^{-10}$ m and $\varepsilon = 1.7 \times 10^{-21}$ J for argon, our model gives

$$E = \frac{72 \times 1.7 \times 10^{-21}}{(3.7 \times 10^{-10})^3} = 2.4 \times 10^9 \text{ Pa}.$$

This value compares very favourably with the measured value of Young's modulus for solid argon, which is 4.8×10^9 Pa. We can expect our value to be an underestimate because we neglected the interaction, i.e. attraction of some of the neighbouring atoms.

Worked example

Obtain an expression for the theoretical value of the ultimate tensile strength of a van der Waals solid in terms of its Young's modulus.

Solution

The maximum restoring force F_{max} that a pair of atoms can provide occurs at the maximum of the F_{ext} curve; see Figure 9.15. Hence, a fracture will occur when the force $f = Fa_0^2/A$ across an individual line of atoms exceeds the maximum restoring force F_{max}. Equating $f = Fa_0^2/A$ with F_{max} then gives the ultimate tensile strength of the solid:

$$\frac{F}{A} = \frac{F_{\text{max}}}{a_0^2}. \tag{9.58}$$

We have from Equation (2.9) that the potential energy between two atoms bound by van der Waals forces can be represented by the Lennard-Jones potential

$$V(r) = \varepsilon\left[\left(\frac{a_0}{r}\right)^{12} - 2\left(\frac{a_0}{r}\right)^{6}\right].$$

Then, for this potential, we have from Equation (9.48),

$$F_{\text{ext}} = \frac{\mathrm{d}V(r)}{\mathrm{d}r} = \varepsilon\left[-12\frac{a_0^{12}}{r^{13}} + 12\frac{a_0^6}{r^7}\right]. \tag{9.59}$$

The maximum restoring force F_{max} occurs at $r = r_{\text{m}}$, where $\mathrm{d}F_{\text{ext}}/\mathrm{d}r = 0$.

$$\frac{\mathrm{d}F_{\text{ext}}}{\mathrm{d}r} = \varepsilon\left[156\frac{a_0^{12}}{r^{14}} - 84\frac{a_0^6}{r^8}\right], \tag{9.60}$$

and equating $\mathrm{d}F_{\text{ext}}/\mathrm{d}r$ to zero gives $r_{\text{m}} = (156/84)^{1/6}a_0$. Then substituting for r_{m} in Equation (9.59), we obtain

$$\begin{aligned} F_{\text{max}} &= 12\varepsilon\left[-\frac{a_0^{12}}{(156/84)^{13/6}a_0^{13}} + \frac{a_0^6}{(156/84)^{7/6}a_0^7}\right] \\ &= \frac{12\varepsilon}{a_0} \times 0.224. \end{aligned}$$

Hence, from Equation (9.58), we have that the ultimate tensile strength

$$\frac{F_{\text{max}}}{a_0^2} = 0.224\frac{12\varepsilon}{a_0^3}.$$

Using $E = 72\varepsilon/a^3$, Equation (9.57), we obtain

$$\frac{\text{Young's modulus}}{\text{Maximum tensile strength}} = \frac{72\varepsilon/a_0^3}{(12\varepsilon/a_0^3)\times 0.224} \approx 27.$$

In practice, this ratio is usually much greater ~100–1000 because of the presence of various imperfections in the crystal lattice.

9.5.2 Bulk modulus

A hydrostatic pressure P that is applied to a solid produces a volume change $-\mathrm{d}V$. The work done by the external force on the crystal is then $-P\mathrm{d}V$. The atoms are pushed closer together and if the work done goes only into increasing that part of the total energy U of the solid that arises from interatomic forces, we have

$$\mathrm{d}U = -P\mathrm{d}V, \tag{9.61}$$

or

$$P = -\frac{\mathrm{d}U}{\mathrm{d}V}. \tag{9.62}$$

We can interpret pressure as energy per unit volume, i.e. as an energy density, as we have noted previously.

If a change in pressure $\mathrm{d}P$ results in a change $\mathrm{d}V$ in the volume of a solid, we can define the bulk modulus as

$$B = -V\frac{\mathrm{d}P}{\mathrm{d}V}. \tag{9.63}$$

From Equation (9.62), $\mathrm{d}P/\mathrm{d}V = -\mathrm{d}^2U/\mathrm{d}V^2$, and therefore,

$$B = V\frac{\mathrm{d}^2U}{\mathrm{d}V^2}. \tag{9.64}$$

Equation (9.64) for the bulk modulus B is expressed in terms of macroscopic quantities, i.e. the energy U of the entire solid and V its volume. We wish to express B in terms of $V(r)$, the potential energy of a pair of atoms and their separation r. To do this, we first need to express $\mathrm{d}^2U/\mathrm{d}V^2$ in terms of r. We have

$$\frac{\mathrm{d}U}{\mathrm{d}V} = \frac{\mathrm{d}U}{\mathrm{d}r}\frac{\mathrm{d}r}{\mathrm{d}V},$$

and

$$\frac{\mathrm{d}^2U}{\mathrm{d}V^2} = \frac{\mathrm{d}}{\mathrm{d}V}\left(\frac{\mathrm{d}U}{\mathrm{d}r}\frac{\mathrm{d}r}{\mathrm{d}V}\right) = \left(\frac{\mathrm{d}r}{\mathrm{d}V}\right)\frac{\mathrm{d}}{\mathrm{d}V}\left(\frac{\mathrm{d}U}{\mathrm{d}r}\right) + \left(\frac{\mathrm{d}U}{\mathrm{d}r}\right)\frac{\mathrm{d}}{\mathrm{d}V}\left(\frac{\mathrm{d}r}{\mathrm{d}V}\right),$$

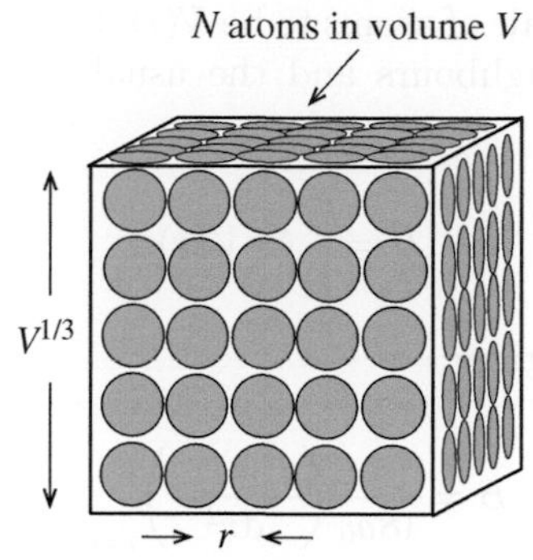

Figure 9.18 A picture of atoms arranged as close-packed spheres with interatomic spacing r. A total of N atoms is accommodated in a cube of volume V. The length of each side of the cube is $V^{1/3}$ and the number of atoms fitting along a side is $N^{1/3}$, giving $V = Nr^3$.

giving

$$\frac{\mathrm{d}^2 U}{\mathrm{d}V^2} = \left(\frac{\mathrm{d}^2 U}{\mathrm{d}r^2}\right)\left(\frac{\mathrm{d}r}{\mathrm{d}V}\right)^2 + \left(\frac{\mathrm{d}U}{\mathrm{d}r}\right)\left(\frac{\mathrm{d}^2 r}{\mathrm{d}V^2}\right). \tag{9.65}$$

If we consider the atoms to be arranged as close-packed spheres with interatomic spacing r, we have the arrangement shown in Figure 9.18. A total of N atoms is accommodated in a cube of volume V. The length of each side of the cube is $V^{1/3}$, and the number of atoms fitting along a side is $N^{1/3}$. Hence $N^{1/3}r = V^{1/3}$, or $V = Nr^3$. Therefore,

$$\frac{\mathrm{d}V}{\mathrm{d}r} = 3Nr^2. \tag{9.66}$$

The solid is in equilibrium when the atoms are at their equilibrium separation when the total energy U of the solid is a minimum. This means that $\mathrm{d}U/\mathrm{d}r = 0$ at $r = a_0$. Taking this into account, Equation (9.65) becomes

$$\frac{\mathrm{d}^2 U}{\mathrm{d}V^2} = \left(\frac{\mathrm{d}^2 U}{\mathrm{d}r^2}\right)_{r=a_0}\left(\frac{\mathrm{d}r}{\mathrm{d}V}\right)^2. \tag{9.67}$$

Substituting for $\mathrm{d}r/\mathrm{d}V$ from Equation (9.66), and taking $r = a_0$, we obtain

$$\frac{\mathrm{d}^2 U}{\mathrm{d}V^2} = \frac{1}{9N^2 a_0^4}\left(\frac{\mathrm{d}^2 U}{\mathrm{d}r^2}\right)_{r=a_0}. \tag{9.68}$$

Then, substituting for $\mathrm{d}^2 U/\mathrm{d}V^2$ in Equation (9.64) and using $V = Na_0^3$, gives

$$B = \frac{Na_0^3}{9N^2 a_0^4}\left(\frac{\mathrm{d}^2 U}{\mathrm{d}r^2}\right)_{r=a_0} = \frac{1}{9Na_0}\left(\frac{\mathrm{d}^2 U}{\mathrm{d}r^2}\right)_{r=a_0}. \tag{9.69}$$

Note that we have assumed that the kinetic energy of the atoms is negligible compared to their potential energy, as occurs at low temperature.

We take the potential energy of a pair of atoms to be $V(r)$. For a total of N atoms, we have ½Nn pairs, where n is the number of nearest neighbours and the usual factor of ½ avoids counting each atom twice. Then,

$$U = \frac{1}{2} NnV(r). \tag{9.70}$$

Substituting for U in Equation (9.69) gives

$$B = \frac{n}{18a_0}\left(\frac{\mathrm{d}^2 V(r)}{\mathrm{d}r^2}\right)_{r=a_0}. \tag{9.71}$$

Again, from Equation (9.12), we have for the particular case of a van der Waals atom:

$$\left(\frac{\mathrm{d}^2 V(r)}{\mathrm{d}r^2}\right)_{r=a_0} = \frac{72\varepsilon}{a_0^2},$$

giving our final result

$$B = \frac{4n\varepsilon}{a_0^3}. \tag{9.72}$$

In Section 3.3, we found an expression for L_0, the latent heat of sublimation of a van der Waals solid in terms of the binding energy ε:

$$L_0 = \frac{1}{2} N_A n\varepsilon,$$

where N_A is Avogadro's number and n is the number of nearest neighbours. We can thus relate the bulk modulus of a solid to its latent heat. Then taking $N_A a_0^3$ to be the molar volume V_m, we find that the isotheral bulk modulus B and the latent heat of sublimation L_0 are related by

$$B = \frac{8L_0}{V_m}. \tag{9.73}$$

This is roughly borne out in practice. For the example of argon, $L_0 = 7.8$ kJ/mol and $V_m = 2.3 \times 10^{-5}$ m^3/mol. Substituting these values in Equation (9.73) predicts $B = 2.7 \times 10^9$ Pa, which is close to the experimental value of 2.5×10^9 Pa.

9.6 The inelastic behaviour of solids

So far, we have confined our attention to small values of strain, for which a solid returns to its original shape when the applied stress is removed, i.e. the solid exhibits elastic behaviour. However, beyond a certain limit, the solid ceases to be elastic. The transition between elastic and inelastic behaviour becomes apparent in a *stress-strain curve.* Such a curve is obtained by stretching a solid and recording the stress it experiences

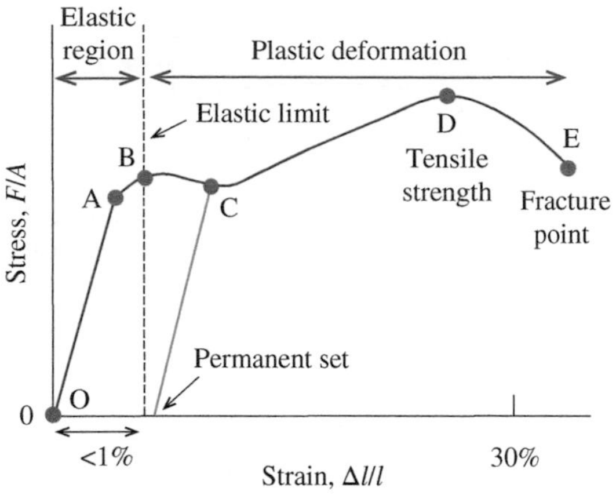

Figure 9.19 A stress-strain curve for a typical metal. Note that the strain axis is expanded below the value of 1% to emphasis the relatively small region over which stress is linearly proportional to strain.

as it is stretched. Taking the example of a long bar of material of length L and cross-sectional area A, the stress-strain curve for a typical metal looks like that shown in Figure 9.19. Note that the strain axis is expanded below the value of 1%.

The first portion of the curve, between O and A, is a straight line showing that the strain is directly proportional to the stress. The slope of the line is equal to Young's modulus of the material. This linearity, however, only occurs for very small values of strain, say less than 1% or even less than 0.01%. The straight-line portion of the curve ends at point A, which is called the *proportionality limit.*

From A to B, the stress is no longer proportional to strain. However, if the stress is removed at any point between O and B, the bar does return to its original length, i.e. the material shows elastic behaviour. Point B is therefore called the *elastic limit.*

Something different happens if the bar is stretched beyond point B, say to point C. In that case, the bar no longer returns to its original length when the stress is removed. Instead, the green line in Figure 9.19 is followed and the bar suffers irreversible deformation. Its length after this deformation is greater than its original length and the bar is said to have acquired a *permanent set.*

The behaviour of the bar from the elastic limit B to the *fracture point* E is called *plastic deformation.* For some materials, a large amount of plastic deformation takes place between these two limits, as in Figure 9.19. Such a material is said to be *ductile.* Copper is an example of a ductile material. Under certain conditions, a bar of copper can be drawn into a long thin wire, so that the original length of the bar increases by several hundred times. For some materials, however, fracture occurs soon after the elastic limit is passed. Such materials are said to be *brittle.* Cast iron is an example of brittle material.

Maximum stress occurs at point D, which corresponds to the tensile strength of the material, the maximum stress that the bar can sustain. When the bar is stretched beyond D, a process called *necking* commences where the cross-sectional area of the bar reduces, as illustrated by Figure 9.20. This means that although the applied stress (F/A), as defined in terms of the applied force F and the original area A, appears to reduce, the actual stress acting at the necking region continues to increase because the cross-sectional area is less than A. Eventually, at fracture point E, the bar fractures.

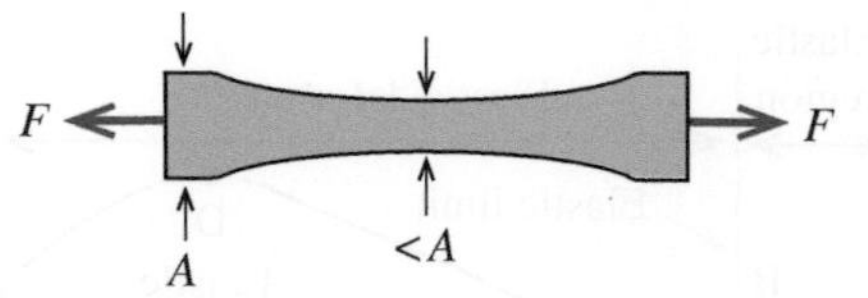

Figure 9.20 The process of necking, where tensile stress causes the cross-sectional area of the bar to reduce. This means that the actual stress acting at the necking region is greater than the applied stress defined in terms of the cross-section area of the un-stretched bar.

9.6.1 Slip

In the elastic region, strain is directly proportional to stress, as explained in terms of the model in which the atoms are connected by Hooke's-law springs; the applied stress causes the interatomic spacing to increase linearly according to Hooke's law. The process occurring in the region of plastic deformation is quite different. This process involves the movement or *slip* of one plane of atoms past another. And this process is not reversible. It is not the case that a whole plane of atoms slips at once. This would require too much energy. Rather, the process occurs one atom at a time. An analogy is the movement of a ripple across a carpet. It takes a much lower force to move the ripple than it takes to move the whole carpet.

To illustrate the process of slip, we first consider two planes of atoms subject to shear stress, as illustrated by Figure 9.21a. As noted earlier, it requires a relatively large shearing force to push a whole layer of atoms past another layer. However, if there is a vacancy in the line of atoms, it takes a relatively small force to push an adjacent atom into that vacancy. Then in a following step, a second atom hops into the newly produced vacancy and so on, as illustrated by Figure 9.21b. Eventually, all the atoms in the top plane have moved one by one to the left by one atomic distance, which is equivalent to the whole of the top plane slipping past the bottom plane. In effect, it is the vacancy that travels through the crystal.

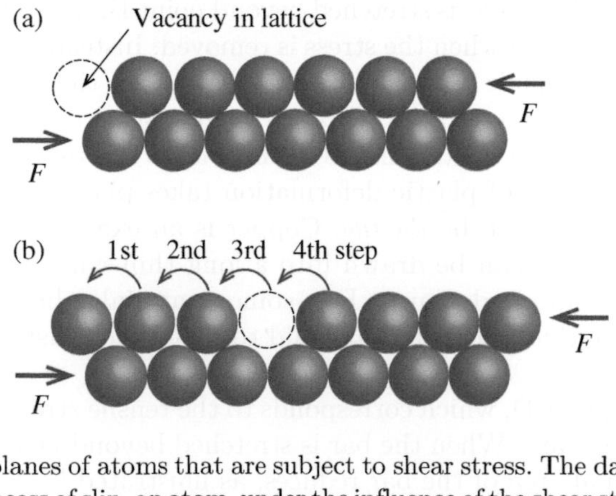

Figure 9.21 (a) Two adjacent planes of atoms that are subject to shear stress. The dashed circle represents a vacancy in the upper plane. (b) In the process of slip, an atom, under the influence of the shear stress hops into the vacancy. Then in a following step, a second atom hops into the newly produced vacancy and so on. Four consecutive steps of the slip process are shown. Eventually, all the atoms in the top plane have moved one by one to the left by one atomic distance, which is equivalent to the whole of the top plane slipping past the bottom plane. In effect, it is the vacancy that travels through the crystal.

In practice, the slipping of crystal planes involves the motion of *dislocations* rather than vacancies through the crystal lattice. An *edge dislocation* is one kind of dislocation, whose nature is illustrated in Figure 9.22. This figure shows an arrangement of atoms in a crystal lattice that is divided into two portions; one above and one below the dashed line. It may be seen that there is an extra half-plane of atoms present in the upper portion. This imperfection in the lattice structure is the edge dislocation. If the upper and lower portions of the crystal lattice are subject to shear stress as indicated, the dislocation can readily move through the crystal.

The mechanism responsible for the mobility of dislocation is also illustrated in Figure 9.22, where two of the atoms have been colour-coded, green and red, respectively. It takes only a relatively small stress to push the atom in green so that it lines up with the line of atoms on its left-hand side, as occurs between panels (a) and (b) of the figure. This essentially moves the half-plane of atoms one atom to the right. Then in the next step, between panels (b) and (c), the atom in red is pushed to line up with the line of atoms on its left. So again, the half-plane of atoms moves one-atom distance to the right. Eventually, the dislocation moves through the whole crystal as, in turn, each atom in the bottom portion moves one atom distance to the left. This is equivalent to the top portion of the crystal lattice slipping past the lower portion.

Dislocations can move freely through a crystal lattice so long as the lattice is otherwise perfect. But they may get stopped if they encounter another imperfection in the lattice. Hence, by incorporating additional imperfections in its crystal structure, a solid acquires greater strength. The most direct way of blocking dislocation motion is to introduce tiny particles of a different atom into the crystal structure. This is exactly what happens in the production of steel. In this case, a small amount, ~1%, of carbon is dissolved into molten iron. The molten iron is then cooled rapidly so that the carbon precipitates out in tiny grains, making many microscopic imperfections in the crystal lattice. These limit the movement of any dislocations, converting relatively soft iron into hard steel.

As we come to the end of our discussion of elastic moduli, we point out that we have assumed that the relationship between stress and strain is independent of the direction of the stress. This is justifiable for

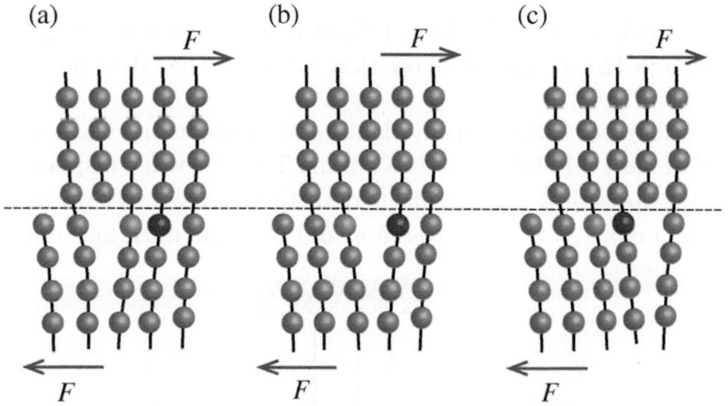

Figure 9.22 The process of slip involving the motion of a dislocation. The dislocation arises from the extra half-plane of atoms present in the upper portion of the crystal lattice. Under the influence of shear stress, the atom shown in green moves to line up with the line of atoms on its left-hand side, as occurs between panels (a) and (b). This essentially moves the half-plane of atoms one atom to the right. Then in the next step, between panels (b) and (c), the atom shown in red moves to line up with the line of atoms on its left. So again, the half-plane of atoms moves one-atom distance to the right. Eventually, the dislocation moves through the whole crystal as, in turn, each atom in the bottom portion moves one atomic distance to the left. This means that the top portion has slipped one atom to the right with respect to the bottom portion.

solids that are amorphous or polycrystalline, which is the case for most structural materials such as concrete and steel. However, it is not the case for single crystals. Because the interatomic bonds in pure crystals are orientated in particular directions within the crystal structure, the magnitude of the strain may depend on the direction of the applied stress. Then different elastic moduli must be assigned to different directions of stress with respect to the crystal structure.

Problems 9

9.1 Young's modulus E for steel is 2×10^{11} Pa and its Poison ratio σ is 0.3. Deduce the shear modulus G and the bulk modulus B of steel.

9.2 A composite wire consists of a copper wire of length 1.5 m and radius 1.0 mm that is connected to a nickel wire of length 2.0 m and radius 1.5 mm to give a composite wire of total length 3.5 m. The top of this wire is attached to a rigid beam and a mass of 10 kg is attached to the bottom end. What is the extension of the composite wire?

9.3 A steel wire of length 4.0 m and diameter 1.0 mm is subject to a progressively increasing tensile stress. Calculate the increase in the energy stored in the wire as the extension of the wire is increased from 3.0 to 4.0 mm. Young's modulus for steel is 2×10^{11} Pa.

9.4 Calculate the thermal fluctuations in the length of a steel rod of length 1.0 m and radius 3.0 mm. Take the temperature to be 300 K. Young's modulus for steel is 2×10^{11} Pa.

9.5 A wire of diameter 1.0 mm is held horizontally between two rigid supports 3.0 m apart. What mass attached to the mid-point of the wire would cause that point to be displaced downwards by 5 cm if Young's modulus for the wire is 11×10^{11} Pa? Ignore the mass of the wire.

9.6 A brass rod of diameter 5.0 mm is heated to a temperature of 300 °C and when it is at this temperature, its ends are firmly clamped. Find the force that must be exerted by the clamps on the rod if it is to be prevented from contracting when it cools to 20 °C. The linear expansion coefficient and Young's modulus for brass are $1.9 \times 10^{-5}\ \mathrm{K}^{-1}$ and 9×10^{11} Pa, respectively.

9.7 A copper wire of diameter 0.5 mm extends by 1 mm when loaded with a mass of 0.5 kg and it twists through 1 rad when a torque of $5.75 \times 10^{-5}\ \mathrm{N \cdot m}$ is applied to its unclamped end. Calculate the value of Poisson's ratio for copper.

9.8 Poisson's ratio for a rod of radius r and length l is defined as the fractional change in radius divided by the fractional change in length, $\sigma = (\Delta r/r)/(\Delta l/l)$. Show that for a rod that does not change its volume when it is stretched, $\sigma = 0.5$.

9.9 A 45 cm length of wire of diameter 1 mm is firmly clamped at its upper end while its lower end is attached to the centre of a disc of metal of mass 0.75 kg and radius 8.0 cm. The disc when displaced from equilibrium performs torsional oscillations in the horizontal plane with a period of 2.6 seconds. What is the shear modulus of the wire? The moment of inertia of a disc $I = ½mR^2$, where m and R are its mass and radius, respectively.

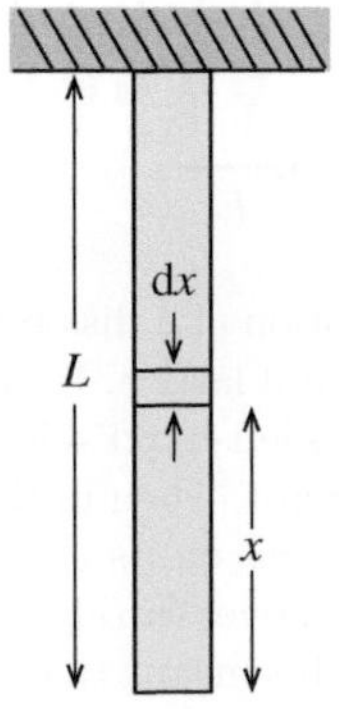

9.10 A wire of length L, cross-section A and density ρ hangs vertically under its own weight as illustrated by the figure. Consider the tensile stress on an elemental slice dx of the wire and the resulting strain on the slice to show that the extension of the wire due to its own weight is $\frac{1}{2}(L^2\rho g)/E$, where E is Young's modulus. Calculate the extension under its own weight of a wire 10 m long suspended freely from its top end if the velocity of sound along the wire is 1000 m/s.

9.11 The latent heat of sublimation of argon is 7.8 kJ/mol and the equilibrium separation of the atoms is 3.7×10^{-10} m. Use these data to obtain a value for the bulk modulus B of solid argon.

9.12 The interatomic potential energy $V(r)$ between two atoms in a solid is given by

$$V(r) = \left(\frac{pq}{p-q}\right)\varepsilon\left[\frac{1}{p}\left[\frac{a_0}{r}\right]^p - \frac{1}{q}\left[\frac{a_0}{r}\right]^q\right],$$

where the symbols have their usual meanings. Show that the bulk modulus of the solid is given by $B = (pqL_s)/9\,V_m$, where L_s is the latent heat of sublimation and V_m is the molar volume.

10

Thermal and transport properties of solids

In this chapter, various thermal and transport properties of solids are discussed. Einstein's theory of the specific heat of solids is described as how its predictions compare to the experimental data. The modified version of the theory due to Debye is described as well. Also discussed is diffusion in solids, which is very different to diffusion in gases. This is because, unlike the molecules in a gas, the atoms in a solid are confined by the forces due to their neighbours. The other transport properties discussed are electrical and thermal conduction in metals, corresponding to the transport of electric charge and thermal energy, respectively. Some of the properties of solids can be described in terms of classical physics. However, as we will see, there are certain properties that require a quantum mechanical description.

10.1 Molar specific heats of solids

We will usually consider one mole of a solid and so deal with molar specific heats. The atoms in a solid are bound to their lattice sites. Consequently, they do not have translational energy but only vibrational energy. So more precisely, we are dealing with the specific heat associated with the vibrations of the atoms in the crystal lattice. Moreover, solids expand by only a small amount when they are heated, and so the difference between the specific heats at constant volume and that at constant pressure can usually be ignored.

An oscillator, such as a mass on a spring, has potential energy and kinetic energy and the total energy E is the sum of the two. For a one-dimensional oscillator moving along the x-axis, we have

$$E = \frac{1}{2}\alpha x^2 + \frac{1}{2}m\left(\frac{\mathrm{d}x}{\mathrm{d}t}\right)^2, \tag{10.1}$$

where α is the spring constant and m is the oscillator mass. According to the classical equipartition theorem, each term on the right-hand side of this equation has a mean energy of $\frac{1}{2}kT$. Hence, the mean total energy of a one-dimensional oscillator is kT.

Suppose that we treat the atoms in a crystal as classical oscillators that can vibrate in three orthogonal directions. The motion in each direction contributes a mean energy of kT and so the mean total energy of an atom is $3kT$. If we have one mole of a monatomic substance, we have N_A atoms, where N_A is Avogadro's number. Therefore, as the total energy of a solid is purely vibrational, we can expect the total energy per mole to be $3N_AkT = 3RT$, where R is the gas constant. This approach based on the equipartition theorem predicts that the molar specific heat C_V for a monatomic solid is

$$C_V = \frac{d}{dT}(3RT) = 3R. \tag{10.2}$$

This classical treatment predicts a constant molar specific heat of $3R$ *at all temperatures*, with a value of about 25 J/mol · K. This is a result that was reached empirically and is known as the law of Dulong and Petit. And indeed, monatomic solids do have a value of C_V close to this value, as shown in Table 10.1.

The values shown in Table 10.1 are obtained at room temperature. If measurements are taken at lower temperatures, it is found, however, that Dulong and Petit's law no longer holds; the value of C_V falls off dramatically with decreasing temperature becoming zero at $T = 0$. Typical experimental results are shown in Figure 10.1, for the example, of copper. Although the classical theory works at room temperature, it clearly fails to match the experimental results at lower temperatures. Any theory about the specific heats

Table 10.1 Molar heat capacities of some solids

Solid	Molar specific heat, C_V (J/mol·K)
Aluminium	24.3
Copper	24.5
Gold	25.6
Lead	26.4
Silver	24.9

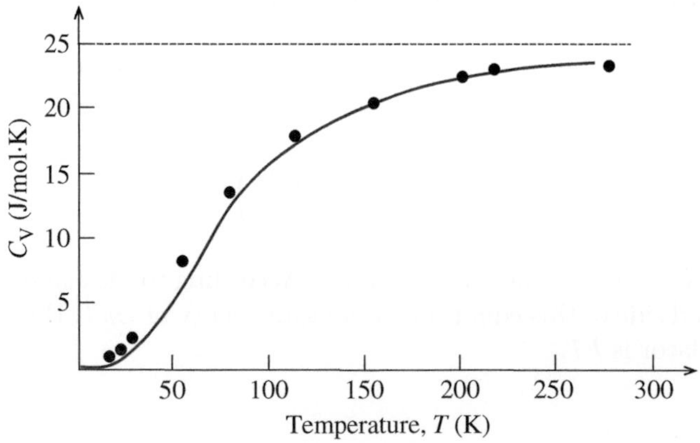

Figure 10.1 The molar specific heat C_V of copper. The dots are the experimental data. The solid curve is a fit to the data of Einstein's theory, where the value of the Einstein temperature $\Theta_E = 240$ K was chosen to give the best fit of the theoretical curve to the data.

of solids must explain the two basic experimental facts: (i) near room temperature, the molar specific heat of most monatomic solids is close to 25 J/mol · K and (ii), at lower temperatures, the heat capacities decrease and indeed vanish at $T = 0$.

10.1.1 The Einstein model

The fundamental step to resolve the discrepancy between classical theory and experiment was taken by Einstein in 1907. He treated the vibrations of the atoms in a solid according to quantum theory, adopting results previously obtained by Max Planck. Planck had sought to explain the intensity distribution of blackbody radiation, i.e. the shape of the blackbody spectrum. Classical physics was unable to explain the shape of the spectrum over its entire frequency range. In the limit of low frequency, the classical spectrum approaches the experimental results. However, as the frequency of the radiation becomes large, the theoretical prediction goes to infinity, which is clearly unphysical. In trying to resolve this discrepancy with the experiment, Planck was led to consider the possibility of a violation of the law of equipartition of energy on which the classical approach was based. This led to Planck's postulate that we may state as follows: A quantum oscillator undergoing simple harmonic motion can only possess discrete values of energy ε_n that satisfy the relation

$$\varepsilon_n = n\frac{h}{2\pi}\omega = n\hbar\omega, \quad n = 0, 1, 2, \ldots, \tag{10.3}$$

where h is Planck's constant and ω is the angular frequency of vibration. Based on this postulate, Planck was able to reproduce the shape of the blackbody spectrum exactly. It was Planck who first put forward the concept of the quantisation of energy.

In Section 3.8 we discussed particles that can have discrete energy levels ε_0, ε_1, ε_2, ε_3.... And for such particles, we saw that Boltzmann's law says that in thermal equilibrium at temperature T, the probability of the particle being in a particular energy level ε_n is proportional to the Boltzmann factor $\mathrm{e}^{-\varepsilon_n/kT}$. From this result, we can deduce that the mean energy of a quantum oscillator of frequency ω is

$$\overline{\varepsilon}(\omega) = \frac{\hbar\omega}{\exp(\hbar\omega/kT) - 1}, \tag{10.4}$$

as shown in the worked example below.

Einstein assumed (i) that all the atoms in a solid vibrate independently of each other and (ii) that all atoms vibrate at the same frequency ω_{E}, where ω_{E}, is known as the *Einstein frequency*. Hence, for N atoms that can vibrate along three orthogonal directions, we have $3N$ independent harmonic oscillators, each of which vibrates at frequency ω_{E}. Einstein then used Equation (10.4) from quantum theory to find the mean total energy U of the $3N$ atoms:

$$U = 3N\left(\frac{\hbar\omega_{\mathrm{E}}}{\exp(\hbar\omega_{\mathrm{E}}/kT) - 1}\right). \tag{10.5}$$

For 1 mol of a solid we have $N = N_{\mathrm{A}}$, for which

$$U = \frac{3N_{\mathrm{A}}\hbar\omega_{\mathrm{E}}}{\exp(\hbar\omega_{\mathrm{E}}/kT) - 1}. \tag{10.6}$$

In the limit of high temperature, as T approaches ∞, this expression reduces to $U = 3N_{\mathrm{A}}kT = 3RT$, in agreement with the classical result.

To determine the specific heat predicted by Einstein's model we differentiate Equation (10.6) with respect to T. This gives

$$C_V = \left(\frac{dU}{dT}\right) = 3N_A\hbar\omega_E \frac{\exp(\hbar\omega_E/kT)}{[\exp(\hbar\omega_E/kT)-1]^2}\frac{\hbar\omega_E}{k},$$
$$= 3R\left(\frac{\hbar\omega_E}{k}\right)^2 \frac{\exp(\hbar\omega_E/kT)}{[\exp(\hbar\omega_E/kT)-1]^2}, \tag{10.7}$$

which can be written as

$$C_V = 3R\left(\frac{\Theta_E}{T}\right)^2 \frac{\exp(\Theta_E/T)}{[\exp(\Theta_E/T)-1]^2}, \tag{10.8}$$

where $\Theta_E = \hbar\omega_E/k$ is known as the *Einstein temperature.* This equation is the basic result of Einstein's theory. It depends on the single parameter Θ_E, which can be chosen differently for each solid. A plot of C_V against T for copper, obtained using Equation (10.8), is shown as the continuous curve in Figure 10.1. A value of $\Theta_E = 240$ K was chosen to give the best fit of the curve to the experimental data. We see that Einstein's model gives qualitative agreement with the experimental data, although the agreement is less good at low temperatures. The value of Θ_E implies a vibrational period $2\pi/\omega_E = 2\pi\hbar/k\Theta_E$ of 2.0×10^{-13} seconds, which is of the correct order of magnitude for atomic vibrations in solids.

To see the high- and low-temperature limits of Equation (10.8), we make the substitution $x = \Theta_E/T$, giving

$$C_V = 3R\frac{x^2 e^x}{(e^x-1)^2}. \tag{10.9}$$

For $T >> \Theta_E$, $x << 1$, and we note that for $x \to 0$,

$$\frac{x^2 e^x}{(e^x-1)^2} = \frac{x^2(1+x+x^2/2+\cdots)}{(x+x^2/2+\cdots)^2} \to 1.$$

Hence, at sufficiently high temperatures, Equation (10.9) reduces to $C_V = 3R$, in agreement with the law of Dulong and Petit. In the low temperature limit, $T \ll \Theta_E$, $x \gg 1$, and we have

$$\frac{x^2 e^x}{(e^x-1)^2} \to \frac{x^2}{e^x}.$$

Using this result in Equation (10.9) gives

$$C_V = 3R\left(\frac{\Theta_E}{T}\right)^2 \exp\left(-\frac{\Theta_E}{T}\right). \tag{10.10}$$

With decreasing temperature, the exponential term $\exp(-\Theta_E/T)$ decreases more rapidly than the quadratic term $(\Theta_E/T)^2$ increases. Hence, the specific heat predicted by the Einstein model decreases with temperature as is observed experimentally. However, the predicted exponential decrease gives a more rapid fall than is observed experimentally, as evidenced by Figure 10.1.

In Section 9.5, we obtained the following relationship between Young's modulus E, the inter-atomic spacing a_0 and the inter-atomic bond strength α of the solid: $E = \alpha / a_0$; see Equation (9.56). For the harmonic oscillators in Einstein's model, $\alpha = M\omega_E^2$, where M is the atomic mass. Then using $\omega_E = k\Theta_E/\hbar$, we obtain

$$E = \frac{M}{a_0}\left(\frac{k\Theta_E}{\hbar}\right)^2. \tag{10.11}$$

For copper of atomic mass 63.5 u, $a_0 = 2.6 \times 10^{-10}$ m, and taking $\Theta_E = 240$ K,

$$E = \frac{63.5 \times 1.66 \times 10^{-27}}{2.6 \times 10^{-10}}\left(\frac{1.38 \times 10^{-23} \times 240}{1.05 \times 10^{-34}}\right)^2 = 4.0 \times 10^{11}\ \text{Pa}.$$

This compares satisfactorily with the actual value of Young's modulus for copper, which is 1.2×10^{11} Pa.

In Section 3.7, we discussed the specific heats of gases. And we saw distinct steps in a plot of specific heat against temperature. These steps arise because the temperature must be sufficiently high for a significant fraction of the molecules to absorb energy and become rotationally or vibrationally excited, i.e. the available thermal energy kT to become comparable to the separations of the quantised energy levels. It is a similar situation for the specific heats of solids. We have from Equation (10.3) that the vibrational energy spacing for a solid is $\hbar\omega_E$. If $kT \ll \hbar\omega_E$ or equivalently $T \ll \Theta_E$, there is little probability of vibrational excitation. On the other hand, if $T \gg \Theta_E$, the probability of vibrational excitation is high. Moreover, when kT becomes much greater than $\hbar\omega_E$, the energy levels may be considered to be continuous. In that case, the classical treatment applies, giving $C_V = 3R$.

The Einstein model is an oversimplification. A solid does not consist of atoms vibrating totally independently of each other. Rather, there exists a strong coupling between them. A mechanical analogue of a two-dimensional crystal lattice consists of billiard balls connected together by identical springs, as illustrated by Figure 10.2. If one ball is set vibrating, say the one labelled A in Figure 10.2, a disturbance is propagated

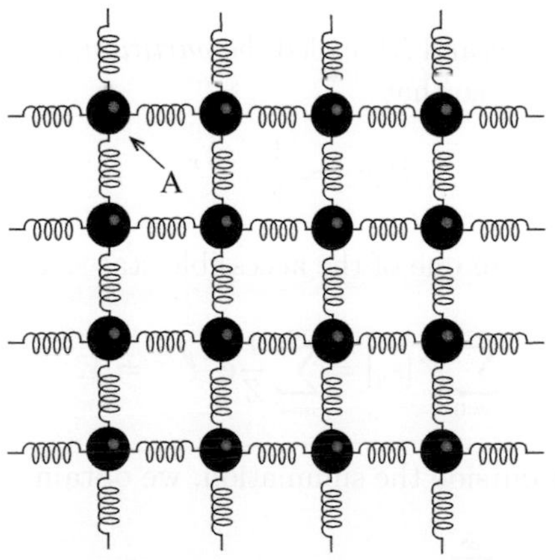

Figure 10.2 A mechanical analogue of a two-dimensional crystal lattice consisting of billiard balls connected together by identical springs. If one ball is set vibrating, say the one labelled A, a disturbance propagates through the whole system. The balls behave as coupled oscillators and the resultant motion involves a wide range of vibrational frequencies.

through the whole system. The balls behave as *coupled oscillators* and they vibrate in a complex manner that involves not just a single frequency but a wide range of frequencies. Similarly, the lattice vibrations of a crystal are really very complicated coupled oscillations of all the atoms, again covering a wide range of frequencies. The individual ways or *normal modes* in which the system vibrates will contribute to the total energy of the solid. And to find the total energy of the solid, these individual contributions must be summed. These aspects are neglected in Einstein's model. He did not expect detailed agreement with experimental measurements and indeed Einstein pointed out the kind of modifications that his model would require.

Worked example

Show that the mean energy of a Planck quantum harmonic oscillator is given by

$$\overline{\varepsilon} = \frac{\hbar\omega/kT}{\left(\mathrm{e}^{\hbar\omega/kT} - 1\right)} kT.$$

Solution

According to Planck's postulate, a quantum harmonic oscillator can only have discrete energies ε_n that satisfy the relation

$$\varepsilon_n = n\hbar\omega, \quad n = 0, 1, 2, \ldots.$$

The probability of it being in state ε_n is given by the Boltzmann law

$$P\left[\varepsilon_n\right] \propto \mathrm{e}^{-\varepsilon_n/kT}.$$

or

$$P\left[\varepsilon_n\right] = \frac{1}{Z}\mathrm{e}^{-\varepsilon_n/kT},$$

where $1/Z$ is the normalization constant and Z is called the *partition function.* It is convenient to change the variable from T to β, where $\beta = 1/kT$, so that

$$P\left[\varepsilon_n\right] = \frac{1}{Z}\mathrm{e}^{-\beta\,\varepsilon_n}.$$

It is certain that the particle must be in one of the accessible states and hence

$$\sum_{n=0}^{\infty} P\left[\varepsilon_n\right] = \sum_{n=0}^{\infty} \frac{1}{Z}\mathrm{e}^{-\beta\,\varepsilon_n} = 1.$$

Taking the normalisation constant outside the summation, we obtain

$$\begin{aligned} Z &= \sum_{n=0}^{\infty} \mathrm{e}^{-\beta\,\varepsilon_n} = \mathrm{e}^{-\beta\,\varepsilon_0} + \mathrm{e}^{-\beta\,\varepsilon_1} + \cdots \\ &= 1 + \mathrm{e}^{-\beta\hbar\omega} + \mathrm{e}^{-\beta 2\hbar\omega} + \cdots \\ &= 1 + r + r^2 + \cdots, \end{aligned}$$

where $r = \mathrm{e}^{-\beta\hbar\omega}$. In the quantum regime in which $\hbar\omega \gg kT$, $\mathrm{e}^{-\beta\hbar\omega} = \mathrm{e}^{-\hbar\omega/kT} \ll 1$. Hence, we have the sum to infinity of a geometric progression:

$$1 + r + r^2 + \cdots = \frac{1}{1-r}.$$

And hence,

$$Z = \frac{1}{1 - \mathrm{e}^{-\beta\hbar\omega}}.$$

The average energy of a quantum oscillator is

$$\begin{aligned}\overline{\varepsilon} &= \varepsilon_0 P[\varepsilon_0] + \varepsilon_1 P[\varepsilon_1] + \cdots \\ &= \sum_{n=0}^{\infty} \varepsilon_n P[\varepsilon_n] = \frac{1}{Z}\sum_{n=0}^{\infty} \varepsilon_n \mathrm{e}^{-\beta\,\varepsilon_n}.\end{aligned}$$

But,

$$\frac{1}{Z}\sum_{n=0}^{\infty} \varepsilon_n \mathrm{e}^{-\beta\,\varepsilon_n} = -\frac{1}{Z}\frac{\mathrm{d}}{\mathrm{d}\beta}\sum_{n=0}^{\infty} \mathrm{e}^{-\beta\,\varepsilon_n} = -\frac{1}{Z}\frac{\mathrm{d}Z}{\mathrm{d}\beta}.$$

Hence,

$$\begin{aligned}\overline{\varepsilon} &= -\frac{1}{Z}\frac{\mathrm{d}}{\mathrm{d}\beta}\left(\frac{1}{1-\mathrm{e}^{-\beta\hbar\omega}}\right) \\ &= -\left(1-\mathrm{e}^{-\beta\hbar\omega}\right)\frac{-\hbar\omega \mathrm{e}^{-\beta\hbar\omega}}{\left(1-\mathrm{e}^{-\beta\hbar\omega}\right)^2} = \frac{\hbar\omega \mathrm{e}^{-\beta\hbar\omega}}{\left(1-\mathrm{e}^{-\beta\hbar\omega}\right)}.\end{aligned}$$

Substituting $\beta = 1/kT$, we obtain

$$\overline{\varepsilon} = \frac{\hbar\omega/kT}{\left(\mathrm{e}^{\hbar\omega/kT} - 1\right)} kT.$$

Note that this is the classical result multiplied by a quantum mechanical factor. Richard Feynman describes this as the first quantum equation that was written down and discussed.

A more exact expression for the energy levels of a quantum harmonic oscillator is

$$\varepsilon_n = \left(n + \frac{1}{2}\right)\hbar\omega + \frac{1}{2}\hbar\omega, \quad n = 0, 1, 2, \ldots.$$

The extra $\hbar\omega/2$ arises from the *uncertainty principle* and is called the *zero point energy*. Taking this into account gives

$$\overline{\varepsilon} = \frac{\hbar\omega/kT}{\left(\mathrm{e}^{\hbar\omega/kT} - 1\right)} kT + \frac{1}{2}\hbar\omega.$$

10.1.2 The Debye model[1]

Peter Debye improved Einstein's model by taking into account the fact that the atoms in a crystal behave as coupled oscillators. From classical mechanics, we know that the vibrations of a system of coupled oscillators can be described in terms of *normal modes of oscillation*.[2] A normal mode is defined as one in which all parts of the system vibrate sinusoidally with the same frequency and with a fixed phase relationship. As an example, Figure 10.3 illustrates the normal modes of a system consisting of three billiard balls connected by springs. The arrows indicate the direction of travel of the balls at a certain instant of time. According to classical mechanics, a system has $3N$ normal modes, where N is the number of entities that the system possesses. Thus for this system of three balls, $N = 3$ and we expect nine normal modes. Three of these normal modes involve longitudinal vibrations, as shown as cases (a), (b), and (c), respectively in Figure 10.3, and six involve transverse vibrations. Three of the six, shown as cases (d), (e), and (f), respectively in Figure 10.3 lie in the x–y plane. The other three transverse vibrations lie in the perpendicular x–z plane, with analogous patterns. The three longitudinal vibrations and the six transverse vibrations make up the total of nine normal modes for the system. Each mode has a well-defined frequency of vibration that is determined by the mass of the balls and the restoring forces due to the connecting springs.

Debye considered the lattice vibrations of N atoms in a solid in terms of $3N$ normal modes of vibration with angular frequencies $\omega_1, \omega_2, \ldots, \omega_{3N}$. Perhaps surprisingly, a normal mode of angular frequency ω has

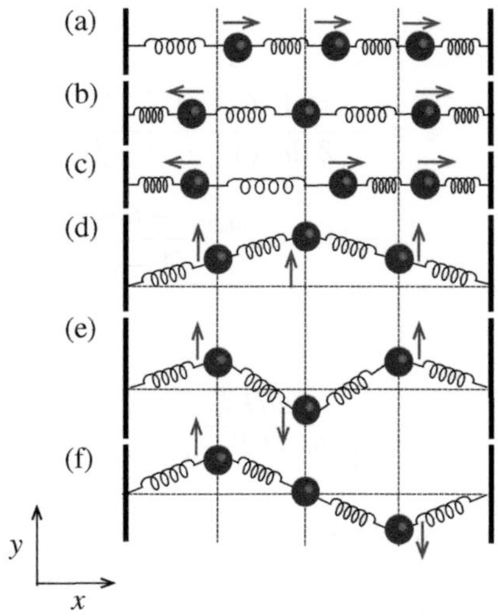

Figure 10.3 The normal modes of a system consisting of three billiard balls connected by springs. The arrows indicate the direction of travel of the balls at a certain instant of time. According to classical mechanics, a system has $3N$ normal modes, where N is the number of entities that the system possesses. Thus, for this system, $N = 3$ and we expect nine normal modes. Three of these normal modes involve longitudinal vibrations, as shown as cases (a), (b), and (c), and six involve transverse vibrations. Three of the six, shown as cases (d), (e), and (f), lie in the x–y plane. The other three transverse vibrations lie in the perpendicular x–z plane, with similar patterns.

[1] This section may be omitted at first reading of the book.

[2] For further discussion of the normal modes of coupled oscillators, see King, G.C. (2009). *Vibrations and Waves*. Wiley.

the same mean energy as a single quantum oscillator of the same frequency, i.e. the mean energy of the ith normal mode of the crystal lattice is given by

$$\bar{\varepsilon}_i = \frac{\hbar\omega_i}{\exp(\hbar\omega_i/kT) - 1}, \tag{10.12}$$

where ω_i is its angular frequency. Then the total energy U of the solid is the sum over all the $3N$ normal modes:

$$U = \sum_{i=0}^{3N} \frac{\hbar\omega_i}{\exp(\hbar\omega_i/kT) - 1}. \tag{10.13}$$

It is possible to evaluate this summation using a computer. Debye, however, took a different approach, and we will outline the essential physics of his approach.

We first consider a one-dimensional model of a crystal and the associated normal modes of oscillation. Figure 10.4 illustrates a row of 15 atoms and some of the possible transverse normal modes lying in the plane of the paper. We see that the wavelength λ of each normal mode is very large compared to the inter-atomic spacing. We can, therefore, ignore the discrete atomic structure and treat the normal modes as those of a continuous elastic medium, just like the standing waves on a string. In analogy with the standing waves on a string, we write

$$n\frac{\lambda_n}{2} = L, \quad n = 1, 2, 3, \ldots, \tag{10.14}$$

where n is the order of the mode, λ_n is the wavelength of the nth mode, and L is the length of the one-dimensional crystal. The wavelength λ and the frequency ν of a normal mode are related by

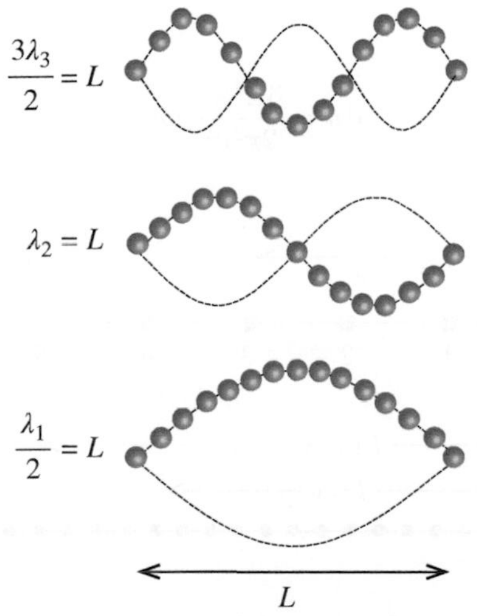

Figure 10.4 The figure illustrates a row of 15 atoms and some of the possible transverse modes lying in the plane of the paper. The wavelength of each normal mode is very large compared to the inter-atomic spacing. Under this condition, we can ignore the discrete atomic structure and treat the normal modes as those of a continuous elastic medium, just like the standing waves on a string.

the familiar expression $\lambda\nu = c$, where c is the wave velocity. This gives the frequency of the nth normal mode to be

$$\nu_n = n\frac{c}{2L}. \tag{10.15}$$

In Figure 10.5a, the integral values of n are plotted as points on a horizontal axis, where the separation of the points is $c/2L$. In this construction, the distance along the axis from the origin, in units of $c/2L$, is equal to the frequency of a given value of n. The particular example of $n = 3$ is shown in Figure 10.5a, where $l_{\nu=3} = 3c/2L = \nu_3$, in accord with Equation (10.15). Then in Figure 10.5b, we have the general case where l is equal to frequency ν and $l + \mathrm{d}l = \nu + \mathrm{d}\nu$. We are interested in the number of points, i.e. the number of normal modes $\mathrm{d}n$ that occur in the frequency range $\mathrm{d}\nu$. To evaluate this quantity, we simply count the number of points on the axis that lie within the distance $\mathrm{d}l$, since

$$(l + \mathrm{d}l) - l = \mathrm{d}l = (\nu + \mathrm{d}\nu) - \nu = \mathrm{d}\nu.$$

Therefore, the number of modes in the frequency range $\mathrm{d}\nu$ is

$$\mathrm{d}n = \frac{\mathrm{d}l}{c/2L} = \frac{\mathrm{d}\nu}{c/2L} = \frac{2L}{c}\mathrm{d}\nu, \tag{10.16}$$

or in terms of angular frequency $\omega = 2\pi\nu$,

$$\mathrm{d}n = \frac{L}{\pi c}\mathrm{d}\omega. \tag{10.17}$$

This equation gives the number of the transverse normal modes in the frequency range $\mathrm{d}\omega$ for our one-dimensional crystal. The problem of counting normal modes, of which Equation (10.17) is a particular example, is of great importance in many branches of physics.

A similar analysis can be applied to the three-dimensional crystal of a solid, although in this case, the points representing particular normal modes lie in three-dimensional space. For the three-dimensional case, the result is

$$\mathrm{d}n = \frac{3L^3\omega^2}{2\pi^2 c^3}\mathrm{d}\omega, \tag{10.18}$$

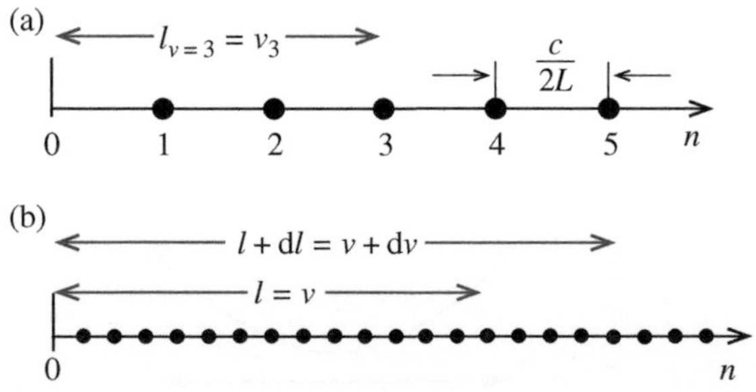

Figure 10.5 (a) A plot of the integral values of n, derived from the condition $\nu_n = nc/2L$, plotted as points on a horizontal axis. The separation of the points is $c/2L$. In this construction, the distance along the axis from the origin, in units of $c/2L$, is equal to the frequency of a given value of n. The particular example of $n = 3$ is shown, where $l = 3c/2L = \nu_3$. (b) The general case, where l is equal to frequency ν and $l + \mathrm{d}l = \nu + \mathrm{d}\nu$. Again, the separation of the points is $c/2L$. The number of normal modes that occur in the frequency range $\mathrm{d}\nu$ is equal to the number of points on the axis that lies within the distance $\mathrm{d}l$.

where we take L to be the length of a cube of the solid. This equation takes into account the fact that a solid consisting of N atoms has $2N$ transverse vibrations and N longitudinal vibrations. The transverse and longitudinal vibrations have different wave velocities. However, for the sake of simplicity, c is taken to be the mean value of these velocities, and all the modes are assumed to have this velocity.

Having the number of modes in the frequency range $\mathrm{d}\omega$ allows the summation in Equation (10.13) to be replaced by an integral, so that

$$U = \int_0^{\omega_{\max}} \left(\frac{\hbar\omega}{\exp(\hbar\omega/kT) - 1} \right) \mathrm{d}n. \tag{10.19}$$

Substituting for $\mathrm{d}n$ from Equation (10.18), we obtain

$$U = \frac{3L^3}{2\pi^2 c^3} \int_0^{\omega_{\max}} \left(\frac{\hbar\omega}{\exp(\hbar\omega/kT) - 1} \right) \omega^2 \mathrm{d}\omega. \tag{10.20}$$

The lower limit of the integral is taken to be zero, which encompasses the lowest frequency mode. To obtain the maximum frequency $\omega_{\max}$, Debye used the fact that the total number of modes must be equal to $3N$. Then from Equation (10.18), we have

$$\int_0^{\omega_{\max}} \frac{3L^3\omega^2}{2\pi^2 c^3} \mathrm{d}\omega = 3N, \tag{10.21}$$

which gives

$$\omega_{\max}^3 = \frac{6\pi^2 c^3 N}{L^3}. \tag{10.22}$$

The 'cut-off' frequency $\omega_{\max}$ arises because, at a sufficiently high frequency, we cannot ignore the atomic nature of the solid. A crystal with interatomic spacing a_0 cannot propagate waves with wavelengths less than $\lambda_{\min} = 2a_0$. This is illustrated for the case of a one-dimensional crystal in Figure 10.6. In this case, neighbouring atoms vibrate in anti-phase. If we consider the atoms in a solid to be arranged as close-packed spheres, we have a total of N atoms in a cube of volume L^3. Thus we can take $L/N^{1/3}$ to be the interatomic spacing a_0. Hence, $\lambda_{\min} \approx a_0 \approx L/N^{1/3}$. Then

$$\omega_{\max} = 2\pi\nu_{\max} = \frac{2\pi c}{\lambda_{\min}} = \frac{2\pi c N^{1/3}}{L},$$

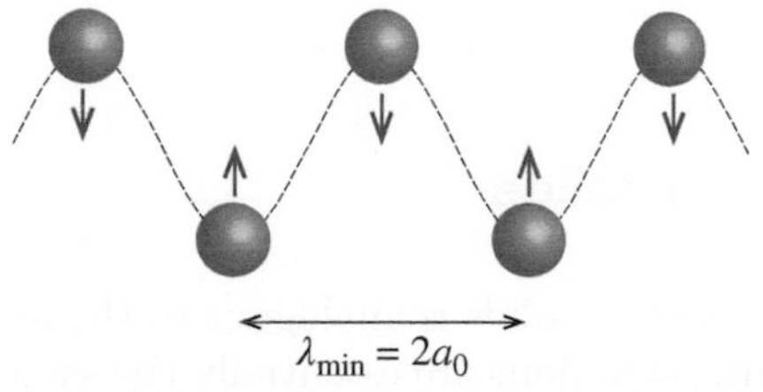

Figure 10.6 A crystal with interatomic spacing a_0 cannot propagate waves with wavelengths less than $\lambda_{\min} = 2a_0$. This is illustrated in the case of a one-dimensional crystal. In this case, neighbouring atoms vibrate in anti-phase.

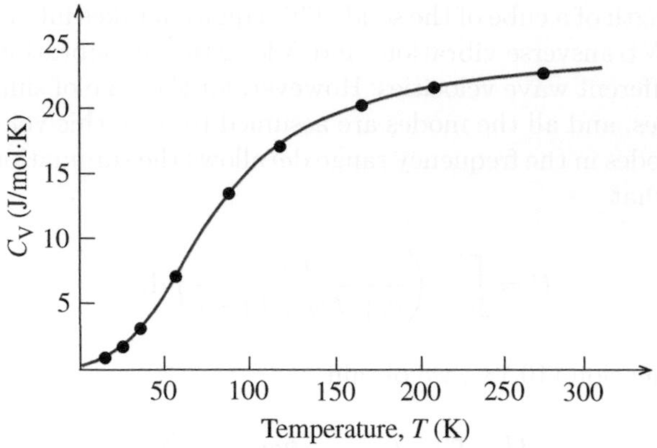

Figure 10.7 The continuous curve shows the molar specific heat curve for copper, calculated using Debye's relation, for a value of $\Theta_D = 315$ K. It is in agreement with the experimental data. In particular, Debye's theory predicts that C_V decreases as T^3 at low temperature, which matches the experimental data.

in essential accord with Equation (10.22). Thus, the limiting frequency in the Debye model corresponds to the shortest possible wavelength being comparable to the interatomic spacing.

Using the value of ω_{max} from Equation (10.22) for the upper limit of the integral in Equation (10.19), an explicit expression for the total energy U of the solid can be obtained. And by differentiating the resulting expression with respect to T, the molar specific heat C_V can be found. When this is done, the final result of the Debye model is

$$C_V = 3Nk\left(\frac{3}{x_D^3}\int_0^{x_D}\frac{x^4 e^x \mathrm{d}x}{(e^x - 1)^2}\right), \tag{10.23}$$

where $x = \hbar\omega/kT$ and $x_D = \hbar\omega_{max}/kT$. x_D can also be written as $x_D = \Theta_D/T$, which defines the Debye temperature Θ_D, in analogy to the Einstein temperature Θ_E. The integral is, of course, just a number and so the specific heat depends only on a single parameter, the Debye temperature Θ_D.

In the high temperature limit, and taking $N = N_A$, Debye's result becomes $C_V = 3N_A k$, in agreement with the law of Dulong and Petit. And in the low temperature limit, it gives $C_V = 0$. Moreover, it predicts that C_V decreases like T^3, at low temperature. The blue curve in Figure 10.7, shows the molar specific heat curve for copper, calculated using Debye's relation with a value of $\Theta_D = 315$ K. It is in good agreement with the experimental data. In particular, it is in accord with experiments at low temperature where the T^3 dependence matches the experimental data, unlike Einstein's result that decreases exponentially, i.e. too fast.

10.2 Thermal conductivity of solids

A discussion of the thermal conductivity of solids is analogous to the discussion of thermal conduction in gases in Section 4.5 and the governing equations are essentially the same. Heat will flow through a slab of material if there is a temperature difference between the faces of the slab. This is illustrated by Figure 10.8,

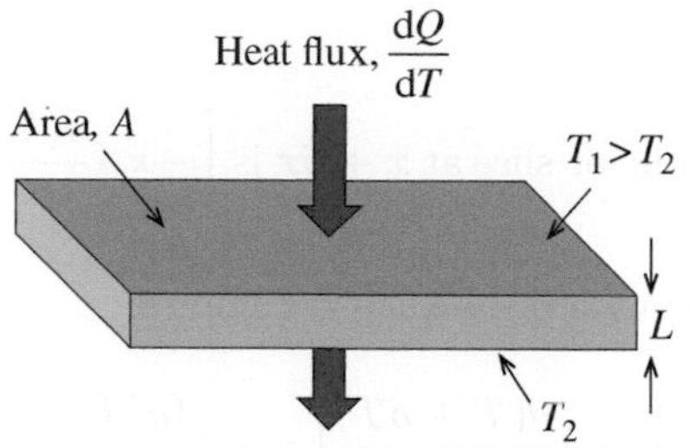

Figure 10.8 Heat will flow through a slab of material if there is a temperature difference between the faces of the slab. For most solids, it is found that the heat conducted per second through the slab is described by $dQ/dt = \kappa A(T_1 - T_2)/L)$, where κ is the thermal conductivity of the solid and A and L are the cross-sectional area and thickness of the slab, respectively.

where the faces of the slab are at temperatures T_1 and T_2, and $T_1 > T_2$. For most solids, it is found that the heat conducted per second through the slab dQ/dt is described by

$$\frac{dQ}{dt} = \kappa A \frac{(T_1 - T_2)}{L}, \tag{10.24}$$

where κ is the *thermal conductivity* of the solid and A and L are the cross-sectional area and thickness of the slab, respectively. The units of κ are $W/m \cdot K$. The differential form of Equation (10.24) is

$$\frac{dQ}{dt} = -\kappa A \frac{dT}{dx}, \tag{10.25}$$

which is the one-dimensional form of Fourier's law of conduction. The negative sign shows that heat flows in the direction of decreasing temperature.

More generally, and for many practical situations, the heat flow is a function of both position and time. Figure 10.9 shows a section of a rod in which the heat flow is a function of both position x and time t. The rod is assumed to be thermally insulated so that there is no loss of heat from its surface. We consider the heat flow into and out of a slice of the rod between x and at $x + \delta x$. Then,

$$\text{heat flow into slice at } x \text{ is } \left[-\kappa A \frac{\partial T}{\partial x}\right],$$

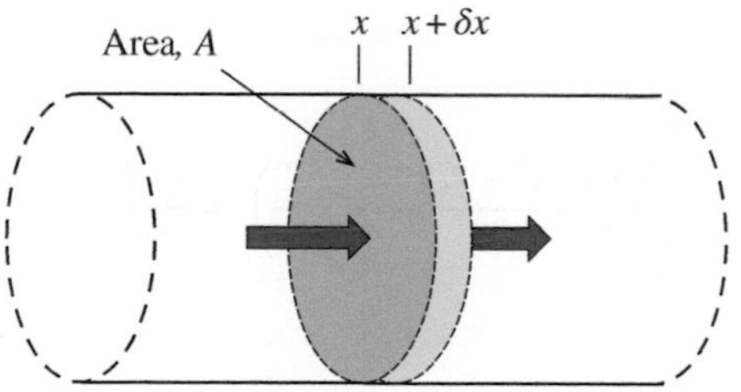

Figure 10.9 The figure shows a section of a rod in which the heat flow is a function of both position x and time t. The rod is assumed to be thermally insulated so that there is no heat loss from its surface. We consider the heat flow into and out of a slice of the rod between x and at $x + \delta x$.

and

$$\text{heat flow out of slice at } x+\delta x \text{ is } \left[-\kappa A\frac{\partial(T+\delta T)}{\partial x}\right].$$

Therefore, the net increase of heat flow into the slice is equal to

$$\left[-\kappa A\frac{\partial T}{\partial x}\right]-\left[-\kappa A\frac{\partial(T+\delta T)}{\partial x}\right]=\kappa A\left[\frac{\partial(T+\delta T)}{\partial x}-\frac{\partial T}{\partial x}\right].$$

This is equal to the rate $\partial Q/\partial t$ at which the heat Q increases in the slice. Hence,

$$\frac{\partial Q}{\partial t}=\kappa A\left[\frac{\partial(T+\delta T)}{\partial x}-\frac{\partial T}{\partial x}\right]. \tag{10.26}$$

When heat energy is added to the slice, the temperature of the slice increases according to

$$\frac{\partial Q}{\partial T}=mc, \tag{10.27}$$

where m is the mass of the slice and c is the specific heat per unit mass. Rearranging this equation and differentiating with respect to t, we obtain

$$\frac{\partial Q}{\partial t}=mc\frac{\partial T}{\partial t}. \tag{10.28}$$

For a slice of thickness δx and density ρ,

$$\frac{\partial Q}{\partial t}=\rho A\delta x c\frac{\partial T}{\partial t}. \tag{10.29}$$

Combining Equations (10.26) and (10.29), we obtain

$$\frac{\partial T}{\partial t}=\frac{\kappa}{\rho c\delta x}\left(\frac{\partial(T+\delta T)}{\partial x}-\frac{\partial T}{\partial x}\right). \tag{10.30}$$

In the limit $\delta x\rightarrow 0$,

$$\frac{1}{\delta x}\left(\frac{\partial(T+\delta T)}{\partial x}-\frac{\partial T}{\partial x}\right)=\frac{\partial^2 T}{\partial x^2}.$$

Hence, we have our final result

$$\frac{\partial T}{\partial t}=\frac{\kappa}{\rho c}\frac{\partial^2 T}{\partial x^2}. \tag{10.31}$$

This is the one-dimensional heat equation and the quantity $\kappa/\rho c$ is the thermal diffusivity of the material.

10.3 Diffusion in solids

We discussed the diffusion of gases in Chapter 4. Diffusion also occurs in solids where atoms diffuse through the crystal lattice, albeit very slowly. And indeed it has a great deal of practical importance. For example, it is exploited in the fabrication of integrated circuits. This involves the evaporation of various impurity atoms onto the surface of a semiconductor wafer, usually silicon that is held at a high temperature of about 1000 °C. The deposited atoms diffuse into the surface of the crystal where they produce the required circuit elements such as resistors and transistors. At this high temperature, the impurity atoms diffuse into the surface on the relatively short time scale of minutes. However, when cooled to room temperature the diffusion rate is reduced enormously so that the diffused impurities are essentially locked in place and the device is stable. This is because, as we shall see, the diffusion rate has an exponential dependence on temperature.

The process of diffusion in solids is very different from that in gases. The predominant mechanism involves vacancies in the crystal lattice. These vacancies can occur when an atom is ejected from its site in the lattice by thermal excitation. An atom from an adjacent site in the lattice can then hop into that vacancy, and the vacancy is thus displaced. This of course leaves another vacancy that can be filled, and the process can repeat. As the vacancies diffuse through the crystal, so too do the atoms. A schematic diagram illustrating this diffusion process is shown in Figure 10.10.

The rate of diffusion depends on the number of vacancies present in the solid, and also on the ability of an atom to 'squeeze' into an adjacent vacancy. For the sake of simplicity, we will consider *self-diffusion* of atoms in a solid. Experimentally, we could coat the surface of a solid with a thin layer of a radioactive isotope of the same atoms in the solid. The radioactive atoms would diffuse through the solid and we could measure their rate of diffusion using radioactive tracer techniques.

Suppose that it takes energy E_V to remove an atom from its site in the lattice and produce a vacancy. The probability that an atom has energy between E and $E + \mathrm{d}E$ at temperature T is given by the Boltzmann distribution

$$P[E]\mathrm{d}E = A\mathrm{e}^{-E/kT}\mathrm{d}E, \tag{10.32}$$

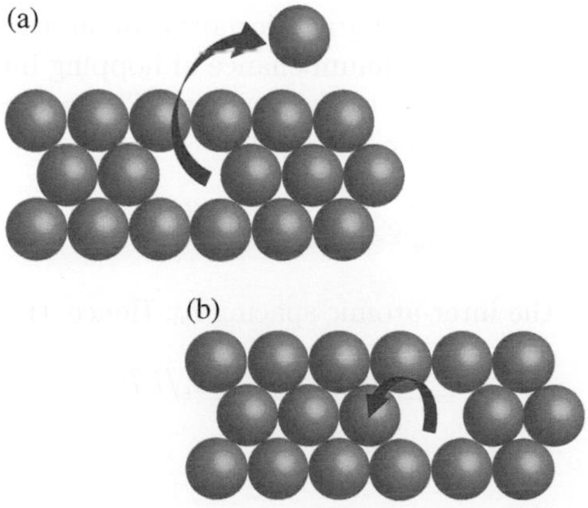

Figure 10.10 A schematic diagram illustrating the predominant diffusion process in solids. (a) Vacancies can occur in a crystal lattice when an atom is ejected from its site by thermal excitation. (b) An atom from an adjacent site in the lattice can then hop into the vacancy, and the vacancy is thus displaced. This leaves another vacancy that can be filled, and the process can repeat.

where A is the normalization constant. We have

$$\int_0^\infty P[E]\mathrm{d}E = A\int_0^\infty \mathrm{e}^{-E/kT}\mathrm{d}E = 1, \tag{10.33}$$

and hence,

$$A = \frac{1}{\displaystyle\int_0^\infty \mathrm{e}^{-E/kT}\mathrm{d}E}. \tag{10.34}$$

Then, the probability that an atom has an energy of *at least* E_V is

$$\frac{\displaystyle\int_{E_V}^\infty \mathrm{e}^{-E/kT}\mathrm{d}E}{\displaystyle\int_0^\infty \mathrm{e}^{-E/kT}\mathrm{d}E} = \mathrm{e}^{-E_V/kT}. \tag{10.35}$$

This is the probability that an atom has enough energy to be ejected from its lattice. Equivalently, the fraction of atoms in the solid that have enough energy to be ejected is $\mathrm{e}^{-E_V/kT}$.

The number of vacancies increases with temperature, thereby increasing the energy of the crystal. It may appear strange that the solid prefers to have a higher energy than one in which there are no vacancies, i.e. a state of lower energy. The reason is that the entropy of disorder is increased by the presence of the vacancies and this compensates for the increase in energy.

Suppose it takes energy E_S for an atom to squeeze into a vacancy. We can reasonably expect E_S to be roughly the same size as E_V. As mentioned earlier, the probability that an atom has at least energy E_S is $\mathrm{e}^{-E_s/kT}$. Then, taking both requirements into account, the probability that an atom has an available adjacent vacancy and that it has the necessary energy to hop into that vacancy is given by

$$\mathrm{e}^{-E_v/kT} \times \mathrm{e}^{-E_s/kT} = \mathrm{e}^{-E_v + E_s/kT}. \tag{10.36}$$

In Einstein's model, the atoms vibrate in simple harmonic motion about their fixed lattice sites at frequency $\omega_E/2\pi$. We can expect an atom to have a maximum chance of hopping into an adjacent vacancy when it reaches its vibrational maxima that lie closest to the vacancy. This occurs at the vibrational frequency. It follows that the average number of hops per second is

$$\frac{\omega_E}{2\pi}\mathrm{e}^{-E_v + E_s/kT}. \tag{10.37}$$

The length of each hop is equal to the inter-atomic spacing a_0. Hence, the average hopping speed v_H is

$$v_H = \frac{a_0\omega_E}{2\pi}\mathrm{e}^{-E_v + E_s/kT}. \tag{10.38}$$

10.3.1 The diffusion coefficient

It is convenient to consider some of the atoms in the solid to be labelled in some way. For example, the solid could contain a certain percentage of atoms that are a radioactive isotope. We can then consider the diffusion of these labelled atoms when there exists a concentration gradient of these atoms in the solid.

In Section 4.4, we considered the number of molecules that cross a unit plane when they diffuse through a gas. We adopt a similar picture for diffusion in a solid. In this case, labelled atoms crossing a unit plane will have hopped there from a distance equal to the interatomic separation a_0 either side of the plane. The densities of the labelled atoms at those distances from the plane are $n - a_0(\mathrm{d}n/\mathrm{d}x)$ and $n + a_0(\mathrm{d}n/\mathrm{d}x)$ respectively, where $\mathrm{d}n/\mathrm{d}x$ is the density gradient. Then the number of labelled atoms that pass through the plane in the positive x-direction is $(1/6)v_{\mathrm{H}}(n - a_0)(\mathrm{d}n/\mathrm{d}x)$, where v_{H} is the hopping speed, and we have used our usual one-sixth model; see Section 4.5. Similarly, the number of labelled atoms that pass through the plane in the opposite direction is $(1/6)v_{\mathrm{H}}(n + a_0)(\mathrm{d}n/\mathrm{d}x)$. The net flux $J(x)$ of these atoms crossing the unit plane per unit time in the positive x-direction is

$$J(x) = \frac{1}{6}v_{\mathrm{H}}(n - a_0)\frac{\mathrm{d}n}{\mathrm{d}x} - \frac{1}{6}v_{\mathrm{H}}(n + a_0)\frac{\mathrm{d}n}{\mathrm{d}x} = -\frac{1}{3}a_0 v_{\mathrm{H}}\frac{\mathrm{d}n}{\mathrm{d}x}. \tag{10.39}$$

This has the same form as Fink's law, $J = -D(\mathrm{d}n/\mathrm{d}x)$, Equation (4.14). Comparing the two expressions, we have $D = -(1/3)a_0 v_{\mathrm{H}}$. And substituting for v_{H} from Equation (10.38), we obtain

$$D = \frac{1}{6\pi}a_0^2\omega_{\mathrm{E}}\mathrm{e}^{-E_{\mathrm{v}} + E_{\mathrm{s}}/kT}.$$

or

$$D = D_0\mathrm{e}^{-E_{\mathrm{v}} + E_{\mathrm{s}}/kT}, \tag{10.40}$$

where the constant of proportionality $D_0 = (1/6\pi)a_0^2\omega_{\mathrm{E}}$.

The most striking feature of Equation (10.40) is the exponential dependence of the rate of diffusion on temperature, which is indeed confirmed by experimental measurements. The diffusion becomes more rapid as the temperature increases. This is in stark contrast to diffusion in gases, where the diffusion rate becomes more rapid as the temperature decreases. Suppose that a surface is coated by depositing atoms on the surface at $T = 1250$ K, allowing these atoms to diffuse into the solid. Taking a typical value of $(E_{\mathrm{V}} + E_{\mathrm{S}})$ to be 2 eV, the ratio of the diffusion rates at $T = 293$ K and at $T = 1250$ K is

$$\frac{\exp(-2 \times 1.6 \times 10^{-19}/1.38 \times 10^{-23} \times 293)}{\exp(-2 \times 1.6 \times 10^{-19}/1.38 \times 10^{-23} \times 1250)} = \frac{\exp(-79)}{\exp(-19)} \approx 1 \times 10^{-26}.$$

We see that the diffusion rate is enormously smaller at room temperature compared to that at the higher temperature, which means the coating remains intact indefinitely at room temperature.

In our discussion of diffusion in gases, we also described diffusion in terms of a random walk process; see Section 4.9. We saw that the mean distance travelled in a medium of diffusion coefficient D in time t is $\sqrt{Dt}$, neglecting numerical constants of order unity. This expression is true in general for diffusion processes and we can use it to estimate the distance d that an atom diffuses in a solid in time t, i.e. $d \sim (Dt)^{1/2}$, where D is the diffusion coefficient of the solid.

Worked example

(a) Determine the fraction of vacancies in a crystal of copper at the temperature of 293 K. (b) Estimate the time it takes an atom of copper to diffuse a distance of 1 mm at the temperature of 1300 K. The values of D_0, E_{V}, and E_{S} for copper are 2.8×10^{-5} m^2/s, 1.45 eV, and 0.75 eV respectively.

Solution

(a) The fraction of vacancies in a crystal at a given temperature is equal to $e^{-E_V/kT}$. Therefore, the fraction of vacancies in the copper crystal at $T = 293$ K is

$$\exp -\left(\frac{1.45 \times 1.6 \times 10^{-19}}{1.38 \times 10^{-23} \times 293}\right) = e^{-57.4} = 1.2 \times 10^{-25}.$$

This result says that if we have a pure copper crystal of mass 1 kg, we would expect there to be

$$\frac{1000}{63.5} \times 6 \times 10^{23} \times 1.2 \times 10^{-25} \sim 1 \text{ vacancy}.$$

At $T = 1300$ K, the fraction is $e^{-13} \sim 2 \times 10^{-6}$, and the corresponding number of vacancies is $\sim 2 \times 10^{19}$.

(b) From Equation (10.40), we have

$$D = 2.8 \times 10^{-5} \exp -\left(\frac{2.2 \times 1.6 \times 10^{-19}}{1.38 \times 10^{-23} \times 1300}\right) = 8.4 \times 10^{-14} \text{ m}^2/\text{s}.$$

Then for diffusion through a distance of 1 mm, we have

$$\left(8.4 \times 10^{-14} \times t\right)^{1/2} \sim 1 \times 10^{-3}$$

which gives $t \sim 1.2 \times 10^7$ seconds $\sim$ 140 days.

10.4 Electrical and thermal conductivities of metals

A detailed understanding of the behaviour of conduction electrons in a metal involves a quantum mechanical approach. However, a simple classical theory that gives a useful qualitative description of the electrical conductivity of a metal is provided by the *free-electron gas* model. This model is attributed to Paul Drude. In this model, a metal is pictured as an array of positive ions in a three-dimensional lattice that exists in a sea of free electrons. In a monovalent metal like sodium, for example, each atom contributes a single free electron with the resulting positive ion being fixed in a lattice site; overall, the metal is electrically neutral. The free electrons are considered to move throughout the volume of the metal, just like molecules do in a gas. In doing so, they make collisions with the positive ions following a random path from one collision to the next with mean speed $\bar{c}$. It is assumed that the electrons do not make collisions with other electrons. The mean distance that an electron travels between collisions is characterised by its mean free path λ, just as for molecules in a gas. Then, the mean time between collisions is $\tau = \lambda/\bar{c}$. Under normal conditions, there is no net movement of electrons in any direction.

If, however, a voltage is applied across the metal, the electrons also acquire a *drift velocity* v_d. The drift velocity arises from the force the free electrons experience from the electric field as they travel between collisions with the ions. It is taken to be responsible for the observed electrical conductance of the metal. The action of the positive ions is to inhibit the flow of the electrons and hence give the metal its resistivity. The drift velocity is superimposed on the mean velocity $\bar{c}$ of the electrons, but is much less than $\bar{c}$ as illustrated by the following worked example.

Worked example

A current of 1.0 A flows down a copper wire of cross-sectional area 1.0 mm^2. Compare the drift velocity v_d of the electrons with their mean speed $\overline{c}$ given by the free electron model. The number of electrons per unit volume n in copper is $8.5 \times 10^{28} m^{-3}$.

Solution

As the electrons have translational energy in three orthogonal directions, each electron has a mean energy of translation equal to $(3/2)kT$, according to the classical equipartition theorem. Taking the mean speed of the electrons to be $\overline{c} \approx c_{rms} = \sqrt{3kT/m}$, where m is the electronic mass, we find at $T = 300$ K,

$$\overline{c} = \sqrt{\frac{3 \times 1.38 \times 10^{-23} \times 300}{9.11 \times 10^{-31}}} = 1.2 \times 10^5 \text{ m/s}.$$

Consider now the current flowing down the copper wire. Figure 10.11 shows a portion of the wire of cross-sectional area A through which current i is flowing. There are n electrons per unit volume with charge e, and these drift down the wire at velocity v_d. Within one second, all those electrons within a distance v_d to the right of the plane P, i.e. in a volume Av_d flow through this plane. This volume contains nAv_d electrons and hence a charge $neAv_d$. Thus, the current $i = neAv_d$ and

$$v_d = \frac{i}{neA} = \frac{1.0}{8.5 \times 10^{28} \times 1.6 \times 10^{-19} \times 1.0 \times 10^{-6}} = 7.3 \times 10^{-5} \text{ m/s}.$$

We see that the drift velocity v_d is exceedingly small compared to the mean speed $\overline{c}$ of the free electrons.

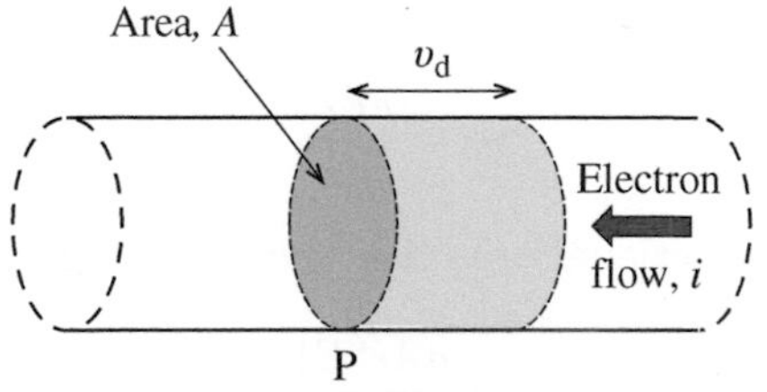

Figure 10.11 The figure shows a portion of copper wire of cross-sectional area A through which current i is flowing. There are n electrons per unit volume with charge e, and these drift down the wire at velocity v_d. Within one second, all those electrons within a distance v_d to the right of the plane P, i.e. in a volume Av_d flow through this plane. The volume contains nAv_d electrons and hence a charge $neAv_d$. Thus, the current $i = neAv_d$.

We assume that when an electron collides with a positive ion, it gives up any velocity it had gained from the electric field. And that an electron leaves each collision with a velocity that may be in any direction with equal likelihood. Hence, the initial velocity of an electron after a collision does not contribute to any net motion. We consider a bar of metal of uniform cross-sectional area A and length L across which a potential V is applied. The electric field experienced by an electron is $E = V/L$. The force on an electron is eE and its resulting acceleration is $a = eE/m$. Therefore, on average, an electron will acquire a drift velocity

$$v_d = at_m = \frac{eEt_m}{m}, \tag{10.41}$$

where t_m is the mean time *since* it made its last collision. To find t_m, we let the probability that an electron survives time t before colliding be $P(t)$ and the probability that the electron survives time $(t+\mathrm{d}t)$ be $P(t+\mathrm{d}t)$. The probability that an electron makes a collision in time $\mathrm{d}t$ is $\mathrm{d}t/\tau$, where τ is the mean time between collisions. Therefore, the probability that an electron does not make a collision in time $\mathrm{d}t$ is $(1-\mathrm{d}t/\tau)$. It follows that

$$P(t+\mathrm{d}t) = P(t)\left(1-\frac{\mathrm{d}t}{\tau}\right),$$

or

$$\frac{P(t+\mathrm{d}t)-P(t)}{\mathrm{d}t} = \frac{\mathrm{d}P(t)}{\mathrm{d}t} = -\frac{1}{\tau}P(t). \tag{10.42}$$

The solution to this differential equation is

$$P(t) = \mathrm{e}^{-t/\tau}. \tag{10.43}$$

The corresponding probability distribution function $P[t]$ is given by

$$P[t]\mathrm{d}t = P(t)\frac{\mathrm{d}t}{\tau}.$$

Then the mean time since the last collision is

$$t_m = \int_0^\infty t\ P[t]\,\mathrm{d}\,t = \frac{1}{\tau}\int_0^\infty t\,\mathrm{e}^{-t/\tau}\mathrm{d}t = \tau. \tag{10.44}$$

We see that t_m is equal to the mean time between collisions τ. Substituting for $t_m = \tau = \lambda/\bar{c}$ in Equation (10.41), we obtain

$$v_d = \frac{eE\lambda}{m\bar{c}}. \tag{10.45}$$

The current i flowing through cross-sectional area A is env_dA, giving

$$i = \frac{nAe^2E\lambda}{m\bar{c}}, \tag{10.46}$$

or

$$i = V\frac{nAe^2\lambda}{Lm\bar{c}}. \tag{10.47}$$

We see that the free electron model correctly predicts Ohm's law; that the current i is directly proportional to the applied voltage V. Then using $R = V/i$, where R is the resistance of the metal bar, we have

$$R = \frac{Lm\bar{c}}{nAe^2\lambda}. \tag{10.48}$$

We also have the expressions $R = \rho L/A$ and $\sigma = 1/\rho$, where ρ is the specific resistivity and σ is the electrical conductivity of the metal, respectively. Hence, we have

$$\rho = \frac{m\bar{c}}{ne^2\lambda}, \tag{10.49}$$

and

$$\sigma = \frac{ne^2\lambda}{m\overline{c}}. \tag{10.50}$$

For copper, $\sigma = 5.8 \times 10^7\,\text{ohm}^{-1}\,\text{m}^{-1}$, using Equation (10.50) gives the value for the mean free path of the electrons to be

$$\lambda = \frac{5.8 \times 10^7 \times 1.72 \times 10^5}{8.5 \times 10^{-31} \times \left(1.6 \times 10^{-19}\right)^2} \approx 3\,\text{nm}.$$

This is of the correct order of magnitude. In general, Equation (10.50) gives good agreement with the experiment when λ is taken to be the order of several atomic spacings.

10.4.1 Thermal conductivity of metals

Good thermal conductors such as metals are usually good electrical conductors too. This suggests that the dominant contribution to thermal conduction in metals arises from the movement of electrons. Again using the free electron model, we imagine that the electrons in a metal behave as a perfect gas transporting thermal energy in an analogous way to that which occurs in heat conduction in a gas. In Section 4.5, we discussed thermal conductivity in gases. For a monatomic gas, we found the following expression for the thermal conductivity κ:

$$\kappa = \frac{1}{2} n\overline{c}\lambda k, \tag{10.51}$$

where n is the number density, $\overline{c}$ is the mean speed of the molecules, λ is their mean free path and k is the Boltzmann constant. We assume that the thermal conductivity of the 'free electron gas' can be described by a similar equation. For the example of copper, we have $n = 8.5 \times 10^{28}\,\text{electrons/m}^3$, and taking $\overline{c} = 1.2 \times 10^5\,\text{m/s}$ and λ to be 3 nm, we find

$$\begin{aligned}\kappa &= \frac{1}{2} \times 8.5 \times 10^{28} \times 1.2 \times 10^5 \times 3 \times 10^{-9} \times 1.38 \times 10^{-23} \\ &= 211\,\text{W/m} \cdot \text{K}.\end{aligned}$$

Considering the simplicity of the model, this value is in satisfactory agreement with the observed value for copper, which is 398 W/m · K.

From Equations (10.50) and (10.51), we can predict the ratio of the thermal to electrical conductivities of a metal:

$$\frac{\kappa}{\sigma} = \frac{1}{2} \frac{n\overline{c}\lambda k m\overline{c}}{ne^2\lambda} = \frac{1}{2} m\overline{c}^2 \frac{k}{e^2}. \tag{10.52}$$

Then using $(1/2)m\overline{c}^2 = (3/2)kT$, we obtain

$$\frac{\kappa}{\sigma} = \frac{3}{2}\left(\frac{k}{e}\right)^2 T. \tag{10.53}$$

Historically, Gustav Wiedemann and Rudolph Franz found empirically that the ratio κ/σ does have approximately the same value for different metals at the same temperature. The dependence of κ/σ on absolute temperature T was later found by Ludvig Lorenz. The quantity $\kappa/\sigma T$ is called the *Lorenz number* L. From Equation (10.53), we can predict its value to be

$$\frac{3}{2}\left(\frac{1.38\times10^{-23}}{1.6\times10^{-19}}\right)^2 \approx 1\times10^{-8}\,\mathrm{W\cdot\Omega/K^2}.$$

This is indeed of the correct order of magnitude. Experimentally determined values of L lie in the range $2\times10^{-8}\,\mathrm{W\cdot\Omega/K^2}$ to $3\times10^{-8}\,\mathrm{W\cdot\Omega/K^2}$.

10.4.2 Successes and failures of the classical free electron model

The classical free electron model succeeds in predicting a range of experimental observations such as Ohm's law and the Wiedemann–Franz law. However, it does have fundamental flaws. Perhaps the most striking is with respect to the specific heats of metals. In the classical free electron model, the electrons have translational energy per mole equal to $(3/2)N_AkT = (3/2)RT$, and thus can be expected to contribute a specific heat of $(3/2)R$ per mole. But as we have seen, the specific heat of monoatomic metals is about 25 J/mol $\approx 3R$, and not about 37 J/mol $\approx (3+3/2)R$. The reason for the discrepancy is that the free electrons do not obey the classical Maxwell–Boltzmann energy distribution. Instead, the free electrons are restricted by the Pauli exclusion principle; the principle we encountered in Chapter 1. This principle requires that no two electrons can be in the same quantum state. As a result, all the energy states up to the highest energy ε_F called the *Fermi energy* are occupied. Typically, the Fermi energy is ~1 eV. The amount of thermal energy available to the electrons at room temperature $\sim kT = 1/40$ eV is much less than this. Consequently, most of the electrons are unable to change their quantum state because all the states within the energy range kT are already occupied. Only those electrons within the small energy interval kT below ε_F can be excited into unoccupied states. Hence, at ordinary temperatures, only a small fraction of the total number of free electrons can absorb thermal energy and consequently their contribution to the specific heat is essentially zero.

Problems 10

10.1 What is the molar specific heat C_V of a monatomic solid at the Einstein temperature?

10.2 Estimate (a) the minimum wavelength of longitudinal waves that can propagate through aluminium and hence the highest frequency of the normal modes of oscillation in aluminium. (b) Use your value of the maximum frequency to estimate the Debye temperature for aluminium. The atomic weight of aluminium is 27 u, its density is $2.7\times10^3\,\mathrm{kg/m^3}$ and its Young's modulus is 6.9×10^{10} Pa.

10.3 Qualitatively, the melting point of a crystalline solid can be defined as the temperature T_M at which the amplitude of simple vibrations of the individual atoms about their lattice sites is about 10% of the interatomic spacing a_0. Using this definition, show that the frequency ν of vibrations of a monatomic crystal satisfies the relation $\nu = (5/\pi a_0)\sqrt{(2kT_M)/m}$, where m is the mass of an atom. Estimate the frequency of vibration of the aluminium atoms. The atomic weight of aluminium is 27 u, its density is $2.7\times10^3\,\mathrm{kg/m^3}$, and its melting point is 660 °C.

10.4 A pond is covered by a layer of ice 5 cm thick. How long will it take before the ice is 10 cm thick if the air temperature remains constant at −7.5 ° C. The latent heat of fusion, thermal conductivity, and density of ice are $L = 3.36\times10^5$ J/kg, $\kappa = 2.1$ W/K · m and $\rho = 920\,\mathrm{kg/m^3}$, respectively.

10.5 The space between two concentric, spherical shells of copper is filled with a thermally insulating material. The radii of the two shells are 5 cm and 10 cm, respectively. An electric heater providing 20 W of power is enclosed within the inner copper shell. It is found that a temperature difference of 50 °C is maintained between the two copper shells. Determine the thermal conductivity κ of the insulating material.

10.6 (a) Use the free electron gas model to obtain a value for the mean free path of an electron in sodium at 293 K. For sodium, the electrical conductivity $\sigma = 2.10 \times 10^7/\Omega/\text{m}$, the density $\rho = 970\ \text{kg/m}^3$, and the atomic mass $= 23$ u.

10.7 Taking the Einstein temperature for silver to be $\Theta_E = 154$ K, obtain a value for its diffusion coefficient D_0. The atomic weight of silver is 108 u, and its density is $10.8 \times 10^3\ \text{kg/m}^3$.

10.8 A thin layer of N atoms is deposited on a surface area A of a solid of the same atoms. These atoms diffuse into the solid. (a) Show that an appropriate solution to the one-dimensional diffusion equation $\partial n/\partial t = D(\partial^2 n/\partial^2 x)$ is $n(x, t) = N\text{e}^{(-x^2/4Dt)}/A(\pi Dt)^{1/2}$, where $n(x, t)$ is the concentration of atoms that have diffused a distance x into the solid at time t and D is the diffusion coefficient. (b) A thick nickel plate is coated with a thin layer of a radioactive isotope of nickel. It is found that the concentration of the radioactive isotope in the solid falls by a factor of 2 at a distance of 5 mm from the surface after one hour. Use these data to obtain a value for the self-diffusion coefficient of nickel.

11

Liquids

In this chapter, we turn our attention to the liquid phase of matter. We have previously discussed the relationships between liquids, gases and solids in terms of the intermolecular forces that operate in the three phases. In all three phases, there exist forces of attraction and repulsion between the constituent molecules. The fact that the molecules are confined within the volume of the liquid demonstrates the attractive force between them. However, the lack of rigidity in a liquid shows that this attractive force is short range. The fact that liquids are essentially incompressible shows the existence of the short-range repulsive force. We have also seen the effects of the interplay in a liquid between the binding energies of the molecules and their kinetic energy due to thermal motion. In a liquid, the kinetic energies of the molecules are comparable to their binding energies. So, although the molecules are confined to the liquid, they are relatively mobile within the liquid. This interplay explains the fact that the liquid phase of a substance occurs over a relatively narrow range of temperature and pressure. As an example, the intermolecular binding energy ε of solid argon is $\sim$10 meV and argon is in its liquid phase between 83.8 and 87.4 °C when $kT \sim 7$ meV.

We describe the structure of liquids, the properties of liquids that relate to the binding energy of the constituent molecules and the flow of liquids. And with respect to liquid flow, we will discuss the important *continuity equation* and *Bernoulli's equation*. As the liquid phase exists between the gaseous and solid phases, we may expect some similarities between gases and liquids and liquids and solids, and indeed we will find this to be the case.

11.1 The structure of liquids

In a solid, there is long-range order extending over $\sim 10^{23}$ atoms in a typical crystal. This long-range order arises because the interatomic forces that operate between the atoms in a solid are also long range. This long-range order is reflected in the diffraction pattern obtained from a solid. The regularity of the crystal structure over such a huge number of atoms means that the diffraction pattern exhibits numerous sharp maxima. In sharp contrast, the molecules in a gas move freely throughout the volume of the gas with a complete absence of spatial order. Consequently, gases do not exhibit a diffraction pattern. In a liquid, there is a degree of spatial order but it extends only a few intermolecular distances because the intermolecular forces

Physics of Matter, First Edition. George C. King.

are short range. And, perhaps surprisingly and despite their fluidity, liquids do exhibit diffraction patterns due to this localised order. However, these diffraction patterns are very different from those of a solid. A diffraction pattern of a liquid typically contains only a few maxima at most and these will be quite diffuse, i.e. not sharp. The structure that a liquid possesses can be described by the *radial distribution function* $g(r)$. This function can be determined from the diffraction pattern of the liquid, and it can also be deduced using theoretical methods.

11.1.1 The radial distribution function

To illustrate the concept of the radial distribution function, Figure 11.1 shows a hypothetical 'snapshot' of a liquid. At the centre of this figure, is a molecule labelled A. Close to molecule A, there can be seen to be a certain degree of spatial order, i.e. some regularity with respect to its nearest neighbours. However, the degree of spatial order between molecule A and molecules further away steadily decreases as their separation increases. This is in stark contrast to the situation in solids where a definite spatial relationship between atoms extends over many atoms. This evolution in the spatial order is exhibited in the radial distribution function $g(r)$, which is defined as

$$g(r) = \frac{n(r)}{n_0}, \tag{11.1}$$

where $n(r)$ is the number of molecules per unit volume at a distance r from a reference molecule and n_0 is the average number density of the bulk liquid, i.e. $n_0 = N_0/V$, where N_0 is the total number of molecules in the liquid of volume V. If there were total disorder in the liquid, just as there is in a gas, $n(r)$ would on average always be equal to n_0 and hence $g(r)$ would always be equal to 1.0. However, if there is a degree of ordering in a liquid, for example, by the grouping of nearest neighbours as in Figure 11.1, then $g(r)$ will be greater than 1.0 in those regions.

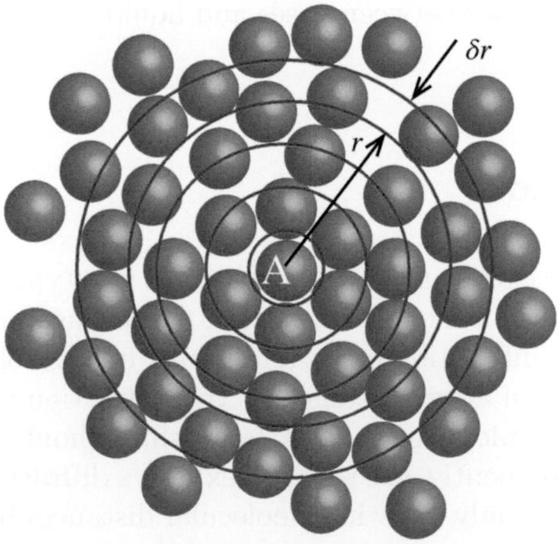

Figure 11.1 The figure shows a hypothetical 'snapshot' of a liquid, where a particular molecule at the centre of the figure and labelled A has been chosen at random. Close to molecule A, there can be seen to be a degree of spatial order, i.e. regularity with respect to its nearest neighbours. However, the degree of spatial order between molecule A and molecules further away steadily decreases as their separation increases.

Suppose a series of concentric spheres centred on molecule A is constructed, each separated by distance δr; see Figure 11.1. Then we can count the number of molecules $N(r)$ whose centres lie within the volume $4\pi r^2 \delta r$ between r and $r + \delta r$. The number density is then

$$n(r) = \frac{N(r)}{4\pi r^2 \delta r}. \tag{11.2}$$

Of course, the molecules in a liquid are continually jiggling about and if we took another 'snapshot', the arrangement of the molecules with respect to molecule A would be slightly different as molecules push past each other. However, if we took many snapshots and then determined $n(r)$ for each one and averaged these data, we would obtain the time-averaged number density, which would still show well-defined maxima and minima.

The radial distribution function for liquid argon is shown in Figure 11.2. Initially, for small values of r, $g(r) = 0$, which simply says that the centre of a second molecule cannot be within the radial extent of molecule A. Then, there is a maximum in $g(r)$ at about 0.28 nm. This maximum gives clear evidence of a 'shell-like' grouping of molecules around molecule A at that distance. Beyond the first maximum, there are further maxima, but these are much smaller in height and much broader. This demonstrates the diminishing spatial correlation between molecule A and molecules further away. Eventually, $g(r)$ converges to the value of 1.0 indicating the absence of any correlation between molecule A and the molecules much far away from it. If there are some regions where the local number density $n(r)$ is larger than the average, bulk density n_0, it follows that there must be some regions where $n(r)$ the local density must be lower: $n(r) < n_0$. This is reflected in the regions where $g(r) < 1.0$. We have chosen one molecule at random. Around any molecule, there will be local order and the net result is that the liquid will show some maxima in a diffraction pattern.

The radial distribution function $g(r)$ of a liquid is determined from the measured diffraction pattern of the liquid in an analogous way to the way the crystal structure of a solid is obtained from its diffraction pattern.

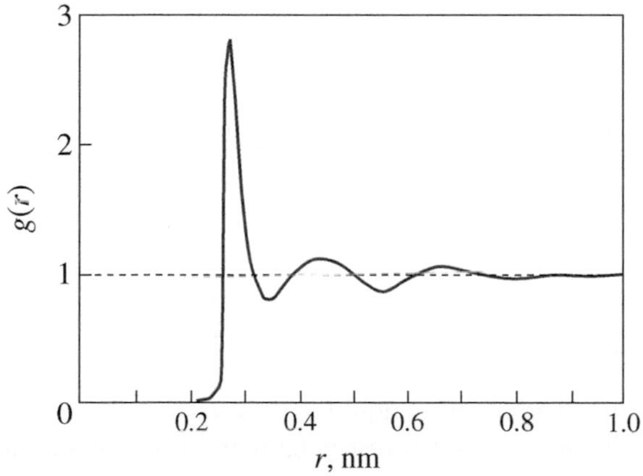

Figure 11.2 The radial distribution function $g(r)$ for liquid argon as a function of radial distance r. Initially, for small values of r, $g(r) = 0$, which simply says that the centre of a second molecule cannot be within the radial extent of molecule A. Then, there is a maximum in $g(r)$ at about 0.28 nm. This maximum gives clear evidence of a 'shell-like' grouping of molecules around molecule A at that distance. Beyond the first maximum, there are further maxima, but these are much smaller in height and much broader. This demonstrates the diminishing correlation between molecule A and molecules further away. Eventually $g(r)$ converges to the value of 1.0 indicating the absence of any correlation between molecule A and the molecules much far away from it. Radial distances where the value of $g(r)$ is less than 1.0 correspond to those regions for which where the local number density $n(r)$ is lower than the average bulk density n_0.

This again uses using the techniques of Fourier analysis. Theoretical techniques can also be used to deduce the radial distribution function, which can then be compared with the one obtained experimentally. These theoretical techniques may employ so-called *monte carlo* methods, which are based on computational algorithms. In these calculations, the interaction between the molecules and their resulting positions are modelled by, for example, the van der Waals interaction. Analysis of the radial distribution function $g(r)$ gives detailed information about the structure of the liquid such as the number of nearest neighbours.

11.2 Physical properties of liquids

Here we consider some of the physical properties of liquids. In particular, we consider properties that we can relate to the behaviour of the constituent molecules and the forces that operate between them.

11.2.1 Latent heats of vapourisation and fusion

Already in Section 2.3, we considered the relationship between the latent heat of vapourisation of liquids (liquid to vapour) and the intermolecular binding energy ε. For a liquid with van der Waals bonding, we obtained the following relationship:

$$\varepsilon \approx \frac{2L_V}{N_A n}, \tag{11.3}$$

where L_V is the latent heat of vapourisation per mole, N_A is Avogadro's number and n is the number of nearest neighbours a molecule has. The usual factor of 2 arises to avoid counting each molecule twice. For a liquid, the value of n is typically 10.

Empirically, it is found that the latent heat of fusion (solid to liquid) is much smaller that the latent heat of vapourisation. For argon, for example:

$$\frac{\text{Latent heat of vapourisation}}{\text{Latent heat of fusion}} = \frac{6.5\,\text{kg/mol}}{1.2\,\text{kg/mol}} = 5.4,$$

and for nitrogen

$$\frac{\text{Latent heat of vapourisation}}{\text{Latent heat of fusion}} = \frac{5.6\,\text{kg/mol}}{0.725.6\,\text{kg/mol}} = 7.8.$$

The large differences arise because in going from liquid to vapour, the change in n is ~10, while in going from solid to liquid, the change in the number of nearest neighbours is much smaller, say from 12 to 10.

11.2.2 Vapour pressure

Suppose that we place a quantity of liquid in a sealed container and evacuate the space above it. Some of the molecules in the liquid will escape into that space and, subsequently, some of those molecules will return to the liquid. Eventually, there will be equilibrium when the number of molecules leaving the liquid is equal to the number returning. The pressure exerted by the vapour under these conditions is known as the *saturated vapour pressure* of the liquid, or simply the *vapour pressure*. Suppose now that the pressure above

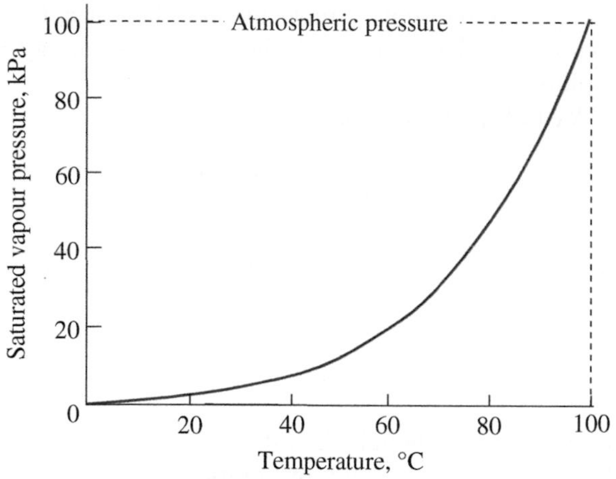

Figure 11.3 The figure shows how the saturated vapour pressure of water varies over the temperature range of its liquid phase. The vapour pressure rises approximately exponentially until at 100 °C, it is equal to atmospheric pressure and the water boils.

the liquid is constant, as, for example, when the liquid surface is open to the atmosphere. As we will see, vapour pressure increases with increasing temperature. Then if the liquid is heated, the vapour pressure rises until eventually the vapour pressure is equal to the external pressure. Bubbles of the vapour appear in the liquid and the liquid is said to be boiling. Figure 11.3 shows how the saturated vapour pressure of water varies over the temperature range of its liquid phase. The vapour pressure rises steadily until at 100 °C, it is equal to atmospheric pressure and the water boils.

The molecular interpretation of vapourisation is as follows. To escape from the liquid, a molecule must have sufficient energy to overcome the attraction of its neighbours. If the latent heat of vapourisation per mole is L_V, the energy required for a molecule to escape is

$$E_v = \frac{L_V}{N_A}, \tag{11.4}$$

where N_A is Avogadro's number. This means that the mean energy of the molecules in the vapour is greater than the mean energy of the molecules in the liquid by the amount E_v. We can apply the Boltzmann distribution to the molecules in the liquid as well as to those in the vapour. It follows then that the relative number densities in the vapour and the liquid, n_V and n_L respectively, are given by

$$\frac{n_V}{n_L} = e^{-E_V/kT}. \tag{11.5}$$

Thus, we have an equation that says how many molecules are in the vapour phase compared to the liquid phase and how the ratio varies with temperature. We make an analogy here with the variation in number density in an isothermal atmosphere. According to Equation (3.52), the number density of molecules at height z is $n(z) = n_0 e^{-E/kT}$, where n_0 is the number density at $z = 0$ and $E = mgz$. Hence, the ratio of number densities at heights z_1 and z_2 is

$$\frac{n_2}{n_1} = e^{-E_2 - E_1/kT} = e^{-\Delta E/kT}, \tag{11.6}$$

where ΔE is the amount of energy required to lift a molecule from height z_1 to z_2. In the case of vapourisation, the molecules in the vapour are less dense than in the liquid because energy E_v must be given to them to release them into the vapour phase.

Substituting for $E_v = L_V/N_A$ in Equation (11.5) gives

$$n_V = n_L e^{-L_V/N_A kT}. \tag{11.7}$$

According to the gas law, $P_{VP}V = N_A kT$ for one mole of vapour, where P_{VP} is the vapour pressure, and $n_V = P_{VP}/kT$. Therefore,

$$P_{VP} = n_V kT = n_L kT e^{-L_V/N_A kT}. \tag{11.8}$$

We can make two simplifications. First, although the ratio n_V/n_L increases with temperature, n_L hardly changes and can be assumed to be constant. Second, by comparison with the exponential term, the linear term kT remains nearly constant throughout the limited temperature range over which a substance remains in the liquid phase. Thus, to a good approximation, we can write

$$P_{VP} = Ae^{-L_V/N_A kT}, \tag{11.9}$$

where A is a constant. The latent heat of vapourisation of water is $L = 4 \times 10^4$ J/mol. Then, for example, increasing the temperature of water from 30 to 50 °C increases the vapour pressure by the factor

$$\exp\left[\frac{4\times 10^4}{8.31}\left(\frac{1}{303} - \frac{1}{323}\right)\right] = 2.7,$$

where we have used $N_A k = R = 8.31$ J/mol. The increase in temperature of 20 °C increases the vapour pressure of water by almost a factor of 3. The exponential dependence of the saturated vapour pressure of water on temperature is apparent in Figure 11.3.

If we differentiate Equation (11.9) with respect to T, we obtain

$$\frac{dP_{VP}}{dT} = A\frac{L_V}{N_A kT^2}e^{-L_V/N_A kT} = \frac{P_{VP}L_V}{N_A kT^2}.$$

Again using $P_{VP} = \frac{N_A}{V}kT$, we obtain

$$\frac{dP_{VP}}{dT} = \frac{L_V}{TV}. \tag{11.10}$$

A more complete derivation of Equation (11.10) that includes the volume of the liquid gives the result

$$\frac{dP_{VP}}{dT} = \frac{L_V}{T(V - V_L)}, \tag{11.11}$$

where V_L is the volume of the liquid. We recognise this equation as the Clausius–Clapeyron equation that we encountered in Section 7.9.

11.2.3 Surface energy and surface tension

In the bulk of a liquid, a molecule is surrounded on all sides by its neighbours. On the other hand, a molecule at the surface has very few neighbours on the vapour side compared with the number on the bulk side. Consequently, it takes less energy to remove a molecule at the surface to infinity than it does to remove a molecule in the bulk to infinity. This means that the molecules at the surface have more potential energy than those in the bulk. The extra energy they possess is called *surface energy*. It is a familiar observation that small drops of liquids tend to have a spherical shape. A sphere is a geometric form that has the smallest surface area for a given volume. Hence, by adopting a spherical shape, the drop minimises its surface area and therefore its surface energy.

Surface tension is the force that is exerted when a surface is cut. It is the force that tends to close the cut. The surface tension γ is thus a force per unit length. Suppose that we have a film of soap supported by a U-shaped frame as shown in Figure 11.4a. The moveable wire can move freely without friction along the frame. Figure 11.4b shows a side view of the film, where the top and bottom surfaces of the film can be seen. A force F must be applied to the wire to maintain static equilibrium. Each surface exerts a force equal to γl, where l is the length of the film and the constant is the surface tension. The total force is thus

$$F = 2\gamma l. \tag{11.12}$$

That the resultant force is due to the two surfaces and not a tension existing throughout the bulk of the liquid is demonstrated by reducing the thickness of the film by, for example, allowing the film to evaporate. If that is done, it is found that the force F remains constant.

If the moveable wire is moved a distance $\mathrm{d}x$ under isothermal conditions so that no heat is exchanged with the surroundings, the work done is $F\mathrm{d}x = 2\gamma l\mathrm{d}x$. As the increased area of the top and bottom surfaces is increased by $2l\mathrm{d}x$, it follows that the surface energy has increased by $2\gamma l\mathrm{d}x/2l\mathrm{d}x = \gamma$ per unit of area. Hence, γ can also be defined as surface energy per unit area, noting that energy/unit area has the same dimensions as force/unit length.

The origin of surface tension lies in the intermolecular forces between the molecules of the liquid. In a simple model, we suppose that the molecules at the surface have neighbours in just one hemisphere and so

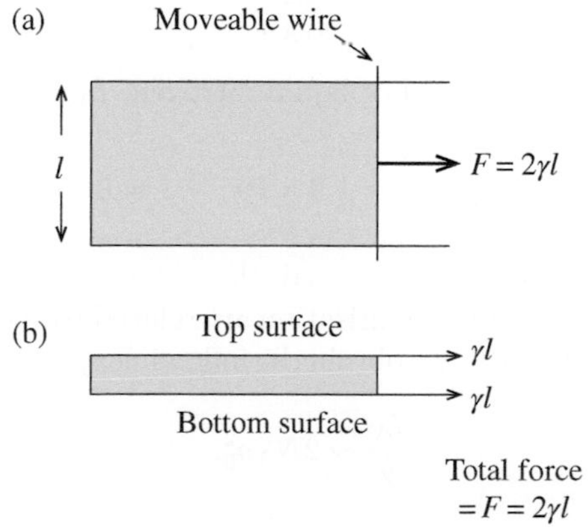

Figure 11.4 (a) The figure shows a film of soap that is supported by a U-shaped frame. The moveable wire can move freely without friction along the frame. (b) A side view of the film, showing the top and bottom surfaces of the film. A force F must be applied to the wire to maintain static equilibrium. Each surface exerts a force equal to γl, where l is the length of the film and γ is the surface tension. The total force is thus $F = 2\gamma l$.

have half the number of bonds of those molecules in the bulk. Hence, if the number of nearest neighbours a molecule has in the bulk liquid is n, we take the number of nearest neighbours that a molecule at the surface has to be $n/2$. If the diameter of a molecule is a_0, the number of atoms per unit area $\sim 1/a_0^2$. And if ε *is* the intermolecular potential energy between two molecules, the total potential energy per unit area, i.e. the surface energy is then

$$\sim \frac{1}{2} \times \frac{n}{2} \times \frac{1}{a_0^2} \times \varepsilon \sim \frac{n\varepsilon}{4a_0^2},$$

where the usual factor of ½ ensures that we do not count each molecule twice. It follows that the surface tension

$$\gamma \sim \frac{n\varepsilon}{4a_0^2}. \tag{11.13}$$

The volume V_m of one mole of substance is equal to M/ρ, where M is the molecular weight and ρ is the mass density. And $V_m \sim N_A a_0^3$, where N_A is Avogadro's number. Substituting for $a_0 = (M/N_A\rho)^{1/3}$ in Equation (11.13), we obtain

$$\varepsilon \approx \frac{4\gamma}{n}\left(\frac{M}{N_A\,\rho}\right)^{2/3}. \tag{11.14}$$

Hence, from a measurement of the surface tension of a liquid, we can obtain a value for the intermolecular potential energy ε. As an example, the surface tension of liquid nitrogen is 4×10^{-3} N/m, the density is 8.0×10^2 kg/m^3, and taking n to be 10 for a liquid, we obtain

$$\varepsilon \sim \frac{4 \times 4 \times 10^{-3}}{10}\left(\frac{28 \times 10^{-3}}{6 \times 10^{23} \times 8.0 \times 10^2}\right)^{2/3} = 2.5 \times 10^{-22}\,\mathrm{J} = 0.016\,\mathrm{eV}.$$

We also have, Equation (11.3),

$$\varepsilon \approx \frac{2L_V}{N_A n},$$

where L_V is the latent heat of vapourisation. For liquid nitrogen, $L_V = 5.6$ kJ/mol, from which

$$\varepsilon \approx \frac{2 \times 5.6 \times 10^3}{6 \times 10^{23} \times 10} = 1.9 \times 10^{-21}\,\mathrm{J} = 0.012\,\mathrm{eV}.$$

Considering the simplicity of the models we have used, these two estimates for ε are in good agreement.

Both surface tension and latent heat of vapourisation are related to the intermolecular potential ε. And by combining Equations (11.3) and (11.13), we obtain the following approximate relation for the ratio L_V/γ:

$$\frac{L_V}{\gamma} \sim 2N_A a_0^2. \tag{11.15}$$

11.2.4 Capillarity

One of the physical effects of surface tension is capillarity. This is observed, for example, when a narrow tube is immersed in water. The water rises in the tube to a height above the surface of the water, as illustrated in

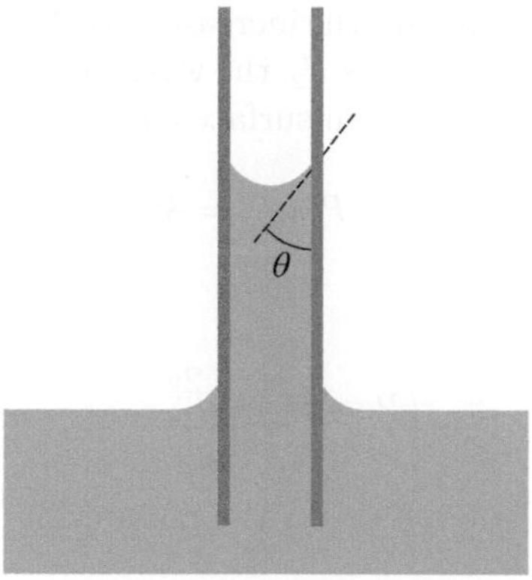

Figure 11.5 When a narrow tube is immersed in water, the water rises in the tube to a height above the surface of the water. The angle θ that the meniscus makes with the tube is called the contact angle that depends on the surface tension of the liquid. The height of the capillary rise depends on the radius of the tube and also the surface tension.

Figure 11.5. The capillary rise depends on the radius of the tube and also on the surface tension of the liquid. And indeed measuring the capillary rise of a liquid provides a method of measuring its surface tension. The angle θ that the meniscus makes with the tube is called the *contact angle.* The value of this angle depends on the surface tension of the liquid. If water is placed on a clean flat glass surface, it is observed that the water 'wets' the glass surface. This is in contrast to say a globule of mercury placed on a glass surface. The globule does not wet the surface but rather retains the shape of a sphere with little contact with the glass. It is found generally that liquids with low surface tension, such as water, readily wet most solids, giving a contact angle of zero. On the other hand, liquids with high surface tension, such as mercury, show a finite contact angle. In molecular terms, we may say that a liquid will wet a surface if the adhesion between the liquid and the solid is greater than the cohesion between the molecules in the liquid. Conversely, if the cohesion between the molecules in the liquid is greater than the adhesion between the liquid and the solid, the liquid will not wet the solid.

To obtain an expression for the capillary rise of a liquid, it is useful to consider first the pressure difference across a liquid–vapour surface. Suppose we have a spherical bubble in a liquid, as in Figure 11.6. The liquid surrounding the vapour in the bubble minimises its surface energy by minimising the surface area of the bubble. If the radius of the bubble is R, its surface area is $4\pi R^2$. Suppose a small expansion of the bubble occurs increasing the radius by $\mathrm{d}R$. Then, the increase in the surface area is

$$4\pi\left[(R+\mathrm{d}R)^2 - R^2\right] = 4\pi\left[R^2 + (\mathrm{d}R)^2 + 2R\mathrm{d}R - R^2\right] \approx 8\pi R\mathrm{d}R,$$

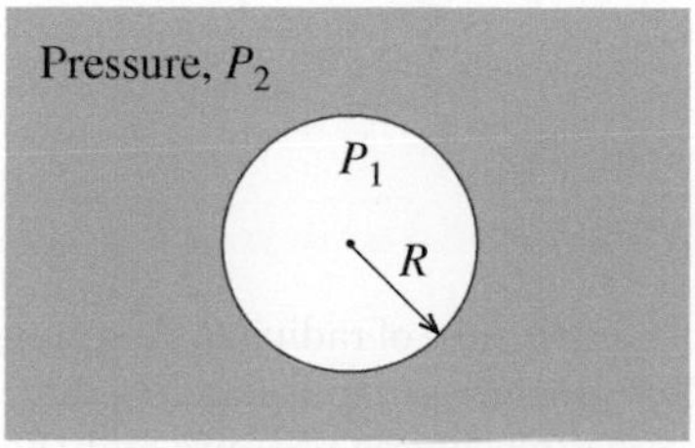

Figure 11.6 The figure shows a spherical bubble of radius R in a liquid. The pressure inside the bubble is P_1 and the pressure in the liquid is P_2. The excess pressure in the bubble over the pressure in the liquid is given by $(P_1 - P_2) = 2\gamma/R$, where γ is the surface tension of the liquid.

taking $(dR)^2$ to be very small. Hence, we can take the increase in surface energy to be $8\pi R dR\gamma$. If the pressure in the bubble is P_1 and the pressure in the liquid is P_2, the work done by the pressure difference $(P_1 - P_2)$ is $4\pi R^2(P_1 - P_2)dR$. Equating this to the increase in surface energy, we have

$$4\pi R^2(P_1 - P_2)dR = 8\pi R dR\gamma,$$

giving

$$(P_1 - P_2) = \frac{2\gamma}{R}.$$

$(P_1 - P_2)$ is the *excess* pressure P in the bubble over the pressure in the liquid. Hence,

$$\text{Excess pressure } P = \frac{2\gamma}{R}. \tag{11.16}$$

Consider now the rise of a liquid in a narrow tube of radius R, as illustrated by Figure 11.7. We assume that the contact angle is zero, which means that the shape of the meniscus is a complete hemisphere. From Equation (11.16), we have that the pressure at point B is less than at point A by the amount $2\gamma/R$, where R is the radius. But the pressure at A is atmospheric pressure and the same as the pressure at point C, at the surface of the liquid. Thus, the pressure at B is less than the pressure of the liquid at point C. Hence, the liquid must rise up the tube to height h such that the pressure at B is less than at A by $2\gamma/R$. This means that

$$\rho g h = \frac{2\gamma}{R}, \tag{11.17}$$

where g is the acceleration due to gravity and ρ is the density of the liquid. This equation shows that h increases as R decreases, i.e. the narrower the tube, the higher the liquid rises.

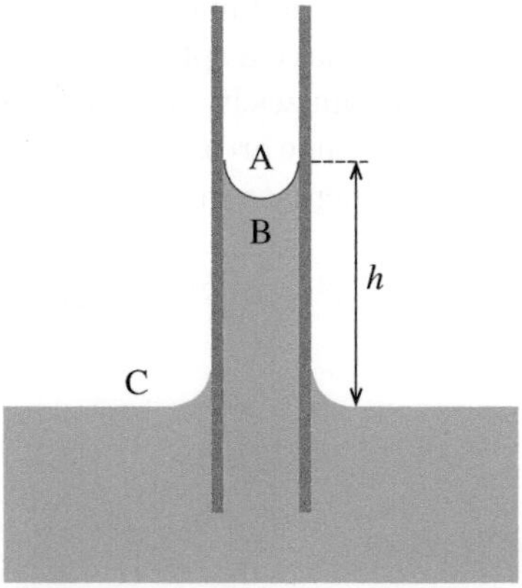

Figure 11.7 The capillary rise of a liquid in a narrow tube of radius R. Assuming that the contact angle is zero, the shape of the meniscus is a complete hemisphere. The pressure at point B is less than at point A by the amount $2\gamma/R$, where γ is the surface tension. But the pressure at A is atmospheric pressure and the same as the pressure at C, the surface of the liquid. Thus the pressure at B is less than the pressure of the liquid at point C. Hence, the liquid must rise up the tube to height h such that the pressure at B is less than at A by $2\gamma/R$. This means that $\rho g h = 2\gamma/R$, where g is the acceleration due to gravity and ρ is the density of the liquid.

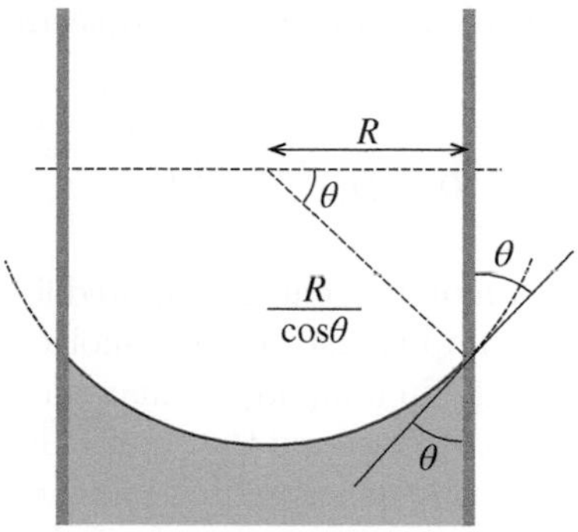

Figure 11.8 This figure shows the situation when the angle of contact of the liquid is greater than zero. Assuming that the radius of the tube is sufficiently small, the meniscus can be taken to be a section of a sphere of radius R. In terms of the radius of the tube, the radius of the spherical surface is $R/\cos\theta$. Then, the height h of the liquid column is given by $h = 2\gamma\cos\theta/\rho gR$.

If the angle of contact of the liquid is greater than zero, we have the situation shown in Figure 11.8. Assuming that the radius of the tube is sufficiently small, the meniscus can be taken to be a section of a sphere with radius R. In terms of the radius of the tube, the radius of the spherical surface is $R/\cos\theta$. Then, the height h of the liquid column is given by

$$h = \frac{2\gamma\cos\theta}{\rho gR}. \tag{11.18}$$

11.2.5 Diffusion

Diffusion occurs in liquids just as it does in gases and solids. And diffusion in liquids can be described using analogous equations to those that apply to gases and solids. For all three phases, diffusion follows Fink's law, which relates the diffusion flux J to the concentration gradient $\mathrm{d}n/\mathrm{d}x$ of a particular species; see Section 4.4. In one dimension, Fink's law is

$$J = -D\frac{\mathrm{d}n}{\mathrm{d}x},$$

where D is the diffusion coefficient. Moreover, in all three cases, diffusion is a random walk process, where the root mean distance travelled by a molecule in a medium of diffusion coefficient D in time t is approximately $\sqrt{2Dt}$. What is different for the three states of matter is their respective rates of diffusion. These rates are very different because of the underlying diffusion mechanisms that occur in each case. In a gas, the molecules do not interact with each other and are free to move freely throughout the volume of the gas. In addition, the length of each step a molecule takes in the diffusion process is $\sim\lambda$, the mean free path, i.e. many molecular diameters. In a liquid, the short-range attraction of a molecule with their neighbours greatly reduces the ability of the molecules to move through the bulk of the liquid. Moreover, the diffusion step is only about one intermolecular distance. Consequently, the diffusion coefficient for a molecule in a liquid solution is several orders of magnitude smaller than for the molecule in the gas phase. In the case of a solid, it takes a finite amount of energy for a molecule to jump into an adjacent site, and we found that the diffusion coefficient for a solid has the form

$$D = D_0 e^{-E_a/kT}, \tag{11.19}$$

where E_a is the *activation energy*, i.e. the amount of energy required for a molecule to jump into an adjacent lattice site. Moreover, the length of the jump is just one intermolecular spacing. For a liquid, we have a somewhat similar situation to that in solids. In a liquid, we may picture the neighbouring molecules that surround a particular molecule to form a 'cage' around that molecule, as in the two-dimensional representation shown in Figure 11.1. The neighbouring molecules tend to confine the molecule and prevent it from making large-scale excursions. However, it may happen that the molecule gains enough energy to break out of its cage. If the amount of energy required to do this is E'_a, the diffusion coefficient for the liquid has the same form as for a solid:

$$D = D_0 e^{-E'_a/kT}. \tag{11.20}$$

An exact derivation of this equation also contains a linear term T in temperature, just as in the relation for the diffusion coefficient of a solid. But again the exponential term dominates the linear term and Equation (11.20) is a good approximation over the temperature range in which a substance is in its liquid phase. And as for a solid, the diffusion step is only about one molecular spacing.

These factors are reflected in the respective values of the diffusion coefficients for the three phases of a substance. A typical diffusion coefficient for the gas phase is in the range 10^{-6} to 10^{-5} m^2/s. In a liquid, a typical diffusion coefficient is in the range 10^{-10} to 10^{-9} m^2/s, while a typical diffusion coefficient for a solid is in the range 10^{-11} to 10^{-10} m^2/s.

11.3 The flow of liquids

Perhaps the most important property that distinguishes liquids is their ability to flow. The flow can be visualised by injecting dye into the liquid through a series of narrow equally spaced channels so that the dye enters parallel to the direction of flow. Figure 11.9 illustrates schematically what would be observed for the flow of a liquid past a cylinder. If the overall pattern of lines does not change with time, as in Figure 11.9a, the dye traces out a set of smooth and continuous lines. This is called *laminar* or *steady flow* and the lines are called *streamlines*. A streamline can be thought of as the path that a particle would take if placed in a flowing liquid. Note that all particles that pass through the same point follow the same path, i.e. the same streamline. Streamlines never intersect; no liquid particle can move across from one streamline to another. At the far left of Figure 11.9a, the streamlines are parallel and horizontal. As they approach the cylinder, the streamlines curve around the cylinder, getting closer together at the top and bottom of the cylinder. Then they spread out again before becoming parallel again at the far right. This flow pattern is typical of laminar flow, where adjacent layers of liquid slide smoothly past each other. At sufficiently high flow rates or when obstacles cause abrupt changes in liquid velocity, the flow can become irregular and chaotic as illustrated by Figure 11.9b. This is called *turbulent flow* and the flow pattern changes continuously.

The condition for turbulent flow was established by Osborne Reynolds, who was a professor of engineering at the University of Manchester. According to Reynolds, if a liquid of density ρ and viscosity η flows past

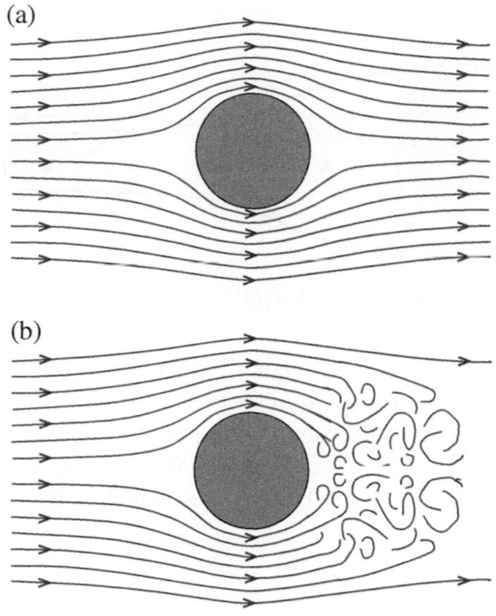

Figure 11.9 The flow of a liquid can be visualised by injecting dye into the liquid through a series of narrow equally spaced channels so that the dye enters parallel to the direction of flow. The figure illustrates schematically what would be observed for the flow of a liquid past a cylinder. (a) If the overall pattern of lines does not change with time, the dye traces out a set of smooth and continuous lines. This is called laminar or steady flow and the lines are called streamlines. (b) At sufficiently high flow rates or when obstacles cause abrupt changes in liquid velocity, the flow can become irregular and chaotic. This is called turbulent flow and the flow pattern changes continuously.

an object of lateral dimension L, there is a critical velocity u_c at which laminar flow changes to turbulent flow. This occurs when

$$u_c = \frac{\mathrm{Re}\,\eta}{\rho L}, \tag{11.21}$$

where Re is a dimensionless quantity called the *Reynolds number*. For practical applications, if Reynolds' number is less than about 2000, the flow will be laminar. If Re is greater than about 4000, the flow will be turbulent. In the range between 2000 and 4000, it is not possible to predict which type of flow exists and therefore this range is called the *critical region*.

We will first consider ideal liquids by which we mean the liquid is (i) incompressible and (ii) has no viscosity. The first assumption is an excellent assumption under normal conditions. The second assumption is less good and in Section 11.4, we will see that viscosity plays a dominant role in the flow of real liquids.

11.3.1 The continuity equation

Imagine that a liquid travels down a pipe of decreasing cross section as shown in Figure 11.10, where the liquid flows from left to right. The volume of liquid that flows through area A_1 in time δt is $A_1 u_1 \delta t$, where u_1 is the velocity of the liquid at A_1. Assuming the liquid to be incompressible, an equal volume of liquid must flow past any point along the pipe during the same time δt. The volume that flows through area A_2 is $A_2 u_2 \delta t$, where u_2 is the velocity of the liquid at A_2. As the two volumes must be equal, we have

$$A_1 u_1 \delta t = A_2 u_2 \delta t \tag{11.22}$$

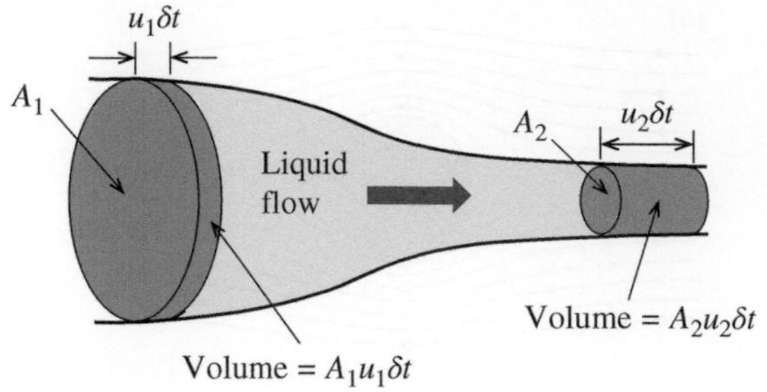

Figure 11.10 The figure shows the flow of a liquid through a pipe of decreasing cross section. The flow is from left to right. The volume of liquid that flows through area A_1 in time δt is $A_1u_1\delta t$, where u_1 is the velocity of the liquid at A_1. Assuming the liquid to be incompressible, an equal volume of liquid must flow past any point along the pipe during the same time δt. The volume that flows through area A_2 is $A_2u_2\delta t$, where u_2 is the velocity of the liquid at A_2. Since the two volumes must be equal, it follows that $A_1u_1 = A_2u_2$, which is the continuity equation.

so that

$$A_1u_1 = A_2u_2 \tag{11.23}$$

This is the *continuity equation* where the quantity $A \times u$ is called the *volume flow rate*. We see that $A \times u$ is a conserved quantity. As $A_2 < A_1$ in Figure 11.10, $u_2 > u_1$. We also have $\delta m = \rho\delta V$, where ρ is the density of the liquid and δm is the mass of liquid in volume δV. Hence, we can write

$$\delta m = \rho A u\delta t, \tag{11.24}$$

giving

$$\frac{\delta m}{\delta t} \approx \frac{\mathrm{d}m}{\mathrm{d}t} = \rho A u. \tag{11.25}$$

As $A \times u$ is a conserved quantity, $\mathrm{d}m/\mathrm{d}t$ is also a conserved quantity for constant density ρ. Hence, the mass flow rate $\mathrm{d}m/\mathrm{d}t$ is the same across any cross sectional area of the pipe.

11.3.2 Bernoulli's equation

In accord with the continuity equation, the velocity of a liquid can vary along its path. The pressure of the liquid can also vary as it depends on the flow velocity, as we shall see. Moreover, the pressure depends on the elevation of the liquid, just as it does for a static liquid. It is Bernoulli's equation that relates the pressure, the flow velocity, and the elevation for an ideal, incompressible liquid, and provides a valuable tool in many practical applications of liquid flow.

Figure 11.11 shows a section of a pipe with changing cross section and changing elevation through which liquid is flowing. We consider the volume element of liquid of density ρ that is initially between points a and b in the pipe. At point a, the cross sectional area of the pipe is A_1, the vertical height above a reference level is h_1, the pressure is P_1, and the flow velocity is u_1. At point b, the corresponding quantities are A_2, h_2, P_2 and u_2.

The liquid behind the volume element pushes the element through the pipe and hence does some work *on* the element. And this volume element pushes against the liquid in front of it and so does some work *on* that

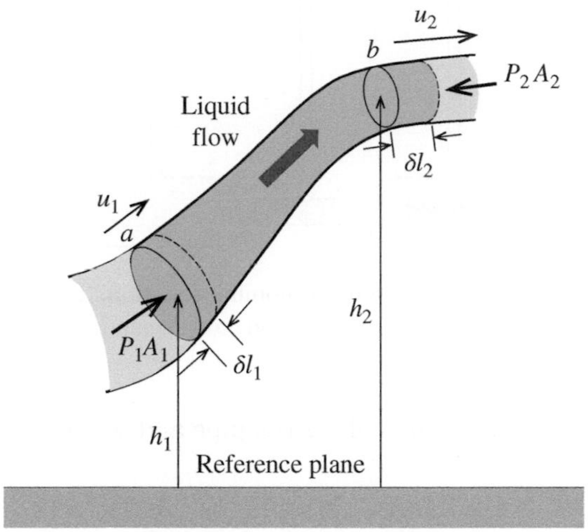

Figure 11.11 The figure shows a section of a pipe of changing cross section and changing elevation through which liquid is flowing. At point a, the cross sectional area of the pipe is A_1, the vertical height above a reference level is h_1, the pressure is P_1, and the flow velocity is u_1. At point b, the corresponding quantities are A_2, h_2, P_2 and u_2. According to Bernoulli's equation $(P_1/\rho) + (1/2)u_1^2 + h_1 g = (P_2/\rho) + (1/2)u_2^2 + h_2 g$.

liquid. In the short time interval δt, the pressure P_1 at a drives the boundary of the volume element through distance $\delta l_1 = u_1 \delta t$. The resultant work done on the element (force × distance) is

$$w_1 = (P_1 A_1) \times \delta l_1 = P_1 A_1 u_1 \delta t.$$

During time interval δt, the boundary of the volume element at b pushes against the pressure P_2 of the liquid in front of it by the distance $\delta l_2 = u_2 \delta t$. And the volume element does work $w_2 = P_2 A_2 u_2 \delta t$ on that liquid. With the assumption that the liquid is incompressible,

$$A_2 u_2 \delta t = A_1 u_1 \delta t = \delta V.$$

Hence, the net work done on the volume element is

$$w_1 - w_2 = (P_1 - P_2)\delta V. \tag{11.26}$$

According to the work-energy theorem, the mechanical energy (kinetic plus potential) of the volume element is increased by this amount of work. We may say that in terms of energy, the flow of the volume element is equivalent to an amount of liquid of mass $\rho \delta V$ having its potential energy increased by the amount $\rho \delta V g(h_2 - h_1)$ and its velocity being increased from u_1 to u_2. Hence,

$$(P_1 - P_2)\delta V = \rho \delta V g(h_2 - h_1) + \frac{1}{2}\rho \delta V\left(u_2^2 - u_1^2\right).$$

Upon rearrangement, this gives

$$\frac{P_1}{\rho} + \frac{1}{2}u_1^2 + h_1 g = \frac{P_2}{\rho} + \frac{1}{2}u_2^2 + h_2 g. \tag{11.27}$$

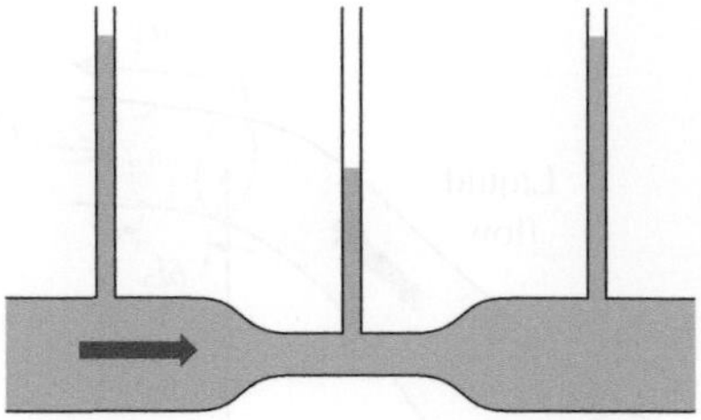

Figure 11.12 The figure provides a vivid illustration of Bernoulli's equation. The flow velocity is higher in the narrower tube, in accordance with the continuity equation, and, consequently, the pressure is lower there.

The subscripts 1 and 2 refer to any two points along the pipe and so we can also write

$$P + \frac{1}{2}\rho u^2 + \rho hg = \text{constant}. \tag{11.28}$$

Equations (11.27) and (11.28) are statements of Bernoulli's equation. In words, it says that the net work done on a volume of liquid is equal to the sum of the changes in kinetic and potential energies of that volume that occur during its flow. Bernoulli's equation was first expressed in words by Daniel Bernoulli in 1738 and was later derived in equation form by Leonard Euler in 1755. Bernoulli's equation is essentially a statement of the conservation of energy, although Bernoulli stated his law long before the conservation of energy was established. Figure 11.12 provides a vivid illustration of Bernoulli's equation. The flow velocity is higher in the narrower tube and, consequently, the pressure is lower there.

We have derived Bernoulli's equation for the case of an ideal liquid, i.e. one that is incompressible. However, we can also treat gases as incompressible if the pressure difference from one region to another is not too great. In this case, we can apply Bernoulli's equation to gases. For example, Bernoulli's equation explains the lift provided by an airplane wing. The wing of an aeroplane is shaped such that the air travels at a greater velocity above the wing than below it. Then, according to Equation (11.28), the pressure above the wing is lower than the pressure below it. Hence, the downward force due to the air on the top side of the wing is less than the upward force of the air on the underside of the wing, and there is a net upward or lift force.

Worked example

Figure 11.13 shows a container of water with a small hole in its side that produces a jet of water. H is the depth of the water and h is the distance of the hole below the surface of the water. The jet of water strikes the floor at a distance x from the side of the container. (a) Determine x if $H = 1.6$ m and $h = 0.40$ m. (b) At what value of h should a second hole be drilled to give the same value of x? (c) What value of h gives the maximum value of x?

Solution

The size of the hole is so small that we can assume that the velocity at which the water level in the container falls is negligible. Also the pressure at the water surface and at the small hole are the same and equal to atmospheric pressure.

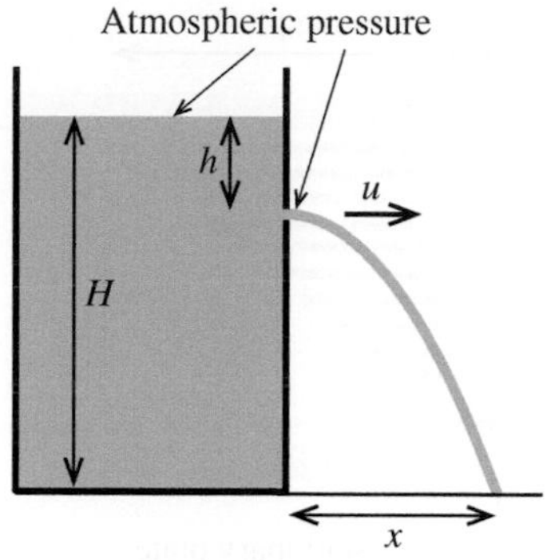

Figure 11.13 The figure shows a container of water with a small hole in its side that produces a jet of water. H is the depth of the water and h is the distance of the hole below the surface of the water. The jet of water strikes the floor at a distance x from the side of the container.

(a) Taking the bottom of the container as the reference level for gravitational potential energy, we have from Bernoulli's equation

$$\frac{P_{\text{atmos}}}{\rho} + \frac{1}{2}(0)^2 + gH = \frac{P_{\text{atmos}}}{\rho} + \frac{1}{2}u^2 + g(H-h),$$

where u is the velocity of the water jet. This gives $u = \sqrt{2gh}$. Interestingly, this is the same result for the final velocity of a body that falls under gravity through a distance h. The time t before the water jet strikes the ground is given by the expression:

$$\frac{1}{2}gt^2 = (H-h),$$

and hence $x = 2\sqrt{h(H-h)}$. For the given values of H and h, $x = 1.4$ m.

(b) Let height h' give the same value of x as $h = 0.40$ m. Then using $x = 2\sqrt{h(H-h)}$, we obtain:

$$2\sqrt{0.40(1.6-0.40)} = 2\sqrt{h'(1.6-h')}.$$

This simplifies to

$$h'^2 - 1.6h' + 0.48 = 0,$$

from which $h' = 1.2$ or 0.40 m.

(c) The distance x will be maximised when ${}^{\mathrm{d}}\!/_{\mathrm{d}h}[h(H-h)] = 0$. This is when $h = H/2$, giving $h = 0.80$ m.

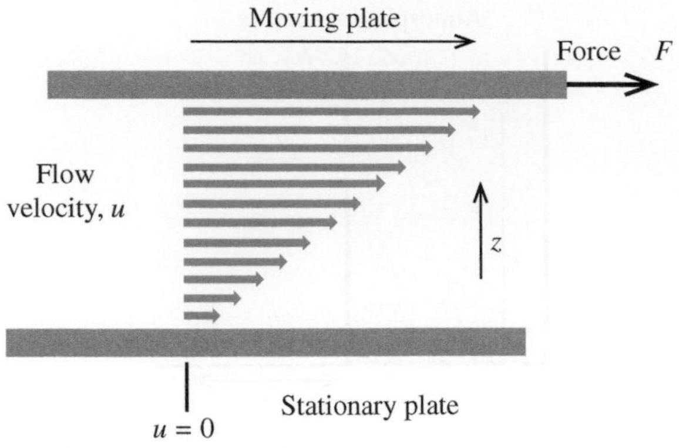

Figure 11.14 A real liquid experiences resistance to flow due to its viscosity. In the figure, the liquid is contained between two plates. The bottom plate is stationary and the upper plate moves along the horizontal direction. The velocity u of the liquid, represented by the green arrows, has a maximum value at the moving plate, and is zero at the stationary plate.

11.4 The flow of real liquids

11.4.1 Viscosity of liquids

A real liquid experiences a resistance to flow due to its viscosity. We discussed the viscosity of gases in Section 4.6, and similar considerations apply to viscous liquids.

Figure 11.14 shows a liquid that is contained between two plates. The bottom plate is stationary and the upper plate moves along the x-direction. The velocity u of the liquid, represented by the green arrows, has a maximum value at the moving plate and is zero at the stationary plate. As for a gas, the force F required to overcome the viscosity is in the direction of the velocity of the moving plate and is proportional to the area A of the plates and the velocity gradient $\mathrm{d}u/\mathrm{d}z$ of the liquid. Hence, we have the following expression for the viscous force F per unit area:

$$F = \eta \frac{\mathrm{d}u}{\mathrm{d}z}, \tag{11.29}$$

where η is the viscosity of the liquid. Fluids that obey this equation are known as Newtonian fluids.

Worked example

We can expect the viscous force that operates on a sphere falling through a liquid to depend on its radius a, its velocity u, and the viscosity η of the liquid. Use the method of dimensions to find an expression for the viscous force in terms of these quantities.

Solution

We let the viscous force $F = Ca^{\alpha}\eta^{\beta}u^{\gamma}$, where C is a constant. The dimensions of F are MLT^{-2}; the dimension of a is L, the dimensions of η are MLT^{-1}, and the dimensions of u are LT^{-1}. Therefore

$$\mathrm{MLT}^{-2} \equiv (\mathrm{L})^{\alpha} \times (\mathrm{ML}^{-1}\mathrm{T}^{-1})^{\beta} \times (\mathrm{LT}^{-1})^{\gamma}.$$

Equating indices of M, L, and T on both sides gives

$$1 = \beta,$$
$$1 = \alpha - \beta + \gamma,$$
$$-2 = -\beta - \gamma.$$

Solving these equations gives $\alpha = 1$, $\beta = 1$ and $\gamma = 1$. Consequently, $F = Ca\eta u$. George Stokes determined that the constant $C = 6\pi$. The resulting equation $F = 6\pi a\eta u$ is known as Stoke's law.

11.4.2 Viscous flow through a pipe

If a liquid had no viscosity, its flow velocity would be uniform across the cross sectional area of the pipe, as illustrated by Figure 11.15a. However, because real fluids do have viscosity, the velocity profile across a cross section of the pipe varies as illustrated by Figure 11.15b. The velocity is zero at the wall of the pipe and has a maximum value at the centre of the pipe.

Figure 11.16 shows a liquid with viscosity η flowing smoothly and steadily through a section of pipe of radius r and length l that is much larger than the radius. The flow is due to a pressure difference $(P_2 - P_1)$ that is maintained across the pipe. It is convenient to consider the flow of liquid in terms of a series of concentric, cylindrical layers of liquid that slide relative to each other. Thus in Figure 11.16 we consider a cylindrical layer of flowing liquid of radius r and width dr, and consider the forces acting on each side of this layer.

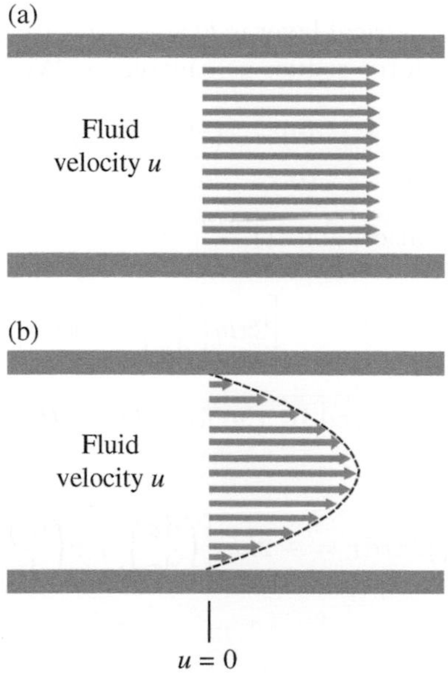

Figure 11.15 (a) If a liquid had no viscosity, its flow velocity would be uniform across the cross section of the pipe. (b) However, because real fluids do have viscosity, the velocity profile varies across the cross section. The velocity is zero at the wall of the pipe and has a maximum value at the centre of the pipe.

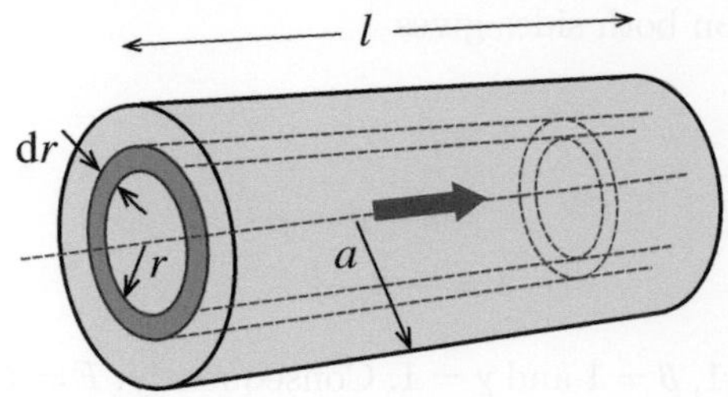

Figure 11.16 The figure shows a liquid of viscosity η flowing smoothly and steadily through a section of pipe of radius a and length l, which is much greater than the radius. The flow is due to a pressure difference $(P_2 - P_1)$ that is maintained across the pipe. It is found that the flow rate V through the pipe is proportional to the pressure difference across the pipe, inversely proportional to the length of the pipe and proportional to the fourth power of the radius of the pipe: $V = \pi(P_2 - P_1)a^4/8\eta l$, which is known as Poiseuille's equation for laminar flow.

The viscous force per unit area on the layer at distance r from the centre of the pipe is given from Equation (11.29) as

$$F = \eta\left(\frac{\mathrm{d}u}{\mathrm{d}r}\right). \tag{11.30}$$

The area of the layer is equal to $2\pi rl$. Hence, the viscous force acting at radius r is

$$F(r) = 2\pi rl\eta\left(\frac{\mathrm{d}u}{\mathrm{d}r}\right). \tag{11.31}$$

The net viscous force acting on the cylindrical layer is $F(r) - F(r + \mathrm{d}r)$. This is a positive quantity because the velocity and therefore the viscous force reduce as r increases. We have

$$\frac{F(r + \mathrm{d}r) - F(r)}{\mathrm{d}r} = \frac{\mathrm{d}F(r)}{\mathrm{d}r}.$$

Hence, substituting for $F(r)$ from Equation (11.31),

$$F(r) - F(r + \mathrm{d}r) = -\left[2\pi l\eta\left(\frac{\mathrm{d}u}{\mathrm{d}r}\right) + 2\pi rl\eta\left(\frac{\mathrm{d}^2u}{\mathrm{d}r^2}\right)\right]\mathrm{d}r. \tag{11.32}$$

In steady, laminar flow, this force is balanced by the force $(P_2 - P_1)2\pi r\mathrm{d}r$ due to the pressure difference across the pipe. Hence,

$$(P_2 - P_1)2\pi r\mathrm{d}r = -2\pi l\eta\left[\left(\frac{\mathrm{d}u}{\mathrm{d}r}\right) + r\left(\frac{\mathrm{d}^2u}{\mathrm{d}r^2}\right)\right]\mathrm{d}r,$$

or

$$(P_2 - P_1) = -l\eta\left[\frac{1}{r}\left(\frac{\mathrm{d}u}{\mathrm{d}r}\right) + \left(\frac{\mathrm{d}^2u}{\mathrm{d}r^2}\right)\right]. \tag{11.33}$$

Assuming a solution of the form $u = A + Br^2$, we obtain $\mathrm{d}u/\mathrm{d}r = 2Br$ and $\mathrm{d}^2u/\mathrm{d}r^2 = 2B$. Substituting for $\mathrm{d}u/\mathrm{d}r$ and $\mathrm{d}^2u/\mathrm{d}r^2$ in Equation (11.33) gives

$$B = -\frac{(P_2 - P_1)}{4l\eta}.$$

Hence,

$$u = A - \frac{(P_2 - P_1)}{4l\eta} r^2.$$

But, $u = 0$ at $r = a$, i.e. at the surface of the pipe, giving

$$A = \frac{(P_2 - P_1)}{4l\eta} a^2.$$

We therefore obtain

$$u(r) = \frac{(P_2 - P_1)}{4l\eta}\left(a^2 - r^2\right). \quad (11.34)$$

This equation shows that the profile of the flow velocity in Figure 11.15b has a parabolic shape.

The volume of liquid emerging from the pipe per second from the layer between r and $r + \mathrm{d}r$ is $2\pi r\mathrm{d}r \times u$, and hence the total volume emerging from the pipe per second is

$$V = \frac{\pi(P_2 - P_1)}{2\eta l}\int_0^a r\left(a^2 - r^2\right)\mathrm{d}r. \quad (11.35)$$

Evaluating the integral gives

$$V = \frac{\pi(P_2 - P_1)a^4}{8\eta l}. \quad (11.36)$$

This is *Poiseuille's equation* for laminar flow. The flow rate is proportional to the pressure difference across the pipe and inversely proportional to the length of the pipe. However, more striking is its dependence on the fourth power of the radius of the pipe. Thus, if the pipe radius is doubled, the throughput of liquid is 16 times greater. We have obtained Poiseuille's equation for the case of liquids, but it also applies to the flow of gases through pipes under normal operating conditions.

Problems 11

11.1 One model of the atomic nucleus is called the *liquid drop model.* In this model, the atomic nucleus is considered as a liquid drop of nucleons with a mean separation of 1 fm $= 1 \times 10^{-15}$ m and a binding energy per nucleon of 10 MeV. Estimate the surface tension of nuclear matter.

11.2 By considering the work required to increase the radius R of a soap bubble by $\mathrm{d}R$, show that the excess pressure within the bubble $P = 4\gamma/R$, where γ is the surface tension of the soap solution. Assume the temperature remains constant. How much work is done in blowing a soap bubble to a radius of 50 mm, if the surface tension of the soap solution is 0.05 N/m?

11.3 Show that the latent heat of vapourisation obeys the following equation:

$$L_V = RT^2 \frac{d}{dT}(\ln P_{VP}),$$

where P_{VP} is the vapour pressure. Assume that the expansion of the liquid can be neglected. The vapour pressure of steam increases from 9.799×10^4 Pa to 10.524×10^4 Pa when the temperature rises from 99 to 101 °C. Use these data to determine the latent heat of vapourisation of water.

11.4 If a drop of liquid is momentarily and slightly deformed, it will oscillate. We can expect the frequency f of this oscillation to depend on the radius r of the drop and the surface tension γ and density ρ of the liquid. Use the method of dimensions to obtain an expression for the frequency of oscillation in terms of these quantities.

11.5 Show by the method of dimensions how the critical velocity u_c for fluid flow through a tube depends on the radius of the tube r, the density ρ, and the viscosity η of the fluid.

11.6 A horizontal tube of diameter 3.0 mm and length 0.35 m is connected to the bottom of a cubical tank of side 1.2 m containing water. If the tank is initially full, how long will it take for the tank to become half full? The viscosity of the water is 8.9×10^{-4} Pa · s.

11.7

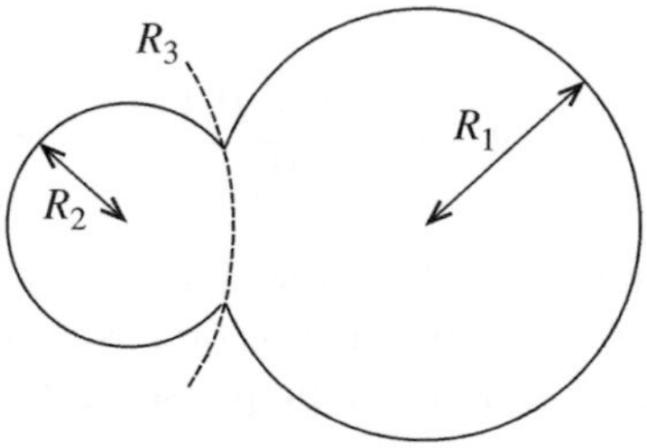

As illustrated by the figure, two soap bubbles of radii $R_1 = 50$ mm and $R_2 = 30$ mm, respectively coalesce so as to have a portion of their surfaces in common. Calculate the radius of curvature R_3 of the common surface.

11.8 The latent heat of vapourisation and the surface tension at 20 °C of water are 2250 kJ/kg and 0.073 N/m, respectively. Use these data to estimate the diameter of a water molecule.

11.9 Compare the excess pressure in a drop of mercury with the excess pressure in a drop of water of the same radius. The surface tensions of mercury and water are 0.485 N/m and 0.073 N/m, respectively.

11.10 Small steel spheres of diameter 2 mm are found to reach a terminal velocity of 6 mm/s when falling through a tall cylinder containing glycerine. Calculate the viscosity of glycerine. The densities of steel and glycerine are 7.85×10^3 kg/m^3 and 1.26×10^3 kg/m^3, respectively.

11.11 Water is supplied to a house through an inlet pipe with an internal diameter of 50 mm. A pipe of internal diameter 20 mm delivers water to a bathroom 6 m above the inlet pipe at the rate of 4 L/min and at a pressure of 2 atm. Calculate the required pressure at the inlet pipe to the house. Atmospheric pressure $= 1.01 \times 10^5$ Pa.

11.12

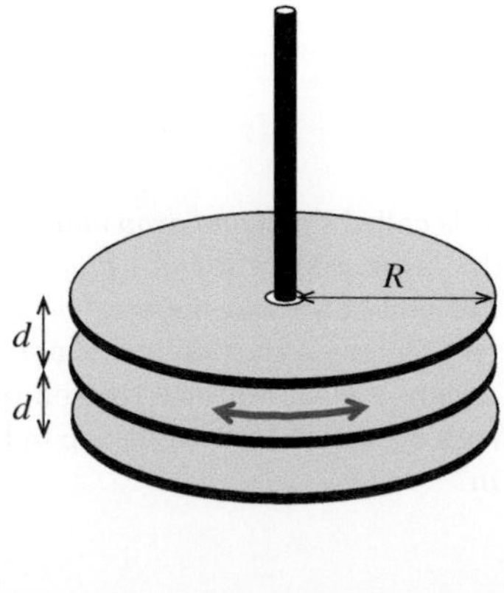

A thin uniform disc of mass 50 g and radius 5 cm is attached to a torsion wire so that it can rotate freely between two fixed metal plates placed on either side of the disc, as shown in the figure. The separation of the disc from both fixed plates is 2 mm. This arrangement is submerged in a tank of water. (a) Show that the equation of motion of the disc is

$$I\frac{\mathrm{d}^2\theta}{\mathrm{d}t^2} + \left(\frac{\pi R^4\eta}{d}\right)\frac{\mathrm{d}\theta}{\mathrm{d}t} + \mu\theta = 0,$$

where I is the moment of inertia of the disc, θ is the angular displacement, R is the radius of the disc, η is the viscosity of water, d is the separation of the disc and fixed plates, and μ is the torsion constant of the torsion wire. (b) When the disc is set into angular oscillation, it is found that the amplitude of oscillation decreases by a factor of 2 after 7.3 seconds. Use these data to determine the viscosity of the water.

For a thin disc of radius R and mass M, $I = (1/2)MR^2$.

12

Liquid crystals

We have discussed gases, liquids, and solids, the three familiar states or phases of matter and we have seen that a substance can exist in one of these three phases depending on temperature and pressure. An obvious example is steam, water, and ice. However, liquid crystals are a class of substances that can exist in more than three different phases. In particular, in going from the liquid to the solid phase, a liquid crystal passes through one (or more) intermediate phases. In the intermediate phase, the liquid crystal has some properties that are normally associated with the solid phase, but it is a true liquid and flows like a liquid; hence the name liquid crystal.

Examples of liquid crystals occur extensively in the natural world and are now ubiquitous in technological applications such as flat-panel displays. However, liquid crystals were only recognised as a distinct phase of matter in 1888. In that year, the botanist Friedrich Reinitzer attempted to measure the melting point of the organic molecule *cholesteryl benzoate* that he had extracted from a plant. To his amazement, he found that this substance appeared to have two melting points. At 145°C, the crystalline solid first melted into a cloudy opaque liquid and then at 178°C, the cloudiness suddenly disappeared, leaving a clear, transparent liquid. And he found that this phenomenon was reversible as the temperature was lowered. To help explain his observations, Reinitzer enlisted the help of physicist Otto Lehmann. Lehmann examined the intermediate, cloudy liquid and recognised that it had a certain degree of order. He proposed that it was a hitherto unknown state of matter, and suggested the name liquid crystal. Despite a number of confirming experiments between 1910 and 1930, the field of liquid crystals remained rather dormant. Then in the mid-1960s, the physicist Pierre-Gilles de Gennes made important contributions to the theoretical understanding of the properties of liquid crystals, particularly their ability to scatter light. For this and for related studies on polymers, de Gennes was awarded the 1991 Nobel Prize in Physics. Interest in liquid crystals really took off after George Gray, at the University of Hull, discovered *cyanobiphenyl* liquid crystals in the 1960s. These materials had all of the properties required for liquid crystal displays and the liquid crystal industry was born.

Physics of Matter, First Edition. George C. King.

12.1 Liquid crystal phases

We recall that a crystalline solid has a well-ordered crystal structure. The constituent molecules have *positional* order and they also have *orientational* order if they are geometrically anisotropic. Figure 12.1 illustrates what happens to a crystalline solid when it melts. As the temperature rises, the thermal motions of the molecules within the lattice increase. Eventually, the vibrations become so intense that the well-ordered structure of the crystal transforms into the random spatial order of a liquid. Remarkably, this transition from the solid to the liquid phase occurs abruptly and at a well-defined temperature. We emphasise that there is a *single* transition between the two phases.

By contrast, Figure 12.2 illustrates what happens in the simplest case where there is only one liquid crystal phase, the *nematic* phase. At low temperatures, a liquid crystal also has positional and

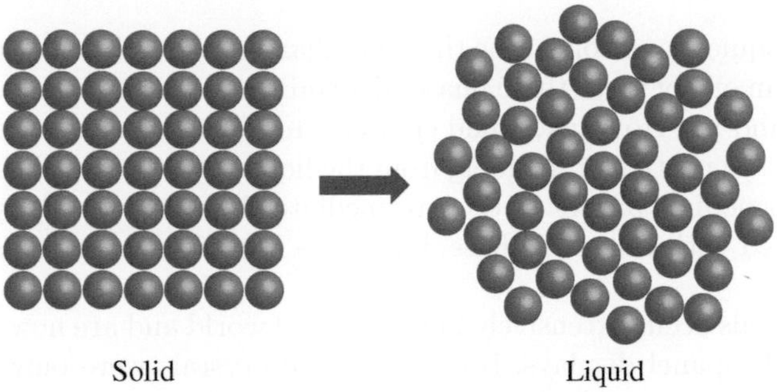

Figure 12.1 When a crystalline solid melts, the well-ordered structure of the crystal transforms into the random spatial order of a liquid. There is a single transition between the two phases and this occurs abruptly at a well-defined temperature.

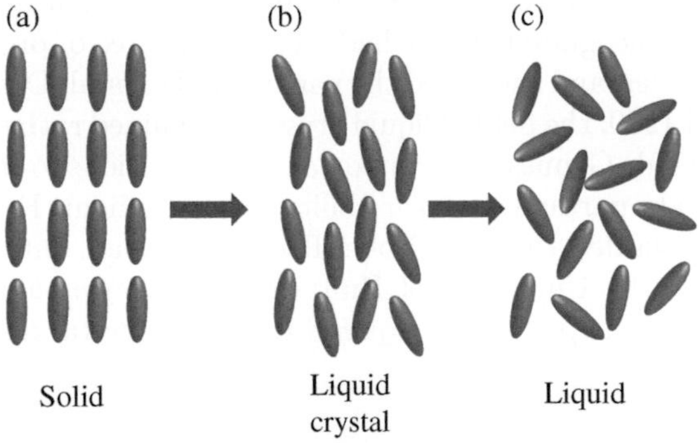

Figure 12.2 This figure illustrates what happens when a liquid crystal changes from the solid to the liquid phase for the simplest case where there is only one liquid crystal phase. (a) At low temperatures, the liquid crystal molecules have positional and orientational order. (b) When the temperature rises sufficiently, the molecules lose their positional order and are able to move around in the liquid state. However, they retain their orientational order with all the molecules tending to point in the same direction. (c) At a still higher temperature, the orientational order breaks down too and the liquid crystal has the random spatial order of a liquid. There are thus two transitions involved when a liquid crystal goes from the solid to the liquid phase, occurring at well-defined temperatures. The intermediate phase between the solid and liquid phases is the liquid crystal phase.

orientational order, as in Figure 12.2a. When the temperature is increased sufficiently, the liquid crystal molecules lose their positional order and are able to move around in the liquid state. However, the molecules retain orientational order, tending to point in the same direction, as in Figure 12.2b. Thus, although an increase in temperature breaks down the positional order, it does not break down the orientational order. This persists until the temperature rises further until the orientational order also breaks down, and the liquid crystal has the random order of a liquid, as in Figure 12.2c. Thus, there are at least two transitions involved when a liquid crystal goes from the solid to the liquid phase. Moreover, these two transitions occur at well-defined temperatures. The intermediate phase is the liquid crystal phase. It is distinctly different from the solid and liquid phases. For example, the liquid crystal phase will usually have a distinctly different appearance; perhaps having a cloudy, opaque appearance while the liquid phase is clear. More fundamentally, it is apparent from Figure 12.2 that there exists a certain amount of molecular order in the liquid crystal phase in that the molecules tend to point in the same direction.

In Figure 12.2, the liquid crystal molecules are depicted as having an anisotropic, cigar-like shape. Such anisotropy is a necessary pre-requisite for a liquid crystal because it is this that gives rise to the orientational order that characterises the liquid crystal phase. So, spherical molecules could not form a liquid crystal state as there is no preferred direction in which they can point. There are several reasons that liquid crystals *self-organise* to give orientational order. There are attractive forces between the liquid crystal molecules, such as van der Waals forces. In addition, the molecules pack together to minimise the total energy. An analogy is the packing of matchsticks in a matchbox. When a bunch of matchsticks is put into a matchbox, they necessarily line up side by side. In a somewhat simplified definition, we may say that a liquid crystal is a substance that has an intermediate phase between its solid and liquid phases. And this arises from the anisotropy of its constituent molecules that endows the intermediate phase with orientational order.

A liquid crystal has properties of both the solid and liquid state; a degree of order but also the ability to flow. So, it is interesting to consider which of these states it most resembles. We can get physical insight into this by comparing the latent heat of transition from solid to liquid crystal to the latent heat of transition from liquid crystal to liquid. Typically, the former is an order of magnitude larger than the latter. This means that it takes much more thermal energy to break down the rigid order of the crystalline solid than it does to break down the orientational order of the liquid crystal, suggesting that the liquid crystal state is more akin to the liquid state.

Liquid crystals can be divided into various classes and sub-classes. Two main classifications are *thermotropic* and *lyotropic* liquid crystals. For the thermotropic class, it is a change in temperature that causes transitions to and from the liquid crystal phase. In a lyotropic liquid crystal, the phase transition again depends on temperature, but in addition, it depends on the concentration of the liquid crystal molecules in a solvent, usually water. Lyotropic liquid crystals occur widely in living systems. For example, many proteins and cell membranes are lyotropic liquid crystals. On the other hand, most technological applications of liquid crystals make use of the thermotropic type and we will devote most of our attention to this type.

12.2 Thermotropic liquid crystal phases

There are various sub-classes of thermotropic liquid crystal phases. The most common phases are the *nematic*, *smectic*, and *chiral* phases. We will see that these are distinguished by the kind of molecular order they possess.

12.2.1 Nematic phase

The nematic phase is the simplest liquid crystal phase, having only orientational order. The name comes from the Greek word *nematos* meaning thread-like and is due to the dark, thread-like appearance of these materials when viewed through a *polarising microscope*. This phase is technologically very important and is widely used in liquid crystal displays. The most common nematic liquid crystals have uniaxial, anisotropic molecular geometries, i.e. are composed of rod-like molecules. In the liquid crystal phase, the molecules self-organise to have orientational order but are completely disordered with respect to their spatial positions. The long molecular axes tend to point in a preferred direction, as illustrated by Figure 12.3. The single degree of molecular order means that nematic liquid crystals have a high degree of fluidity, similar to that of ordinary liquids. This fluidity combined with the anisotropic shape of the molecules means that their orientation can be easily manipulated by an external electric field as we discuss in Section 12.6.

The orientational order of a nematic liquid crystal can be characterised in terms of the preferred direction in which the molecules tend to point and their degree of order. The preferred direction is described in terms of a unit vector called the *director* $\mathbf{n}$: see Figure 12.3. The degree of order is denoted by the *order parameter* P_2, where

$$P_2 = \left\langle \frac{3\cos^2\theta - 1}{2} \right\rangle, \tag{12.1}$$

and θ is the angle that the long axis of a molecule makes with respect to the director. The orientation of individual molecules will change over time, and the angled brackets denote that the function is averaged over many molecules at the same time or the average over time of a single molecule. P_2 conveniently takes a value between 0 for a completely disordered phase and 1 for a completely ordered phase. This parameter is highly dependent on temperature. Figure 12.4 illustrates qualitatively this temperature dependence for a nematic liquid crystal. T_c is the temperature of the transition between the liquid crystal phase and the isotropic liquid phase. At low temperature, most nematic liquid crystals have order parameters that do not go much above 0.7. As the temperature rises, it falls to a value between ~0.3 and 0.4 just below the liquid

Figure 12.3 Nematic liquid crystals have an anisotropic shape, and typically are composed of rod-like molecules. In the liquid crystal phase, the molecules self-organise to have orientational order. The long molecular axes, on average, point in a preferred direction, which is described in terms of a unit vector called the director $\mathbf{n}$.

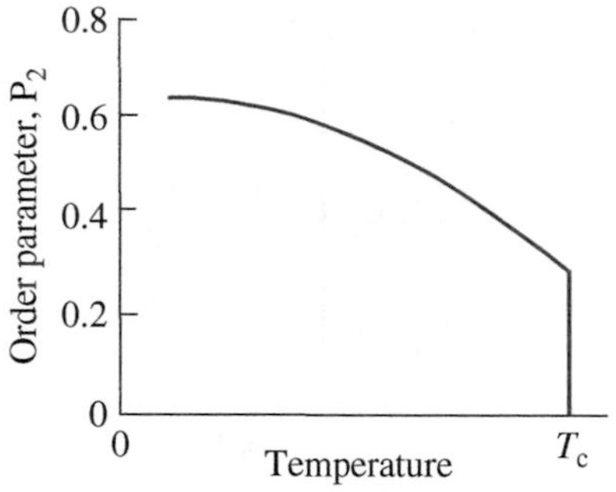

Figure 12.4 A typical plot of the temperature dependence of the order parameter P_2 of a nematic liquid crystal. T_c is the temperature of the transition between the liquid crystal and liquid phases. P_2 drops abruptly to zero at the liquid crystal to liquid transition.

crystal to liquid transition before dropping abruptly to 0 at the transition. The abrupt change in the order parameter is just a reflection of the breakdown of the orientational order.

A single direction of order is not maintained throughout the bulk of a nematic liquid crystal. Instead, the constituent molecules group together to form small *domains*, and the order parameter is the same across the individual domains. Discontinuities exist between the domains and when viewed under a polarising microscope, these give rise to the appearance of the dark threads noted earlier. Moreover, these domains scatter incident radiation giving rise to the milky, opaque appearance of nematic materials.

It is not only rod-like molecules that give rise to nematic liquid crystal phases. Other molecular shapes are capable of generating such phases. Perhaps, the most significant of these are *discotic* liquid crystals, which have a disc-like structure, as illustrated schematically by Figure 12.5. Discotic liquid crystals self-assemble so that their short axes are on average parallel to each other.

12.2.2 Smectic phase

In the smectic phase, the liquid crystals have orientational order, just as for the nematic phase. In addition, however, they also have a degree of spatial order. The smectic phase is thus more ordered than the nematic phase. The additional order arises because the molecules are arranged in layers, as illustrated by Figure 12.6. The thickness of a layer is about the length of a molecule. Within an individual layer, the molecules have orientational order but no positional order; i.e. the local order in the layers is nematic-like.

Figure 12.5 It is not only rod-like molecules that give rise to nematic liquid crystal phases. So too do discotic nematic liquid crystals that have a disc-like structure and self-assemble so that their short axes are on average parallel to each other. As for the rod-like nematic liquid crystals, their spatial arrangement is random.

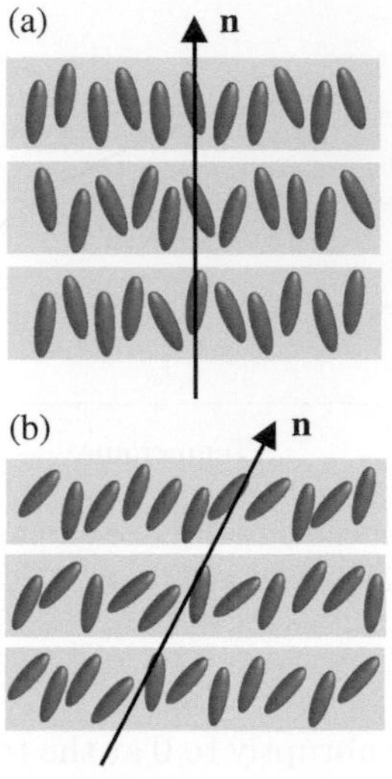

Figure 12.6 In the smectic phase, the liquid crystal molecules are arranged in layers. Within an individual layer, the molecules have orientational order but no positional order. A subdivision between smectic phases is made depending upon the orientation of the liquid crystal molecules with respect to the layer normal. (a) In the Smectic A phase, the director is oriented along the layer normal. (b) In the Smectic C phase, the director lies at an angle with respect to the layer normal.

The lateral forces between the liquid crystal molecules within a layer are stronger than the interactions between the layers, i.e. the side-to-side interactions between molecules are stronger than the interactions between the ends of two molecules. The result is that individual layers can slip over each other in a manner similar to that of soap. Indeed, the word smectic originates from the Latin word *smecticus*, meaning having soap-like properties.

A subdivision between smectic phases is made depending upon the orientation of the liquid crystal molecules in the layers with respect to the layer normal. In the *Smectic A* phase, the director is oriented along the layer normal as illustrated in Figure 12.6a, whereas in the *Smectic C* phase, it lies at an angle with respect to the layer normal as shown in Figure 12.6b.

12.2.3 Chiral liquid crystal phases

The special characteristic of the chiral nematic phase, also called the *cholesteric* phase, is that the direction of the director **n** varies in a helical fashion as one proceeds through the liquid crystal. This is the liquid crystal phase that Reinitzer discovered in 1888. It is convenient to imagine that the molecules are arranged in parallel layers, as illustrated by Figure 12.7. In each layer, all molecules tend to point in the same direction, as in the nematic phase. However, from one layer to the next, there exists a spiral twisting of the director **n** about an axis that is normal to the layers, as illustrated by Figure 12.7. This helical structure can be seen by tracing out the tips of the directors on progressing from one layer to the next. This helical structure occurs when the liquid crystal molecules have the property of *chirality*. A molecule is chiral if it cannot be superimposed on its mirror image. This is illustrated by Figure 12.8, which is a representation of the chiral molecule amino acid *alanine*, $NH_2CH(CH_3)COOH$. Many naturally occurring liquid crystals are chiral. However, a liquid crystal can be made to be chiral by inserting a chiral molecular group into its structure.

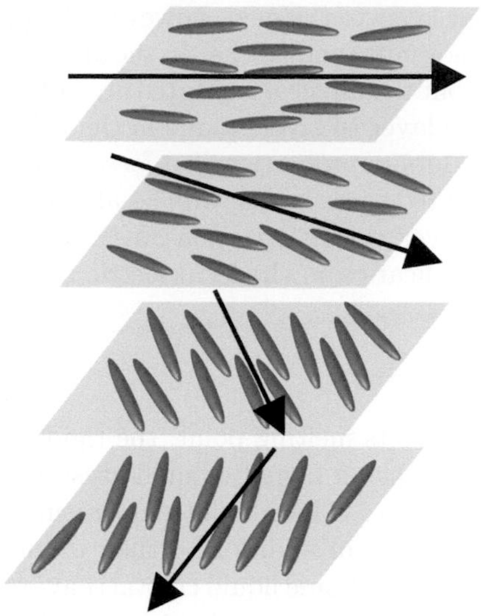

Figure 12.7 A representation of the chiral nematic liquid crystal phase. It is convenient to imagine that the molecules are arranged in parallel layers, as illustrated. However, these layers are imaginary and do not imply any kind of layered structure. In each layer, the molecules tend to point in the same direction, as in the nematic phase. However, there exists a spiral twisting of the director **n** about an axis that is normal to the layers. This helical structure can be seen by tracing out the tips of the directors on progressing from one layer to the next.

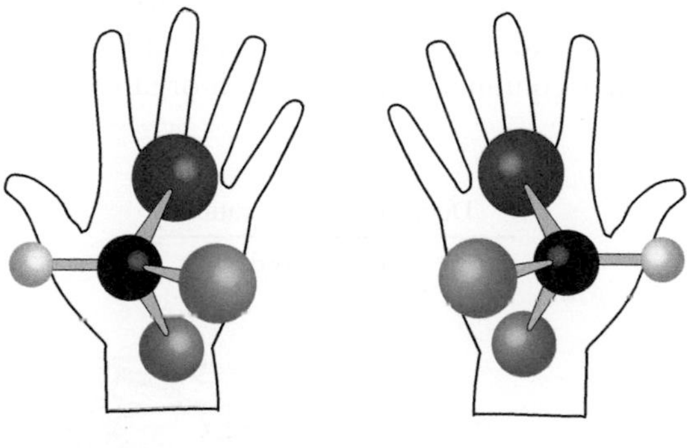

Figure 12.8 A molecule is chiral if it cannot be superimposed on its mirror image. This is illustrated by this representation of the chiral molecule amino acid *alanine*, $NH_2CH(CH_3)COOH$.

The chiral nematic phase is also called the cholesteric phase because many liquid crystals that were discovered were derived from cholesteric. The *pitch* P of a chiral nematic liquid crystal refers to the distance over which the director undergoes a full 360° twist. In some chiral nematic liquid crystals, the pitch P is of the same order as the wavelength of visible light. This gives rise to several important optical properties, which we will discuss in Section 12.4.

In addition to the chiral nematic phase, there are also chiral Smectic phases. The chiral Smectic C phase is the chiral analogue of the Smectic C phase shown in Figure 12.6b. And again the molecules are arranged in layers. In each layer, the molecules have orientational order and are tilted at an angle with respect to the normal to the layer. In going from one layer to the next, however, the molecular chirality causes a gradual change in the direction in which the molecules point. That is, the angle that the director makes with respect to the layer normal remains the same but the director is rotated about the normal from one layer to the next. This gradual change in direction from layer to layer results in the helical structure of the director.

The chiral Smectic-C phase can exhibit a property called *ferroelectricity.* This is because the particular arrangement of the molecules can lead to the molecular dipoles aligning to give a bulk polarisation, known as the *spontaneous polarisation.* Moreover, this polarisation can be reversed by the application of an applied electric field. Ferroelectricity finds application in the ferroelectric liquid crystal display.

12.2.4 Molecular order and temperature

When a crystalline solid is heated above its melting point, the rigid order of the crystal breaks down to become the random order of a liquid. When a nematic liquid crystal is heated, the positional order first breaks down leaving only orientational order. Then as the temperature rises still further, the orientational order breaks down. Smectic liquid crystals, in addition to their orientational order, have an additional degree of order due to their layered structure. Some liquid crystals have both a nematic and a smectic phase. As noted previously, the inter-molecular interactions that give the liquid crystal orientational order are stronger than those that give the layered structure. Thus, it is found that as the temperature is steadily increased, the layered structure first breaks down leaving only orientational order. Then at higher temperature, the orientational order beaks down, i.e. the smectic phase exists at a lower temperature than the nematic phase. We see a general principle that as temperature increases, the degree of order possessed by a liquid crystal decreases. A further example is provided by discotic liquid crystals. The nematic phase of discotic liquid crystals has only orientational order, as described earlier; see Figure 12.9a. However, the molecules of some discotic liquid crystals can self-organise to produce other phases with additional kinds of order. They can sit on top of each other, just like a pile of coins, as in Figure 12.9b. This is called the *columnar* discotic phase. Furthermore, these columns may further self-organise to form more complicated ways of

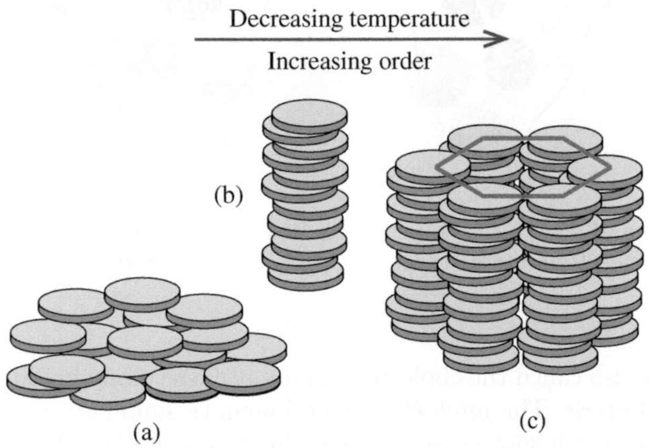

Figure 12.9 The molecules of some discotic liquid crystals can self-organise to produce other liquid crystal phases with different kinds of order. (a) The nematic phase, with just orientational order. (b) The columnar discotic phase in which the molecules sit on top of each other, just like a pile of coins. (c) The columns may further self-organise to form more complicated ways of stacking. As indicated, the phases with higher degrees of order occur as the temperature decreases.

stacking as shown in Figure 12.9c. The particular phase in which a discotic liquid crystal exists is determined by temperature. And as indicated by Figure 12.9, the phases with higher degrees of order occur as the temperature decreases. Again we have the familiar interplay and competition between thermal, i.e. kinetic energy and binding energy that we have seen in all phases of matter.

12.2.5 Molecular structure of liquid crystals

There are thousands of different liquid crystals with a wide range of molecular structures. However, we can see strong similarities between, for example, the rod-like molecules that are found in nematic and smectic liquid crystals. Figure 12.10 is a schematic diagram to illustrate the general structure or template of such rod-like molecules. To have a rod-like structure, they must have a certain amount of rigidity. This is provided by the *core* of the structure, which typically consists of two benzene-like ring systems that have planar geometry. The two ring systems may be directly linked together or by a linking group, such as –CH=N–. The linking group increases the rigidity of the molecule and also usefully extends its length. However, the molecular structure must have a certain amount of flexibility as this ensures the transition to the liquid phase occurs at a reasonably low temperature. It must also give the liquid crystal mobility, which is important when the molecules are to be manipulated by electric fields. Flexibility and mobility are provided by terminal molecular groups that are attached to the core. In addition, terminal or side groups may be attached to give the liquid a permanent electric dipole or to make the liquid crystal chiral. With such considerations in mind, it is possible to design and synthesise a liquid crystal to have the desired properties for a particular application, for example, in a liquid crystal display. And this is what is done in practice. A typical example of a nematic liquid crystal is pentylcyanobiphenyl (5CB), which has the chemical formula

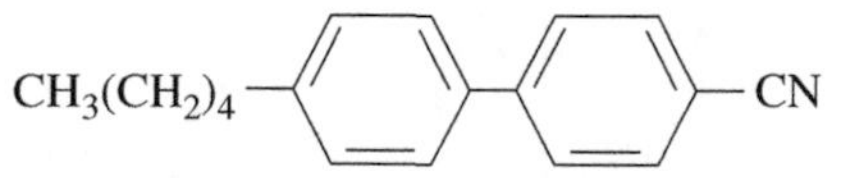

This molecule is about 2 nm long. It undergoes a phase transition from a crystalline state to a nematic state at 22.5°C and it goes from a nematic to an isotropic state at 35°C.

The geometrical anisotropy of the constituent molecules of a liquid crystal means that its physical properties will differ from one direction to another. This includes its relative permittivity, refractive index, elastic moduli, and viscosity. The most important of these, at least for technological applications, are the optical properties of a liquid crystal, such as their *birefringence*. These applications usually involve the manipulation of polarised light and so it is useful to first discuss some properties of polarised light.

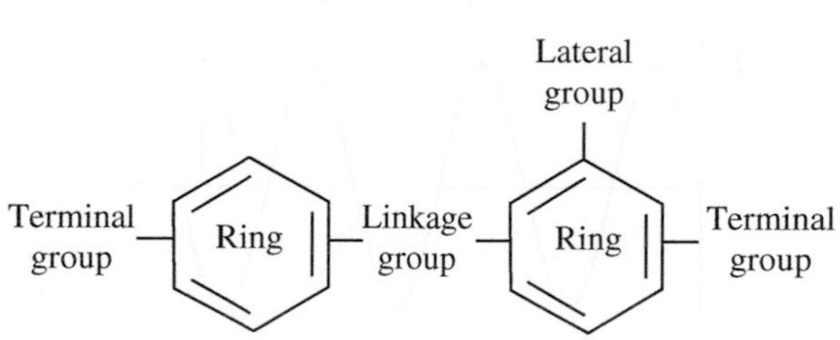

Figure 12.10 A schematic diagram to illustrate the general structure or template of the rod-like molecules that are found in nematic and smectic liquid crystals.

12.3 Polarised light

Light, we recall, is an electromagnetic wave consisting of propagating electric and magnetic fields. The electric and magnetic fields both point in directions that are perpendicular to the direction of propagation. The orientations of the electric and magnetic fields determine the *polarisation state* of the light. However, as the magnetic field is perpendicular to the electric field at all times, it is the convention to specify the polarisation state in terms of the orientation of the electric field alone. Figure 12.11 shows an electromagnetic wave that propagates in the $+z$-direction. The electric field vector associated with this wave is contained in a single plane, and so the light wave is referred to as *linearly polarised.*

The electric field of the wave oscillates in both time and space. Hence, if a snapshot is taken of the wave, the electric field oscillates as a function of the propagation direction, as illustrated by Figure 12.11, where E_0 is the amplitude of the electric field. The distance it takes for the field to repeat itself is the wavelength λ. Alternatively, at any fixed value of z, the electric field varies sinusoidally as in Figure 12.12. The time it takes for the electric field variation to repeat itself is the period T of the wave. In vacuum, the wavelength and period are related to the velocity of light c by the expression

$$c = \frac{\lambda}{T}. \tag{12.2}$$

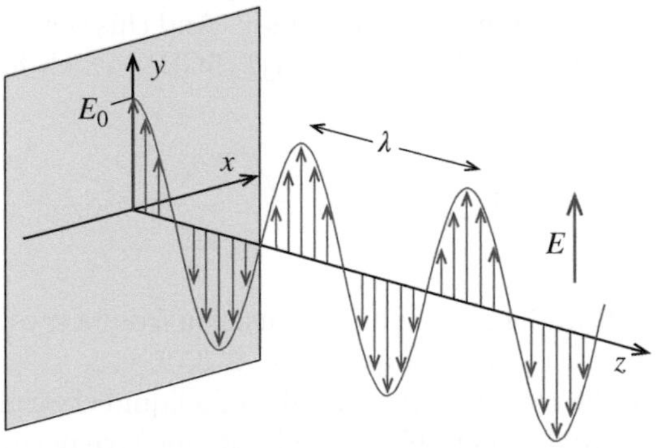

Figure 12.11 An electromagnetic wave that propagates in the $+z$-direction. The electric field vector of this wave is contained in a single plane, and so the light wave is referred to as linearly polarised. The electric field of amplitude E_0 oscillates as a function of the propagation direction. The distance it takes for the field to repeat itself is the wavelength λ.

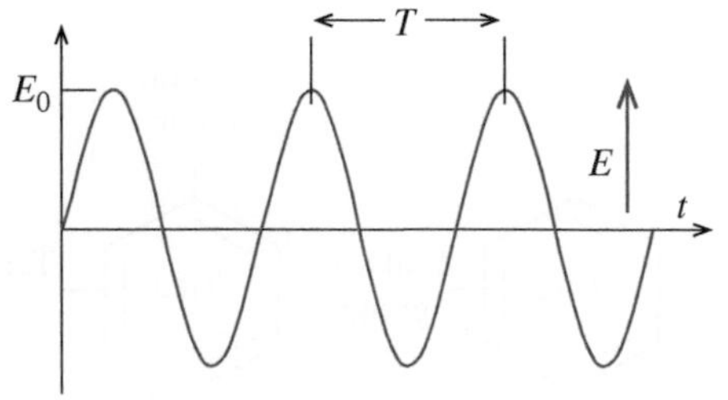

Figure 12.12 At any fixed value of z, the electric field of the electromagnetic wave shown in Figure 12.11 varies sinusoidally. The time it takes for the electric field variation to repeat itself is the period T of the wave.

It is, however, often convenient to replace λ and T by the variables

$$k = \frac{2\pi}{\lambda}; \quad \omega = \frac{2\pi}{T},$$

where k is the wave number and ω is the angular frequency, giving $c = \omega/k$.

The orientation of the plane of polarisation of a linearly polarised wave is given by the ratio of the x- and y-components of the electric field. For the linearly polarised wave shown in Figure 12.13, the two components have equal amplitudes E_0 and may be written as

$$E_x(x, t) = E_0 \sin(\omega t - kz),$$
$$E_y(y, t) = E_0 \sin(\omega t - kz).$$

The x-component varies sinusoidally along the x-axis and the y-component varies sinusoidally along the y-axis and is in phase with the x-component. It follows that the resultant wave has amplitude

$$\sqrt{E_x(x, t)^2 + E_y(y, t)^2} = \sqrt{2}E_0,$$

and the wave may be written as

$$E = \sqrt{2}E_0 \sin(\omega t - kz).$$

The electric field vector of this polarised wave makes an angle θ with respect to the x-z plane where $\tan\theta = E_y/E_x = 1$, giving $\theta = \pi/4$, or 45°. The inset in the figure shows the orientation of the two components.

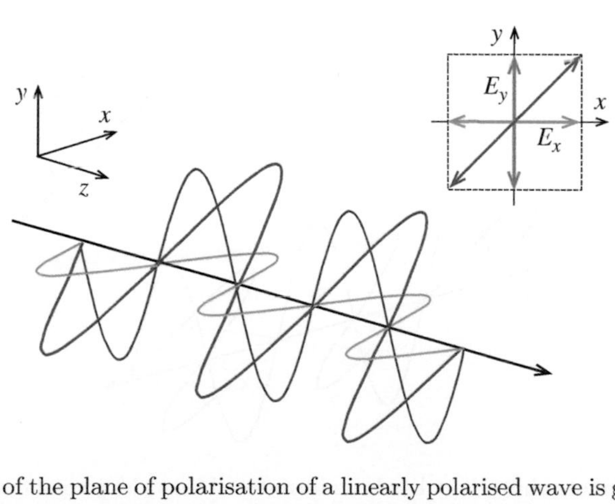

Figure 12.13 The orientation of the plane of polarisation of a linearly polarised wave is given by the ratio of the x- and y-components of the electric field. For the linearly polarised wave shown, the two components are in phase and have equal amplitudes. This gives the plane of polarisation to be at an angle of 45° with respect to the x-axis. The inset in the figure shows the orientation of the two components.

Suppose now that we have a wave in which there is a phase difference π between the x- and y-components such that

$$E_x = E_0 \sin(\omega t - kz),$$

$$E_y = E_0 \sin(\omega t - kz + \pi).$$

This is shown in Figure 12.14, where the y-component has 'slid' a distance of $\lambda/2$ in the $-z$-direction with respect to the x-component. Using the identity

$$\sin(A + B) = \sin A \cos B + \cos A \sin B,$$

$$E_y = E_0 \sin(\omega t - kz + \pi) = -E_0 \sin(\omega t - kz).$$

The amplitude of the resultant wave is

$$\sqrt{E_x^2 + E_y^2} = \sqrt{2}E_0,$$

as before, while the electric field vector makes an angle θ with the x–z plane, where

$$\tan\theta = \frac{E_y}{E_x} = -1,$$

giving $\theta = -\pi/4$, or $-45°$. The inset in the figure shows the orientation of the two components. Comparing Figures 12.13 and 12.14, we see that introducing a phase difference of π has caused the direction of the electric field vector, i.e. the polarisation plane to be rotated through 90°.

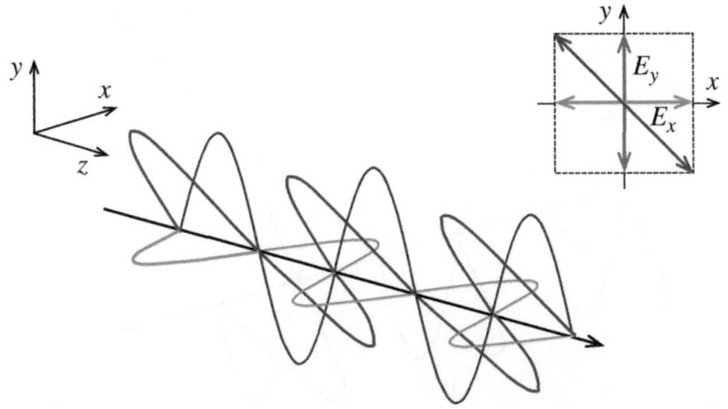

Figure 12.14 For this wave, a phase shift of $+\pi$ has been introduced into the y-component of the wave shown in Figure 12.13. This causes the y-component to 'slide' a distance equal to half a wavelength in the $-z$-direction, with respect to the x-component. The inset in the figure shows the orientation of the two components. Comparison of Figures 12.13 and 12.14 shows that this results in the plane of polarisation of the light being rotated through 90°.

Finally, suppose that we have a wave in which there is a phase difference of $\pi/2$ between the x- and y-components such that

$$E_x = E_0 \sin\left(\omega t - kz + \frac{\pi}{2}\right),$$
$$E_y = E_0 \sin(\omega t - kz).$$

This case is shown in Figure 12.15. We have

$$\sin\left(\omega t - kz + \frac{\pi}{2}\right) = \cos(\omega t - kz),$$

giving,

$$E_x = E_0 \cos(\omega t - kz).$$

The y-component varies sinusoidally along the y-axis and the x-component varies co-sinusoidally along the x-axis, while the direction of the electric field varies according to

$$\tan\theta = \frac{E_y}{E_x} = \frac{\sin(\omega t - kz)}{\cos(\omega t - kz)} = \tan(\omega t - kz),$$

or

$$\theta = (\omega t - kz).$$

At a fixed value of z, $kz =$ constant, and hence, $\theta = \omega t +$ constant. This means that at any fixed value of z, the electric field vector rotates about the propagation axis at angular frequency ω, as illustrated by the inset in Figure 12.15. For an observer looking towards the incoming wave, the electric field vector rotates

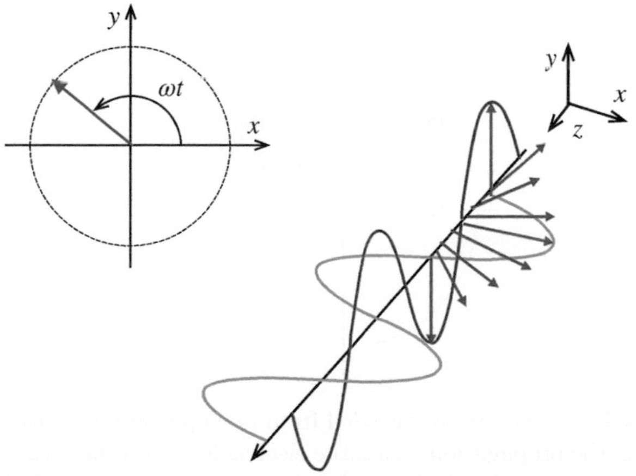

Figure 12.15 A phase shift of $\pi/2$ between the x- and y-components of the electric field of a wave produces circularly polarised light. At any point on the z-axis, the electric field vector of the wave rotates about the z-axis at angular frequency ω, with constant amplitude, as shown in the figure inset. In this particular case, the electric field vector rotates in a counter-clockwise direction and the wave is called left-circularly polarised.

in a counter-*clockwise* direction and the wave is called *left-circularly polarised.* If the phase angle were $-\pi/2$ instead of $+\pi/2$, the wave would be *right-circularly polarised.* At any instant of time and at any point on the z-axis, the amplitude of the x-component of the wave is $E_0 \cos(\omega t - kz)$ and the amplitude of the y-component of the wave is $E_0 \sin(\omega t - kz)$. It follows that the amplitude of the rotating electric field vector is

$$\sqrt{E_0^2 \cos^2(\omega t - kz) + E_0^2 \sin^2(\omega t - kz)} = E_0.$$

We see that a circularly polarised wave is obtained from a combination of two linearly polarised waves of equal amplitude. It is also the case that a linearly polarised wave can be obtained from the combination of a right-circularly polarised wave and a left-circularly polarised wave of the same amplitude. We can see this graphically in Figure 12.16. At any fixed point on the propagation axis z, the electric field vectors of the circularly polarised waves rotate about the propagation axis; one in the clockwise direction and one in the anti-clockwise direction. The x-components of the two electric field vectors are equal in amplitude but always point in opposite directions. They, therefore, combine destructively so that the x-component of the resultant wave is zero. On the other hand, the y-components always point in the same direction and add constructively. The combination of the two helical waves, therefore, is a linearly polarised wave lying in the y–z plane, with an amplitude that is twice the amplitude of the individual circularly polarised waves. From these examples, we see that we can change the polarisation state of a wave by changing the phase angle between x- and y-components of the electric field.

Linearly and circularly polarised waves are two special cases of polarised light. More generally, the x- and y-components are not equal and there is an arbitrary phase difference between them, neither π nor $\pi/2$, nor a multiple of these values. Then the polarisation of the wave is referred to as *elliptically polarised.* In this case, the tip of the electric field vector traces out an ellipse in a plane perpendicular to the propagation direction.

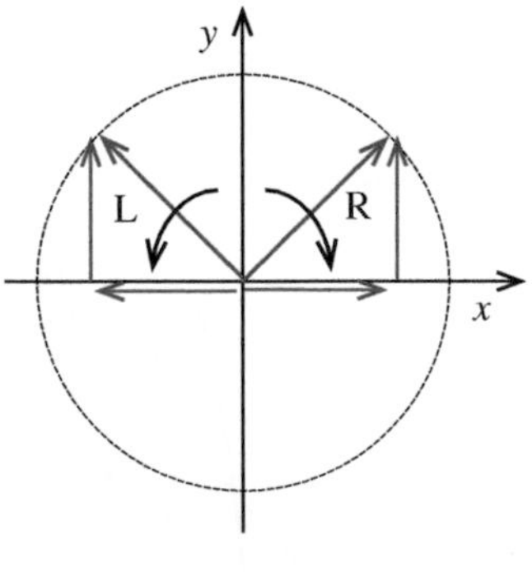

Figure 12.16 A linearly polarised wave can be synthesised from two oppositely polarised circular waves of equal amplitude. At any fixed point on the propagation axis z, the electric field vectors of the two circularly polarised waves rotate about the propagation axis; one in the clockwise direction and one in the anti-clockwise direction. The x-components of the two electric field vectors are equal in amplitude but always point in opposite directions. They, therefore, combine destructively. On the other hand, the y-components always point in the same direction and combine constructively. The combination of the two helical waves, therefore, is a linearly polarised wave lying in the y–z plane, with an amplitude that is twice the amplitude of the individual circularly polarised waves.

The light that is emitted by most light sources, for example, fluorescent and incandescent lamps and light-emitting diodes is unpolarised. We can obtain linearly polarised light by passing the light through a *polariser*, such as a sheet of Polaroid, as illustrated in Figure 12.17. An ideal polariser passes 100% of the incident light that is polarised in the direction of the polariser's *polarisation axis*, but completely blocks all light that is perpendicular to this axis. The light waves from an unpolarised light source are a random mix of waves whose electric fields oscillate in all possible transverse directions. We can resolve each of these oscillations into a component that is parallel to the polarisation axis and a component that is perpendicular to the axis. Only the parallel components are transmitted by the polariser; the perpendicular components are not. If the incident waves are randomly orientated, the sum of the parallel and perpendicular components is equal. Hence, when unpolarised light is incident on a polariser, exactly half of the light is transmitted.

If, as in Figure 12.17, a second polariser is placed so that its polarisation axis is at 90° with respect to the first, in a so-called *crossed-polariser* arrangement, no light will pass through the second polariser. This crossed-polariser arrangement is often used in liquid crystal displays, as we will see. More generally, the polarisation axes of the two polarisers may be at an arbitrary angle θ with respect to each other. Suppose the polarisation of the first polariser is in the vertical direction so that the plane of polarisation of the transmitted light is also in the vertical direction. When this light, of electric field amplitude E_0, meets the second polariser, it will have a component $E_0 \cos\theta$ parallel to the polarisation axis of the second polariser; see Figure 12.18. This component is transmitted by the second polariser. There is also a component E_0 $\sin\theta$ that is perpendicular to the polarisation axis, which is not transmitted. Hence, the intensity I of the light transmitted by the second polariser is proportional to $E_0^2 \cos^2\theta$. The intensity is zero at $\theta = 90°$, as earlier, and the maximum intensity $I_0 = E_0^2$ is at $\theta = 0°$. Hence, the transmitted intensity I at angle θ is

$$I = I_0 \cos^2\theta. \tag{12.3}$$

Equation (12.3) is known as the law of Malus.

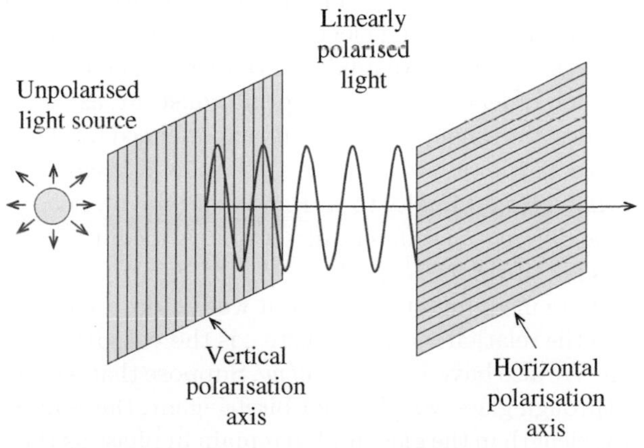

Figure 12.17 Linearly polarised light can be obtained from an unpolarised source by passing the light through a polariser. An ideal polariser passes 100% of the incident light that is polarised in the direction of the polarisation axis, but completely blocks all light that is perpendicular to this axis. If a second polariser is placed so that its polarisation axis is at 90° with respect to the first, in a crossed polariser arrangement, no light will pass through the second polariser.

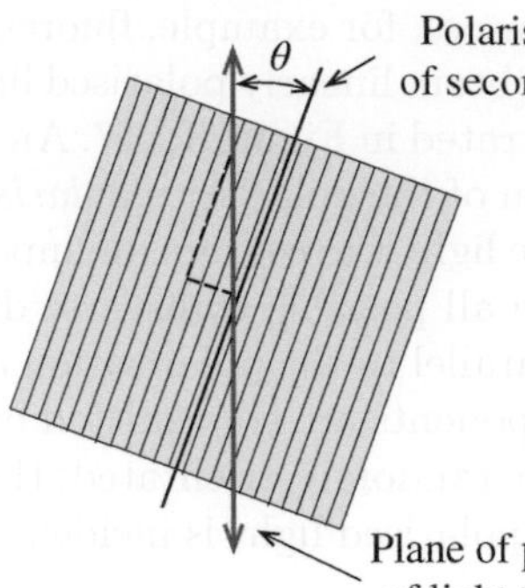

Figure 12.18 If the polarisation axes of two polarisers are at an angle θ with respect to each other, the transmitted intensity I at angle θ is given by the Malus law: $I = I_0\cos^2\theta$, where I_0 is the maximum transmitted intensity, which occurs at $\theta = 0°$.

12.4 Optical properties of liquid crystals

The huge interest in liquid crystals is partly explained by their special and unique optical properties. Here we describe some of these properties and how they are used in technological applications.

12.4.1 Birefringence

Light propagates through a transparent medium by exciting the electrons within the medium. The oscillating electric field of the light causes the electrons to oscillate and these electrons emit secondary wavelets at the same frequency. These secondary wavelets recombine and the resulting refracted wave carries on through the medium. The velocity of the wave depends on the difference between the frequency of the light wave and the natural frequency of the oscillating electrons. This frequency depends on the strength of the bond the electrons experience in the medium. An analogy is a mass oscillating on the end of a spring, where the natural frequency of oscillation is dependent on the spring constant, as well as the mass. If a medium is composed of anisotropic molecules, the strength of the bond experienced by the electrons will be different in different directions. For example, in the case of a rod-like molecule, it is easier to move the electrons along the long axis of the molecule than at right angles to it. Hence, the wave velocity and therefore the index of refraction of the medium will be different in different directions. A medium in which the refractive index is different in different directions is called *birefringent*.

The frequency ν of light *within* a medium is the same as it was before it entered the medium. However, its velocity v changes according to the relation $v = c/n$, where c is the velocity of light in vacuum and n is the refractive index of the medium. We also have $\lambda = v/\nu = c/n\nu$. Suppose that we have a linearly polarised light wave. If the light propagates through glass, which is not birefringent, the x- and y-components of the wave have the same velocity and wavelength in the glass and so remain in phase as they pass through the glass, as illustrated in Figure 12.19a. Now, suppose that this light wave passes through a birefringent medium for which the refractive index for the x-component of the light is different from the refractive index for the y-component. The two components have the same frequency but travel at different velocities and therefore have different wavelengths within the medium. In Figure 12.19b, the y-component travels more slowly than

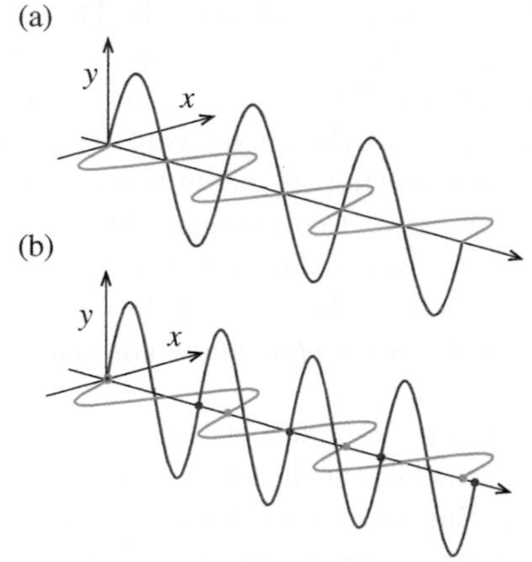

Figure 12.19 (a) If a linearly polarised light wave propagates through glass, which is not birefringent, the x- and y-components of the wave have the same velocity and wavelength in the glass. They remain in phase as they pass through the glass, and are in phase after leaving the glass. (b) In a medium that is birefringent, for which the refractive index for the x-component of the light is different from the refractive index for the y-component, the two components have the same frequency but travel at different velocities. They, therefore, have different wavelengths within the medium. In this figure, the y-component travels more slowly than the x-component and, therefore, has a shorter wavelength. The x-component completes three full wavelengths while the y-component completes almost four. It follows that the two components go steadily out of phase.

the x-component and therefore has a shorter wavelength. Inspection of Figure 12.19b shows that the x-component completes three full wavelengths while the y-component completes almost four over the same distance. It follows that the two components go steadily out of phase.

Suppose the length L of a birefringent medium is chosen so that the resultant phase difference between the x- and y-components of a linearly polarised light wave is exactly π after the light has passed through the medium. This situation is illustrated by Figure 12.20. As the light wave entered the medium, at $z = 0$, its

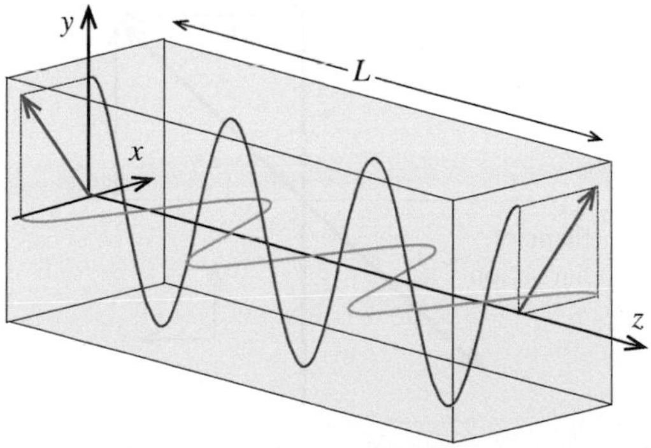

Figure 12.20 In this figure, the length L of the birefringent medium is chosen so that the resultant phase difference between the x- and y-components of an incident, linearly polarised light wave is exactly π after the light has passed through the medium. The effect of the birefringence is to rotate the plane of polarisation of the linearly polarised light through 90°. An optical component that produces such a rotation of 90° is called a half-wave plate.

electric field vector was at angle of 135° with respect to the x-axis. This is consistent with the x-component of the electric field vector having the value of $-E_0$ and the y-component of the electric field vector having the value of $+E_0$. But as the light wave left the medium, at $z = L$, the phase shift of π results in the electric field vector being at an angle of 45° with respect to the x-axis, consistent with the x- and y-components both having a value of $+E_0$. The effect of the birefringent medium is to rotate the plane of polarisation of the linearly polarised light through 90°. An optical component that produces a rotation of 90° is called a *half-wave plate*. So if a half-wave plate were placed between the two crossed polarisers shown in Figure 12.17, the light would be transmitted by the second polariser. Similarly, an optical component that produces a rotation of 45° is called a *quarter-wave plate* and an optical component that produces a rotation of 180° is called a *full-wave plate*.

Being composed of anisotropic molecules, a liquid crystal is naturally birefringent. In particular, the index of refraction for light polarised parallel to the director is different from the index of refraction for light polarised perpendicular to the director. Suppose that a linearly polarised wave enters a liquid crystal such that its plane of polarisation makes a finite angle with respect to the director, say 45° for the sake of simplicity. This is illustrated in Figure 12.21, where the director is parallel to the y-axis. In this case, the wave has electric field components

$$E_x = E_0 \sin(\omega t - kz),$$
$$E_y = E_0 \sin(\omega t - kz).$$

These two components travel at different velocities in the liquid crystal according to the two different indices of refraction. This means that the two components will go steadily out of phase. The result is that the light emerging from the liquid crystal will be elliptically polarised.

However, suppose that the plane of polarisation of the wave is parallel to the director. In that case, the wave has only a y-component of electric field and therefore the wave propagates in accord with a single refractive index; that for propagation parallel to the director. As there is no x-component, the wave maintains the orientation of its plane of polarisation as it propagates through the crystal. Similarly, a linearly polarised wave whose polarisation plane is perpendicular to the director has only an x-component of electric field, and propagates in accord with a single refractive index; that for propagation perpendicular to the

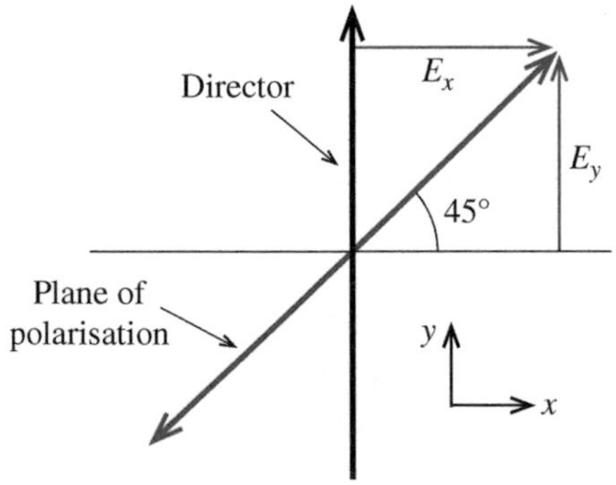

Figure 12.21 If a linearly polarised wave enters a liquid crystal such that its plane of polarisation makes an angle of say 45° with respect to the director, the wave will have x- and y-components. These travel at different velocities according to the two different indices of refraction. This means that the two components will go steadily out of phase. The result is that the light emerging from the liquid crystal will be elliptically polarised. However, if the polarisation plane is parallel or perpendicular to the director, the orientation of the plane of polarisation is unchanged.

director. And again, the wave maintains the orientation of its plane of polarisation as it propagates. Thus, the orientation of the plane of polarisation of a linearly polarised wave is unchanged as it propagates through a liquid crystal if the plane is parallel or perpendicular to the director.

Worked example

In a liquid crystal, light polarised parallel to the director propagates according to the index of refraction n_1, while light polarised perpendicular to the director propagates according to the index of refraction n_2. In a particular liquid crystal $n_1 = 1.723$ and $n_2 = 1.532$. Consider a beam of linearly polarised light of wavelength 0.535 μm, whose plane of polarisation is at an angle of 45° with respect to the director so that we can consider two equal components of the light, one traveling parallel to the director and one travelling perpendicular to it. What is the minimum length of liquid crystal for which the phase difference between the two components after passing through the liquid crystal is π rad?

Solution

We have, $\lambda_0 \nu = c$, where λ_0 is the wavelength of the light in vacuum, ν is its frequency and c is the velocity of light in vacuum. In the liquid crystal, we have $\lambda \nu = v$, where λ is the wavelength of light in the liquid crystal and $v = c/n$ where n is the refractive index. Hence, $\lambda = \lambda_0/n$, from which

$$\lambda_1 = \frac{\lambda_0}{n_1}, \quad \text{and} \quad \lambda_2 = \frac{\lambda_0}{n_2}.$$

If a light wave travels a distance L in vacuum, the phase change incurred is $(2\pi/\lambda_0)L$, as the change in phase is 2π per wavelength. In the liquid crystal with refractive index n_1, the corresponding phase change is $(2\pi/\lambda_1)L$. The phase difference between the two light waves after distance L, called the *relative phase retardation*, is then

$$2\pi L\left(\frac{1}{\lambda_1} - \frac{1}{\lambda_0}\right) = \frac{2\pi L}{\lambda_0}(n_1 - 1).$$

Similarly, the phase retardation for the refractive index n_2, is $2\pi L/\lambda_0(n_2 - 1)$.

Subtracting these two results we find that the retardation in phase between the two components in the liquid crystal is $2\pi L/\lambda_0(n_1 - n_2)$. We want this to be π rad. Hence,

$$L = \frac{\lambda_0}{2(n_1 - n_2)} = \frac{0.535}{2(1.723 - 1.532)} = 1.40\ \mu\text{m}.$$

12.4.2 Selective reflection

In Section 12.2.3, we described the chiral nematic liquid crystal phase and its helical structure. This helical structure, which is repeated throughout the liquid crystal, is illustrated schematically by Figure 12.22a, which shows the rotation of the liquid crystal molecules along the axis of the helix. The helical shape leads to a periodicity in the structure of the liquid crystal. When light is reflected from a chiral liquid crystal, this periodicity leads to constructive interference of the reflected light waves. The effect, which is called *selective reflection*, has similarity to the Bragg reflection of X-rays from a crystal; see Figure 12.22b.

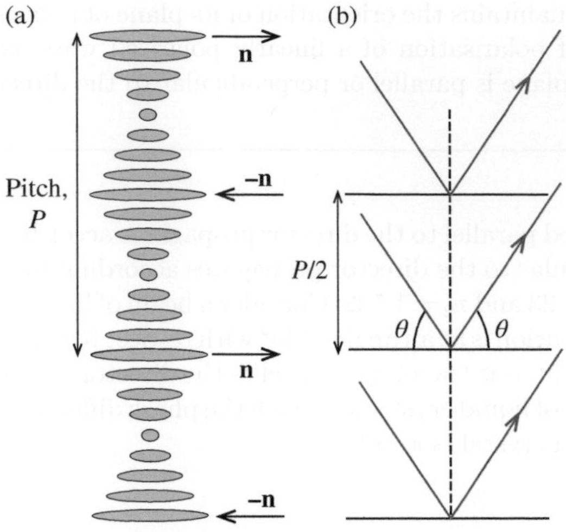

Figure 12.22 (a) A schematic representation of the helical structure of a chiral nematic liquid crystal. The pitch P is the distance along the axis over which the director rotates by 360°. Because the director $-\mathbf{n}$ is equivalent to the director $+\mathbf{n}$, the repeat distance of the structure causing the constructive interference is equal to $P/2$. (b) Reflected waves that emanate from parts of the liquid crystal separated by $P/2$ are in phase and produce constructive interference when $2(P/2)n\sin\theta = m\lambda$, where λ is the wavelength of the light, n is the refractive index of the material, and $m = 0, 1, 2, \ldots$.

The pitch P is the distance along the axis of a nematic liquid crystal over which the *director* rotates by 360°. However, because the director $-\mathbf{n}$ is equivalent to the director $+\mathbf{n}$, the repeat distance of the structure causing the constructive interference is equal to $P/2$, as in Figure 12.22b. Hence, reflected waves emanating from parts of the liquid crystal separated by $P/2$ are in phase and produce constructive interference when the following condition is obtained:

$$2\left(\frac{P}{2}\right)n\sin\theta = m\lambda, \tag{12.4}$$

where θ is the incident glancing angle, λ is the wavelength of the light, $m = 1, 2, \ldots$, and n is the average of the two refractive indices of the chiral nematic liquid crystal, which is birefringent. Usually, m is taken to be 1.

If white light is incident on the liquid crystal, then at any particular angle of reflection, only light of a narrow range of wavelengths, according to Equation (12.4), will be reflected, i.e. the liquid crystal will appear to be coloured. Clearly, the colour of light reflected by a chiral liquid crystal depends on the pitch P of its helical structure. This structure winds and unwinds as the temperature changes. This is because as the temperature increases, the molecules have greater thermal energy, which means that the angle at which the director changes between layers becomes greater. In turn, this means that the repeat distance for the director to complete a full circle becomes shorter, i.e. the pitch becomes smaller. According to Equation (12.4), a longer pitch, corresponding to a low temperature, results in the reflection of red light by the liquid crystal, while a shorter pitch, corresponding to a high temperature, results in the reflection of blue light. Figure 12.23 illustrates an example of how the colour of light reflected from a chiral nematic liquid crystal varies with temperature. Such liquid crystals find wide application in medical applications to monitor temperature.

Although this selective reflection of light can be thought of in the same way as Bragg reflection, the helical structure of the chiral liquid crystal has some complex properties. The helical structure of a

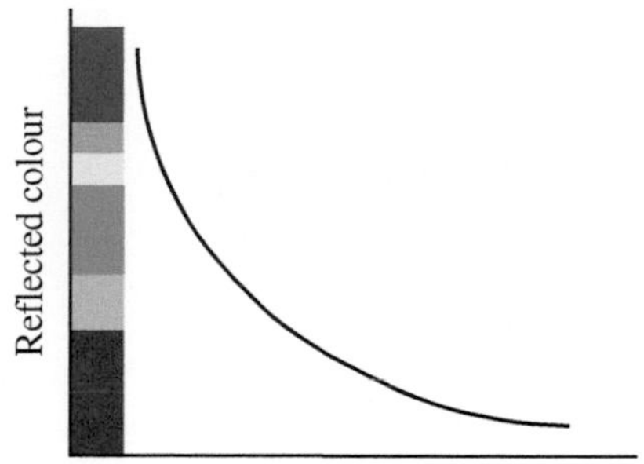

Figure 12.23 The figure illustrates how the colour of light reflected from a chiral nematic liquid crystal varies with temperature.

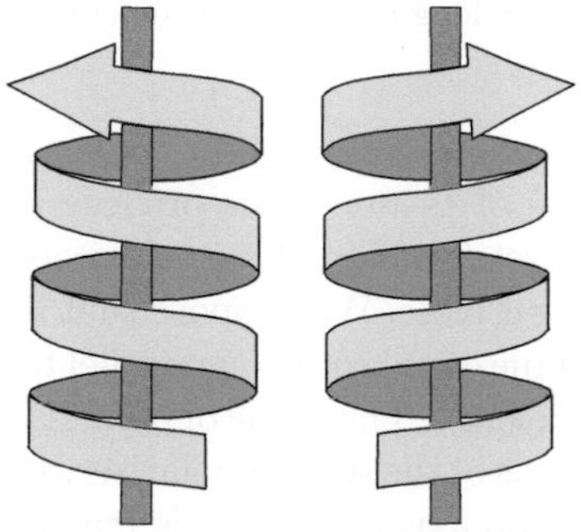

Figure 12.24 Left- and right-handed helices.

particular liquid crystal may be either left-handed or right-handed, as defined by Figure 12.24a, b, respectively. The helical structure selectively reflects *circular polarised* light of the handedness that matches the handedness of the structure. The other handedness of the circular polarised light is transmitted by the helical structure even though the wavelength may match the Bragg condition, Equation (12.4). Thus, a chiral liquid crystal structure will selectively reflect light of a particular colour but only half of the incident light will be reflected. The rest will pass through the material unaffected.

12.4.3 Waveguide regime

There is a special case of propagation of linearly polarised light along the helical axis of a chiral nematic liquid crystal when the pitch of the helical structure is much greater than the wavelength of the light. Here, the polarisation direction of the light, i.e. the direction of the electric field vector, follows the rotation of the director as it propagates through the liquid crystal. This is described as the *waveguide regime*. To help explain this effect, we first consider an analogous situation where a stack of polarisers is used to rotate the plane of polarisation of a linearly polarised beam of light.

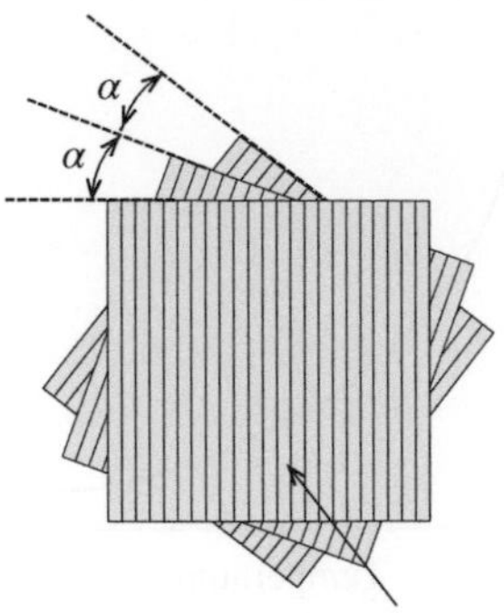

Figure 12.25 The figure shows the first three of a stack of a large number of polarisers where the polarisation axis of each polariser is at an angle α greater than that of its immediate predecessor in the stack. The effect of the stack is to rotate the plane of polarisation of incident linearly polarised light. Even if the first and last polarisers are crossed, the output light intensity is equal to the input intensity so long as there are enough intermediate polarisers.

We consider a stack consisting of a large number $N+1$ of polarisers where the polarisation axis of each polariser is at an angle α greater than that of its immediate predecessor in the stack. Thus, the last polariser is at an angle $\theta = N\alpha$ with respect to the first. The first three polarisers are shown in Figure 12.25. We neglect any losses due to reflection at the many surfaces. We suppose that linearly polarised light of intensity I_0 is incident on the first polariser with the plane of polarisation parallel to the polarisation axis. The light passes through the first polariser with 100% efficiency. The plane of polarisation of this transmitted light makes an angle α with respect to the polarisation axis of the second polariser; see also Figure 12.25. The component of the light that is parallel to the polarisation axis and hence is transmitted by the polariser is $I_0 \cos\alpha$. The component of the light that is perpendicular to the polarisation axis is not transmitted by the polariser. It follows that the intensity of the light transmitted by the second polariser is just

$$I_1 = I_0 \cos^2\alpha.$$

Similarly, the intensity of the light transmitted by the third polariser is

$$I_2 = I_1 \cos^2\alpha = I_0 \left(\cos^2\alpha\right)^2,$$

and the intensity of the light transmitted by the Nth polariser is $I_0(\cos^2\alpha)^N$. If N is a very large number, the angle α will be small. We have the identity

$$\cos^2\alpha = 1 - \sin^2\alpha,$$

and the expansion

$$\sin\alpha = \alpha - \frac{\alpha^3}{3!} + \frac{\alpha^5}{5!} - \cdots.$$

Retaining only the first term, as α is small, we have that the light intensity transmitted by the Nth polariser is

$$I_0\left(1-\alpha^2\right)^N \approx I_0\left(1-N\alpha^2\right) = I_0\left(1-\frac{\theta^2}{N}\right).$$

This result says that even if $\theta = 90°$, so that the first and last polarisers are crossed, the output intensity is equal to the input intensity so long as we have enough intermediate polarisers, i.e. N is very large. And note that the plane of polarisation is rotated 90° about the propagation direction, as the polarising axis of the last polariser is at an angle of 90° with respect to the first. Thus, we have a way of 'gently' rotating the plane of polarisation of the light.

An analogous effect occurs in a chiral nematic liquid crystal, in which the director rotates about the axis of the crystal. In this case, the rotation of the plane of polarisation occurs because of birefringence. To understand what happens to the light, a sample of the chiral nematic liquid crystal is considered to be divided into a large number of narrow slices. And each slice has a director that is at an angle of α with respect to the proceeding slice. When a linearly polarised wave leaving the previous slice is incident on the following slice, it finds that the director is at an angle of α with respect to its electric field vector. Again, we can resolve the wave into two orthogonal components with respect to the director. One component oscillates along the direction of the director and the other oscillates along a direction that is perpendicular to the director. These two components travel at different velocities according to the two respective refractive indices. This introduces a phase difference between the two phases. Strictly speaking, the light transmitted by the slice is elliptically polarised. However, because both the width of the slice and the angle α between successive directors are small, the polarisation state of the wave is essentially unchanged, i.e. the action of the slice is to slightly rotate the electric vector of the light wave but to maintain its state of linear polarisation. Moreover, detailed analysis shows that the angle through which the electric field vector rotates is just α, the same as the angle between successive directors. Hence, when light propagates through the chiral nematic liquid crystal, the light remains linearly polarised but the direction of polarisation follows the director of the helical structure.

12.4.4 Optical polarising microscopy

It is important for both academic reasons and practical applications to be able to identify the types of phases that a given liquid crystal exhibits. There are several different techniques for doing this. A widely used technique is *optical polarising microscopy*. In this technique, a thin sample of the liquid crystal is sandwiched between two glass microscope slides. This assembly, whose temperature may be varied, is positioned between crossed polarisers and illuminated by white light. The liquid crystal is then viewed with a microscope. What is observed through the microscope is a so-called *optical texture*. An example is shown in Figure 12.26.

A thin film of liquid crystal will effectively be a half-wave plate for some wavelength, completely transmitting that colour and a full-wave plate for a different wavelength, not transmitting that colour. Every other wavelength will be elliptically polarised and so partially transmitted by the second polariser. That means that a uniform film of a particular thickness will appear as one colour, related to the birefringence and sample thickness. Where different colours are seen, as in the photograph, there is usually an out-of-plane angle adopted by the director. The effect of this is to reduce the effective refractive index for light polarised parallel to the director. Hence, the difference in refractive indices of the liquid crystal and hence the birefringence is reduced. Different tilts in different parts of the sample will give different birefringence colours. Optical textures are valuable in identifying the phase of a given liquid crystal. Moreover, if the

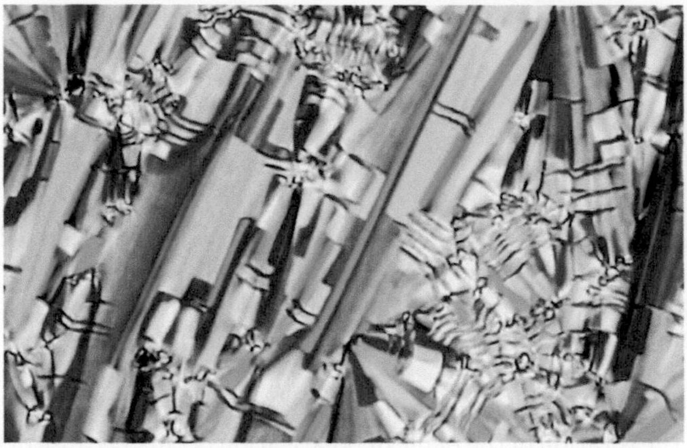

Figure 12.26 An example of an optical texture of a liquid crystal obtained using the technique of optical polarising microscopy. Ingo Dierking/2009 Wiley-VCH Verlag GmbH & Co. KGaA, Weinheim 1617-9437/09/0404-27 Physik Journal 8 (2009) Nr. 4.

liquid crystal has more than one phase, the order in which they appear as the temperature of the sample is varied gives further information about the phases.

12.5 Liquid crystal displays

There are various types of liquid crystal displays, and these involve the manipulation of the liquid crystal molecules by the application of an external electric field.

12.5.1 Reorientation of liquid crystals in an electric field

The molecules of a particular liquid crystal may have a permanent electric dipole moment, where one end of the molecules has a net positive charge and the other end a net negative charge. However, even molecules that have no permanent dipole moment will become *polarised* in the presence of an electric field. The electric field causes a slight charge separation in the molecule creating an electric dipole. The action of an electric field $\boldsymbol{E}$ on a polarised molecule is to exert a torque $\boldsymbol{\tau}$ on it tending to align the molecule along the direction of the electric field. The torque depends on the electric field $\boldsymbol{E}$ and the electric dipole moment $\boldsymbol{p}$ of the molecules according to

$$\boldsymbol{\tau} = \boldsymbol{E} \times \boldsymbol{p}. \tag{12.5}$$

Suppose that we have an arrangement in which the polarised molecules of a liquid crystal all point in a horizontal direction, as shown in Figure 12.27a. It is quite possible to arrange for this to happen, as we shall see. When the electric field is applied in a direction perpendicular to the molecules, they experience a torque that tends to point them along the direction of the field. And for a sufficiently strong applied field, the molecules will completely align with the electric field direction, as in Figure 12.27b. Thus, by applying an electric field, the orientation of the liquid crystal molecules can be changed and controlled.

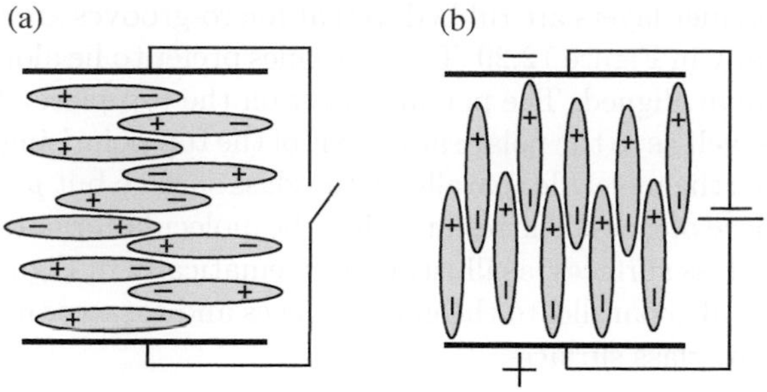

Figure 12.27 (a) An arrangement of polarised liquid crystal molecules where all the molecules are aligned along the horizontal direction. (b) The action of an electric field on a polarised molecule is to exert a torque on it tending to align the molecule along the direction of the electric field. For a sufficiently strong applied field, the molecules align completely with the electric field direction.

12.5.2 The twisted nematic liquid crystal display

The heart of this display is a nematic liquid crystal cell, the basic parts of which are shown in Figure 12.28. The liquid crystal is contained between two pieces of glass. On the outside of the top piece of glass and the outside of the bottom piece of glass are polarising films. The polarisation axes of these two polarising films are perpendicular to each other so that they form a pair of crossed polarisers. The inside surface of the top piece of glass is coated with a thin, transparent layer of indium-tin-oxide (ITO), which acts as an electrical electrode. Similarly, a transparent conducting film is coated on the inside of the bottom piece of glass. With this arrangement, a voltage can be applied across the liquid crystal.

Although the molecules in a nematic liquid crystal tend to point in the same direction, this direction must be defined for the liquid crystal cell to operate. This is achieved by coating each of the two ITO layers

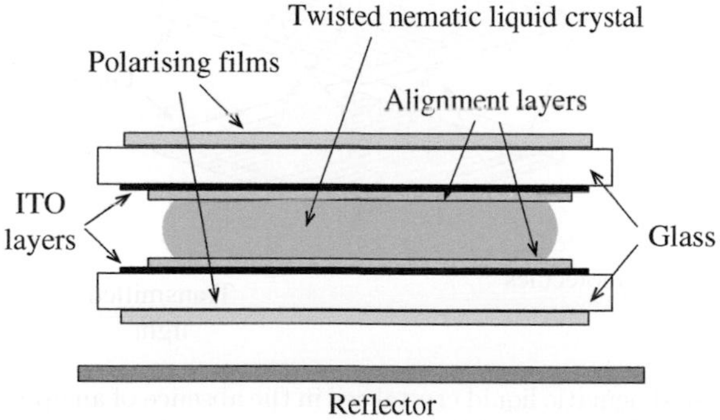

Figure 12.28 The basic parts of a twisted nematic liquid crystal cell. The liquid crystal is contained between two pieces of glass. On the outside of the top piece of glass and the outside of the bottom piece of glass are placed polarising films. The polarisation axes of these two polarising films are perpendicular to each other so that they form a pair of crossed polarisers. The inside surface of the top piece of glass is coated with a thin, transparent layer of indium-tin-oxide (ITO), which acts as an electrical electrode. Similarly, a transparent conducting film is coated on the inside of the bottom piece of glass. With this arrangement, a voltage can be applied across the liquid crystal. Each of the two ITO layers is coated with a polymer layer. The polymer layers are rubbed so that micro-grooves are formed in the polymer surface, which align the liquid crystal molecules. The reflector may be used to reflect light back into the cell.

with a polymer layer. The polymer layers are rubbed so that micro-grooves are formed on the polymer surface, as illustrated schematically in Figure 12.29. The molecules prefer to lie along the micro-grooves rather than across them and so become aligned. The polymer layer on the top piece of glass aligns the molecules parallel to the glass surface as well as to the polarisation axis of the top polarising film. The polymer layer on the bottom piece of glass aligns the molecules parallel to the glass surface but *perpendicular* to the direction of the alignment on the top piece of glass. The result is that the molecular orientation is twisted through 90° between the top and bottom glass surfaces, as illustrated schematically in Figure 12.30. It follows that the director of the liquid crystal is also parallel to the glass surfaces and rotates through 90° in going from the top glass surface to the bottom glass surface.

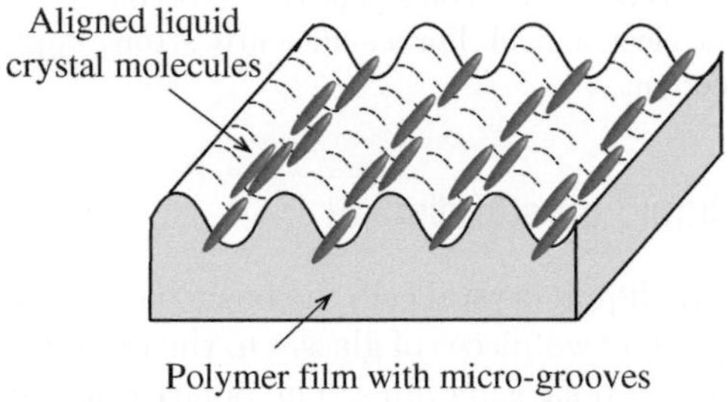

Figure 12.29 The figure illustrates the micro-grooves that are formed on a polymer surface. The liquid crystal molecules prefer to lie along the micro-grooves rather than across them and this aligns the molecules.

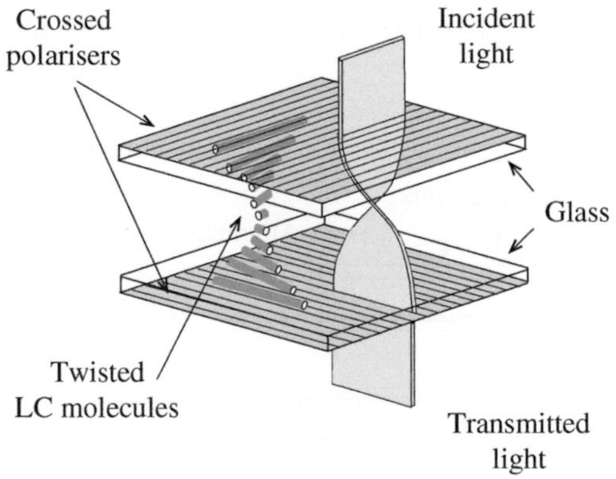

Figure 12.30 The action of a twisted nematic liquid crystal cell in the absence of an applied voltage. The polymer layer on the top piece of glass aligns the molecules parallel to the glass surface and parallel to the polarisation axis of the top polarising film. The polymer layer on the bottom piece of glass aligns the molecules parallel to the glass surface but perpendicular to the direction of the alignment on the top piece of glass. The result is that the orientation of the molecules is twisted through 90° between the top and bottom glass surfaces. It follows that the director of the liquid crystal is also parallel to the glass surfaces and rotates through 90° in going from the top glass surface to the bottom glass surface. Ambient light that is incident on the cell is plane polarised by the top polariser. The plane of polarisation of this light follows the director of the twisted nematic liquid crystal in accord with the waveguide regime, and is rotated by 90° as it passes through the crystal. As the polarisation axis of the bottom polarising film is perpendicular to that of the top polarising film, light passes through the cell.

Ambient light that is incident on the cell is linearly polarised by the top polarising film, with its plane of polarisation parallel to the director of the liquid crystal. The plane of polarisation follows the rotation of the director as the light passes through the liquid crystal according to the waveguide regime and is rotated by 90°. As the polarisation axis of the bottom polarising film is perpendicular to that of the top polarising film, the light passes through the cell.

This situation in which the liquid crystal molecules are aligned by the micro-grooves on the glass surfaces is the case when there is no voltage applied to the cell. However, when a voltage is applied, the molecules change their orientation to become aligned with the electric field direction, as illustrated by Figure 12.31. Then, the direction of the molecules and hence the director become perpendicular to the glass surfaces. Now, the plane of polarisation of the light is parallel to the director. Accordingly, the orientation of the plane of polarisation does not change as the light passes through the liquid crystal. Consequently, the light does not pass through the bottom polarising film and the light is extinguished by the cell.

The operation of the device is clear. In the absence of the applied voltage, the liquid crystal adopts a twisted structure that guides the light so that it passed through the device. However, when the voltage is applied, the molecules re-orientate to point in the direction of the electric field and the twist is destroyed. In the absence of the twist, the light is not guided through the crossed polarisers and the display appears dark. When the voltage is removed, the molecules relax back to the original twisted state and the display again transmits light. We have a light switch.

There are two modes of operation for the display. In one mode, a reflector is placed behind the liquid crystal cell, as shown in Figure 12.28. In that case, the light transmitted by the cell is reflected towards the cell. If no voltage is applied, this reflected light is again rotated through 90° and passes through the polarising film on the top piece of glass. If a voltage is applied, the molecules re-orientate, the light is not rotated, and the display appears black. Such a liquid crystal display produces black numbers or letters on a silvery background in ambient light. In a second mode of operation, a diffuse light source is placed behind the liquid crystal cell and transmitted light is viewed from the front of the display.

Another type of display exploits the birefringence of nematic liquid crystals in a slightly different way. It uses a liquid crystal cell, essentially like that in Figure 12.28. The glass and polarising layers are still set up as in Figure 12.28. Now, however, the polymer layers align the liquid crystal molecules *perpendicular* to the glass surfaces, as illustrated by Figure 12.32a. In this case, the director is perpendicular to the glass surfaces.

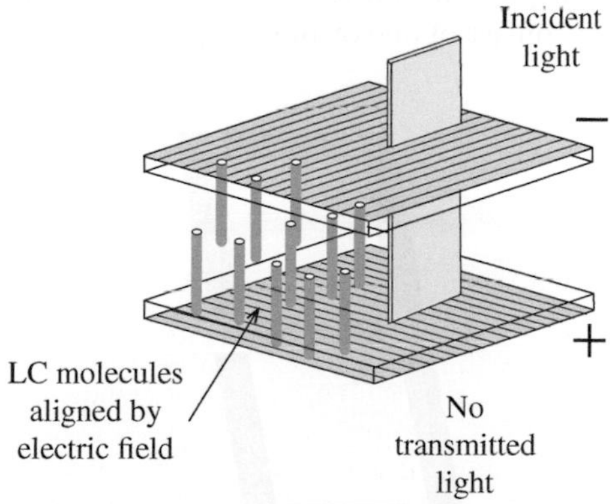

Figure 12.31 When a voltage is applied to the cell, the liquid crystal molecules change their orientation to become aligned with the electric field. Now, the director becomes perpendicular to the glass surfaces and parallel to the plane of polarisation of the light. Accordingly, the orientation of the plane of polarisation is unchanged as the light passes through the liquid crystal, and the light does not pass through the bottom polarising film.

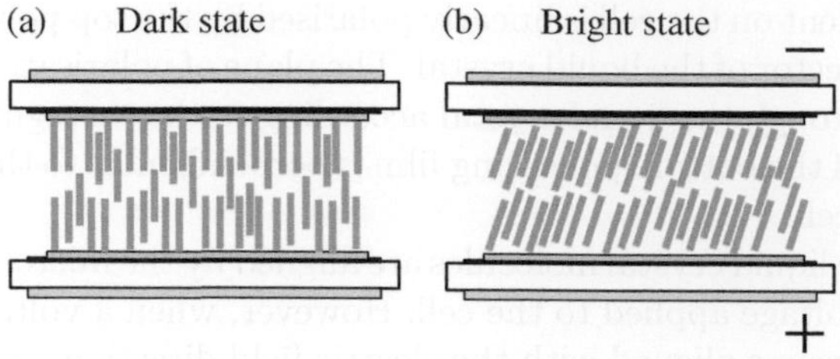

Figure 12.32 Here, the polymer layers of the liquid crystal cell align the molecules perpendicular to the glass surfaces. (a) With no voltage applied, the plane of polarisation of the light transmitted by the top polarising film is parallel to the director. Consequently, the orientation of the polarisation plane is unchanged as the light passes through the liquid crystal, which means that it is not transmitted by the bottom polarising film. The display appears black. (b) When a voltage is applied, the electric field causes the polarised molecules to tilt away from the normal to the glass surfaces. In that case, the director is no longer perpendicular to the glass surfaces, and the linearly polarised light now has components that are not parallel or perpendicular to the director. This causes light to become elliptically polarised. This elliptically polarised light will have a component that is parallel to the polarisation axis of the bottom polarising film and is transmitted by it. And, the display appears bright.

And it is parallel to the plane of polarisation of the light from the top polarising film. Consequently, the orientation of the polarisation plane is unchanged as it passes through the liquid crystal. And this means that it is not transmitted by the bottom polarising film. The display appears black.

However, the application of an electric field causes the polarised molecules to tilt away from the normal to the glass surfaces, as in Figure 12.32b. In that case, the director is no longer perpendicular to the glass surfaces, and the linearly polarised light now has components that are not in line or perpendicular to the director. This results in phase retardation of the light as it propagates through the liquid crystal. The resultant elliptically polarised light will have a component that is parallel to the polarisation axis of the bottom polarising film and is transmitted by it. And the display appears bright. The amount of molecular tilt depends on the strength of the electric field. This means that the phase retardation and hence the brightness of the display also depends on the strength of the applied field. So, this type of display can be used to produce intensities between its brightest state and its dark state.

One of the simplest formats is to display the numbers 0–9 in the form of seven bar segments as illustrated by Figure 12.33. Each segment may consist of one or more liquid crystal cells. By appropriate switching of

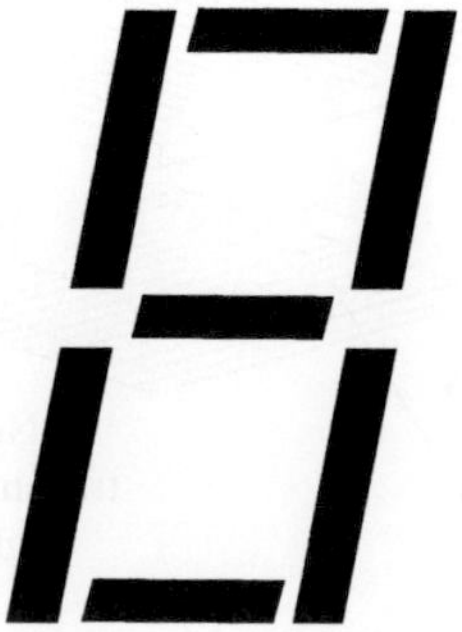

Figure 12.33 A display format to display the numbers 0–9 in the form of seven bar segments. Each segment consists of one or more liquid crystal cells. By appropriate switching of the applied voltages to the cells of the segments, the various numbers can be displayed.

the applied voltages to the cells in the segments, the various numbers can be displayed. Essentially, no current flows between the two ITO electrodes and so the display consumes extremely little power. Moreover, the required operational voltage, $\sim$1.5 V, is readily obtainable from a battery. A flat screen for a laptop computer, for example, requires a matrix of a large number of liquid crystal cells. Each cell corresponds to a pixel on the screen. Colour is achieved by putting coloured filters in front of each cell. The colours chosen are red, green, and blue from which any other colour can be made.

12.6 Liquid crystals in nature

Liquid crystals are ubiquitous in our technological world. However, before finishing our discussion of liquid crystals, it is important to emphasise that they are widespread in the natural world too. Many examples of liquid crystals appear in plants and animals, and many of these exhibit the chiral nematic, i.e. cholesteric phases. Indeed the helical organisation that we saw in chiral liquid crystals is involved in most of the molecules that are essential to life. This is notably the case for DNA, cell wall cellulose in plants and fruits, collagen in bones, cornea or fish scales and chitin in the exoskeletons of insects and crustaceans. The helical organisation of biological molecules endows living matter with a variety of crucial physical properties. It enables the optimisation of DNA packing in the nucleus of cells and strengthens the shells of arthropods, our bones, and fish scales. It is also the reason for the iridescent colours of certain fruits and beetles. For example, biological liquid crystals give an intense electric blue colour to the fruits of *Pollia condensata*, a plant from Africa. Their iridescent colour results from the cholesteric arrangement of liquid crystals in the cellulose strands of the skin of the fruit, and thus not from pigments that selectively absorb light, as is frequently the case for colours in the natural world. We have seen how the colour reflected from a cholesteric liquid crystal depends on the twist of the helix and the angle of observation. These properties account, for example, for the iridescent and shimmering colours of the *Chrysina gloriosa* scarab beetle whose cuticle of cholesteric chitin displays bright green and silver stripes.

The other main class of liquid crystals are the lyotropic liquid crystals, which are formed on the dissolution of the liquid crystals in a solvent. For example, the slime that forms in a soap dish is a liquid crystal phase formed by the dissolution of soap in water. Of particular importance is the occurrence of lyotropic liquid crystal phases in biological membranes. In particular, biological membranes and cell membranes are a form of liquid crystal. Accordingly, life itself critically depends on lyotropic liquid crystal phases. Many other biological structures exhibit lyotropic liquid-crystal behaviour. For example, the material that is extruded by a spider to make silk is a lyotropic phase. The molecular order in the silk is crucial to its renowned strength.

Problems 12

12.1 Blue light of wavelength 465 nm is Bragg reflected in first order ($m = 1$) from a certain chiral liquid crystal at a glancing angle of 19.4°. At what glancing angle would red light of wavelength 659 nm be reflected? If the temperature of the liquid crystal changed so that the red light was observed at the glancing angle of 19.4°, would the liquid crystal be hotter or colder than before?

12.2 Two polarisers initially have their polarisation directions parallel. Through what angle must one of the polarisers be rotated so that the intensity of the transmitted light is reduced to a third of the originally transmitted intensity?

12.3 The figure shows an arrangement in which a polariser is inserted between a pair of crossed polarisers such that its polarisation direction is at an angle of 45° with respect to the polarisation directions of both crossed polarisers. Explain the observation that some light is transmitted by this arrangement. What fraction of incident unpolarised light intensity is transmitted? The red arrows in the figure indicate the polarisation directions of the polarisers.

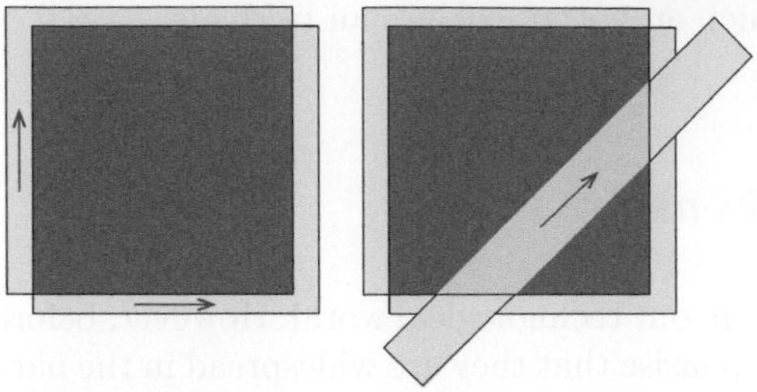

12.4 A polariser is situated between a pair of crossed polarisers. This middle polariser is rotated at angular frequency ω about the direction of an incident light beam. Show that the intensity I of the light emerging from this arrangement of polarisers is modulated at four times the rotational frequency ω according to $I = I_1/8(1 - \cos 4\omega t)$, where I_1 is the intensity of the light emerging from the first polariser.

12.5 Show that a linear wave may be considered as the combination of a right-circularly polarised wave and a left-circularly polarised wave of the same amplitude.

12.6 Circularly polarised light of intensity I_0 is incident on a linear polariser. Show that the intensity of the light transmitted by the polariser is $I_0/2$.

12.7 (a) Show that the minimum thickness for a quarter-wave plate is

$$d = \frac{\lambda_0}{4(n_1 - n_2)},$$

where λ_0 is the wavelength in vacuum and n_1 and n_2 are the indices of refraction for the two perpendicular components of linearly polarised light passing through it.

(b) The two refractive indices n_1 and n_2 for the nematic liquid crystal pentylcyanobiphenyl are 1.6173 and 1.4924, respectively at the wavelength of 589.3 nm. Calculate the thickness for a quarter-wave plate of the liquid crystal at this wavelength.

(c) With the aid of a sketch, show what orientation of the quarter–wave plate is required to convert linearly polarised light into circularly polarised light.

12.8 A liquid-crystal computer screen of dimensions 28 cm by 20 cm is viewed from a distance of 40 cm. Assuming that the size of each pixel matches the diffraction limit of the human eye, calculate the total number of pixels on the screen. Take the diameter of the pupil of an eye to be 8 mm.

Solutions to problems

Problems 1

1.1 (a) Assuming the oil layer is 1 molecule thick, find that the size of the oil molecules is ~ 1 nm. (b) The depth of tread on a car tyre that is used before the tyre needs changing is ~ 15 mm. The radius of a tyre is ~ 200 mm and let its width be w.

This gives $d \sim 6 \times 10^{-10}$ m ~ 0.6 nm.

1.2 $(1/2)Mv^2 = V_{acc}Ze$. As $t = L/v$, $t = L[M/(2V_{acc}Ze)]^{1/2}$. This gives $t = 5.6 \times 10^{-5}$ seconds.

1.3 In the electrolysis process, 1 mol of water is converted into 1 mol of H_2 and 0.5 mol of O_2. 5 L of water has a mass of 5 kg. Therefore, number of moles of water = 278 mol. Therefore, 278 mol of H_2 and 139 mol of O_2 are produced.

1.4 1 mol of potassium contains N_A atoms and weighs 0.039 kg, and 1.0 m^3 of potassium has a mass of 860 kg. Therefore, 1.0 m^3 of potassium contains $(860/0.039) \times 6.0 \times 10^{23}$ atoms. If these have radius R, find $R \sim 0.22$ nm.

1.5 Assuming a molecular weight of 30 u, 1 mol of air weighs 0.030 kg and occupies 22.4 L. Therefore 1 m^3 of air has mass $0.030/(22.4 \times 10^{-3}) = 1.3$ kg.

1.6 The surface area of the earth $= 5 \times 10^{14}$ m^2. $F = Mg$, and $P = F/A$, giving $M = 5 \times 10^{18}$ kg.

1.7 For hydrogen, we have $r_n = n^2 a_0$. Therefore, $r_{150} = 150^2\, a_0 = 1.2$ μm.

As the difference in n between $n = 150$ and $n = 151$ is small compared to n, we can write, $\mathrm{d}E_n = 2 \times 13.6(1/n^3)\,\mathrm{d}n$, giving, $\Delta E = 2 \times 13.6(1/n^3)\Delta n$, with $\Delta n = 1$. Therefore, $\Delta E = 8.1 \times 10^{-6}$ eV.

1.8 (a) For hydrogen, we have $E_n = -13.6(1/n^2)$ eV. Therefore, it takes 10.2 eV to remove the electron. (b) Li^{++} is hydrogenic, but with one electron orbiting about a nucleus with charge $Z = 3$. Therefore,

$$E_n = -13.6 \times Z^2 \frac{1}{n^2} = -13.6 \times 3^2 \frac{1}{1^2} = -122\,\text{eV}.$$

Hence, the energy required to ionise the Li^{++} ion is 122 eV.

1.9 He^{+} is hydrogenic, but with one electron orbiting about a nucleus with charge $Z = 2$. Therefore,

$$E_n = -13.6 \times Z^2 \frac{1}{n^2} = -54.4\,\text{eV}.$$

Therefore, the total energy required is $24.6 + 54.4 = 79$ eV.

1.10 (a) The Bohr radius for muonic hydrogen is $a_0/207 = (0.53 \times 10^{-10})/207 = 2.6 \times 10^{-13}$ m.

(b) For muonic hydrogen, $E_n = -[(m_\mu e^4)/(4\pi\varepsilon_0)^2 2\hbar^2](1/1^2) = 207 \times 13.6 = -2815$ eV.

(c) $E_1 - E_2 = \dfrac{m_\mu e^4}{(4\pi\varepsilon_0)^2 2\hbar^2}\left(\dfrac{1}{2^2} - \dfrac{1}{1^2}\right) = 2815 \times -0.75 = -2111$ eV.

Hence, the energy difference between the two levels is 2.1 keV.

Physics of Matter, First Edition. George C. King.

1.11 $E_1 = -[(m_e/2)e^4/(4\pi\varepsilon_0)^2 2\hbar^2](1/1^2)$, where $[m_e e^4/(4\pi\varepsilon_0)^2 2\hbar^2] = 13.6\,\text{eV}$.

Therefore, the ionisation energy of positronium is equal to $-E_1 = (13.6/2) = 6.8\,\text{eV}$.

Then, $E_1 - E_2 = -5.1\,\text{eV}$. Therefore, $\lambda = (hc)/E = 244\,\text{nm}$.

1.12 Theoretically, the spatial resolution is the order of the de Broglie wavelength of the electrons. The energy of the electrons is 2500 eV. Therefore, de Broglie wavelength $\lambda = 0.25 \times 10^{-10}\,\text{m}$. In practice, aberrations in the magnetic lenses used in a conventional electron microscope limit the spatial resolution to a value of an order of magnitude larger than the above result.

1.13 $E = p^2/2m$. Then using $\Delta p \Delta x \sim \hbar$, and taking $\Delta x = D$, where D is the diameter of the nucleus and $p = \Delta p$,

$$E = \frac{1}{2m}\left(\frac{\hbar}{D}\right)^2 = 8.4 \times 10^{-13}\,\text{J} \sim 5\,\text{MeV}.$$

1.14 (a) Energy of the particle $E = p^2/2m = V(r) = bx$, $p^2 = 2mbx$.

Taking $p \sim \Delta p$ and $\Delta x \sim x$, the relationship $\Delta p \Delta x \sim \hbar$ gives

$$\Delta p \sim (2mb\Delta x)^{1/2} \sim \left(\frac{2mb\hbar}{\Delta p}\right)^{1/2}, \quad \text{giving,} \quad \Delta p \sim (2mb\hbar)^{1/3}.$$

Therefore,

$$E = \frac{p^2}{2m} \sim \frac{(2mb\hbar)^{2/3}}{2m} \sim \frac{1}{2^{1/3}}\left(\frac{b^2\hbar^2}{m}\right)^{1/3}.$$

(b) $E = p^2/2m = V(r) = kx^2$, gives $p^2 = 2mkx^2$. Taking $p \sim \Delta p$ and $\Delta x \sim x$, the relationship $\Delta p \Delta x \sim \hbar$ gives $(\Delta p)^4 = 2mk\hbar^2$.

Therefore,

$$E = \frac{p^2}{2m} \sim \frac{(2mk)^{1/2}\hbar}{2m} \sim \frac{1}{\sqrt{2}}\left(\frac{k}{m}\right)^{1/2}\hbar \sim \frac{1}{\sqrt{2}}\hbar\omega.$$

Problems 2

2.1 $\varepsilon = \dfrac{2L}{nN_A}$. Taking $n = 10$, $\varepsilon = 12\,\text{meV}$.

2.2 (a) Electrostatic force $F_C = (1/4\pi\varepsilon_0) \cdot e^2/r^2$. Gravitational force $F_G = G(m_e m_e/r^2)$.

$$\frac{F_C}{F_G} = \frac{1}{4\pi\varepsilon_0 G}\left(\frac{e}{m_e}\right)^2 = 4.2 \times 10^{42}.$$

(b) Energy due to gravitational attraction of two helium atoms at separation $2.9 \times 10^{-10}\,\text{m}$ is $G(M_{He}^2/r_0^2) = 3.5 \times 10^{-44}\,\text{J} = 2.2 \times 10^{-25}\,\text{eV}$, which is a factor of 4×10^{21} smaller than the binding energy of the van der Walls bond.

2.3 (a) 1 kg of argon has a volume $1/\rho\,\mathrm{m}^3$. Number N of atoms in 1 kg of argon is $\sim 1/\rho a_0^3$. Using $\varepsilon = 2L_V/Nn$ with $n = 10$, $\varepsilon = \left(2L_V\rho a_0^3\right)/n$. Substituting $\gamma = n\varepsilon/4a_0^2$, we obtain $a_0 = 2\gamma/\rho L_V = 1.2 \times 10^{-10}$ m, and (b) $\varepsilon = 7.5 \times 10^{-23}\,\mathrm{J} \sim 5 \times 10^{-4}\,\mathrm{eV}$.

2.4 $\mathrm{d}V/\mathrm{d}r = 0$ gives $(12/a)(a/r_{\min})^{13} = (6/a)(a/r_{\min})^7$, and $r_{\min} = \sqrt[6]{2}a$.
Substituting for $r_{\min} = \sqrt[6]{2}a$, gives $V(r) = 4\varepsilon[(a^{12}/2^2a^{12}) - (a^6/2a^6)] = -\varepsilon$.
Hence, $-\varepsilon$ is the binding energy of the two atoms.

2.5 $\mathrm{d}V/\mathrm{d}r = 0$ $\exp[-2(r_0 - r_e)/a] = \exp[-(r_0 - r_e)/a]$, and $r_0 = r_e$.
Substituting for $r_0 = r_e$ gives $V(r) = -D_e$, which is the depth of the well.

2.6 $\mathrm{d}V/\mathrm{d}r = 0$ at equilibrium gives $B = \left(pAr_0^{q-p}\right)/q$. Substituting for B, $\varepsilon = [(q-p)A]/\left(qr_0^p\right)$.
Therefore, $A = \left(q\varepsilon r_0^p\right)/(q-p)$. Using this result gives $B = \left(p\varepsilon r_0^q\right)/(q-p)$. Substituting for A and B gives

$$V = \frac{pq\varepsilon}{q-p}\left[-\frac{1}{p}\left(\frac{r_0}{r}\right)^p + \frac{1}{q}\left(\frac{r_0}{r}\right)^q\right].$$

2.7 The restoring force $F = -\mathrm{d}V/\mathrm{d}r$ is

$$\frac{pq\varepsilon}{(q-p)r_0}\left[-\left(\frac{r_0}{r}\right)^{p+1} + \left(\frac{r_0}{r}\right)^{q+1}\right],$$

which at $r = r_0 + \delta$ is

$$\frac{pq\varepsilon}{(q-p)r_0}\left[-\frac{r_0^{p+1}}{(r_0+\delta)^{p+1}} + \frac{r_0^{q+1}}{(r_0+\delta)^{q+1}}\right].$$

Using binomial expansion,

$$\frac{1}{(r_0+\delta)^{p+1}} = \frac{1}{r_0^{p+1}(1+\delta/r_0)^{p+1}} \cong \frac{1}{r_0^{p+1}}[1 - (p+1)\delta/r_0 + \cdots].$$

Then to first order in δ/r_0,

$$F = \frac{pq\varepsilon}{(q-p)r_0}\left\{-\left[1-(p+1)\frac{\delta}{r_0}\right] + \left[1-(q+1)\frac{\delta}{r_0}\right]\right\} = -\frac{pq\varepsilon\delta}{r_0^2}.$$

The equation of motion of each hydrogen atom is

$$m\frac{\mathrm{d}^2}{\mathrm{d}t^2}\left(\frac{r_0+\delta}{2}\right) = -\frac{pq\varepsilon\delta}{r_0^2}.$$

Therefore, $(m/2)\left(\mathrm{d}^2\delta/\mathrm{d}t^2\right) = -(pq\varepsilon\delta)/r_0^2$, which gives SHM with frequency $\nu = 1/2\pi\sqrt{(2pq\varepsilon)/\left(r_0^2 m\right)}$.
Therefore, $pq = \left(2\pi^2\nu^2 r_0^2 m\right)/\varepsilon = 4.4$.

2.8 The electric field at r due to the charge $+2q$ is $2q/(4\pi\varepsilon_0 r^2)$. The electric field at r due to each of the $-q$ charges is $q/[4\pi\varepsilon_0(r^2 + d^2)]$. The two charges have field components parallel to the quadrupole which cancel. The resultant field of these two charges, which is perpendicular to the quadrupole, is

$$\frac{2q}{4\pi\varepsilon_0\left(r^2+d^2\right)} \times \frac{r}{\left(r^2+d^2\right)^{1/2}} = \frac{2q}{4\pi\varepsilon_0\left(r^2+d^2\right)^{3/2}}.$$

Therefore, the resultant electric field at r is

$$\frac{2q}{4\pi\varepsilon_0}\left[\frac{1}{r^2} - \frac{r}{\left(r^2+d^2\right)^{3/2}}\right] = \frac{2q}{4\pi\varepsilon_0 r^2}\left[1 - \left(1 + \frac{d^2}{r^2}\right)^{-3/2}\right].$$

For $d \ll r$, $[1+(d^2/r^2)]^{-3/2} \approx 1-[(3d^2)/2r^2]$. Hence, $E(r) = (3qd^2)/(4\pi\varepsilon_0 r^4)$.

This means that the electrical potential of a quadrupole varies as $1/r^5$. We recall that it varies as $1/r$ for a point charge and as $1/r^4$ for an electric dipole.

2.9 (a) $F(r) = \mathrm{d}V/\mathrm{d}r = -(1/\rho)Ae^{r/\rho} + [(\alpha e^2)/(4\pi\varepsilon_0 r^2)]$. At equilibrium $r = a_0$, $F(r) = 0$.

Therefore, $(1/\rho)Ae^{-a_0/\rho} = (\alpha e^2)/(4\pi\varepsilon_0 a_0^2)$, and $A = (\alpha\rho e^2 e^{a_0/\rho})/(4\pi\varepsilon_0 a_0^2)$.

Therefore,

$$V(r) = \frac{\alpha e^2}{4\pi\varepsilon_0}\left[\frac{\rho e^{(a_0-r)/\rho}}{a_0^2} - \frac{1}{r}\right].$$

Therefore, binding energy ε at $r = a_0$

$$= -\frac{\alpha e^2}{4\pi\varepsilon_0 a_0}\left[\frac{\rho}{a_0} - 1\right], \quad \text{and lattice energy} \quad = -N_A\varepsilon = \frac{N_A\alpha e^2}{4\pi\varepsilon_0 a_0}\left[1 - \frac{\rho}{a_0}\right].$$

(b) $(N_A\alpha e^2)/(4\pi\varepsilon_0 a_0) = 8.63 \times 10^5$. Therefore, $[1-(\rho/a_0)]8.63 \times 10^5 = 763 \times 10^3$, giving $\rho = 0.32 \times 10^{-10}$ m.

2.10 (a) No. At room temperature, $kT = 1/40$ eV $= 4 \times 10^{-21}$ J is much greater than the binding energy ε, and hence any molecules formed would soon be broken up by collisions with the atoms. (b) We have $(3/2)kT = D_e$. Therefore, $T = 2D_e/3k = 3.4 \times 10^4$ K.

2.11 (a) Energy of a photon is $E = h\nu = hc/\lambda$. Therefore, $\lambda = 283$ nm, which lies in the ultraviolet region of the electromagnetic spectrum.

Problems 3

3.1 $\Delta\lambda = 2[(2\ln 2kT)/m]^{1/2}(\lambda_0/c)$ gives $\Delta\lambda = 2.8 \times 10^{-4}$ nm.

3.2 (a) $\int_0^{20} P[t]\mathrm{d}t = A\int_0^{20}\mathrm{d}t = 1$, gives $A = 1/20 = 0.05$.

(c) Zero

(d) (i) $\int_0^5 P[t]\mathrm{d}t = [At]_0^5 = 0.25$. (ii) $\int_{15}^{20} P[t]\mathrm{d}t = [At]_{15}^{20} = 0.25$.

(e) $\bar{t} = \int_0^{20} tP[t]\mathrm{d}t = \frac{1}{20}\int_0^{20} t\mathrm{d}t = \frac{1}{20}\left[\frac{t^2}{2}\right]_0^{20} = 10$ min.

(f) Because $P[t]$ is a constant, equal to 1/20, see your sketch in part (b), all times are equally probable.

3.3 In order to escape to infinity, the initial kinetic energy of a molecule should be equal to the magnitude of the gravitational potential energy: $(1/2)mv_e^2 = G[(M_E m)/r]$, where v_e is the escape velocity. This gives, $v_e = 11.2$ km/s.

For helium, $v_{rms} = \sqrt{\overline{v^2}} = [(3kT)/m]^{1/2} = 1.35$ km/s. Although the rms velocity of helium is a factor of ~ 8 lower than the escape velocity, the high-energy tail of the velocity distribution gives a small but significant probability of helium atoms having enough energy to escape. Oxygen and nitrogen have much higher masses and therefore a much smaller probability for escape.

3.4 Height of Coulomb barrier, $V_{Coul} = (1/4\pi\varepsilon_0)[e^2/(r_D + r_T)]$, where $r_D = 1.5 \times 10^{-15}$ m; $r_T = 1.7 \times 10^{-15}$ m. This gives $V_{Coul} = 7.2 \times 10^{-14}$ J $= 0.45$ MeV. In a head-on collision, each deuterium ion must have $\sim 3.6 \times 10^{-14}$ J.

Thermal energy of ions is $(3/2)kT$ giving $T = 1.7 \times 10^9$ K. In practice, the temperature does not need to be as high as this because of two effects: *quantum mechanical tunnelling* where the nuclei tunnel through the Coulomb barrier, and the fact that there are more energetic ions in the high-energy tail of the Maxwell–Boltzmann distribution.

3.5 $A\int_0^\infty e^{-mgz/kT}dz = 1$ gives $A = mg/kT$.

Mean height $\bar{z} = (mg/kT)\int_0^\infty z\,e^{-mgz/kT}dz = kT/mg$, and mean potential energy $mg\bar{z} = kT = 1/40$ eV.

(b) $(3/2)kT = mgz$, giving $z = (3/2)(kT/mg) = 12.7$ km. (c) Zero.

3.6 The thermal energy of the molecules.

3.7 (a) Effective mass $m' = (\rho - \rho_0)V = [(\rho - \rho_0)/\rho]m$. For mass m' rotating with angular frequency ω, at a distance r from the axis, force $F(r) = m'r\omega^2$.

(b) Potential energy $V(r) = -\int m'r\omega^2 dr$, giving $V(r) = -[(m'r\omega^2)/2] + \text{constant}$.

$$V(r = 0) = 0, \text{gives } V(r) = -\frac{m'r\omega^2}{2}.$$

(c) Boltzmann gives probability distribution $P[r] \propto -\exp[-(m'r^2\omega^2)/2kT]$. Therefore, particle density $n(r) = A\exp[(m'r^2\omega^2)/2kT]$.

(d) The ratio of concentrations at r and $r + dr$ is given by

$$\frac{n(r + dr)}{n(r)} = \frac{A\exp\left[m'(r + dr)^2\omega^2/2kT\right]}{A\exp[m'r^2\omega^2/2kT]} = \frac{1000}{1}.$$

Neglecting terms in $(dr)^2$, $\exp[m'2rdr\omega^2/2kT] = 1000$. This gives $dr = 0.95$ mm.

3.8 $(d/dc)P[c] = Ae^{-mc^2/2kT}(2c - [(2mc)/2kT]c^2) = 0$. Therefore, the most probable speed $c_m = \sqrt{(2kT)/m} = 276$ m/s.

3.9 (a) $\bar{c} = \dfrac{2}{\sqrt{\pi}}\left(\dfrac{2kT}{m}\right)^{1/2} = 599$ m/s.

(b) $c_{rms} = \left(\dfrac{3kT}{m}\right)^{1/2} = 517$ m/s.

(c) Maximum of graph should be at $c_m = [(2kT)/m]^{1/2} = 422$ m/s.

(d) (i) $P[c]dc = 4\pi[m/(2\pi kT)]^{3/2}c^2e^{-mc^2/2kT}dc$. At $T = 300$ K, $P[c] = 0.00143$ s/m at $c = 600$ m/s. Then, $P[c]\,dc = 0.00143 \times 2 = 0.0029$ or 0.29%. (ii) By plotting a graph on squared graph paper and counting squares, find that the fraction of molecules with speed between 600 and 900 m/s is 0.25 or 25%.

3.10 (a) Kinetic energy $K = (1/2)mc^2$, giving $dK = mcdc$.

Therefore, $P[K] = P[c](dc/dK) = P[c](1/mc)$. Substituting for $P[c]$ gives $P[K] = (4\pi/mc)\,[m/(2\pi kT)]^{3/2}c^2e^{-mc^2/2kT}$. Then substituting $c = (2K/m)^{1/2}$ gives $P[K]dK = [4/(\pi k^3T^3)]^{1/2}K^{1/2}e^{-K/kT}dK$.

(b) $(d/dK)P[K] = 0$ gives the most probable kinetic energy $= (1/2)kT$.

(c) Mean kinetic energy $\overline{K} = \displaystyle\int_0^\infty K\,P[K]dK = \left[4/(\pi k^3T^3)\right]^{1/2}\int_0^\infty K^{3/2}e^{-K/kT}dK$.

Substitution $K = x^2$ gives $\overline{K} = \left[4/(\pi k^3T^3)\right]^{1/2}\displaystyle\int_0^\infty 2x^4e^{-x^2/kT}dx$. Using the standard integral, obtain $\overline{K} = (3/2)kT$.

3.11 (a) $(d/dc)P[c] = 0$ gives $(3c^2 - c^4[2m/(2kT)]) = 0$, and $c = \sqrt{(3kT)/m}$.

(b) $K = (1/2)mc^2$, with $dK = mc\,dc$. Substituting for c and dc gives $P[K]dK = A(2K/m^2)e^{-K/kT}dK$. $(d/dK)P[K] = 0$ gives $K_m = kT$.

(c) $A\int_0^\infty c^3 e^{-mc^2/2kT}dc = 1$. Using given integral, $A = 2[m/(2kT)]^2$.

$$\overline{c^2} = 2\left(\frac{m}{2kT}\right)^2 \int_0^\infty c^5 e^{-mc^2/2kT}dc.$$

Obtain $\overline{c^2} = (4kT)/m$, and $K = 2kT$.

3.12 (a) Fraction of molecules with energy greater than ε is equal to $(1/kT)^2 \int_\varepsilon^\infty K e^{-K/kT}dK$. Using a given integral, this is equal to $[1 + (\varepsilon/kT)]e^{-\varepsilon/kT}$.

(b) Fraction F_1 at temperature $T_1 = [1 + (\varepsilon/kT_1)]e^{-\varepsilon/kT_1}$, where $T_1 = 293$ K. $kT_1 = 1/40$ eV $\ll \varepsilon = 0.7$ eV. Hence, $F_1 \approx (\varepsilon/kT_1)e^{-\varepsilon/kT_1}$. Similarly, $F_2 \approx (\varepsilon/kT_2)e^{-\varepsilon/kT_2}$, where $T_1 = 303$ K. Reaction rate $\propto F$. Therefore, ratio of reaction rates $= F_2/F_1 = T_1/T_2 e^{\varepsilon\Delta T/kT_1T_2}$. $\Delta T \ll T_1T_2$. Therefore, ratio of reaction rates ≈ 1.9.

3.13 The total energy of a pendulum is the sum of its kinetic and potential energies $E = (1/2)mv^2 + (1/2)(mg/l)x^2$, where v is its velocity and x is its distance from its equilibrium position. It follows that $(1/2)(mg/l)\overline{x^2}. = (1/2)kT$. Therefore, $\sqrt{\overline{x^2}} = \sqrt{(kTl)/mg} = 2.2 \times 10^{-11}$ m. This is an exceedingly small amount of jiggle considering an atom has a diameter of $\sim 1 \times 10^{-10}$ m. However, such thermal motions are relevant in the extremely sensitive apparatus used at the *Laser Interferometer Gravitational Wave Observatory* for the detection of gravitational waves.

3.14 We have $n_1 = n_0 e^{-\varepsilon/kT}$. In the limit $T \to 0$, $n_1 = 0$, and all the particles are in the ground state. In the limit $T \to \infty$. $n_1 = n_0$, and the populations in the two energy levels are the same.

3.15 (a) Rotational energy $\varepsilon_{rot} \sim \hbar^2/I \sim [\hbar^2(m_1 + m_2)/m_1m_2d^2] = 2\hbar^2/(M_H d^2) = 2.45 \times 10^{-21}$ J ~ 15 meV. Rotational excitation will occur when $kT \sim \varepsilon_{rot}$. Therefore, $T \sim \varepsilon_{rot}/k \sim 180$ K.

(b) $\varepsilon_{vib} \sim h\nu = (hc)/\lambda = 8.3 \times 10^{-20}$ J ~ 0.5 eV. Vibrational excitation will occur when $kT \sim \varepsilon_{vib}$. Therefore, $T \sim \varepsilon_{vib}/k \sim 6000$ K.

Problems 4

4.1 (i) $n = P/kT = 2.44 \times 10^{25}$ mol/m^3. $\lambda = 1/(\sqrt{2}\, n\pi d^2) = 7.53 \times 10^{-8}$ m. $\lambda \propto 1/n$ and therefore to $\propto 1/P$. Therefore, (ii) $\lambda = 7.63 \times 10^{-2}$ m, (iii) $\lambda = 76.3$ m.

4.2 Roughly speaking, mean free path λ should be 10 times larger than the distance to the anode, i.e. $\lambda \sim 250$ mm. Probability of travelling $l = \lambda/10$ without a collision $= e^{-\lambda/10\lambda} \approx 90\,\%$. Hence, with a mean free path of 250 mm, 90% of the electrons would travel to the anode without suffering a collision.

$$n = \frac{1}{\sqrt{2}\,\lambda\pi d^2} = \frac{P}{kT}, \quad \text{from which} \quad P = 3 \times 10^{-2}\,\text{Pa} = 3 \times 10^{-4}\,\text{mbar}.$$

4.3 Consider an area A of the crystal surface. The number of molecules that would fit on that surface $\sim A/d^2$, where d is the molecular diameter. The rate at which the molecules strike area A is $(1/4)An\bar{c}$, and the rate at which molecules stick to the surface is $(1/4)An\varepsilon\bar{c}$. Therefore, the time t required to cover area A with molecules is $\sim (A/d^2)/[(An\varepsilon\bar{c})/4] \sim [4/(n\varepsilon\bar{c}d^2)]$. $n = 2.6 \times 10^{13}$ molecules/m^3. $\bar{c} = (8kT/\pi m)^{1/2} = 440$ m/s.

Therefore, $t \sim 5700$ seconds ~ 100 minutes. At atmospheric pressure, a monolayer of molecules would build up 10^{12} times faster and this example demonstrates the need for ultra-high vacuum in studies of the surface properties of pure materials.

4.4 The probability $P[l]\mathrm{d}l$ of a free path being within the range l to $l+\mathrm{d}l$ is $P[l]\mathrm{d}l = \dfrac{\mathrm{e}^{-l/\lambda}}{\lambda}\mathrm{d}l$. The probability function $P[l]$ has a maximum at $l = 0$. Hence the most probable value of free path is zero.

4.5 Number of molecules absorbed in time $\mathrm{d}t$ is $N = \dfrac{1}{4}n\bar{c}A\varepsilon\mathrm{d}t$.

where A is the total area of the plate. Therefore, $\mathrm{d}n = \dfrac{N}{V} = \dfrac{n\bar{c}A\varepsilon\mathrm{d}t}{4V}$, where V is the volume of the chamber. Then, $n(t+\mathrm{d}t) = n(t) - \dfrac{n\bar{c}A\varepsilon\mathrm{d}t}{4V}$, or, $\dfrac{n(t+\mathrm{d}t)-n(t)}{\mathrm{d}t} = \dfrac{\mathrm{d}n}{\mathrm{d}t} = -\dfrac{n\bar{c}A\varepsilon}{4V}$. The solution is $n = n_0 e^{-\frac{\bar{c}A\varepsilon t}{4V}}$. It follows that $t = \dfrac{4V}{\bar{c}A\varepsilon}\ln\left(\dfrac{n_0}{n}\right)$. Taking $\bar{c} = \dfrac{2}{\sqrt{\pi}}\left(\dfrac{2kT}{m}\right)^{1/2} = 460\ \mathrm{m/s}$, giving $t = 0.1$ s.

4.6 $n = P/kT = 2.44 \times 10^{25}\ \mathrm{mol/m^3}$. $\lambda = 1/2\pi n d^2 = 5.3 \times 10^{-8}\ \mathrm{m}$. Mean speed of the molecules is $(2/\sqrt{\pi})(2kT/m)^{1/2} = 476\ \mathrm{m/s}$. (i) The diffusion coefficient $D = (1/3)\bar{c}\lambda = 1.15 \times 10^{-5}\ \mathrm{m^2/s}$. (ii) The viscosity $\eta = (1/3)nm\bar{c}\lambda = 1.3 \times 10^{-5}$ Pa·s. (iii) The molar specific heat of N_2 is $5R/2$. Hence, $\kappa = (1/3)n\bar{c}\lambda(C_V/N_A) = 9.7 \times 10^{-3}$ W/m·K.

4.7 Considering a cylindrical shell of width $\mathrm{d}r$ and radius r, with $r_w > r < r_c$, Hence, $(Q/2\pi l)\displaystyle\int_{r_w}^{r_c}(1/r)\mathrm{d}r = -\tau\int_{T_w}^{T_c}\mathrm{d}T$, giving $\kappa = 1.6 \times 10^{-2}\ \mathrm{W/m\cdot K}$.

4.8 $F/A = \eta(\mathrm{d}v/\mathrm{d}y)$. $F = \tau/r_b$; $A = 2\pi r_b l$; $v = r_a\Omega$; $\mathrm{d}v/\mathrm{d}y = \Omega r_a/(r_a - r_b)$.
Hence, $\eta = (\tau/r_b)(1/2\pi r_b l)[(r_a - r_b)/\Omega r_a]$, giving $\eta = 2.25 \times 10^{-5}\ \mathrm{Pa\cdot s}$.

4.9 $\dfrac{F}{A} = \eta\dfrac{\mathrm{d}v}{\mathrm{d}r} = \eta\dfrac{\mathrm{d}(r\omega)}{\mathrm{d}r}$. $\dfrac{\mathrm{d}(r\omega)}{\mathrm{d}r} = \omega + r\dfrac{\mathrm{d}\omega}{\mathrm{d}r}$.
The first term on the right-hand side represents the angular motion the layer would have if there were no viscous slip, and hence can be omitted. The second term is responsible for the viscous stress. Hence, $\tau/(r \times 2\pi r l) = \eta r(\mathrm{d}\omega/\mathrm{d}r)$, which gives

$$\frac{\tau}{2\pi r^3 l}\int_{r_b}^{r_a}\frac{\mathrm{d}r}{r} = \eta\int_0^{\Omega}\mathrm{d}\omega, \quad \text{from which} \quad \eta = 2.22 \times 10^{-5}\ \text{Pa·s}.$$

4.10 Concentration gradient is n_v/h, and the number crossing any plane per second is $-(DAn_v)/h$, where A is the cross sectional area of the test tube. Therefore the mass crossing any plane per second is $-(DAn_v m)/h$. Mass of liquid that evaporates per second is $\rho(\mathrm{d}V/\mathrm{d}t) = \rho A(\mathrm{d}h/\mathrm{d}t)$. Therefore, $\rho A(\mathrm{d}h/\mathrm{d}t) = -(DAn_v m)/h$, which upon integration gives $h^2 = [(2Dn_v m)/\rho]t$.

$$n = \frac{P}{kT} = 7.9 \times 10^{23}\ \text{molecules/m}^3. \quad t = \frac{h^2\rho}{2Dn_v m} = 4.4 \times 10^5\ \text{seconds} \approx 5\ \text{days}.$$

Problems 5

5.1 At the Boyle temperature T_B, $[P + (a/V^2)](V - b) = RT_B$. In addition, for an ideal gas, $PV = RT_B \Rightarrow P = RT_B/V$. Substituting for P gives $[(RT_B/V) + (a/V^2)](V - b) = RT_B$.
Therefore, $RT_B - (ab/V^2) + (a/V) - (RbT_B/V) = RT_B$. giving $T_B = (a/Rb)[1 - (a/V)]$.
Assuming $b \ll V$, $T_B = a/Rb = 1011$ K.

5.2 $T_C = \dfrac{8a}{27bR} = \dfrac{8 \times 0.359}{27 \times 4.27 \times 10^{-5} \times 8.31} = 300$ K.

5.3 (i) $P = nRT/V = 7.48 \times 10^6$ Pa. (ii) $[P + (n^2a/V^2)](V - nb) = nRT$ gives $P = 7.55 \times 10^6$ Pa. (iii) $PV/nRT = 1 + (B/V) + (C/V^2)$ gives $P = 7.67 \times 10^6$ Pa.

5.4 $b = (2/3)N_A\pi d^3$ gives $d = 2.9 \times 10^{-10}$ m.

5.5 For a gas of rigid spheres, there is no attractive force and van der Waals equation becomes $P(V - b) = RT$, where $b = (2/3)N_A\pi d^3$. Then $P = RT/(V - b) = (RT/V)[1 + (b/V)]$ for $b \ll V$. $PV/RT = 1 + (b/V) = [1 + (2/3V)N_A\pi d^3]$, and hence, $B = (2/3)N_A\pi d^3$.

5.6 Making the substitutions obtain $(P_r P_C + [a/(V_r V_C)^2])(V_r V_C - b) = RT_r T_C$.

Substituting $V_C = 3b$, $P_C = a/27b^2$ and $T_C = 8a/27bR$, obtain $[P_r + (3/V_r^2)](3V_r - 1) = 8T_r$.

Note that the van der Waals constants a and b have conveniently disappeared. Hence, this *reduced equation of state* should apply to all substances. Thus, although the pressures and volumes may be different, two gases are said to be in *corresponding states* if the reduced pressure, volume, and temperature are the same.

5.7 The mathematical definition of the critical point is $(\partial P/\partial V)_{T_C} = (\partial^2 P/\partial V^2)_{T_C} = 0$. $(\partial P/\partial V)_{T_C} = 0$ gives $1/(V-b) = (a/RT)/V^2$. Differentiating this equation with respect to V and equating to zero gives $1/(V-b)^2 = (2a/RT)/V^3$. Substituting $1/RT = V^3/[2a(V-b)^2]$ gives $V_C = 2b$. Substituting $V_C = 2b$ gives $T_C = a/4bR$. Then, substituting for $V_C = 2b$ and $T_C = a/4bR$ in Dieterici's equation gives $P_C = a/4bR = a/4e^2b^2$.

5.8 $C = b^2$ gives $b = 18.7 \times 10^{-6}\ \mathrm{m^3/mol}$.

$B = b - (a/RT)$ gives $a = RT(b - B) = 0.114 \times 10^{-1}\ \mathrm{m^6 \cdot Pa/mol^2}$.

5.9 $P_{vdw} = \dfrac{RT}{V}\left[1 + \dfrac{1}{V}\left(b - \dfrac{a}{RT}\right) + \dfrac{b^2}{V^2}\right]$. $P_i = \dfrac{RT}{V}$.

$$\Delta P = P_{vdw} - P_i = \frac{RT}{V}\left[\frac{1}{V}\left(b - \frac{a}{RT}\right) + \frac{b^2}{V^2}\right].$$

$$\frac{\Delta P}{P_i} = \frac{P_{vdw} - P_i}{P_i} = \left[\frac{1}{V}\left(b - \frac{a}{RT}\right) + \frac{b^2}{V^2}\right].$$

(i) $\Delta P/P_i = -0.0026$, i.e. a 0.26% difference. (ii) $\Delta P/P_i = -0.22$, i.e. a 22% difference.

5.10 $C_V = (3/2)R$. $C_P - C_V = R[1 + (2a/RTV)] = 1.34R$. Hence, $C_P = 2.84R$.

Problems 6

6.1 Number of molecules $= 2.99 \times 10^{23}$. Nitrogen has five degrees of freedom. Therefore, internal energy of the gas at (i) 20 °C is 3.02 kJ, and (ii) at 80 °C is 3.64 kJ.

6.2 The temperature of the gas does not change during the free expansion and so, using $PV =$ constant, final pressure $= P_0/4$.

6.3 The temperature θ_p in °C on the platinum resistance thermometer scale is given by $\theta_p = [(R_\theta - R_0)/(R_{100} - R_0)] \times 100$, where R_θ, R_0, R_{100} are the respective resistances at the temperatures concerned, at 0 and 100 °C. Therefore

$$\theta_p = \frac{R_0(1 + 200\alpha + 200^2\beta) - R_0}{R_0(1 + 100\alpha + 100^2\beta) - R_0} \times 100 = 197\ ^\circ\mathrm{C}.$$

6.4 First law: $Q = \Delta U + W$. Therefore, $W/Q = (Q - \Delta U)/Q = 1 - (\Delta U/Q)$.

Suppose we have n moles of helium and that the temperature increases from T_i to T_f.

$$\Delta U = n\frac{1}{2}f\,R(T_f - T_i), \text{and } Q = nC_P(T_f - T_i), \quad \text{with } f = 3 \text{ and } C_P = \frac{5}{2}.$$

Therefore,

$$\frac{W}{Q} = 1 - n\frac{3}{2}\,R(T_f - T_i)\frac{2}{5}\,\frac{1}{n(T_f - T_i)} = 1 - \frac{3}{5}, \quad \text{giving } W = \frac{2}{5}Q.$$

6.5 Suppose that the mass is displaced at a distance x. Then the volume of the upper vessel increases to $(V_0 + xA)$ and the reduced pressure P_U is given by $P_0 V_0^\gamma = P_U(V_0 + xA)^\gamma$. Hence, $P_U = P_0/[(1 + xA/V_0)^\gamma]$. As $xA \ll V_0$, $P_U = P_0(1 - \gamma xA/V_0)$.

Similarly, increased pressure in the lower vessel is $P_L = P_0(1 + \gamma xA/V_0)$.

From Newton's second law, $m(\mathrm{d}^2x/\mathrm{d}t^2) = A[P_0(1 - \gamma xA/V_0) - P_0(1 + \gamma xA/V_0)]$, giving $\mathrm{d}^2x/\mathrm{d}t^2 = -[(2\gamma P_0 A^2)/mV_0]x$. This motion is SHM with period $\tau = 2\pi\sqrt{mV_0/(2\gamma P_0 A^2)}$.

From the given data, $\gamma = (2\pi^2 m V_0)/(\tau^2 P_0 A^2) = 1.4$.

6.6 Amount of oxygen gas is 5 mol. (a) At constant volume, $Q_V = nC_V\Delta T$, where $C_V = (5/2)R$. Therefore, $Q_V = 8.31$ kJ. (b) At constant pressure, $Q_P = nC_P\Delta T$, where $C_P = [(5/2) + 1]R$. Therefore, $Q_P = 11.63$ kJ. (c) The internal energy depends on the temperature of the gas: $\Delta U = nC_V\Delta T = n(5/2)\Delta T$. Therefore, for both processes, $\Delta U = 8.31$ kJ. (d) From first law, $W = Q - \Delta U$. Therefore, $W = 3.32$ kJ.

6.7 (a), (i) $W = P\int_{V_1}^{V_2} P\mathrm{d}V = nRT\int_{V_1}^{V_2}\mathrm{d}V/V = nRT\ln(V_2/V_1) = 1.70$ kJ. (ii) As T constant, $\Delta U = 0$, and $Q = W = 1.7$ kJ. (b), (i) $T_1V_1^{\gamma-1} = T_2V_2^{\gamma-1}$, with $\gamma = 1.4$. Therefore, $T_2 = T_1(V_1/V_2)^{\gamma-1} = 224$ K. (ii) $\Delta U = nC_V\Delta T$. Therefore, $\Delta U = -1.48$ kJ. Hence, $W = 1.48$ k J.

6.8 The work done by the gas $W = (P_2 - P_1)(V_2 - V_1)$, i.e. the area enclosed by the solid lines. There is no change in the internal energy of the gas because it returns to its original state: $\Delta U = 0$. Therefore, $Q = W$.

6.9 During step $a \rightarrow b$, work done by gas $W_{ab} = P_1(V_2 - V_1) = 455$ J. During step $b \rightarrow c$, no change in V and hence no work done by gas. During step $c \rightarrow d$, work done by gas $W_{cd} = P_1(V_2 - V_1) = -152$ J. During step $d \rightarrow a$, no change in V and hence no work done by gas. Therefore, total work done by gas $=(455 - 152) = 303$ J.

$Q = \Delta U + W = W = 303$ J, as $\Delta U = 0$. At point a, $P_1 = 3$ atm and $V_1 = 1$ l, gives $T_a = PV/nR = 365$ K. Similarly, $T_b = 912$ K, $T_c = 304$ K and $T_d = 122$ K.

6.10 (i) $\Delta H = \Delta U + P\Delta V$. $P\Delta V = P(V_2 - V_1) = 3.03$ kJ.

(ii) $\Delta U = \Delta H - P\Delta V$. Therefore, increase in the internal energy of water $\Delta U = 3.8$ kJ.

6.11 (a) For an isothermal process, $PV = nRT$. Then, $P\mathrm{d}V + V\mathrm{d}P = nR\mathrm{d}T = 0$, giving $P = -\mathrm{d}P/(\mathrm{d}V/V)$. Substituting for B gives $P = B = v^2\rho$. Hence, $v = (P/\rho)^{1/2}$.

Recalling that $P = (N/V)m\overline{u_x^2}$, where N is the number of molecules in volume V, we have $P = (N_A/V_m)m\overline{u_x^2} = \rho\overline{u_x^2}$, where V_m is the molar volume. Hence, $v = \sqrt{\overline{u_x^2}}$.

(b) For an adiabatic process, PV^γ = constant giving $V^\gamma\mathrm{d}P + P\gamma V^{\gamma-1}\mathrm{d}V = 0$, and $\gamma P = -\mathrm{d}P/(\mathrm{d}V/\mathrm{d}V)$. Substituting for B gives $\gamma P = B = v^2\rho$. Hence,$v = \sqrt{\gamma P/\rho} = \sqrt{\gamma\overline{u_x^2}}$.

(c) Using $(1/2)m\overline{u_x^2} = (1/2)kT$, gives $\sqrt{\overline{u_x^2}} = \sqrt{kT/m} = 285$ m/s.

$v = \sqrt{\overline{u_x^2}}$ gives velocity of sound =285 m/s. $v = \sqrt{\gamma\overline{u_x^2}}$ gives velocity of sound =337 m/s. Results indicate propagation of sound in air is an adiabatic process.

6.12 At constant enthalpy, $\mu = \mathrm{d}T/\mathrm{d}P$. Hence,

$$\Delta T = \int_{P_1}^{P_2}(a - bP)\mathrm{d}P = \left[aP - \frac{b}{2}P^2\right]_{P_1}^{P_2}.$$

Therefore, $\Delta T = a(P_2 - P_1) - (b/2)(P_2^2 - P_1^{20}) = -4.9$ K. The air cools down by 4.9 K.

Problems 7

7.1 $\varepsilon = 40\,\%$, $W = 800$ J, and discarded heat $Q_C = 1200$ J.

7.2 (a) $\Delta S = C_V\int_{T_i}^{T_f}\mathrm{d}T/T = C_V\ln(T_f/T_i) = +3.6$ J/K. For constant pressure, $\Delta S = C_P\int_{T_i}^{T_f}\mathrm{d}T/T = C_P\ln(T_f/T_i) = +6.0$ J/K. (b) Now, $\Delta S = m \times C\int_{T_i}^{T_f}\mathrm{d}T/T = +111$ J/K.

7.3 The entropy does increase, even though $Q = 0$, because the process is dissipative and hence irreversible. At constant V, $\Delta S = nR\ln(T_f/T_i) = 1.88$ J/K. As $Q = 0$, first law gives $W = U_f - U_i = (5/2)R(T_f - T_i) = 582$ J.

7.4 The heat gained by the ice cube is 1.17×10^4 J, giving $\Delta S_{\text{ice cube}} = +(1.17 \times 10^4)/273 = +42.7$ J/K and $\Delta S_{\text{water}} = -(1.17 \times 10^4)/273 = -42.7$ J/K.

Overall, $\Delta S_{\text{overall}} = \Delta S_{\text{water}} + \Delta S_{\text{ice cube}} = 0$. This is a reversible process.

7.5 There are three stages in this process: (i) the warming of the ice cube from −10 to 0 °C, (ii) the melting of the ice and (iii) the warming of the ice-cube water from 0 to 15 °C. (i) $\Delta S_1 = \int \mathrm{d}Q/T = mc_{\text{ice}}\ln(T_f/T_i) = +2.90$ J/K. (ii) $\Delta S_2 = mL/T = +42.7$ J/K. (iii) $\Delta S_3 = mc_{\text{water}}\ln(T_f/T_i) = +7.84$ J/K. Therefore, total change of entropy of ice cube plus melted water is +53.43 J/K. Temperature of bucket water in bucket does not change significantly. In each of the above three stages, this water loses heat: (i) $Q_1 = -mc_{\text{ice}}(T_f - T_i) = -777$ J, (ii) $Q_2 = -mL = -1.165 \times 10^4$ J, (iii) $Q_1 = -mc_{\text{icewater}}(T_f - T_i) = -2.2 \times 10^3$ J. Therefore, total heat lost from water in bucket $= 1.463 \times 10^4$ J, and its change in entropy = − 50.80 J/K. Therefore, overall entropy change = + 2.63 J/K.

7.6 During melting, $\Delta S = Q/T = 61.0$ J/K. During heating of melted water, $\Delta S = m \times c\ln(T_f/T_i) = 14.8$ J/K.

$$\Delta S_{\text{reservoir}} = -\frac{Q}{T_{\text{reservoir}}} = -\frac{m \times c \times \Delta T}{293} = -14.3\ \text{J/K}.$$

Hence, change in entropy of ice cubes plus melted water =75.8, change in reservoir = − 14.3, and change in universe =61.5 J/K.

7.7 $\Delta S_{\text{water}} = mc\int_{T_1}^{T_2} \mathrm{d}T/T = mc\ln(T_2/T_1)$. Therefore,

$$\sum \Delta S_{\text{water}} = mC\left[\ln\left(\frac{T_2}{T_1}\right) + \ln\left(\frac{T_{23}}{T_2}\right) + \cdots\right] = 782\ \text{J/K}.$$

$\Delta S_{\text{bath}} = -Q/T = -mc(\Delta T/T)$. Therefore, $\sum \Delta S_{\text{bath}} = -770$ J/K.

Therefore, the total energy change of the water plus the six heat baths is +12 J/K.

7.8 Taking $T_H = 300$ K, $T_C = 280$ K, $\varepsilon = 1 - (T_C/T_H) = 0.067$, i.e. about 7%.

For, $W = 5$ kW, $\mathrm{d}Q_H/\mathrm{d}t = 5/\varepsilon = 75$ kW. Therefore, $\mathrm{d}Q_C/\mathrm{d}t = 75 - 5 = 70$ kW.

7.9

Step	Q (J)	W (J)	ΔU (J)
$a \to b$	1440	1440	0
$b \to c$	0	2078	−2078
$c \to d$	−863	−863	0
$d \to a$	0	−2078	+2078
Total	578	578	0

Hence, $Q_H = 1441$ J and total work W is 578 J, giving $\varepsilon = 0.40 = 40\,\%$. We can compare this result with $\varepsilon = (T_H - T_C)/T_H = 0.40 = 40\,\%$.

7.10 (a) $\varepsilon_1 = 1 - (Q_M/Q_H)$; $\varepsilon_2 = 1 - (Q_C/Q_M)$. Therefore, $Q_M = (1 - \varepsilon_1)Q_C$.

Therefore, $Q_C/Q_H = (1 - \varepsilon_2)(1 - \varepsilon_1)$. $Q_C = Q_H - (W_1 + W_2)$, or $Q_C/Q_H = 1 - [(W_1 + W_2)/Q_H] = 1 - \varepsilon_{\text{net}}$. which gives $\varepsilon = \varepsilon_1 + \varepsilon_2(1 - \varepsilon_1)$.

(b) $\varepsilon_1 = 1 - (T_M/T_H)$; $\varepsilon_2 = 1 - (T_C/T_M)$. Therefore,

$$\varepsilon = \left(1 - \frac{T_M}{T_H}\right) + \left(1 - \left[1 - \frac{T_M}{T_H}\right]\right)\left(1 - \frac{T_C}{T_M}\right),$$

which simplifies to $\varepsilon = 1 - (T_C/T_H)$.

7.11 Along the paths $(b \to c)$ and $(d \to a)$, there is no work done, and hence, $Q_H = nC_V(T_c - T_b)$, $Q_C = nC_V(T_d - T_a)$, where n is the number of moles of the working substance and C_V is its heat capacity. Total work W done by the working substance over the cycle is $W = Q_H - Q_C = nC_V(T_c - T_b) - nC_V(T_d - T_a)$.

Therefore, efficiency $\varepsilon = W/Q_H = [(T_c - T_b) - (T_d - T_a)]/(T_c - T_b)$. For the adiabatic compression and expansion: $T_b(V_1)^{\gamma-1} = T_a(V_1)^{\gamma-1}$, $T_c(V_1)^{\gamma-1} = T_d(V_1)^{\gamma-1}$. Substituting for T_b and T_c, $\varepsilon = (T_d r^{\gamma-1} - T_a r^{\gamma-1} - T_d + T_a)/(T_d r^{\gamma-1} - T_a r^{\gamma-1}) = 1 - (1/r^{\gamma-1})$. For $r = 8$, the maximum efficiency $\varepsilon = 0.46$, i.e. 46%. This is an overestimate because there are various heat losses from the engine. More typically, the efficiency ~35%.

7.12 Heat engine: $\varepsilon = (1 - Q_C/Q_H)$. In one cycle, $\Delta S_{universe} = -(Q_H/T_H) + (Q_C/T_C) \geq 0$.
Hence, $(Q_C/Q_H) \geq (T_C/T_H)$, so that $\varepsilon \leq (1 - T_C/T_H)$. For heat pump, $CoP = (1 - Q_C/Q_H)^{-1}$. In one cycle, $\Delta S_{universe} = +(Q_H/T_H) - (Q_C/T_C) \geq 0$. Hence, $Q_H/Q_C \geq T_H/T_C$. so that $CoP \leq (1 - T_C/T_H)^{-1}$.

7.13 (a) $n_2/n_1 = e^{-\varepsilon/kT}$, and $n_1 + n_2 = N$.
Therefore, $n_1 = N/(e^{-\varepsilon/kT} + 1)$, $n_2 = Ne^{-\varepsilon/kT}/(e^{-\varepsilon/kT} + 1) = N/(e^{\varepsilon/kT} + 1)$. (i) As $T \to 0$, $n_1 \to N$, $n_2 \to 0$. (ii) As $T \to \infty$, $n_1 \to N/2$ and $n_2 \to N/2$, $n_1 \to N$, i.e. at sufficiently high temperature, the populations become equal.

(b) $E = n_2\varepsilon = N\varepsilon/(e^{\varepsilon/kT} + 1)$. (i) As $T \to 0$, $E \to 0$. (ii) As $T \to \infty$, $E \to N\varepsilon/2$.

(c) $E = N\varepsilon/(e^{\varepsilon/kT} + 1)$. Letting $u = \varepsilon/kT$, gives, $du = -(\varepsilon/kT^2)dT$. Then, $E = N\varepsilon[1/(e^u + 1)]$, and $(dE/du)N\varepsilon [-e^{\varepsilon/kT}/(e^{\varepsilon/kT} + 1)^2]$. Therefore,

$$\frac{dE}{dT} = \frac{dE}{du}\frac{du}{dT} = \left(\frac{N\varepsilon^2}{kT^2}\right)\frac{e^{\varepsilon/kT}}{(e^{\varepsilon/kT} + 1)^2}.$$

At high T, $e^{\varepsilon/kT} \to 1$ and $dE/dT = N\varepsilon^2/2kT^2$.

7.14 $dU = TdS - PdV$. Therefore, $dS = (dU/T) + (P/T)dV$. For van der Waals gas, $dU = (3/2)RdT + (a/V^2)dV$. Therefore, $dS = (3/2)R(dT/T) + (1/T)[P + (a/V^2)]dV$. Then, using van der Waals equation of state, $dS = (3/2)R(dT/T) + [R/(V - b)]dV$.

7.15 Clausius-Clapeyron equation: $dP/dT = L/(T\Delta V)$. 1 mol of grey tin occupies 2.064×10^{-5} m^3, and 1 mol of white tin occupies 1.626×10^{-5} m^3. Therefore, $\Delta V = V_{white} - V_{grey} = -4.38 \times 10^{-6}$ m^3. For 1 mol, $dP/dT = -(2.2 \times 10^3)/(291 \times 4.38 \times 10^{-6})$ J/m$^3 \cdot$ K. Therefore, $dT = -2.9$ K. Therefore, transition temperature at 50 atm = 15.1 °C.

7.17 (i) There are five macrostates, (ii) with multiplicities, 1, 4, 6, 4, and 1, respectively. (iii) Using $S = k \ln w$, where w is the multiplicity, the entropies are 0, 1.39 k, 1.79 k, 1.39 k, and 0, respectively.

Problems 8

8.1 For CsBr, $r_1/r_2 = 0.167/0.196 = 0.85$, which is larger than the critical value of 0.723.
Therefore, we would expect CsBr to adopt the CsCl structure, which it does.

8.2 Each carbon atom has four bonds and each bond connects two atoms. Hence, the number of bonds per mole is $2N_A$. Therefore, the binding energy of the covalent bond is $[(715 \times 10^3)/2N_A]$ J = 3.7 eV.

8.3 (a) Each cube has 4Na and 4F ions, i.e. four NaF molecules. Each corner is shared by eight neighbouring cubes. Hence, each cube of volume a^3 contains on average two NaF molecules. Therefore, the molar volume occupied by N_A molecules is $V_m = 2N_A a^3$. (b) M_m for NaF = 42×10^{-3} kg. $\rho = M_m/V_m$, i.e., $N_A = M_m/(\rho \times 2a^3) = 6.1 \times 10^{23}$/mol.

8.4

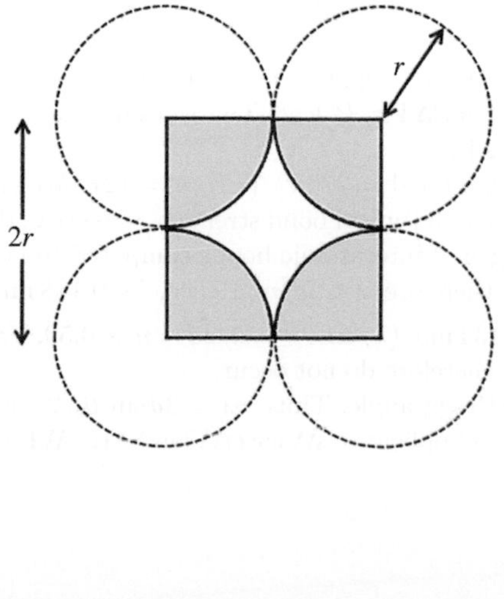

Total volume of spheres within the cube = $8 \times (1/8) \times (4/3)\pi r^3$. Therefore, fractional packing =$8 \times (1/8) \times (4/3)\pi r^3 \div 8r^3 = \pi/6 = 0.524$.

8.5 (i) @@

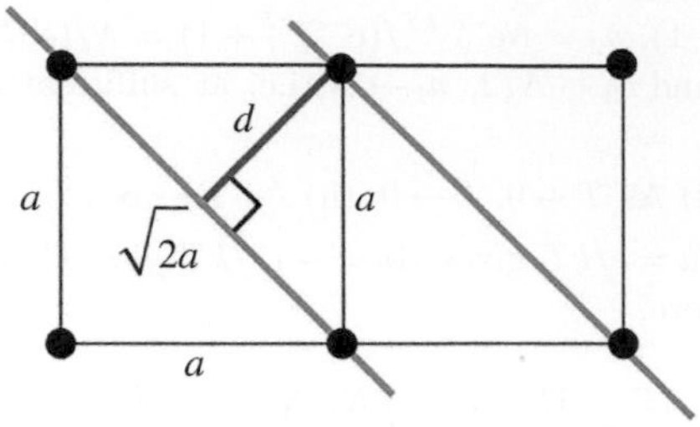

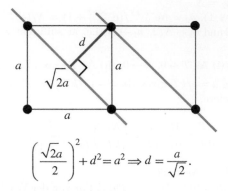

$$\left(\frac{\sqrt{2}a}{2}\right)^2 + d^2 = a^2 \Rightarrow d = \frac{a}{\sqrt{2}}.$$

(ii) Looking in the plane of the body diagonal at the edges of the planes:

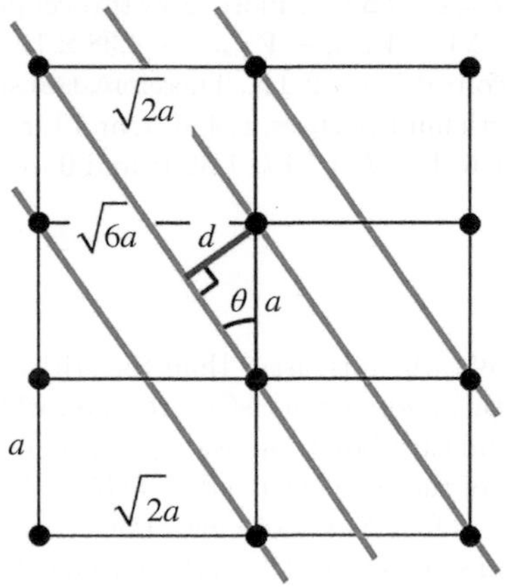

$$\sin\theta = \frac{\sqrt{2}a}{\sqrt{6}a} = \frac{d}{a} \Rightarrow d = \frac{a}{\sqrt{3}}.$$

8.6 As it has a face-centred cubic structure, copper has $n = 12$ nearest neighbours.

The work done by the pressure is $P\Delta V = P^2 V/B$. This work increases the potential energy of the inter-atomic bonds. Total number of bonds in 1 mol = $(1/2)N_A n$.

Therefore, increase in energy per bond =$(2P^2V)/(BN_A n) = (2P^2M)/(BN_A n\rho) = 1.7 \times 10^{-27}$ J, or 1.1×10^{-8} eV. This increase is a tiny fraction of a typical bond strength $\sim$ few eV, despite the very large applied pressure. The result emphasises the strength of interatomic bonds compared to laboratory scale forces.

8.7 (a) $2d\sin\theta = m\lambda$. $\sin\theta < 1$ and, therefore, $\lambda < 2d/m$. Hence, $\lambda < 0.188$ nm.

(b) (i) $d = (1 \times \lambda)/(2\sin\theta) = 0.134$ nm. (ii) $\sin\theta = m\lambda/2d = m \times 0.59$. Values of $m = 2$ or greater give values of $\sin\theta$ greater than unity and therefore do not occur.

8.8 Scattering angle θ_s is twice the Bragg angle. Thus, $n\lambda = 2d\sin(\theta_s/2)$, with $n = 1$. Therefore, $d_1 = 0.224$ nm. Similarly, %XED_Home%\Server\Applications\Viewer\ViewXML_WB.exe.config and $d_3 = 0.129$ nm.

$$\text{Hence,}\quad d_1 : d_2 : d_3 \approx 1 : \frac{1}{\sqrt{2}} : \frac{1}{\sqrt{3}}.$$

8.9 (a) $\lambda = h/p = h/\sqrt{2mE}$. Therefore, $\lambda_e = h/\sqrt{2m_e eV} = \lambda_n = h/\sqrt{2m_n kT}$, giving, $m_e eV = m_n kT$. Therefore, $V = (m_n kT)/(m_e e) = 46$ V. (b) Taking energy of a neutron to be 2 MeV, $\lambda = h/p = h/\sqrt{2m_n E} = 2 \times 10^{-14}$ m. The wavelength of the X-rays should be about the same order as the spacing of the lattice planes $\sim 10^{-10}$ m. The above de Broglie wavelength is several orders of magnitude too small. This shows the need for energetic neutrons to first be slowed down for crystallographic studies.

8.10 The de Broglie wavelength λ must be ~ 1 fm $= 1 \times 10^{-15}$ m.

Then, $\lambda = hc/\sqrt{K(K + 2mc^2)} \cong hc/K$ for $K \gg mc^2$. This gives $K \cong 1.99 \times 10^{-10}$ J $= 1.24$ GeV. This value is much greater than the rest mass energy of an electron.

8.11 (a) $\nu = c/\lambda = 5 \times 10^{12}$ Hz. (b) In SHM we have, frequency $\nu = (1/2\pi)\sqrt{\alpha/m}$, where α is the force constant and m is the mass, i.e. the mass of a sodium atom. Therefore, $\alpha = 37.7$ N/m. (c) Potential energy $V(r) = (1/2)\alpha r^2$. Therefore, $V(r) = 18.8\, r^2$ J. (d) Boltzmann gives $P[r]\mathrm{d}r = A\mathrm{e}^{-V(r)/kT}\mathrm{d}r = A\mathrm{e}^{-\alpha r^2/2kT}\mathrm{d}r$. Using standard integral gives $A = \sqrt{\alpha/(2\pi kT)}$, and hence, $P[r] = \sqrt{\alpha/(2\pi kT)}\mathrm{e}^{-\alpha r^2/2kT}$. (e) From equipartition of energy theorem, $\overline{V(r)} = (1/2)\alpha\overline{r^2} = (1/2)kT$.

Therefore, $\sqrt{\overline{r^2}} = \sqrt{kT/\alpha} = \sqrt{(1.38 \times 10^{-23} \times 400)/37.7} = 1.2 \times 10^{-11}$ m.

8.12 We have

$$V(a_0 + x) = V(a_0) + x\left(\frac{\mathrm{d}V(r)}{\mathrm{d}r}\right)_{r=a_0} + \frac{x^2}{2!}\left(\frac{\mathrm{d}^2 V(r)}{\mathrm{d}r^2}\right)_{r=a_0} + \frac{x^3}{3!}\left(\frac{\mathrm{d}^3 V(r)}{\mathrm{d}r^3}\right)_{r=a_{00}}.$$

ΔV is the increase in potential energy with respect to the value at the minimum:

$$\Delta V = V(a_0 + x) - V(a_0),$$

and $(\mathrm{d}V(r)/\mathrm{d}r)_{r=a_0} = 0$ as the potential is a minimum at $r = a_0$.

$$\text{Hence,}\quad \Delta V = \frac{x^2}{2!}\left(\frac{\mathrm{d}^2 V(r)}{\mathrm{d}r^2}\right)_{r=a_0} + \frac{x^3}{3!}\left(\frac{\mathrm{d}^3 V(r)}{\mathrm{d}r^3}\right)_{r=a_{00}}.$$

Writing,

$$V(r) = 12\varepsilon\left[\frac{1}{12}\left(\frac{a_0}{r}\right)^{12} - \frac{1}{6}\left(\frac{a_0}{r}\right)^{6}\right],$$

$$\frac{\mathrm{d}V(r)}{\mathrm{d}r} = \frac{12\varepsilon}{a_0}\left[\left(\frac{a_0}{r}\right)^{13} - \left(\frac{a_0}{r}\right)^{7}\right] = 0, \quad \text{at } r = a_0,$$

$$\frac{\mathrm{d}^2 V(r)}{\mathrm{d}r^2} = \frac{12\varepsilon}{a_0^2}\left[13\left(\frac{a_0}{r}\right)^{14} - 7\left(\frac{a_0}{r}\right)^{8}\right] = \frac{72\varepsilon}{a_0^2}, \quad \text{at } r = a_0,$$

and

$$\frac{\mathrm{d}^3 V(r)}{\mathrm{d}r^3} = -\frac{12\varepsilon}{a_0^3}\left[13 \times 14\left(\frac{a_0}{r}\right)^{15} - 7 \times 8\left(\frac{a_0}{r}\right)^{9}\right] = -\frac{1512\varepsilon}{a_0^3}, \text{at } r = a_0.$$

$$\text{Hence,}\quad \Delta V = \frac{x^2}{2!}\left(\frac{72\varepsilon}{a_0^2}\right) + \frac{x^3}{3!}\left(-\frac{1512\varepsilon}{a_0^3}\right) \quad \text{or} \quad \Delta V = 36\varepsilon\left(\frac{x}{a_0}\right)^2 - 252\varepsilon\left(\frac{x}{a_0}\right)^3.$$

Problems 9

9.1 $G = 7.7 \times 10^{10}$ Pa, $B = 1.7 \times 10^{11}$ Pa.

9.2 The force mg is constant throughout the wire. Therefore, the stress in the copper wire $= mg/A$, and the resulting strain $= mg/EA$. Therefore, extension of the copper wire $= mgL/EA = 4.3 \times 10^{-4}$ m. Similarly, the extension of nickel wire $= 1.3 \times 10^{-4}$ m.

Therefore, total extension of the composite wire $= 5.6 \times 10^{-4}$ m $= 0.56$ mm.

Young's modulus is 11×10^{10} Pa for copper and 21×10^{10} Pafor nickel.

9.3 The extension x of a wire of length l and cross-section A is related to the tensile stress F by $F = [(EA)/l]x$. Work done in increasing the extension by $\mathrm{d}x$ is $F\mathrm{d}x = [(EA)/l]x\,\mathrm{d}x$.

Therefore, work done in increasing the extension from x_1 to x_2 and hence the stored energy is $[(EA)/l]\int_{x_1}^{x_2} x\,\mathrm{d}x = (1/2)[(EA)/l]\left(x_2^2 - x_1^2\right) = 0.14$ J.

9.4 The potential energy stored in a wire of length l and cross-section A is $(1/2)[(EA)/l](\Delta l)^2$, where Δl is the extension. Thus, the potential energy is proportional to the square of Δl. According to the equipartition theorem, the average value of Δl is then given by $(1/2)[(EA)/l]\overline{(\Delta l)^2} = (1/2)kT$. Therefore, $\sqrt{\overline{(\Delta l)^2}} = \sqrt{kTl/EA} = 5.4 \times 10^{-14}$ m. This compares with a typical atomic diameter of $\sim 1 \times 10^{-10}$ m.

9.5

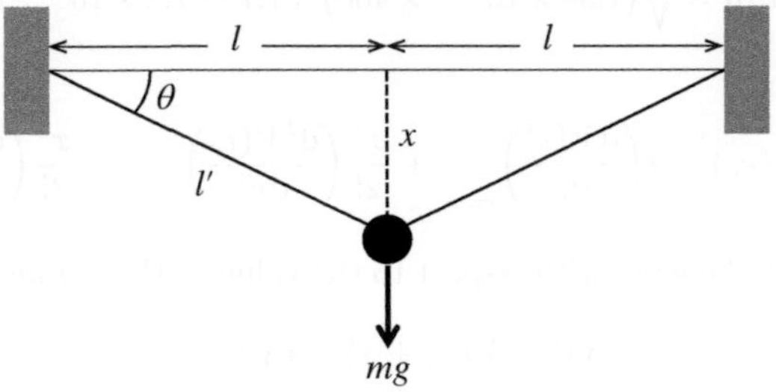

Tension in each wire

$$T = \frac{mg}{2\sin\theta} = \frac{mg}{2(x/l')} \cong \frac{mg}{2(x/l)} = \frac{mgl}{2x},$$

giving stress $= mgl/2Ax$.

$$l + \Delta l \cong l\left[1 + \frac{1}{2}\left(\frac{x^2}{l^2}\right)\right].$$

Therefore, $\Delta l \cong (1/2)(x^2/l)$, and strain $\Delta l/l \cong (1/2)(x/l)^2$.

Therefore, $mgl/2Ax \cong E(1/2)(x/l)^2$, giving $m \cong AE/g(x/l)^3 = 0.21$ kg.

9.6 $l = l_0(1 + \alpha\Delta T)$, giving $\Delta l = l - l_0 = l_0\alpha\Delta T$ and strain $= \Delta l/l_0 = \alpha\Delta T$.

Therefore, force $F = AE(\Delta l/l_0) = AE\alpha\Delta T = 9.4 \times 10^4$ N.

9.7 We have,

$$G = \frac{2L\tau}{\pi r_0^4\phi}; \quad E = \frac{FL}{A\Delta L}; \quad G = \frac{E}{2(1+\sigma)}.$$

Therefore, $E/G = \left(Fr_0^2\phi\right)/(2\Delta L\tau) = 2.66$. Using $\sigma = (E/2G) - 1$ gives $\sigma = 0.33$.

9.8 Volume of rod $V = \pi r^2 l$. Then, $\mathrm{d}V = \pi r^2\mathrm{d}l - 2\pi rl\mathrm{d}r$. As $\mathrm{d}V = 0$, we have for small changes in r and l, $\pi r^2\Delta l = 2\pi rl\Delta r$, giving $\Delta l/l = (2\Delta r)/r$. Therefore, $\sigma = (\Delta r/r)/(\Delta l/l) = (1/2) = 0.5$.

9.9 $T = 2\pi\sqrt{I/c}$, where $c = (\pi G r_0^2)/2L$. Therefore,

$$G = \frac{4\pi^2 I \times 2L}{\pi T^2 r_0^2} = \frac{4\pi \times 2L}{T^2 r_0^2}\frac{1}{2}mR^2.$$

which gives $G = 6.4 \times 10^{10}$ Pa.

9.10 The force on the elemental slice is $F = Ax\rho g$, resulting in tensile stress $= x\rho g$.

Using stress/strain $= E$, the strain of the elemental slice $= x\rho g/E$, it follows that the extension of the slice of width dx is $= (x\rho g/E)/dx$. Therefore, the total extension ΔL of the wire is $\Delta L = (\rho g)/E\int_0^L x dx = (1/2)[(L^2\rho g)/E]$. Velocity of sound waves $v = \sqrt{E/\rho}$, where ρ is the density of wire. Therefore, $\Delta L = (1/2)(L^2 g/v^2) = 0.49$ mm.

9.11 We have $B = 8L_s/V_m$, where V_m is the molar volume. Taking $V_m = N_A a_0^3$, $B = 8L_s/(N_A a_0^3) = 2.0 \times 10^9$ Pa.

9.12

$$B = \frac{n}{18a_0}\left(\frac{d V^2(r)}{dr^2}\right)_{r=a_0}.$$

$$\frac{d V(r)}{dr} = \left(\frac{pq}{p-q}\right)\varepsilon\left[-\frac{a_0^p}{p}p\left(\frac{1}{r^{p+1}}\right) + \frac{a_0^q}{q}q\left(\frac{1}{r^{q+1}}\right)\right],$$

$$\frac{d^2 V(r)}{dr^2} = \left(\frac{pq}{p-q}\right)\varepsilon\left[\frac{a_0^p}{p}p(p+1)\left(\frac{1}{r^{p+2}}\right) - \frac{a_0^q}{q}q(q+1)\left(\frac{1}{r^{q+2}}\right)\right].$$

For $r = a_0$,

$$\frac{d^2 V(r)}{dr^2} = \varepsilon\left(\frac{pq}{{a_0}^2}\right).$$

Therefore, $B = (n/18a_0)(pq\varepsilon/a_0^2)$. $V_m = N_A a_0^3$ and $L_s = (1/2)N_A n\varepsilon$. Substituting $n\varepsilon/a_0^2 = (2L_s/V_m)a_0$ gives $B = (pqL_s)/9V_m$.

Problems 10

10.1 We have

$$C_V = 3R\left(\frac{\Theta_E}{T}\right)^2\frac{\exp(\Theta_E/T)}{[\exp(\Theta_E/T)-1]^2}.$$

At Einstein, temperature $T = \Theta_E$. Therefore, $C_V = 2.76R$.

10.2 (a) One mole of aluminium occupies $0.027/(2.7 \times 10^3) = 1.0 \times 10^{-5}\,\text{m}^3$. Average spacing of atoms is therefore $\sim[(1.0 \times 10^{-5})/(6.0 \times 10^{23})]^{1/3} = 2.6 \times 10^{-10}$ m. Therefore, $\lambda_{min} \sim 5.2 \times 10^{-10}$ m. Velocity $v = \sqrt{E/\rho} = 5.1 \times 10^3$ m/s, and $\nu_{max} = 0.97 \times 10^{13}$ Hz.

(b) $\Theta_D = (\hbar\omega_{max})/k = (\hbar 2\pi v_{max})/k = 464$ K. This is in satisfactory agreement with the actual value of the Debye temperature of aluminium, which is 433 K.

10.3 The energy of a simple harmonic oscillator is $E = (1/2)m\omega^2 A^2$, where we take m as the mass of an atom and ω as the angular frequency of oscillation $= 2\pi\nu$. At the melting temperature T_M, the energy of the atom is kT_M. Therefore,

$$\frac{1}{2}m(2\pi\nu)^2\left(\frac{a_0}{10}\right)^2 = kT_M.$$

Rearrangement gives

$$\nu = \frac{5}{\pi a_0}\sqrt{\frac{2kT_{\mathrm{M}}}{m}}.$$

For aluminium,

$$a_0 = \left(\frac{0.027}{6.0\times 10^{23}\times 2.7\times 10^{3}}\right)^{1/3} = 2.6\times 10^{-10}\ \mathrm{m}^3,$$

which gives $\nu \sim 5\times 10^{12}$ Hz. Considering the simplicity of the models used, this is in reasonably good agreement with the value obtained in Problem 10.2.

10.4 Suppose that the thickness of the ice is x. Then the rate at which heat is conducted through the ice is $\mathrm{d}Q/\mathrm{d}t = \kappa A(\Delta T/x)$, where $\Delta T = 7.5\,°\mathrm{C}$. This heat is supplied by the latent heat of fusion of the water. The heat released when water forms a layer of ice of thickness $\mathrm{d}x$ is $\rho A \mathrm{d}x L$. Therefore,

$$\rho AL\frac{\mathrm{d}x}{\mathrm{d}t} = \frac{\mathrm{d}Q}{\mathrm{d}t} = \kappa A\frac{\Delta T}{x}.$$

Hence,

$$t = \frac{\rho L}{\kappa\Delta T}\int_{0.05}^{0.1} x\mathrm{d}x = \frac{\rho L}{2\kappa\Delta T}\left[x^2\right]_{0.05}^{0.1}.$$

This gives $t = 7.34\times 10^5$ seconds $\sim$ 20 hours.

10.5 Considering a spherical shell of the insulating material between r and $r + \mathrm{d}r$.

$$\kappa 4\pi r^2\frac{\mathrm{d}T}{\mathrm{d}r} = -\frac{\mathrm{d}Q}{\mathrm{d}t} = -20\ \mathrm{W}.$$

Therefore,

$$\frac{4\pi\kappa}{20}\int_{T_1}^{T_2}\mathrm{d}T = -\int_{r_1}^{r_2}\frac{\mathrm{d}r}{r^2}.$$

This gives $\kappa = 0.32\ \mathrm{W/K\cdot m}$.

10.6 We have $\lambda = (m_{\mathrm{e}}\bar{c}\sigma)/(ne^2)$. $\bar{c} \approx \sqrt{(3kT)/m_{\mathrm{e}}} = 1.15\times 10^5$ m/s. Molar volume $= 0.023/970 = 2.37\times 10^{-5}\ \mathrm{m}^3$. Therefore, the volume of sodium atom $= 3.95\times 10^{-29}\ \mathrm{m}^3$. Assuming each sodium atom contributes one electron, electron number density $n \sim 2.53\times 10^{28}\ \mathrm{m}^{-3}$. This gives $\lambda = 3.4$ nm.

10.7 $D_0 = (1/6\pi)a_0^2\omega_{\mathrm{E}}$, where the Einstein frequency $\omega_{\mathrm{E}} = (k\Theta_{\mathrm{E}})/\hbar$.

Therefore, $D_0 = \left(a_0^2 k\Theta_{\mathrm{E}}\right)/(6\pi\hbar)$. For silver, $a_0 = 2.6\times 10^{-10}$ m, giving $D_0 = 7.3\times 10^{-8}\ \mathrm{m}^2/\mathrm{s}$.

10.8 (a)
$$\frac{\partial n}{\partial t} = \frac{N}{A(\pi D)^{1/2}}\left[-\frac{1}{2t^{3/2}} + \left(\frac{x^2}{4Dt^{5/2}}\right)\right]\mathrm{e}^{(-x^2/4Dt)}.$$

$$\frac{\partial^2 n}{\partial x^2} = \frac{N}{A(\pi Dt)^{1/2}D}\left[-\frac{1}{2t^{3/2}} + \frac{x^2}{4Dt^{5/2}}\right]\mathrm{e}^{(-x^2/4Dt)}$$

Therefore,

$$D\frac{\partial^2 n}{\partial x^2} = \frac{N}{A(\pi Dt)^{1/2}}\left[-\frac{1}{2t^{3/2}} + \frac{x^2}{4Dt^{5/2}}\right]\mathrm{e}^{(-x^2/4Dt)} = \frac{\partial n}{\partial t}.$$

(b) $\dfrac{n(x=x')}{n(x=0)} = \dfrac{Ne^{(-x'^2/4Dt)}}{A(\pi Dt)^{1/2}} \cdot \dfrac{A(\pi Dt)^{1/2}}{Ne^0} = e^{(-x'^2/4Dt)}$.
Therefore, $(x'^2)/(4Dt) = \ln 2$, which gives $D = x'^2/(4t\ln 2) = 2.5 \times 10^{-9}\,\mathrm{m^2/s}$.

Problems 11

11.1 Using similar arguments as for a normal liquid, the surface tension of an atomic nucleus is $\gamma \sim n\varepsilon/4d^2$, where n is the number of nearest neighbours a nucleon has, ε is the binding energy between two nucleons, and d is the diameter of a nucleon.
Total binding energy per nucleon $\sim (1/2)n\varepsilon$, giving $n\varepsilon = 2 \times 10\,\mathrm{MeV}$. Therefore,

$$\gamma \sim \frac{n\varepsilon}{4d^2} \sim 5\,\mathrm{MeV/fm^2} \quad \text{or} \quad 8 \times 10^{17}\,\mathrm{J/m^2}.$$

11.2 The work done in increasing the radius of the soap bubble from R to $R + \mathrm{d}R$ is $P\mathrm{d}V$, where $\mathrm{d}V$ is the increase in the volume of the bubble. As $V = (4/3)\pi R^3$, $\mathrm{d}V = 4\pi R^2\mathrm{d}R$. Therefore, work done $=4\pi PR^2\mathrm{d}R$. Surface area A of the soap bubble $= 2 \times 4\pi R^2$, as the two surfaces of the soap bubble are involved. Therefore, the increase in the surface area of the bubble $\mathrm{d}A = 16\pi R\mathrm{d}R$. This involves an increase in the surface energy of $\gamma\mathrm{d}A$. It follows that $\gamma\mathrm{d}A = 16\pi\gamma R\mathrm{d}R = 4\pi PR^2\mathrm{d}R$, and $P = 4\gamma/R$.
Work done in blowing the bubble is

$$4\pi\int_0^R PR^2\mathrm{d}R = 16\pi\gamma\int_0^R R\mathrm{d}R = 8\pi\gamma R^2.$$

Therefore, for $R = 50 \times 10^{-3}\,\mathrm{m}$, work done is $3.1 \times 10^{-3}\,\mathrm{J}$.

11.3 The Clausius–Clapeyron equation $\mathrm{d}P_{\mathrm{VP}}/\mathrm{d}T = L_{\mathrm{V}}/[T(V - V_{\mathrm{L}})]$ becomes $\mathrm{d}P_{\mathrm{VP}}/\mathrm{d}T = L_{\mathrm{V}}/TV$ when changes in V_{L} can be neglected. Assuming the vapour behaves like an ideal gas, $P_{\mathrm{VP}} = RT/V$. Substituting for V gives $L_{\mathrm{V}} = RT^2(1/P_{\mathrm{V}})(\mathrm{d}P_{\mathrm{VP}}/\mathrm{d}T)$, which can be written as $L_{\mathrm{V}} = RT^2(\mathrm{d}/\mathrm{d}T)\ln P_{\mathrm{VP}}$. Using $L_{\mathrm{V}} = RT^2(1/P_{\mathrm{V}})(\mathrm{d}P_{\mathrm{VP}}/\mathrm{d}T)$, obtain $L_{\mathrm{V}} = 41.5\,\mathrm{kJ/mol} = 2306\,\mathrm{kJ/kg}$.

11.4 We let $f \propto r^\alpha\gamma^\beta\rho^\gamma$. Considering the dimensions of all these quantities, we have $[\mathrm{T}]^{-1} \equiv [\mathrm{L}]^\alpha[\mathrm{MT^{-2}}]^\beta[\mathrm{M\,L^{-3}}]^\gamma$. This gives for T: $-1 = -2\beta$, L: $0 = \alpha - 3\gamma$, M: $0 = \beta + \gamma$. Solving these equations, $f = k\left[\left(\gamma^{1/2}\right)/\left(r^{3/2}\rho^{1/2}\right)\right] = k\sqrt{\gamma/r^3\rho}$, where k is a constant.

11.5 We let $u_{\mathrm{c}} \propto r^\alpha\rho^\beta\eta^\gamma$. Considering the dimensions of all the above quantities, we have $[\mathrm{LT^{-1}}] \equiv [\mathrm{L}]^\alpha[\mathrm{M\,L^{-3}}]^\beta[\mathrm{M\,L^{-1}T^{-1}}]^\gamma$. This gives for L: $1 = \alpha - 3\beta - \gamma$, T: $-1 = -\gamma$, M: $0 = \beta + \gamma$. Solving these three equations gives $u_{\mathrm{c}} = k(\eta/r\rho)$, where k is a constant. The constant k is just the Reynolds number Re.

11.6 Flow rate of water leaving the pipe is $V = (\pi\Delta Pr^4)/8\eta l$, where $\Delta P = \rho gh$, and h is the height of the water level. Therefore, the volume leaving the pipe in time $\mathrm{d}t = [(\pi\rho ghr^4)/8\eta l]\mathrm{d}t$. If the height h of the water level in the tank falls by $\mathrm{d}h$ in time $\mathrm{d}t$, the change in the volume of the water $= -A\mathrm{d}h$, where A is the base area of the tank. Therefore, $[(\pi\rho ghr^4)/8\eta l]\mathrm{d}t = -A\mathrm{d}h$, giving, $\mathrm{d}t = -(8\eta lA\mathrm{d}h)/(\pi\rho ghr^4)$. Integrating from $h = h_0$ to $h = h_0/2$, we obtain $t = [(8\eta lA)/(\pi\rho gr^4)]\ln 2$. Hence, $t = 1.6 \times 10^4$ seconds $= 4.4$ hours.

11.7 The excess pressures above atmospheric pressure of the two bubbles is, respectively, $4\gamma/R_1$ and $4\gamma/R_2$. Therefore, the excess pressure in the smaller bubble with respect to the larger bubble is $(4\gamma/R_1) - (4\gamma/R_2)$. This excess pressure must be equal to $4\gamma/R_3$.

$$\text{Hence,} \quad \frac{4\gamma}{R_3} = \frac{4\gamma}{R_1} - \frac{4\gamma}{R_2}, \quad \text{and} \quad R_3 = 75\,\mathrm{mm}.$$

11.8 $L_V/\gamma \sim 2N_A a_0^2$, where L_V is latent heat per mole and a_0 is the molecular diameter, giving $a_0 \sim \sqrt{(L/M)/(2N_A\gamma)}$, where L is latent heat per kg. Therefore, $a_0 \sim 12 \times 10^{-10}$ m.

11.9 Excess pressure of a drop scales linearly with surface tension γ. Hence, the excess pressure in a drop of mercury is ~ 7 times greater than in a drop of water.

11.10 By Stoke's law, viscous force $= 6\pi\eta rv$. At the terminal velocity, these balance.
Therefore, $(4/3)\pi r^2(\rho_s - \rho_g)g = 6\pi\eta rv$, which gives $\eta = 2.4$ Pa · s.

11.11 Bernoulli: $(P_1 - P_2) = (1/2)\rho(u_2^2 - u_1^2) + \rho g(h_2 - h_1)$. At the bathroom, we have, $u_2 = (4 \times 10^{-3})/[60 \times \pi \times (10 \times 10^{-3})^2] = 0.21$ m/s. Hence, $u_1 = 3.4 \times 10^{-2}$ m/s.
Then taking into account the difference in vertical height, $P_1 = 2.6 \times 10^5$ Pa $= 2.6$ atm.

11.12 Considering one surface of the plate, and an annulus of the plate between radii r, and $r + \mathrm{d}r$, the viscous force acting on this annulus of area $2\pi r\mathrm{d}r$ is $\mathrm{d}F = (\eta v/d) \times 2\pi r\mathrm{d}r = (\eta r\omega 2\pi r\mathrm{d}r)/d$, where ω is the angular velocity. The contribution to the couple acting on the disc is $\mathrm{d}G = [(\eta r\omega 2\pi r\mathrm{d}r)/d] \times r = (2\pi\eta\omega r^3\mathrm{d}r)/\mathrm{d}$, and hence the couple due to one surface is $G = [(2\pi\eta\omega)/d]\int_0^R r^3\mathrm{d}r = [(\pi\eta\omega)/2d]R^4$. As there are two surfaces, the total couple acting on the disc is $(\pi\eta\omega R^4)/d = [(\pi\eta R^4)/d](\mathrm{d}\theta/\mathrm{d}t)$. The equation of motion for the disc is then $I(\mathrm{d}^2\theta/\mathrm{d}t^2) + [(\pi R^4\eta)/d](\mathrm{d}\theta/\mathrm{d}t) + \mu\theta = 0$. Comparing this equation with the equation of motion for damped harmonic oscillations $(\mathrm{d}^2\theta/\mathrm{d}t^2) + \gamma(\mathrm{d}\theta/\mathrm{d}t) + \mu\theta = 0$, we see that the damping factor $\gamma = (\pi R^4\eta)/Id$, giving $\eta = (\gamma Id)/\pi R^4$. The oscillations decay according to $\theta = \theta_0 \mathrm{e}^{-\gamma t/2}$. For $\theta = (\theta_0/2)$, $t = (2\ln 2)/\gamma$, giving $\gamma = (2\ln 2)/t$. Thus, $\eta = (\gamma Id)/(\pi R^4) = (Id2\ln 2)/(\pi R^4 t)$. This gives $\eta = 1.2 \times 10^{-3}$ Pa · s.

Problems 12

12.1 Using, $2(P/2)n\sin\theta = m\lambda$, with $m = 1$, obtain reflection for 659 nm wavelength at 28 °. To observe red reflected light at 19.4°, we have $Pn = 659/\sin(19.4°) = 1984$ nm. Assuming that refractive index remains approximately constant, the pitch length P must increase. Cooling a chiral liquid crystal increases the length of the pitch P. Therefore, the liquid crystal would be cooler than before.

12.2 (a) From Malus law, $I = I_0\cos^2\theta$, find $\theta = \cos^{-1}\sqrt{1/3} = 54.7°$.

12.3 The plane of polarisation of the light transmitted by the first polariser lies in the vertical plane. This transmitted light has half the intensity of the incident unpolarised light. It has two orthogonal components that each lie at 45° with respect to the horizontal. Each component has half the intensity of the light transmitted by the first polariser. The component that is parallel to the polarisation direction of the intermediate polariser is transmitted by it. This transmitted light has a component in the vertical direction that is transmitted by the third polariser. And it has half the intensity of the light transmitted by the intermediate polariser. Thus, the intensity of the light transmitted by this combination of three polarisers is $1/2 \times 1/2 \times 1/2 = 1/8$ that of the incident unpolarised light.

12.4 Suppose the polarisation direction of the middle polariser is at angle θ with respect to the polarisation direction of the first polariser. The intensity of the light transmitted by the middle polariser will be $I_1\cos^2\theta$. The plane of polarisation of this transmitted light is at angle $(90 - \theta)°$ with respect to the polarisation direction of the third polariser. Hence, the intensity of the light transmitted by the third analyser is $I_1\cos^2\theta\cos^2(90 - \theta) = (I_1/4)\sin^2 2\theta$. $\cos 4\theta = 1 - 2\sin^2 2\theta$, giving $\sin^2 2\theta = (1/2)(1 - \cos 4\theta)$. Therefore, $I = (I_1/8)(1 - \cos 4\theta)$. $\omega = \mathrm{d}\theta/\mathrm{d}t$, from which $\theta = \omega t$, assuming that $\theta = 0$ at $t = 0$. Then, $I = (I_1/8)(1 - \cos 4\omega t)$.

12.5 We may write for the two circularly polarised waves

$$E_x = E_0\sin(\omega t - kz): \quad E_y = E_0\sin\left[(\omega t - kz) + \frac{\pi}{2}\right],$$

$$E_x{}' = E_0\sin(\omega t - kz): \quad E_y{}' = E_0\sin\left[(\omega t - kz) - \frac{\pi}{2}\right].$$

Expanding the terms in square brackets: $E_y = E_0\cos(\omega t - kz)$, $E_y{}' = -E_0\cos(\omega t - kz)$.
Hence, $E_y + E_y{}' = 0$ and $E_x + E_x{}' = 2E_0\sin(\omega t - kz)$. These components represent a linearly polarised wave of amplitude $2E_0$.

12.6 Circularly polarised light of intensity I_0 can be considered to be two linearly polarised beams of equal intensity $I_0/2$, with planes of polarisation parallel to say the x- and y-directions. Suppose the polarisation direction of the linear polariser is at an arbitrary angle θ with respect to the y-direction. The transmitted intensity of the linear polarised beam whose plane of polarisation is parallel to the y-direction will be $(I_0/2)\cos^2\theta$. Similarly, the transmitted intensity of the linear polarised beam whose plane of polarisation is perpendicular to the y-direction will be $(I_0/2)\sin^2\theta$. The total transmitted intensity is then $(I_0/2)\cos^2\theta + (I_0/2)\sin^2\theta = I_0/2$.

12.7 (a) We have, $\lambda_1 = \lambda_0/n_1$, and $\lambda_2 = \lambda_0/n_2$. Phase changes incurred in travelling distance $d = 2\pi(d/\lambda_1)$ and $2\pi(\delta/\lambda_2)$, respectively. Therefore, the phase difference between the two light waves after distance d is $2\pi d[(1/\lambda_1) - (1/\lambda_2)] = (2\pi d/\lambda_0)(n_1 - n_2)$. For a quarter-wave plate, the phase difference corresponding to a quarter of a cycle or $\lambda_0/4$ or $\pi/2$ rad. Therefore, $(2\pi d/\lambda_0)(n_1 - n_2) = \pi/2$. This gives, $d = \lambda_0/4(n_1 - n_2)$.

(b) For the liquid crystal $d = \lambda_0/4(n_1 - n_2) = 1.18 \times 10^{-6}\,\text{m} = 1.18\,\mu\text{m}$.

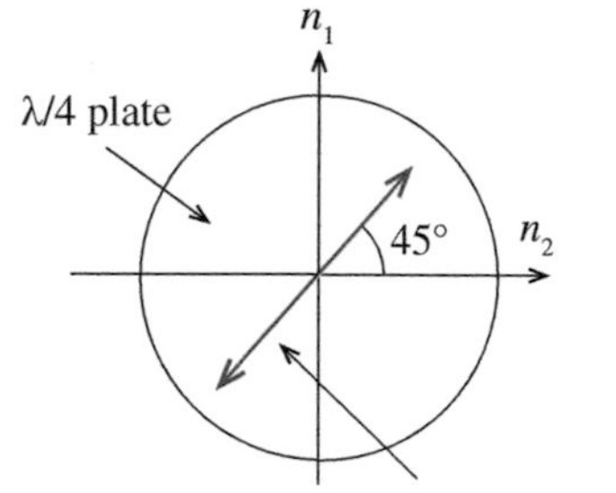

(c) The figure illustrates the two perpendicular axes of a quarter-wave plate; one is chosen to be in the vertical plane and the other in the horizontal plane. A light wave whose plane of polarisation is parallel to the vertical direction experiences a refractive index n_1, while a light wave whose plane of polarisation is parallel to the horizontal direction experiences a refractive index n_2. To convert linearly polarised light into circularly polarised light, the orientation of the quarter-wave plate must be such that the plane of polarisation of the incident linearly polarised light is at 45° with respect to the vertical. This means that its two mutually perpendicular components, in the vertical and horizontal directions respectively, have the same amplitude. This means that they combine to produce circularly polarised light after travelling through the quarter-wave plate. Other angles would produce the more general case of elliptically polarised light.

12.8 Angular diffraction limit $= 1.22(\lambda/D)$, where λ is the wavelength and D is diameter of the eye. Therefore, pixel size on screen corresponding to this diffraction limit is $= 1.22(\lambda/D)d$, where d is distance to screen. Therefore, minimum number of pixels is $A/[1.22(\lambda/D)d]^2$, where A is area of screen. Taking $\lambda = 550$ nm, the minimum number of pixels is 5×10^5 pixels.

Index

Physics of Matter, First Edition. George C. King.

D

E

H

I

J

K

L

M

X

Y

Z

Printed and bound by CPI Group (UK) Ltd, Croydon, CR0 4YY
26/01/2024
08227847-0001